Outstanding Contributions to Logic

Volume 4

For further volumes:
http://www.springer.com/series/10033

Guram Bezhanishvili
Editor

Leo Esakia on Duality in Modal and Intuitionistic Logics

Springer

Editor
Guram Bezhanishvili
New Mexico State University
Las Cruces, NM
USA

ISSN 2211-2758 ISSN 2211-2766 (electronic)
ISBN 978-94-017-8859-5 ISBN 978-94-017-8860-1 (eBook)
DOI 10.1007/978-94-017-8860-1
Springer Dordrecht Heidelberg New York London

Library of Congress Control Number: 2014936452

Printed on acid-free paper

Springer is part of Springer Science+Business Media (www.springer.com)

Preface

This volume is dedicated to Leo Esakia's contributions to the theory of modal and intuitionistic systems. Leo Esakia was one of the pioneers in developing duality theory for modal and intuitionistic logics, and masterfully utilizing it to obtain some major results in the area. The volume consists of 10 chapters, written by leading experts, that discuss Leo's original contributions and consequent developments that have shaped the current state of the field.

I would like to express sincere gratitude to the authors as well as to the referees without whose outstanding job the volume would not have been possible. It is my belief that the volume will serve as an excellent tribute to Leo Esakia's pioneering achievements in developing algebraic and topological semantics of modal and intuitionistic logics, which have paved the way for the next generations of researchers interested in this area.

Guram Bezhanishvili

Contents

Introduction

Leo Esakia's lifelong passion for modal and intuitionistic logics started to develop in the 1960s. Soon after it became apparent that Kripke semantics [28], although very attractive and intuitive, was not adequate for handling large classes of modal logics (the phenomenon of Kripke incompleteness). It was already understood that Kripke frames provide a nice representation for modal algebras, but a modal algebra can in general be realized only as a subalgebra of the modal algebra arising from a Kripke frame.

Leo's main interest at the time was Gödel's translation [23] of the intuitionistic propositional logic **Int** into Lewis' modal system **S4**, and the corresponding classes of Heyting algebras and **S4**-algebras. Influenced by the work of Stone [38, 39], Tarski (and his collaborators McKinsey and Jónsson) [26, 27, 30–32], and Halmos [25], Leo realized that the missing link between the algebraic and relational semantics of these systems is topology. This yielded the notion of what we now call *(quasi-ordered) Esakia spaces* (namely quasi-ordered Stone spaces with additional properties) and the representation of **S4**-algebras as the algebras of clopen subsets of Esakia spaces. This representation extends to full duality between the categories of **S4**-algebras and Esakia spaces. In his discussions with Sikorski, Leo also realized an apparent need for duality for Heyting algebras. He was able to obtain such a duality as a particular case of his duality for **S4**-algebras, thus obtaining a powerful machinery to study modal logics over **S4** and superintuitionistic logics (extensions of **Int**). These ground-breaking results were published in Esakia's 1974 paper [10], which remains one of the most cited papers by Leo.

Around the same time (mid 1970s), Goldblatt and Thomason came to the same realization, and developed what later became known as the *descriptive frame semantics* for modal logic. These findings were published in Goldblatt [21, 22]. Note that although Esakia worked with quasi-ordered Stone Spaces, replacing a quasi-order with an arbitrary binary relation in an Esakia space yields the descriptive frame semantics of Goldblatt and Thomason.

The machinery Leo developed was powerful in many respects. In particular, what we now call the *Esakia lemma* was a consequence of his duality (in fact, Leo developed the lemma to obtain the morphism correspondence of his duality). As was shown by Sambin and Vaccaro [35] it plays a crucial role in developing the Sahlqvist completeness and correspondence in modal logic. Subsequently, many generalizations of Sahlqvist's theorem have been obtained that utilize Esakia's lemma.

The volume opens with the chapter by Mai Gehrke which discusses Esakia duality for **S4**-algebras, and how to derive Esakia duality for Heyting algebras from it. Gehrke provides a more general setting for this approach, which also yields the celebrated Priestley duality for bounded distributive lattices [33, 34]. All this is done utilizing the theory of canonical extensions, a very active field of research of today. Gehrke also discusses Esakia's lemma and gives a modern account of how to construct free finitely generated Heyting algebras and their Esakia duals.

The dual description of free finitely generated Heyting algebras and **S4**-algebras was initiated by Esakia and his student Grigolia in the mid 1970s. They developed the so-called *coloring technique* [19, 20] which became very useful in describing "upper-parts" of the dual spaces of the free finitely generated Heyting and modal algebras. This important topic was further developed in the 1980s by Shehtman, Rybakov, Grigolia, and Belissima. In the 1990s, Ghilardi published a series of papers which gave a novel perspective on the topic. This paved the way for the follow-on papers by N. Bezhanishvili, A. Kurz, M. Gehrke, D. Coumans, S. van Gool, and others. An up-to-date survey of this topic is given in the chapter by Nick Bezhanishvili, Silvio Ghilardi, and Mamuka Jibladze.

Over the years, several generalizations of Esakia duality have been developed. To name a few, Leo himself generalized his duality to the setting of bi-Heyting algebras and temporal algebras [11, 13] (see also F. Wolter [40]), G. Bezhanishvili generalized Esakia duality to monadic Heyting algebras [1], S. Celani and R. Jansana generalized it to weak Heyting algebras [9], and G. Bezhanishvili and R. Jansana to implicative semilattices [3]. The chapter by Sergio Celani and Ramon Jansana discusses Esakia duality for Heyting algebras and its generalizations to weak Heyting algebras and implicative semilattices. It also discusses how to obtain the duals of maps between Heyting algebras that only preserve part of the Heyting algebra structure. These turn out to be partial Esakia morphisms that play a crucial role in developing Zakharyaschev's canonical formulas, which provide an axiomatization of superintuitionistic logics (as well as transitive modal logics).

Esakia spaces are closely related to the celebrated Vietoris construction. In fact, originally Esakia defined his spaces by means of the Vietoris space of a Stone space [10]. This was the precursor of the coalgebraic semantics for modal logic. This topic and the related recent developments are discussed in the chapter by Yde Venema and Jacob Vosmaer.

Another important result in modal logic associated with Esakia's name is the so-called *Blok-Esakia theorem*. It establishes that the lattice of normal extensions of Grzegorczyk's modal system **S4.Grz** and the lattice of superintuitionistic logics are isomorphic. The modal system **S4.Grz** was introduced by Grzegorczyk [24], who proved a topological completeness of **S4.Grz**, and showed that the Gödel embedding of **Int** into **S4** also embeds **Int** into **S4.Grz**. In Esakia's terminology, both **S4** and **S4.Grz** are *modal companions* of **Int**. Grezegorczyk's modal system **S4.Grz** was one of the favorite modal systems of Leo. He investigated it in great detail. In particular, Esakia showed that **S4.Grz** is the largest modal companion of **Int**. He also showed that each superintuitionistic logic L has the largest modal companion, obtained by adding the Grzegorczyk axiom to the Gödel translation of L. This yields the Blok-Esakia theorem, which was obtained independently by Blok [8] and Esakia [12]. Several generalizations of the Blok-Esakia theorem were obtained by A. Kuznetsov and A. Muravitsky [29], F. Wolter and M. Zakharyaschev [41–43], F. Wolter [40], and G. Bezhanishvili [2].

The chapter by Frank Wolter and Michael Zakharyaschev is dedicated to the Blok-Esakia theorem, while the chapter by Alexei Muravitsky provides an outline of the intuitionistic modal logic **KM** which is closely related to the Gödel-Löb provability logic **GL**. In particular, it discusses the generalization of the Blok–Esakia isomorphism to an isomorphism between the lattices of all normal extensions of **KM** and **GL**, respectively. This isomorphism has a further generalization. Namely, in [18] Esakia introduced the *modalized Heyting calculus***mHC** and announced that the isomorphism between the lattices of all normal extensions of **KM** and **GL** extends to an isomorphism between the lattices of all normal extensions of **mHC** and **K4.Grz**—the modal system obtained by adding to the well-known modal system **K4** a version of Grzegorczyk's axiom. The syntax and semantics of the intuitionistic modal logic **mHC** are discussed in the chapter by Tadeusz Litak. The chapter also proves the isomorphism between the lattices of all normal extensions of **mHC** and **K4.Grz** announced in [18], and discusses the important related issues of well-foundedness, scatteredness, and constructive fixed point theorems, as well as interpretations of constructive modalities in scattered topoi.

Leo Esakia was also one of the pioneers in developing the topological semantics for modal logic. In the 1970s he proved that if we interpret modal diamond as the derivative of a topological space, then **GL** is the modal logic of all scattered spaces [14, 15]. This result was obtained independently and slightly earlier by Simmons [36]. Whether or not a given space is scattered depends on whether or not the assembly (i.e., the frame of nuclei) of the frame of opens of the space is Boolean [37]. This and related issues about the assembly tower of a given frame are discussed in the chapter by Harold Simmons in the setting of point-free topology. The topological semantics of the provability logic **GL** and the polymodal provability logic **GLP** is reviewed in the chapter by Lev Beklemishev and David Gabelaia, who also point out interesting connections between the topological semantics of **GLP**, large cardinals, and consistency issues in set theory.

As we pointed out, Esakia and Simmons were the first who developed the topological semantics for the provability logic **GL**. In [16], Esakia introduced a weakening of the modal system **K4**, which he termed *weak***K4** and denoted by **wK4**. He showed that when interpreting modal diamond as the derivative of a topological space, **wK4** is the modal logic of all topological spaces, and that **K4** is the modal logic of all spaces satisfying the so-called *T_d-separation axiom* (a lower separation axiom properly situated between T_0 and T_1, asserting that each point is locally closed). These results were originally obtained by Leo in the 1970s, but were published for the first time only in 2001. Further results in this direction were obtained by Leo and his collaborators in the follow-up papers [17, 4–7], as well as by Shehtman and his school, Joel Lucero-Bryan, and others. The last chapter in the volume by Andrey Kudinov and Valentin Shehtman is dedicated to the derivational semantics of modal logic and other related issues.

References

1. Bezhanishvili G (1999) Varieties of monadic Heyting algebras, II. Duality theory. Stud Logica 62(1):21–48
2. Bezhanishvili G (2009) The universal modality, the center of a Heyting algebra, and the Blok-Esakia theorem. Ann Pure Appl Logic 161(3):253–267
3. Bezhanishvili G, Jansana R (2013) Esakia style duality for implicative semilattices, Appl Categorical Struct 21:181–208
4. Bezhanishvili G, Esakia L, Gabelaia D (2005) Some results on modal axiomatization and definability for topological spaces. Stud Logica 81(3):325–355
5. Bezhanishvili G, Esakia L, Gabelaia D (2010) K4.Grz and hereditarily irresolvable spaces. In: Feferman S, Sieg W, Kreinovich V, Lifschitz V, de Queiroz R (eds) Proofs, Categories and Computations. Essays in honor of Grigori Mints. College Publications, London, pp 61–69
6. Bezhanishvili G, Esakia L, Gabelaia D (2010) K4.Grz, The modal logic of Stone spaces: diamond as derivative. Rev Symb Log 3(1):26–40
7. Bezhanishvili G, Esakia L, Gabelaia D (2011) K4.Grz, Spectral and T_0-spaces in d-semantics. In: Bezhanishvili N, Löbner S, Schwabe K, Spada L (eds) Lecture notes in artificial intelligence. Springer, Berlin, pp 16–29
8. Blok W (1976) Varieties of interior algebras. Ph.D. thesis, University of Amsterdam
9. Celani S, Jansana R (2005) Bounded distributive lattices with strict implication. MLQ Math Log Q 51(3):219–246
10. Esakia L (1974) Topological Kripke models. Soviet Math Dokl 15:147–151
11. Esakia L (1975) The problem of dualism in the intuitionistic logic and Browerian lattices. In: Fifth International Congress of Logic, Methodology and Philosophy of Science, Canada, pp 7–8
12. Esakia L (1976) On modal "companions" of superintuitionistic logics. VII Soviet Symposium on Logic, Kiev, pp 135–136 (Russian)
13. Esakia L (1978) Semantical analysis of bimodal (tense) systems. Logic, Semantics and Methodology. Metsniereba, Tbilisi, pp 87–99 (Russian)
14. Esakia L (1978) Diagonal constructions, Löb's formula and Cantor's scattered spaces. Modal and Intensional Logics, Moscow, pp 177–179 (Russian)
15. Esakia L (1981) Diagonal constructions, Löb's formula, and Cantor's scattered spaces. Studies in logic and semantics, Metsniereba, Tbilisi, pp 128–143 (Russian)
16. Esakia L (2001) Weak transitivity—a restitution, Logical investigations, vol 8. Nauka, Moscow, pp 244–255 (Russian)
17. Esakia L (2004) Intuitionistic logic and modality via topology. Ann Pure Appl Logic 127(1–3):155–170
18. Esakia L (2006) The modalized Heyting calculus: a conservative modal extension of the intuitionistic logic. J Appl Non-Classical Logics 16(3–4):349–366

19. Esakia L, Grigolia R (1975) Christmas trees. On free cyclic algebras in some varieties of closure algebras. Bull Sect Logic 4(3):95–102
20. Esakia L, Grigolia R (1977) The criterion of Brouwerian and closure algebras to be finitely generated. Bull Sect Logic 6(2):46–52
21. Goldblatt R (1976) Metamathematics of modal logic. Rep Math Logic 6:41–77
22. Goldblatt R (1976) Metamathematics of modal logic, II. Rep Math Logic 7:21–52
23. Gödel K (1933) Eine Interpretation des intuitionistischen Aussagenkalkülus. Ergebnisse eines mathematischen Kolloquiums 4:39–40
24. Grzegorczyk A (1967) Some relational systems and the associated topological spaces. Fund Math 60:223–231
25. Halmos P (1962) Algebraic logic. Chelsea Publishing Co., New York
26. Jónsson B, Tarski A (1951) Boolean algebras with operators, I. Amer J Math 73:891–939
27. Jónsson B, Tarski A (1952) Boolean algebras with operators, II. Amer J Math 74:127–162
28. Kripke S (1963) Semantical analysis of modal logic, I. Normal modal propositional calculi. Z Math Logik Grundlagen Math 9:67–96
29. Kuznetsov A, Muravitsky A (1986) On superintuitionistic logics as fragments of proof logic extensions. Stud Logica 45(1):77–99
30. McKinsey J, Tarski A (1944) The algebra of topology. Ann Math 45:141–191
31. McKinsey J, Tarski A (1946) On closed elements in closure algebras. Ann Math 47(2):122–162
32. McKinsey J, Tarski A (1948) Some theorems about the sentential calculi of Lewis and Heyting. J Symbolic Logic 13:1–15
33. Priestley H (1970) Representation of distributive lattices by means of ordered Stone spaces. Bull London Math Soc 2:186–190
34. Priestley H (1972) Ordered topological spaces and the representation of distributive lattices. Proc London Math Soc 24(3):507–530
35. Sambin G, Vaccaro V (1989) A new proof of Sahlqvist's theorem on modal definability and completeness. J Symbolic Logic 54(3):992–999
36. Simmons H (1974/75) Topological aspects of suitable theories. Proc Edinburgh Math Soc (2) 19(4):383–391
37. Simmons H (1981) Spaces with Boolean assemblies. Colloq Math 43 (1980), 1:23–39
38. Stone M (1936) The theory of representations for Boolean algebras. Trans Amer Math Soc 40(1):37–111
39. Stone M (1937) Topological representation of distributive lattices and Brouwerian logics. Časopis pešt mat fys 67:1–25
40. Wolter F (1998) On logics with coimplication. J Philos Logic 27(4):353–387
41. Wolter F, Zakharyaschev M (1995) Intuitionistic modal logic, Logic and foundations of mathematics (Florence), Synthese Lib, vol 280. Kluwer Academic Publishers, Dordrecht, pp 227–238
42. Wolter F, Zakharyaschev M (1997) On the relation between intuitionistic and classical modal logics. Algebra Logic 36(2):73–92
43. Wolter F, Zakharyaschev M (1999) Intuitionistic modal logics as fragments of classical bimodal logics. Logic at work, Studies in Fuzziness Soft Computing, vol 24. Physica, Heidelberg, pp 168–183

Chapter 1
Esakia's Biography and Bibliography

Esakia's Biography

1934	Born on 14 November. Named after his father, who was a famous movie director in Georgia. His mother was an actress.
1953	Entered Tbilisi State University. Majored in Physics.
1958	Graduated from Tbilisi State University with the degree in Physics.
1958	Joined Institute of Physics of the Georgian Academy of Sciences.
1963	Moved to the newly founded Institute of Cybernetics of the Georgian Academy of Sciences. Stayed at the institute for 40 years.
1960s	Main scientific ideas started to form.
1970s	Started the famous *Esakia seminar*. Esakia run the seminar for 40 years until his death. The seminar would be on Wednesdays, and it would last the entire day. The seminar continues to this day, and is now run by Mamuka Jibladze and David Gabelaia.
Late 1960s–early 1970s	Obtained dualities for the categories of Heyting algebras and closure algebras by means of ordered Stone spaces, which later were coined *Esakia spaces*.
1974	"*Topological Kripke models*" appeared in Soviet Math. Doklady. The paper develops Esakia duality for closure algebras and Heyting algebras. *Esakia's lemma* also appears in the paper. Later it became a primary tool in proving Sahlqvist type correspondence results. The paper is one of the most cited papers of Esakia.
1976	The Blok-Esakia theorem on the isomorphism between the lattices of superintuitionistic logics and normal extensions

G. Bezhanishvili (ed.), *Leo Esakia on Duality in Modal and Intuitionistic Logics*,
Outstanding Contributions to Logic 4, DOI: 10.1007/978-94-017-8860-1_1,

	of Grzegorczyk's modal system is established independently by Blok and Esakia.
1985	The monograph "*Heyting Algebras. Duality Theory*" appears in Russian.
Late 1980s–1990s	Worked on extending the correspondence between the intuitionistic propositional calculus, Grzegorczyk's logic, and the Gödel-Löb logic to the predicate case.
1990s	The severe economic hardship period in Georgia after the collapse of the Soviet Union. The Esakia seminar continued to thrive, but more often than not, the seminar was held without electricity or heat.
1995	The first *International Tbilisi Symposium on Logic, Language and Computation* (**TbiLLC**) was held in Gudauri, Georgia. Esakia was instrumental in organizing the symposium, as well as making it an extremely successful bi-anneal conference series. The tenth TbiLLC was held in 2013.
2000s	Worked on topological semantics of modal logic, provability logic, and related intuitionistic modal logics.
2002	Moved from Institute of Cybernetics to Institute of Mathematics of the Georgian Academy of Sciences. Remained at the institute until his death.
2003	The first *International Conference on Topological and Algebraic Methods in Non-Classical Logics* was held in Tbilisi, Georgia. The conference was organized by Leo's group and Department of Mathematics of New Mexico State University, and was funded by the Georgian-US bilateral grant. This became one of the main conference series in the area, and is now known as **TACL** (Topology, Algebra and Categories in Logic). The sixth TACL was held in 2013.
2008	The first *International Conference on Topological Methods in Logic* (**TOLO**) was held in Tbilisi, Georgia. The conference was organized by Leo's group, and was funded by the Georgian National Science Foundation. This also became a very successful bi-anneal conference series, which continues to thrive. The fourth TOLO will be held in the summer of 2014.
2010	Esakia's health started to fail. In spite of this, Esakia remained active. Several days before his death he conducted the last of his famous seminars, where he discussed his latest findings and new ideas. Esakia passed away on November 15. He is survived by a wife, two daughters, and three grandchildren.

Esakia's work and personality have been a constant source of inspiration for a logic community in general, and for the many generations of researchers he has raised in particular. The 2011 installments of TACL and OGLAL (Ordered Groups and Lattices in Algebraic Logic) were dedicated to Esakia's memory, as well as the 2012 special issue of Studia Logica (Volume 100, Numbers 1–2). More biographical notes on Leo Esakia can be found in the foreword to that issue.

Esakia's Bibliography

1960s

1. L. Esakia, *Functioning of finite automata*, III Soviet Meeting in Probability Theory and Mathematical Statistics, Tbilisi, 1963, 53–66 (Russian).
2. L. Esakia, *Modal extensions of superintuitionistic logics*, II Symposium in Cybernetics, Tbilisi, 1965, 173 (Russian).
3. L. Esakia, *A tense propositional calculus*, II Symposium in Cybernetics, Tbilisi, 1965, 174–175 (Russian).
4. L. Esakia, *The Δ-system and algebras with derivation*, Georgian Math. Dokl. **40** (1965), 537–543 (Russian).
5. L. Esakia, *On the structure of varieties of Brouwerian algebras*, VIII Soviet Algebraic Colloquium, Riga, 1967, 144–146 (Russian).
6. L. Esakia, *On covers of Brouwerian algebras*, IX Soviet Algebraic Colloquium, Gomel, 1968, 222–223 (Russian).
7. L. Esakia, *On semigroups generated by the operations of closure and pseudo-complementation*, IX Soviet Algebraic Colloquium, Gomel, 1968, 224–225 (Russian).
8. L. Esakia, *On a class of topological structures*, X Soviet Algebraic Colloquium, Volume II, Novosibirsk, 1969, 137–138 (Russian).
9. L. Esakia, *Brouwerian and their dual topological structures*, II Tiraspol Symposium in General Topology and Its Applications, Chisinau, 1969, 102–103 (Russian).

1970s

10. L. Esakia, *On the structure of quasi-closed elements in closure algebras*, Institute of Applied Mathematics, Tbilisi State University, Tbilisi, 1970, 24–25 (Russian).
11. L. Esakia, *Topological representation of some Brouwerian algebras*, Conference in Algebra and Mathematical Logic, Ivanovo, 1970, 134–135 (Russian).
12. L. Esakia, *On Lukasiewicz algebras*, IV Inter. Congress of Logic, Methodology and Philosophy of Science, Romania, 1971, 25–26.
13. L. Esakia, *On subdirect decomposability of Lukasiewicz algebras*, Institute of Applied Mathematics, Tbilisi State University, Tbilisi, 1971, 25–26 (Russian).
14. L. Esakia, *On the category of quasi-ordered Stone spaces*, VI Soviet Conference, Tbilisi, 1971, 136–137 (Russian).

15. L. Esakia, *On the category of perfect Kripke models*, II Soviet Conference in Mathematical Logic, Moscow, 1972, 54 (Russian).
16. L. Esakia, *Topological Kripke models*, Soviet Math. Dokl. **15** (1974), 147–151.
17. L. Esakia and S. Meskhi, *On five critical systems*, Theory of Logical Entailment, Part I, Moscow, 1974, 76–79 (Russian).
18. L. Esakia, *On some new results in the theory of modal and syperintuitionistic systems*, Theory of Logical Entailment, Part II, Moscow, 1974, 173–184 (Russian).
19. L. Esakia, *On the theory of topological Kripke models*, III Soviet Conference in Mathematical Logic, Novosibirsk, 1974, 233–235 (Russian).
20. L. Esakia, *The problem of dualism in the intuitionistic logic and Browerian lattices*, V Inter. Congress of Logic, Methodology and Philosophy of Science, Canada, 1975, 7–8.
21. L. Esakia and R. Grigolia, *Christmas trees. On free cyclic algebras in some varieties of closure algebras*, Bull. Sect. Logic **4** (1975), 95–102.
22. L. Esakia, *On the construction of fibered sums and coproducts in some classes of distributive structures and closure algebras*, V Georgian Math. Conference, Tbilisi, 1975, 29–30 (Russian).
23. L. Esakia, *On the Dummett-Lemmon hypothesis*, IV Soviet Conference in Mathematical Logic, Chisinau, 1976, 160 (Russian).
24. L. Esakia, *On modal "companions" of superintuitionistic logics*, VII Soviet Symposium in Logic and Methodology of Science, Kiev, 1976, 135–136 (Russian).
25. L. Esakia and R. Grigolia, *The criterion of Brouwerian and closure algebras to be finitely generated*, Bull. Sect. Logic **6** (1977), 46–52.
26. L. Esakia and S. Meskhi, *Five critical modal systems*, Theoria **43** (1977), 52–60.
27. L. Esakia, *On coproducts in the category of topological Kripke models*, VII Georgian Math. Conference, Tbilisi, 1977, 47–48 (Russian).
28. L. Esakia, *On the category of Pierce algebras*, XIV Soviet Conference in Algebra, Part II, Novosibirsk, 1977, 127 (Russian).
29. L. Esakia, *Semantical analysis of bimodal (tense) systems*, Logic, Semantics and Methodology, Metsniereba Press, Tbilisi, 1978, 87–99 (Russian).
30. L. Esakia, *Diagonal constructions, Löb's formula and Cantor's scattered spaces*, Modal and Intensional Logics, Moscow, 1978, 177–179 (Russian).
31. L. Esakia, *A note on the category of Stone spaces*, IX Georgian Math. Conference, Tbilisi, 1979, 67–68 (Russian).
32. L. Esakia, *On weak filtered products of Boolean algebras*, V Soviet Conference in Mathematical Logic, Novosibirsk, 1979, 169 (Russian).
33. L. Esakia, *Duality for almost direct products of Boolean, Heyting and closure algebras*, VI Inter. Congress of Logic, Methodology and Philosophy of Science, West Germany, 1979, 103–106.
34. L. Esakia, *On the theory of modal and superintuitionistic systems*, Logical Inference, Nauka Press, Moscow, 1979, 147–172 (Russian).
35. L. Esakia, *On the variety of Grzegorczyk algebras*, Studies in Non-Classical Logics and Set Theory, Nauka Press, Moscow, 1979, 257–287 (Russian).

1980s

36. L. Esakia and R. Grigolia, *The variety of Heyting algebras is balanced*, XVI Soviet Algebraic Conference, Part II, Leningrad, 1981, 37–38 (Russian).
37. L. Esakia, *On Hausdorff reducible spaces*, X Georgian Math. Conference, Batumi, 1981, 199 (Russian).
38. L. Esakia, *Diagonal constructions, Löb's formula and Cantor's scattered spaces*, Studies in Logic and Semantics, Metsniereba Press, Tbilisi, 1981, 128–143 (Russian).
39. L. Esakia, *On two modal systems related to Peano Arithmetic*, VIII Soviet Conference in Logic and Methodology of Science, Moscow, 1982, 130–133 (Russian).
40. L. Esakia, *Algebras of logic (Survey, 1973–1977)*, Ordered Sets and Lattices, Saratov Univ. Press, Saratov, 1983, 15–26 (Russian).
41. L. Esakia, *Weak decomposition of Heyting and Boolean algebras*, VII Inter. Congress of Logic, Methodology and Philosophy of Science, Austria, 1983, 71–73.
42. L. Esakia, *On a locally finite variety of Heyting algebras*, XVII Soviet Algebraic Conference, Part II, Minsk, 1983, 280–281 (Russian).
43. L. Esakia, *On the variety of Grzegorczyk algebras*, Sel. Sov. Math. **3** (1984), 343–366.
44. L. Esakia, *Intuitionistic logic and the weak law of Pierce*, Intensional Logics and Logical Structure of Theories, IV Soviet-Finnish Colloquium in Logic, Metsniereba Press, Tbilisi, 1985, 170–171 (Russian).
45. L. Esakia, *Heyting Algebras. Duality Theory*, Metsniereba Press, Tbilisi, 1985, 104 pages (Russian).
46. L. Esakia, *On a classification of elements in closure algebras*, Methods for Research in Logic, Metsniereba Press, Tbilisi, 1986, 48–54 (Russian).
47. L. Esakia, *A classification of elements in closure algebras: Hausdorff residues*, IX Soviet Conference in Logic and Philosophy of Science, Moscow, 1986, 172–173 (Russian).
48. L. Esakia, *A note on polynomials in monadic diagonalizable algebras*, VIII Soviet Conference in Mathematical Logic, Moscow, 1986, 219 (Russian).
49. M. Abashidze and L. Esakia, *Cantor's scattered spaces and the logic of provability*, Baku International Topological Conference, Part I, Baku, 1987, 3.
50. M. Abashidze and L. Esakia, *A single adequate model for Intuitionistic Logic and the modal systems of Gödel-Löb and Grzegorczyk*, IX Soviet Conference in Mathematical Logic, Leningrad, 1988, 3 (Russian).
51. L. Esakia, *Provability logic with quantifier modalities*, Intensional Logics and the Logical Structure of Theories, Metsniereba Press, Tbilisi, 1988, 4–9 (Russian).

1990s

52. L. Esakia, *Glivenko's theorem in the context of diagonalizable algebras*, X Soviet Conference in Mathematical Logic, Almaty, 1990, 172 (Russian).
53. L. Esakia, *McKinsey's modal system half a century later*, Methodology and Philosophy of Science, Vol. II, RGNF Press, Moscow–Obninsk, 1995, 95–98 (Russian).

54. L. Esakia, *Amended intuitionistic logic and formal provability*, X Inter. Congress of Logic, Methodology and Philosophy of Science, Italy, 1995, 40.
55. L. Esakia, *An outline for the amended intuitionistic logic and the kindred modal systems*, Human Communication Research Center, Univ. of Edinburgh, Edinburgh, 1995, 1–3.
56. L. Esakia, *The problem of quantification in intuitionistic logic in light of provability interpretation*, Russian Academy of Sciences, Moscow, 1997, 28–31 (Russian).
57. L. Esakia, *Provability interpretations of intuitionistic logic*, Logical Investigations, No. 5, Nauka Press, Moscow, 1998, 19–24 (Russian).
58. L. Esakia, *Quantification in intuitionistic logic with provability smack*, Bull. Sect. Logic **27** (1998), 26–28.
59. L. Esakia, *Creative and critical points in intuitionistic Kripke models and adjoint modalities*, Russian Academy of Sciences, Moscow, 1999, 78–82 (Russian).

2000s

60. L. Esakia, M. Jibladze, D. Pataraia, *Scattered toposes*, Ann. Pure Appl. Logic **103** (2000), 97–107.
61. L. Esakia, *Synopsis of fronton theory*, Logical investigations, No. 7, Nauka Press, Moscow, 2000, 137–147 (Russian).
62. L. Esakia, *The Gödel-Löb modal system—an addendum*, Russian Academy of Sciences, Moscow, 2001, 77–79 (Russian).
63. L. Esakia, *Weak transitivity—a restitution*, Logical investigations, No. 8, Nauka Press, Moscow, 2001, 244–255 (Russian).
64. L. Esakia, *A modal version of Gödel's second incompleteness theorem and the McKinsey system*, Logical investigations, No. 9, Nauka Press, Moscow, 2002, 292–300 (Russian).
65. L. Esakia, *A topological interpretation of the provability predicate in Peano Arithmetic and Zermelo-Fraenkel Set Theory*, Modern Algebra and Its Applications, Tbilisi, 2002, 16–17.
66. L. Esakia, *Dissident as an accessibility relation in the possible worlds semantics*, VINITI, Moscow, 2003, 8–11 (Russian).
67. L. Esakia, *The Gödel embedding of the intuitionistic calculus to modal logic: recent and new observations*, Russian Academy of Sciences, Moscow, 2003, 72–75 (Russian).
68. L. Esakia, *Intuitionistic logic and modality via topology*, Ann. Pure Appl. Logic **127** (2004), 155–170.
69. G. Bezhanishvili, L. Esakia, D. Gabelaia, *Some results on modal axiomatization and definability for topological spaces*, Studia Logica **81** (2005), 325–355.
70. L. Esakia, *The modalized Heyting calculus: a conservative modal extension of the intuitionistic logic*, J. Appl. Non-Classical Logics **16** (2006), 349–366.
71. L. Esakia and R. Grigolia, *Formulas of one propositional variable in intuitionistic logic with the Solovay modality*, Logic Log. Philos. **17** (2008), 111–127.
72. L. Esakia, *Around provability logic*, Ann. Pure Appl. Logic **161** (2009), 174–184.

2010s

73. G. Bezhanishvili, L. Esakia, D. Gabelaia, *The modal logic of Stone spaces: diamond as derivative*, Rev. Symb. Log. **3** (2010), 26–40.
74. G. Bezhanishvili, L. Esakia, D. Gabelaia, *K4.Grz and hereditarily irresolvable spaces*, Proofs, Categories and Computations. Essays in honor of Grigori Mints. Solomon Feferman, Wilfried Sieg, Vladik Kreinovich, Vladimir Lifschitz, and Ruy de Queiroz, editors. College Publications, 2010, 61–69.
75. G. Bezhanishvili, L. Esakia, D. Gabelaia, *Spectral and T_0-spaces in d-semantics*, Lecture Notes in Artificial Intelligence. N. Bezhanishvili, S. Löbner, K. Schwabe, and L. Spada, editors. Springer, 2011, 16–29.
76. L. Esakia and B. Löwe, *Fatal Heyting algebras and forcing persistent sentences*, Studia Logica **100** (2012), 163–173.

Chapter 2
Canonical Extensions, Esakia Spaces, and Universal Models

Mai Gehrke

In memory of Leo Esakia

Abstract In this chapter we survey some recent developments in duality for lattices with additional operations paying special attention to Heyting algebras and the connections to Esakia's work in this area. In the process we analyse the Heyting implication in the setting of canonical extensions both as a property of the lattice and as an additional operation. We describe Stone duality as derived from canonical extension and derive Priestley and Esakia duality from Stone duality for maps. In preparation for this we show that the categories of Heyting and modal algebras are both equivalent to certain categories of maps between distributive lattices and Boolean algebras. Finally we relate the N-universal model of intuitionistic logic to the Esakia space of the corresponding Heyting algebra via bicompletion of quasi-uniform spaces.

Keywords Heyting algebra · Booleanization · Canonical extension · Eskia duality · Universal model

2.1 Introduction

With his study of duality for Heyting algebras and modal algebras, Esakia was one of the first to study duality for lattices with additional operations [12]. Jónsson and Tarski studied duality for Boolean algebras with additional operations in the form of

---

This work was partially supported by ANR 2010 BLAN 0202 02 FREC

M. Gehrke (✉)
LIAFA, CNRS, Université Paris Diderot, Paris 7, France
e-mail: mai.gehrke@liafa.univ-paris-diderot.fr

G. Bezhanishvili (ed.), *Leo Esakia on Duality in Modal and Intuitionistic Logics*, Outstanding Contributions to Logic 4, DOI: 10.1007/978-94-017-8860-1_2,

canonical extensions very early on [34, 35]. In this chapter we highlight some of the main results of central importance for these works of Esakia, and Jónsson and Tarski, and indicate some further developments in duality theory for lattices with additional operations, mainly as they pertain to Heyting algebras.

The theory of canonical extensions, which recasts duality in a purely algebraic setting of lattice completions, was initiated by Jónsson and Tarski in their two papers [34, 35]. At first sight, one might think the purpose was to get rid of the topological nature of Stone duality, but this aspect is actually still very much present, though in a form more similar to the later developed point-free approach to topology. The main purpose for Jónsson and Tarski in recasting duality in algebraic form was to make it easier to identify what form the dual of an additional algebraic operation on a lattice should take. Thus in the first paper [34] they proved a general theorem about the extension of maps and preservation of equational properties, and the second paper [35] then specified the ensuing dual structures in which one can represent the original algebras.

The theory of canonical extensions has moved forward significantly since the seminal work of Jónsson and Tarski and is now applicable far beyond the original Boolean setting. It has the advantage of allowing a uniform and relatively transparent treatment of duality issues concerning additional operations. Since Esakia's work on duality for Heyting and S4 modal algebras may be seen as special instances of duality for lattices with additional operations, the theory of canonical extensions has in fact allowed the generalisation of many of Esakia's results and methods to a much wider setting. In this chapter we will show this while focussing mainly on Heyting algebras.

In Sect. 2.2 we give a brief introduction to canonical extension and show that it provides a point-free approach to duality for Heyting algebras. To this end, we show, in a constructive manner, that the canonical extension of any distributive lattice is a complete $\bigvee$-distributive lattice and thus also a Heyting algebra. Further, we show that the canonical embedding of a lattice in its canonical extension is conditionally Heyting and thus a Heyting algebra embedding if the original distributive lattice is a Heyting algebra. In Sect. 2.3 we explain how additional operations are treated in the theory of canonical extensions and illustrate this in the particular example of Heyting implication, viewed as an additional operation on a lattice.

Section 2.4 is purely algebraic and is a preparation for the duality results for Heyting and S4 modal algebras. Here we discuss Booleanisation of distributive lattices. In modern terms Booleanisation is the fact that the inclusion of the category of Boolean algebras in the category of distributive lattices has a left adjoint. This means, among other, that for any distributive lattice D, there is a unique Boolean algebra D^- that contains D as a sublattice, and that is generated by D as a Boolean algebra. This implies that the category of distributive lattices is equivalent to the category of lattice inclusions $D \hookrightarrow D^-$ with commutative diagrams for which the maps between the domains are lattice maps and the maps between the codomains are Boolean algebra maps. One can then see Heyting algebras as the (non full) subcategory of those inclusions $D \hookrightarrow D^-$ which have an upper adjoint $g\colon D^- \to D$ and with commutative diagrams that also commute for the adjoint maps. Finally S4 modal algebras also live inside the category of lattice embeddings from distributive lattices to Boolean algebras as those lattice embeddings $e: D \to B$ that have an upper adjoint, and the maps are, as for Heyting algebras, the commutative diagrams that commute both for the embeddings and for their upper adjoints. That is, in this setting, one may see the category of Heyting algebras as the intersection of the category of distributive lattices and the category of S4 modal algebras. This point of view on distributive lattices, Heyting algebras, and S4 modal algebras allows one to see all of them as certain *maps* between distributive lattices and Boolean algebras. Now applying Stone duality to these maps yields Priestley duality for distributive lattices, and Esakia duality for Heyting algebras and modal algebras.

In Sect. 2.5 we show how Stone duality may be derived from the canonical extension results by 'adding points' in the sense of point-free topology. Further, we give an algebraic account of the duality for operators and the corresponding notion of bisimulation. This last topic is treated further in Sect. 2.7 where we discuss Esakia's lemma and its generalisation as obtained in our paper [22] with Bjarni Jónsson. Finally, in Sect. 2.6, we derive both Priestley and Esakia duality from Stone duality with the help of the results of Sect. 2.4.

In Sect. 2.8, we briefly discuss set representations of distributive lattices and in particular the representation of the free N-generated Heyting algebra in the so-called N-universal model. In particular, we outline recent results from [19] which show that the Stone or Priestley space of a distributive lattice is the bicompletion of *any* set representation of the lattice, viewed as a quasi-uniform space.

I would like to thank the anonymous referee as well as Dion Coumans and Sam van Gool for carefully reading a draft of the chapter and making many useful comments which have improved the chapter greatly. I also want to thank Guram Bezhanishvili for detailed historical comments pointing out how my view point of Esakia duality here closely corresponds to Leo Esakia's own derivation of the duality. These historical comments have been incorporated in Sects. 2.4 and 2.6 as appropriate. Finally I would like to thank Dick de Jongh for helpful discussions about the universal models for intuitionistic logic.

2.2 Canonical Extension

Canonical extensions were first introduced by Jónsson and Tarski [34] in order to deal with additional operations such as modalities and relation algebraic operations in the setting of Stone duality. The idea is the following: A topological space is a pair, $(X, \mathcal{O})$ where X is a set, and $\mathcal{O}$ is a collection of downsets of X (in the order on X induced by $\mathcal{O}$). Since the poset X and the complete lattice, $\mathcal{D}(X)$, of all downsets of X are dual to each other via the discrete duality between posets and downset lattices, the information in giving $(X, \mathcal{O})$ is exactly the same as the information in the embedding $\mathcal{O} \hookrightarrow \mathcal{D}(X)$. Further, in the case of Stone duality, since $\mathcal{O}$ is generated by the dual lattice D, the data $\mathcal{O} \hookrightarrow \mathcal{D}(X)$, in turn, amounts to giving an embedding $D \hookrightarrow \mathcal{D}(X)$. This latter formulation brings the entire duality within the setting of lattice theory, making the proper translation of additional structure such as operations on the lattice more transparent. The key insight needed here is that this embedding may be uniquely characterised among the completions of D.

Definition 1. Let L be a lattice, a *canonical extension* of L is a lattice completion $L \hookrightarrow L^\delta$ of L with the following two properties:

density: L is $\bigvee\bigwedge$- and $\bigwedge\bigvee$-dense in L^δ, that is, every element of L^δ is both a join of meets and a meet of joins of elements of L;

compactness: given any subsets S and T of L with $\bigwedge S \leq \bigvee T$ in L^δ, there exist finite sets $S' \subseteq S$ and $T' \subseteq T$ such that $\bigwedge S' \leq \bigvee T'$ in L.

The fundamental facts about canonical extensions are the following.

Theorem 1. *Every bounded lattice has a canonical extension and it is unique up to an isomorphism which commutes with the embedding of the original lattice in the extension.*

Theorem 2. *For any bounded distributive lattice D, the map $\eta\colon D \to \mathscr{D}(X, \leq)$ into the lattice of all downsets of the dual space X of D which sends each element $d \in D$ to the corresponding clopen downset $\eta(d)$ is a canonical extension of D.*

In the original approach of Jónsson and Tarski for Boolean algebras [34] and Gehrke and Jónsson for bounded distributive lattices [22], Theorem 2 provided the existence part of Theorem 1. However, the canonical extension may also be obtained directly from the lattice without the use of the axiom of choice. This was first identified by Ghilardi and Meloni in the case of Heyting algebras in their work on intermediate logics [31]. A similar choice-free approach was used in [20] where Theorem 1 was first proved in the setting of arbitrary (i.e., not-necessarily-distributive) bounded lattices. There, it was shown that the lattice of Galois closed sets of the polarity $(\mathscr{F}, \mathscr{I}, R)$, where $\mathscr{F}$ is the collection of lattice filters of L and $\mathscr{I}$ is the collection of lattice ideals of L, and FRI if and only if $F \cap I \neq \emptyset$, yields a canonical extension of L. This is actually part of a more general representation theorem for so-called Δ_1-completions of a lattice. These are the completions satisfying the density condition in Definition 1. In [21] it was shown that any such completion may be obtained as the Galois closed subsets of a certain kind of polarity between a closure system of filters and a closure system of ideals of the original lattice. A different choice-free approach to the existence of canonical extensions for lattices via dcpo presentations was given in [26].

It follows from Theorem 2 that the canonical extension of any bounded distributive lattice is a downset lattice and therefore a complete Heyting algebra. We show here, in a choice-free manner, that the canonical extension of any bounded distributive lattice is a frame and thus also a complete Heyting algebra. Further, we show that the embedding of a bounded lattice in its canonical extension is conditionally Heyting, meaning that it preserves the implication whenever defined. Thus canonical extension also provides a constructive approach to Esakia duality.

We first need a few facts about canonical extensions. In working with topological spaces, the closed and the open subsets, obtained for a Boolean space by taking arbitrary intersections and arbitrary unions of basic clopens, respectively, play a very important role. For canonical extensions, the basic clopens are replaced by the image of the embedding $L \hookrightarrow L^\delta$ and the closures under infima and suprema play roles similar to those of closed and open subsets in topology. Also, from a lattice theoretic perspective, the density condition that is part of the abstract definition of canonical extension makes it clear that the meet and the join closures of L in L^δ play a central role.

Definition 2. Let L be a lattice, and L^δ a canonical extension of L. Define

$$F(L^\delta) := \{\, x \in L^\delta \mid x \text{ is a meet of elements from } L \,\},$$

$$I(L^\delta) := \{\, y \in L^\delta \mid y \text{ is a join of elements from } L \,\}.$$

We refer to the elements of $F(L^\delta)$ as *filter elements* and to the elements of $I(L^\delta)$ as *ideal elements*.

The reason for this nomenclature is that the poset $F(L^\delta)$ of filter elements of L^δ is reverse order isomorphic to the poset **Filt**(L) of lattice filters of L via the maps $x \mapsto (\uparrow x) \cap L$ and $F \mapsto \bigwedge_{L^\delta} F$, and, dually, the poset $I(L^\delta)$ of ideal elements of L^δ is order isomorphic to the poset **Idl**(L) of lattice ideals of L via the maps $y \mapsto (\downarrow y) \cap L$ and $I \mapsto \bigvee_{L^\delta} I$. Establishing these isomorphisms is in fact the first step in proving the uniqueness of the canonical extension, see e.g. [21, Theorem 5.10]. Note that now we can reformulate the density condition for canonical extensions by saying that $F(L^\delta)$ is join dense in L^δ and $I(L^\delta)$ is meet dense in L^δ.

We are now ready to prove that the canonical extension of a bounded distributive lattice satisfies the join-infinite distributive law.

Theorem 3. *Let D be a bounded distributive lattice. Then D^δ is $\bigvee$-distributive.*

Proof. As a first step we want to show for $x \in F(D^\delta)$ and $X \subseteq F(D^\delta)$ that

$$x \wedge \bigvee X \leq \bigvee \{\, x \wedge x' \mid x' \in X \,\}.$$

To this end, let $z \in F(D^\delta)$ with $z \leq x \wedge \bigvee X$ and $y \in I(D^\delta)$ with $\bigvee\{x \wedge x' \mid x' \in X\} \leq y$. By the join density of $F(D^\delta)$ and the meet density of $I(D^\delta)$, it suffices to show that we must have $z \leq y$.

The condition on y implies that, for each $x' \in X$, we have $x \wedge x' \leq y$, and thus, by compactness, there are $a, b \in D$ with $x \leq a$, $x' \leq b$ and $a \wedge b \leq y$. As a consequence we have

$$\bigvee X \leq \bigvee \{\, b \in D \mid \exists x' \in X\, \exists a \in D \text{ with } x \leq a,\, x' \leq b, \text{ and } a \wedge b \leq y \,\}.$$

This inequality, combined with $z \leq x \wedge \bigvee X \leq \bigvee X$ and compactness, now implies that there are $b_1, \ldots, b_n \in D$, there are $x'_1, \ldots, x'_n \in X$ with $x'_i \leq b_i$, and there are $a_1, \ldots, a_n \in D$ with $x \leq a_i$ and $a_i \wedge b_i \leq y$ and $z \leq b_1 \vee \ldots \vee b_n$. Let $a = a_1 \wedge \ldots \wedge a_n$ then, since $z \leq x \wedge \bigvee X \leq x$, we have

$$\begin{aligned} z &\leq x \wedge (b_1 \vee \ldots \vee b_n) \\ &\leq a \wedge (b_1 \vee \ldots \vee b_n) \\ &= (a \wedge b_1) \vee \ldots \vee (a \wedge b_n) \\ &\leq (a_1 \wedge b_1) \vee \ldots \vee (a_n \wedge b_n) \leq y. \end{aligned}$$

It follows that $x \wedge \bigvee X \leq \bigvee \{ x \wedge x' \mid x' \in X \}$ as desired. In order to prove that L^δ is $\bigvee$-distributive, it is enough to consider suprema of collections X of filter elements since the filter elements are join dense in L^δ. However, we need to know that for any $u \in L^\delta$, we have $u \wedge \bigvee X \leq \bigvee \{ u \wedge x' \mid x' \in X \}$. To this end we have

$$\begin{aligned} u \wedge \bigvee X &= \bigvee \{ x \in F(L^\delta) \mid x \leq u \wedge \bigvee X \} \\ &\leq \bigvee \{ x \wedge \bigvee X \mid u \geq x \in F(L^\delta) \} \\ &\leq \bigvee \{ \bigvee \{ x \wedge x' \mid x' \in X \} \mid u \geq x \in F(L^\delta) \} \\ &= \bigvee \{ \bigvee \{ x \wedge x' \mid u \geq x \in F(L^\delta) \} \mid x' \in X \} \\ &\leq \bigvee \{ u \wedge x' \mid x' \in X \}. \end{aligned}$$

This completes the proof. □

Note that, by order duality, it follows that the canonical extension of a distributive lattice also is $\bigwedge$-distributive, but this is not our focus here. Next we prove that the canonical embedding is conditionally Heyting.

Proposition 1. *Let L be a bounded lattice. The canonical extension $\eta : L \to L^\delta$ preserves any existing relative pseudocomplements.*

Proof. Let $a, b \in L$ and suppose $a \to_L b = \max\{c \in L \mid a \wedge c \leq b\}$ exists. Let $x \in F(L^\delta)$ with $a \wedge x \leq b$, then, by compactness, there is $c \in L$ with $a \wedge c \leq b$ and $x \leq c$. Thus $x \leq a \to_L b$. Since $F(L^\delta)$ is join dense in L^δ, it follows that $a \to_L b = \max\{u \in L^\delta \mid a \wedge u \leq b\}$ and thus $a \to_{L^\delta} b$ exists and is equal to $a \to_L b$. □

Corollary 1. *Let A be a Heyting algebra. The canonical extension of A as a bounded lattice is a Heyting algebra embedding.*

2.3 Implication as an Additional Operation

In the previous section, we saw that canonical extension, or equivalently, the representation given by topological duality for bounded lattices, restricts to Heyting algebras giving Heyting algebra representations. The theory of canonical extensions was developed to deal with additional operations that may not be determined by the order of the underlying lattice. As such, canonical extension for distributive lattices with additional operations may be seen as a generalisation (in algebraic form) of Esakia duality.

In this section, we give the general definitions of extensions of maps and relate these, in the case of Heyting algebras, to the Heyting implication on the canonical extension.

Definition 3. Let K and L be lattices, $f\colon K \to L$ any function. We define maps f^σ and f^π from K^δ into L^δ by

$$f^\sigma(u) := \bigvee \left\{ \bigwedge\{ f(a) \mid a \in K \text{ and } x \leq a \leq y\} \mid F(K^\delta) \ni x \leq u \leq y \in I(K^\delta)\right\},$$

$$f^\pi(u) := \bigwedge \left\{ \bigvee\{ f(a) \mid a \in K \text{ and } x \leq a \leq y\} \mid F(K^\delta) \ni x \leq u \leq y \in I(K^\delta)\right\}.$$

The above definition, for arbitrary maps, was first given in the setting of distributive lattices and it was shown that these are in fact upper and lower envelopes with respect to certain topologies [23]. This is not always true in the general lattice setting, but it is still true for mono- and antitone maps and for arbitrary maps on lattices lying in finitely generated lattice varieties. For details, see Sect. 4 of [25] and the Ph.D. thesis of Jacob Vosmaer [47].

The two above extensions of a map f are not always equal, but for maps that are join or meet preserving, or that turn joins into meets or vice versa, the two extensions agree and we say such maps are *smooth*, see e.g. [23, Corollary 2.25]. However, for binary operations, coordinate-wise preservation, or reversal, of join and/or meet is not sufficient to imply smoothness. Example 1 below shows that implication, viewed as an additional binary operation on the lattice underlying a Heyting algebra, need not be smooth.

In the example we will make use of two basic facts about canonical extensions of lattices which are useful when dealing with additional operations: First of all, the canonical extension of a finite product is, up to isomorphism, the product of the canonical extensions of the individual lattices. This allows one to compute coordinate-wise. Secondly, the operation $L \mapsto L^\partial$ which yields for each lattice the order dual lattice also commutes with canonical extension and the filter elements of L^δ are precisely the ideal elements of $(L^\partial)^\delta = (L^\delta)^\partial$ and vice versa.

Looking at the definitions of the extensions of maps, note that they are self dual in the order *on the domain* of the map. Thus we can take the order dual of the domain, or of any coordinate of the domain, and still obtain the same extension. Further note that, if the map is order preserving, then the upper bounds of the intervals on which we are taking meets, and the lower bounds of the intervals on which we are taking joins play no role. Accordingly, for $f : K \to L$ order-preserving we have

$$f^\sigma(u) = \bigvee \left\{ \bigwedge\{ f(a) \mid x \leq a \in K\} \mid F(K^\delta) \ni x \leq u\right\},$$

and, in particular, $f^\sigma(x) = \bigwedge\{ f(a) \mid x \leq a \in K\}$ for filter elements $x \in F(K^\delta)$.

Example 1. Let A be the Heyting algebra consisting of a countable decreasing chain with a bottom added. Since A is a chain it is a Heyting algebra in which the implication is given by

$$f(a, b) = \begin{cases} 1 \text{ if } a \le b, \\ b \text{ if } a > b. \end{cases}$$

Since f is order-preserving in its second coordinate and order-reversing in its first, it will be convenient to regard f as a map from $A^\partial \times A$ to A; we then have an order-preserving map to work with. We label the canonical extensions of the two chains as in the figures below.

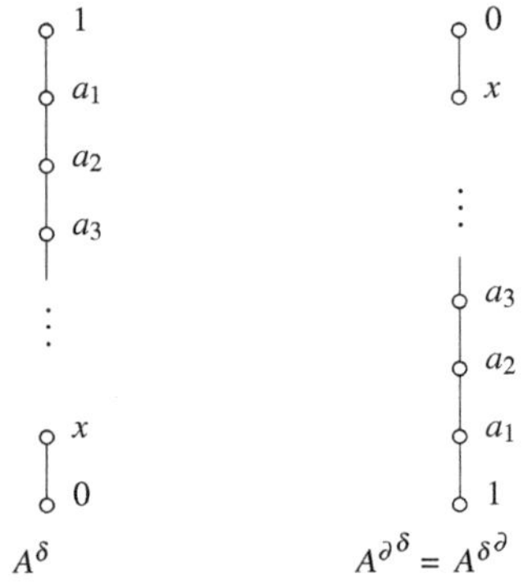

In A^δ there is a single element which is not a lattice element, namely the filter element x. The canonical extension of $A^\partial \times A$ is the product shown below. The only element which is neither a filter element nor an ideal element of the product is (x, x) and this is where the two extensions take different values.

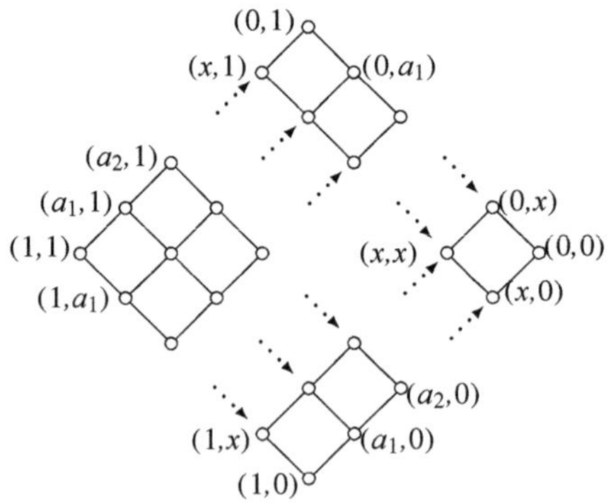

The value of $f^\sigma(x, x)$ is calculated by approaching (x, x) from below with filter elements, and in the second coordinate x is itself a filter element, while it is an ideal element of the dual lattice A^∂. Also, in A^∂, the only filter elements below x are the actual lattice elements $a \in A^\partial$. Thus we get:

$$\begin{aligned}
f^{\sigma}(x,x) &= \bigvee\{\, f^{\sigma}(a,x) \mid A^{\partial} \ni a \leq^{\partial} x \,\} \\
&= \bigvee\left\{\bigwedge\{\, f(a,b) \mid x \leq b \in A \,\} \mid A^{\partial} \ni a \leq^{\partial} x \right\} \\
&= \bigvee\left\{\bigwedge\{b \mid x \leq b \in A, b < a\,\} \mid x \leq a \in A \right\} \\
&= x.
\end{aligned}$$

The value of $f^{\pi}(x,x)$ is calculated by approaching (x,x) from above

$$\begin{aligned}
f^{\pi}(x,x) &= \bigwedge\{\, f^{\pi}(x,b) \mid x \leq b \in A \,\} \\
&= \bigwedge\left\{\bigvee\{f(a,b) \mid A^{\partial} \ni a \leq^{\partial} x\,\} \mid x \leq b \in A \right\} \\
&= \bigwedge\{1 \mid x \leq b \in A\,\} \\
&= 1.
\end{aligned}$$

We conclude that f is not smooth.

It is clear from the above computation, that, if one of the extensions of the implication on A is equal to the Heyting implication that exists on A^{δ}, then it must be the π-extension, and that is indeed the case in general. This is actually just a special instance of the fact that, for an order preserving map f, if g is the upper adjoint of f with respect to some coordinate, then g^{π} is the upper adjoint of f^{σ} with respect to the same coordinate [24, Proposition 4.2].

Proposition 2. *Let* $(A, \to)$ *be a Heyting algebra, then* $(A^{\delta}, (\to)^{\pi})$ *is a Heyting algebra.*

Proof. Let $x, x' \in F(A^{\delta})$, and $y \in I(A^{\delta})$. Using the fact that $\to$ is order preserving as a map from $A^{\partial} \times A$, we have

$$x' \to^{\pi} y = \bigvee\{a' \to b \mid x' \leq a' \in A \ni b \leq y\}.$$

Using the compactness property and the definition of filter and ideal elements we then obtain the following string of equivalences

$$\begin{aligned}
x \wedge x' \leq y &\iff \exists a, a', b \in A\ (x \leq a, x' \leq a', b \leq y \text{ and } a \wedge a' \leq b) \\
&\iff \exists a, a', b \in A\ (x \leq a, x' \leq a', b \leq y \text{ and } a \leq a' \to b) \\
&\iff x \leq x' \to^{\pi} y.
\end{aligned}$$

Now let $u, v, w \in A^{\delta}$, then, using the density property, the fact that A^{δ} is $\bigvee$-distributive, and the definition of extension for additional operations we have the following string of equivalences

$$\begin{aligned}
& u \wedge v \leq w \\
\iff & \forall x, x' \in F(A^\delta)\, \forall y \in I(A^\delta)\left[\, (x \leq u, x' \leq v, \text{and}\, w \leq y) \Rightarrow x \wedge x' \leq y \,\right] \\
\iff & \forall x, x' \in F(A^\delta)\, \forall y \in I(A^\delta)\left[\, (x \leq u, x' \leq v, \text{and}\, w \leq y) \Rightarrow x \leq x' \to^\pi y \,\right] \\
\iff & u \leq v \to^\pi w.
\end{aligned}$$

□

2.4 The Connection Between Heyting Algebras and S4 Modal Algebras

The well-known equivalence between Heyting algebras and certain S4 modal algebras plays a fundamental role in Esakia duality, and in order to clarify the relationship of Esakia's duality to Stone and Priestley duality and to canonical extension, we need a purely algebraic and categorical description of this classical connection. This is the purpose of the current section.

McKinsey and Tarski [38] initiated the rigorous study of the connection between Heyting algebras and S4 modal algebras. They worked with closed sets instead of opens and thus with closure algebras and what we now call co-Heyting algebras. In the 1950s, Rasiowa and Sikorski worked further in this area. Their work may be found in their influential monograph [42]. They are the ones who switched to working with the interior and open sets as it is done now. Another paper in the area that was important to Leo Esakia was the 1959 paper by Dummet and Lemmon [11]. The next period of activity occurred in the 1970s with the work of Esakia and Blok, the most relevant and important publications being [4, 5, 12–14]. In particular, one may find a treatment of the results presented in this section in Sect. 5 of Chap. II of Leo Esakia's 1985 book [14]. This book is written in Russian, but it may soon be available in English translation. Esakia called the S4-algebras corresponding to Heyting algebras stencil algebras. These were also already studied by McKinsey-Tarski and Rasiowa-Sikorski. Blok (and Dwinger) also payed special attention to these. Finally, the fully categorical description of the relationship between S4 and Heyting may be found in the paper [37] by Makkai and Reyes from 1995.

This equivalence between the category of Heyting algebras and the category of what was called stencil S4-algebras is also at the heart of the Blok-Esakia theorem. This theorem states that the lattice of subvarieties of the variety of Heyting algebras is isomorphic to the lattice of subvarieties of the variety of Grzegorczyk algebras. In fact, the reason that the Blok-Esakia theorem is true is that all varieties of Grzegorczyk algebras are generated by the stencil algebras that they contain. The Blok-Esakia theorem is treated in detail in the chapter by Frank Wolter and Michael Zakharyaschev in this same volume.

Here we start from Booleanisation for distributive lattices in general, a theory which dates back to Peremans' 1957 paper [39]. This is already responsible for the fact that distributive lattices alternatively may be seen as embeddings of distributive lattices into Boolean algebras that they generate. This fact is the algebraic counterpart to Priestley duality. In categorical terms, Booleanisation is the left adjoint of the inclusion of the category of Boolean algebras as a full subcategory of the category of bounded distributive lattices. More concretely, given a bounded distributive lattice D, its Booleanisation D^- is the unique, up to isomorphism, Boolean algebra containing D as a bounded sublattice and generated as a Boolean algebra by D. It may be obtained algebraically by a free construction [15, 39] or via duality (or otherwise) by embedding D in the power set of its dual space (or in any other Boolean algebra) and generating a Boolean algebra with the image. The inclusion homomorphism $e_D : D \to D^-$ is the unit of the adjunction and thus the Booleanisation of a bounded lattice homomorphism $h : D \to E$ commutes with the inclusions e_D and e_E so that h^- extends h. Note that the elements of D^- can be written in the form $\bigwedge_{i=1}^n (\neg a_i \vee b_i)$ where the a_is and the b_is all belong to D.

Next, Heyting algebras may be seen as those distributive lattices for which $e_D : D \to D^-$ has a left adjoint, and this extends to a categorical duality.

Proposition 3. *A bounded distributive lattice A is the reduct of a Heyting algebra if and only if the inclusion $e : A \to A^-$ of A in its Booleanisation has an upper adjoint $g : A^- \to A$. Furthermore, a lattice homomorphism $h : A_1 \to A_2$ is a Heyting algebra homomorphism if and only if the following diagram commutes*

$$\begin{array}{ccc} A_1^- & \xrightarrow{h^-} & A_2^- \\ \downarrow{\scriptstyle g_1} & & \downarrow{\scriptstyle g_2} \\ A_1 & \xrightarrow{h} & A_2 \end{array}$$

where $g_i : A_i^- \to A_i$ is the upper adjoint of the embedding $e_i : A_i \to A_i^-$ for $i =$ 1 and 2.

Proof. Suppose $e : A \to A^-$ has an upper adjoint $g : A^- \to A$, and let $a, b, c \in A$. We have

$$\begin{aligned} & a \wedge b \leq c \\ \iff \quad & e(a) \wedge e(b) \leq e(c) \\ \iff \quad & e(a) \leq \neg e(b) \vee e(c) \\ \iff \quad & a \leq g(\neg e(b) \vee e(c)). \end{aligned}$$

Thus A is a Heyting algebra with $b \to c := g(\neg e(b) \vee e(c))$.

Conversely, if A is a Heyting algebra, the following string of equivalences, toggling carefully between the algebras A and A^-, shows that the adjoint does exist and it gives an explicit way of calculating it. Let $a \in A$ and $u = \bigwedge_{i=1}^n (\neg e(b_i) \vee e(c_i)) \in A^-$

where $b_i, c_i \in A$ for each $i \in \{1, \ldots, n\}$, then we have

$$\begin{array}{lrll}
 & e(a) & \leq \bigwedge_{i=1}^{n}(\neg e(b_i) \vee e(c_i)) & \\
\iff & e(a) & \leq \neg e(b_i) \vee e(c_i) & \text{for all } i \in \{1, \ldots, n\} \\
\iff & e(a) \wedge e(b_i) & \leq e(c_i) & \text{for all } i \in \{1, \ldots, n\} \\
\iff & a \wedge b_i & \leq c_i & \text{for all } i \in \{1, \ldots, n\} \\
\iff & a & \leq b_i \to c_i & \text{for all } i \in \{1, \ldots, n\} \\
\iff & a & \leq \bigwedge_{i=1}^{n}(b_i \to c_i) &
\end{array}$$

Finally, given the formulas relating the upper adjoint of the inclusion and the Heyting implication, and using the fact that h^- extends h, it is a simple calculation to see that the statement about morphisms is true. □

On the other hand, this is closely related to S4 modal algebras via the following observation.

Proposition 4. *The category of S4 modal algebras is equivalent to the following category: The objects of the category are adjoint pairs $e : D \leftrightarrows B : g$ where D is a bounded distributive lattice, B is a Boolean algebra, and the lower adjoint $e$is a lattice embedding; The morphisms of the category are pairs (h, k), where $h \colon B \to B'$ is a homomorphism of Boolean algebras, $k \colon D \to D'$ is a bounded lattice homomorphism, and the resulting squares commute both for the upper and lower adjoints.*

Proof. Given an S4 modal algebra $(B, \Box)$, it is easy to check that $\mathrm{Im}(\Box)$ is a bounded distributive sublattice of B, and that the map $g \colon B \to \mathrm{Im}(\Box)$ defined by $b \mapsto \Box(b)$ is upper adjoint to the inclusion map $e \colon \mathrm{Im}(\Box) \to B$. Conversely, given an object $e \colon D \leftrightarrows B \colon g$ in the category as described above, it is also easy to see that $(B, e \circ g)$ is an S4 modal algebra, as well as that the compositions of these two assignments bring us back to an object equal or isomorphic to the one we started with. For the morphisms the pertinent diagram is

$$\begin{array}{ccc}
B_1 & \xrightarrow{h} & B_2 \\
{\scriptstyle e_1}\uparrow\downarrow{\scriptstyle g_1} & & {\scriptstyle e_2}\uparrow\downarrow{\scriptstyle g_2} \\
D_1 & \xrightarrow{k} & D_2
\end{array}$$

Suppose the square commutes both up and down. We have

$$
\begin{aligned}
h \circ (e_1 \circ g_1) &= (h \circ e_1) \circ g_1 \\
&= (e_2 \circ k) \circ g_1 \\
&= e_2 \circ (k \circ g_1) \\
&= e_2 \circ (g_2 \circ h) \\
&= (e_2 \circ g_2) \circ h
\end{aligned}
$$

so that h is a homomorphism for the corresponding modal algebras. Conversely, suppose that $h\colon B_1 \to B_2$ satisfies $h \circ (e_1 \circ g_1) = (e_2 \circ g_2) \circ h$. We define $k\colon D_1 \to D_2$ by $k = g_2 \circ h \circ e_1$. Then we have

$$k \circ g_1 = (g_2 \circ h \circ e_1) \circ g_1 = g_2 \circ (e_2 \circ g_2 \circ h) = g_2 \circ h$$

and

$$e_2 \circ k = e_2 \circ (g_2 \circ h \circ e_1) = \mathrm{id}_{D_2} \circ (h \circ e_1) = h \circ e_1$$

and thus both diagrams commute. □

Combining Proposition 3 and Proposition 4 we obtain the following corollary which is the algebraic counterpart of the famous Gödel translation in logic.

Corollary 2. *The category of Heyting algebras is equivalent to the full subcategory of those S4 modal algebras* $(B, \Box)$ *for which* $\mathrm{Im}(\Box)$ *generates* B.

2.5 From Canonical Extensions to Stone Duality

Historically Jónsson and Tarski obtained canonical extension as an algebraic description of Stone duality. However, in retrospect, canonical extension can be obtained directly and in a choice-free manner, and then the duality can be obtained from it by adding points (via Stone's Prime Filter Theorem). This point of view is particularly advantageous when one wants to understand additional operations on lattices.

Given a distributive lattice D, the canonical extension D^δ is a complete distributive lattice, and, using Stone's Prime Filter Theorem, one can prove that it has enough completely join prime elements. For completeness we give the argument here. Before we do this, note that completely join and meet prime elements of a complete lattice C come in splitting pairs (p, m) satisfying

$$\forall u \in C \qquad (p \nleq u \iff u \leq m)$$

and thus the correspondence between completely join and meet prime elements is given by $p \mapsto \kappa(p) = \bigvee\{u \in C \mid p \nleq u\}$. We denote the poset of completely join prime elements by $J^\infty(C)$ and the poset of completely meet prime elements by $M^\infty(C)$. It then follows that $\kappa : J^\infty(C) \to M^\infty(C)$ is an isomorphism of posets.

Proposition 5. *Let D be a bounded distributive lattice and D^δ its canonical extension. Then every element of D^δ is a join of completely join prime elements and a meet of completely meet prime elements.*

Proof. It suffices to show that for $u, v \in D^\delta$ with $u \not\leq v$ there is a splitting pair (p, m) as described above with $p \leq u$ and $v \leq m$. By the density condition for canonical extensions, $u \not\leq v$ implies that there are $x \in F(D^\delta)$ and $y \in I(D^\delta)$ with $x \leq u$, $v \leq y$, and $x \not\leq y$. Let $F_x = (\uparrow x) \cap D$ be the filter of D corresponding to x and $I_y = (\downarrow y) \cap D$ be the ideal corresponding to y. If $a \in F_x \cap I_y$, then

$$x = \bigwedge F_x \leq a \leq \bigvee I_y = y$$

which contradicts the choice of x and y. Thus, by Stone's Prime Filter Theorem, there is a prime filter $\mathfrak{p}$ of D with $F_x \subseteq \mathfrak{p}$ and $I_x \subseteq D \setminus \mathfrak{p}$. Now, letting $p = \bigwedge \mathfrak{p}$ and $m = \bigvee (D \setminus \mathfrak{p})$, where the extrema are taken in D^δ, we see that $p \leq x \leq u$ and $v \leq y \leq m$.

It remains to show that (p, m) is a splitting pair. To this end, suppose $u \in D^\delta$ and $p \not\leq u$, then there is $y \in I(D^\delta)$ with $u \leq y$ but $p \not\leq y$. Now $p \not\leq y$ means that $\mathfrak{p} \cap I_y = \emptyset$ where $I_y = (\downarrow y) \cap D$. Thus we have $I_y \subseteq D \setminus \mathfrak{p}$ and $u \leq y \leq m$ as required. Conversely, since $\mathfrak{p} \cap (D \setminus \mathfrak{p}) = \emptyset$, we have $p \not\leq m$, and thus $u \leq m$ implies $p \not\leq u$. □

Given the canonical extension of a bounded distributive lattice D, the Stone space of D may be obtained by topologising the set $X = J^\infty(D^\delta)$ with the topology given by the 'shadows' of the ideal elements on X, that is, by the sets $y \mapsto \{p \in J^\infty(D^\delta) \mid p \leq y\}$ for $y \in I(D^\delta)$. Since $I(D^\delta)$ is closed under finite meets and arbitrary joins and the elements of X are completely join prime, it follows that these sets form a topology. One can then show that the sets $\hat{a} = \{p \in J^\infty(D^\delta) \mid p \leq a\}$ for $a \in D$ are precisely the compact open subsets of this space and they generate the topology. This yields a compact sober space in which the compact-open sets form a basis, which is closed under finite intersection. These spaces are known as Stone spaces or spectral spaces. In case the lattice is Boolean, all join primes are atoms and the corresponding space is a compact Hausdorff space with a basis of clopen sets. These spaces are (unfortunately) *also* known as Stone spaces or, for some authors, as Boolean spaces. We call the spaces for distributive lattices *Stone spaces* and the ones for Boolean algebras *Boolean spaces*. As mentioned above, the elements of D correspond to the compact open subsets of the Stone space, the ideal elements of D^δ correspond to the open subsets. Order dually, the filter elements of D^δ correspond, again via the assignment $x \mapsto \{p \in J^\infty(D^\delta) \mid p \leq x\}$, to the closed sets in the Stone topology for the lattice D^∂ that is order dual to D. One can also understand these sets directly relative to the Stone space of D itself. For this purpose some concepts from stably compact spaces are needed (see [36, Sect. 2] for further details): A subset S of a space X is called *saturated* provided it is an intersection of opens (this will yield precisely the downsets of $X = J^\infty(D^\delta)$). Then the sets $\{p \in J^\infty(D^\delta) \mid p \leq x\}$ for $x \in X$ are precisely the compact saturated subsets of of the Stone space of D.

The specialisation order of a topology on a set X is usually defined by $x \leq y$ if and only if $x \in \overline{\{y\}}$ if and only if for every open subset $U \subseteq X$, we have $x \in U$ implies $y \in U$. In this setting this yields the reverse order to the order on $J^{\infty}(D^{\delta})$ as inherited from D^{δ}. Since it is more convenient to work with the order that fits with D^{δ}, we work with the dual definition of specialisation order: $x \leq y$ if and only if $y \in \overline{\{x\}}$ if and only if for every open subset $U \subseteq X$, we have $y \in U$ implies $x \in U$. Thus opens are downsets here rather than upsets.

Given a modality $\Box\colon D \to E$ (that preserves 1 and $\wedge$), the extension $\Box^{\sigma} = \Box^{\pi}\colon D^{\delta} \to E^{\delta}$ (which we call $\Box^{\delta}$) is completely meet preserving, see [23, Theorem 2.21]. Accordingly, it is completely determined by its action on the completely meet prime elements of D^{δ}. This action is encoded using pairs from the Cartesian product of $X_E = J^{\infty}(E^{\delta})$ and $X_D = J^{\infty}(D^{\delta})$ via the relation

$$xSy \iff \kappa(x) \geq \Box^{\delta}(\kappa(y))$$

The relations thus obtained are characterised by three properties:

(B1) $\geq \circ\, S \circ \geq\, = S$;
(B2) $S[x] = \{y \in X_D \mid xSy\}$ is compact saturated for each $x \in X_E$;
(B3) $\Box_S(U) := (S^{-1}[U^c])^c = \{x \in X_E \mid \forall y \in X_D\ (xSy \implies y \in U\}$ is compact open for each compact open subset $U \subseteq X_D$.

The first property is clearly satisfied, the second corresponds to the fact that $\Box^{\delta}$ sends completely meet prime elements to ideal elements, and the third property corresponds to the fact that $\Box^{\delta}$ restricts to a map from D to E (we give the details of the correspondence below in the order dual case of join and 0 preserving modality).

Recovering the modal operator from the relation is easily seen to work just as in Kripke semantics. In fact, this approach via canonical extension makes clear why the box operation given by a Kripke relation should be defined the way it is:

$$\begin{aligned}
\widehat{\Box^{\delta}(a)} &= \big\{x \in X_E \mid x \leq \Box^{\delta}(a)\big\} \\
&= \Big\{x \in X_E \mid \bigwedge\{\Box^{\delta}(\kappa(y)) \mid a \leq \kappa(y)\} \nleq \kappa(x)\Big\} \\
&= \big\{x \in X_E \mid \forall y \in X\ (y \nleq a \implies \Box^{\delta}(\kappa(y)) \nleq \kappa(x))\big\} \\
&= \big\{x \in X_E \mid \forall y \in X\ (xSy \implies y \in \hat{a})\big\} \\
&= \Box_S(\hat{a}).
\end{aligned}$$

Dual statements of course hold for a modality $\Diamond\colon D \to E$ (that preserves 0 and $\vee$). In particular, $\Diamond^{\sigma} = \Diamond^{\pi}\colon D^{\delta} \to E^{\delta}$, which we call $\Diamond^{\delta}$, is completely join preserving and the dual relation is given by $R = \{(x, y) \in X_E \times X_D \mid x \leq \Diamond^{\delta}(y)\}$. This relation satisfies

(D1) $\leq \circ\, R \circ \leq\, = R$;
(D2) $R[x] = \{y \in X_D \mid xRy\}$ is closed for each $x \in X_E$;

(D3) $\Diamond_R(U){:}=R^{-1}[U]=\{x\in X_E\mid \exists y\ (xRy \text{ and } y\in U)\}$ is compact open for each compact open $U\subseteq X_D$.

Finally, we recover the operation from a relation R with these properties via

$$\widehat{\Diamond^\delta(a)}=\Diamond_R(\hat{a})=\big\{x\in X_E\mid \exists y\in X_D\ (xRy \text{ and } y\in\hat{a})\big\}=R^{-1}[\hat{a}].$$

The conditions given here are well-known to duality theorists and may be found, e.g. in [32], but we give an algebraic derivation here based on the canonical extension. To this end first note that it is a simple fact from discrete (Birkhoff) duality that $\bigvee$-preserving maps on downset lattices, $f\colon \mathscr{D}(X)\to\mathscr{D}(Y)$, are in one-to-one correspondence with relations $R\subseteq Y\times X$ satisfying $\leq\circ R\circ\leq\,=R$. Also, the canonical extension of an operation preserving finite joins is completely join preserving [23, Theorem 2.21]. Thus, it suffices to show that the extensions $\Diamond^\delta\colon D^\delta\to E^\delta$ of $\vee$ and 0 preserving maps $\Diamond\colon D\to E$ are characterised within the $\bigvee$-preserving maps from D^δ to E^δ by the conditions (D2) and (D3).

Proposition 6. *Let $k\colon D\to E$ be an order preserving map. Then $k^\sigma\colon D^\delta\to E^\delta$ sends filter elements to filter elements and consequently, if k^σ has an upper adjoint, then this upper adjoint sends ideal elements to ideal elements.*

Proof. By definition

$$k^\sigma(u)=\bigvee\Big\{\bigwedge\{k(a)\mid a\in D \text{ and } x\leq a\leq y\}\mid F(D^\delta)\ni x\leq u\leq y\in I(D^\delta)\Big\}.$$

Thus for $u=x\in F(D^\delta)$ this definition reduces to

$$k^\sigma(x)=\bigvee\Big\{\bigwedge\{k(a)\mid a\in D \text{ and } x\leq a\leq y\}\mid x\leq y\in I(D^\delta)\Big\},$$

and since k is order preserving and $D\hookrightarrow D^\delta$ is compact this is the same as

$$k^\sigma(x)=\bigwedge\{k(a)\mid a\in D \text{ and } x\leq a\}.$$

Thus $k^\sigma(x)\in F(D^\delta)$. Now suppose $g\colon E^\delta\to D^\delta$ is an upper adjoint to k^σ and that $x\in J^\infty(E^\delta)$ and $y\in I(D^\delta)$, then we have

$$\begin{aligned}
x\leq g(y) &\iff k^\sigma(x)\leq y\\
&\iff \bigwedge_{x\leq a\in D} k(a)\leq y\\
&\iff \exists a\in D\ (x\leq a \text{ and } k(a)\leq y)\\
&\implies \exists a\in D\ (x\leq a\leq g(y)).
\end{aligned}$$

Now since $g(y)$ is the join of all the $x \in J^\infty(E^\delta)$ below it, it follows that it is the join of all the $a \in D$ below it and thus it is an ideal element. □

Theorem 4. *Let $f: D^\delta \to E^\delta$ be a $\bigvee$-preserving map. Then $f = \Diamond^\delta$ for some $\Diamond: D \to E$ if and only if the following conditions are met:*

1. *The upper adjoint of f sends completely meet prime elements to ideal elements;*
2. *f sends elements of D to elements of D.*

Proof If $f = \Diamond^\delta$ for some $\Diamond: D \to E$, then it follows from Proposition 6 and the fact that $M^\infty(D^\delta) \subseteq I(D^\delta)$ that the upper adjoint of f sends completely meet prime elements to ideal elements. The second condition is clearly also satisfied as f restricted to D is $\Diamond$.

For the converse, suppose $f: D^\delta \to E^\delta$ is $\bigvee$-preserving and satisfies the two conditions in the theorem. Define $\Diamond: D \to E$ by $\Diamond(a) := f(a)$ for $a \in D$. Then certainly $\Diamond^\delta = f$ on D. Now let $x \in F(D^\delta)$. Then

$$\Diamond^\delta(x) = \bigwedge\{\Diamond a \mid x \le a \in D\} = \bigwedge\{f(a) \mid x \le a \in D\} \ge f(x)$$

since f is order preserving. On the other hand, if $m \in M^\infty(D^\delta)$ and $f(x) \le m$, then $x \le g(m)$ where $g: E^\delta \to D^\delta$ is the upper adjoint of f. Now since, by the first condition, $g(m)$ is an ideal element, there is $a \in D$ with $x \le a \le g(m)$. Thus $f(a) \le m$ and now, as $\Diamond$ is order-preserving

$$\Diamond^\delta(x) \le \Diamond(a) = f(a) \le m.$$

By the meet density of $M^\infty(D^\delta)$ in D^δ, it follows that $\Diamond^\delta(x) = f(x)$ and thus $\Diamond^\delta = f$ on $F(D^\delta)$. Finally since both functions are $\bigvee$-preserving, it now follows that $\Diamond^\delta = f$ on all of D^δ. □

We are now ready to verify that the two conditions in the theorem correspond dually to the conditions (D2) and (D3) given above.

Proposition 7. *Let D and E be bounded distributive lattices with dual spaces X_D and X_E, respectively. Let $f: D^\delta \to E^\delta$ be a $\bigvee$-preserving map and $R \subseteq X_E \times X_D$ the corresponding dual relation. Then the following hold:*

1. *The upper adjoint of f sends completely meet prime elements to ideal elements if and only if R satisfies condition* (D2);
2. *f sends elements of D to elements of D if and only if R satisfies condition* (D3).

Proof. Since the compact open subsets of X_D are precisely the downsets in X_D of elements of D, and since f is obtained from R as the map $U \mapsto R^{-1}[U]$ on downsets, or in other words as the map $u \mapsto \bigvee R^{-1}[{\downarrow}u \cap J^\infty(D^\delta)]$ on the canonical extensions, it is clear that f sends elements of D to elements of D if and only if R satisfies condition (D3). In order to prove the first equivalence, let $m \in M^\infty(E^\delta)$ and take $x' \in J^\infty(E^\delta)$ with $m = \kappa(x')$, then we have

$$
\begin{aligned}
\{x \in X_D \mid x \leq g(m)\} &= \{x \in X_D \mid x \leq g(\kappa(x'))\} \\
&= \{x \in X_D \mid f(x) \leq \kappa(x')\} \\
&= \{x \in X_D \mid x' \nleq f(x)\} = (R[x])^c.
\end{aligned}
$$

Thus $(R[x])^c$ is the downset in X_D of $g(m)$ and this set is open if and only if $g(m)$ is an ideal element. □

The duality for homomorphisms is derivable in a similar manner. The pertinent facts are the following. Let $h: D \to E$ be a map between bounded distributive lattices. Then the following statements are equivalent:

1. h is a bounded lattice homomorphism;
2. $h^\sigma = h^\pi$ is a complete lattice homomorphism;
3. $h^\sigma = h^\pi$ has a lower adjoint which sends completely join primes to completely join primes.

The dual of a bounded lattice homomorphism $h: D \to E$ is the map

$$(h^\delta)^\flat \restriction X_E : X_E \to X_D$$

and it is characterised by the property that, under this map, the pre-image of a compact open is always compact open. Such maps are usually called *spectral maps* or *Stone maps*. The dual of a spectral map $f: X_E \to X_D$ is given by inverse image and so is the canonical extension of the dual map.

If a lattice homomorphism also preserves an additional operation on the lattice, then one can derive, in the same way as we've done above, that the dual maps will satisfy bisimulation conditions with respect to the relation corresponding to the additional operation. We finish this section by considering this situation.

Consider a diagram

$$
\begin{array}{ccc}
E_1 & \xrightarrow{h_1} & E_2 \\
\Big\downarrow{\scriptstyle \Diamond_1} & & \Big\downarrow{\scriptstyle \Diamond_2} \\
D_1 & \xrightarrow{h_2} & D_2
\end{array}
$$

where E_i and D_i are bounded distributive lattices, the h_i are lattice homomorphisms, and the maps $\Diamond_i$ are operators (the argument is similar for n-ary operators). We want to obtain the dual condition to the diagram commuting. If $D_1 = D_2$, and h_1 is equal to h_2, this is simply the statement that it is a homomorphism with respect to the diamonds. To this end one may first show that the following statements are equivalent:

1. $h_2 \circ \Diamond_1 = \Diamond_2 \circ h_1$;
2. $h_2^\delta \circ \Diamond_1^\sigma = \Diamond_2^\sigma \circ h_1^\delta$;
3. $\forall x \in X_{E_1}\ h_2^\delta(\Diamond_1^\sigma(x)) = \Diamond_2^\sigma(h_1^\delta(x))$

4. $\forall x \in X_{E_1} \forall z \in X_{D_2}\ (z \leq h_2^\delta(\Diamond_1^\sigma(x)) \iff z \leq \Diamond_2^\sigma(h_1^\delta(x)))$.

The equivalence of (1) and (2) follows from the fact that the first map is an operator and the second one order preserving on either side of the equality since this implies that $(h_2 \circ \Diamond_1)^\sigma = h_2^\delta \circ \Diamond_1^\sigma$ and $(\Diamond_2 \circ h_1)^\sigma = \Diamond_2^\sigma \circ h_1^\delta$, see [22, Theorem 4.3]. The equivalence of (2) and (3) follows because all the extended functions are completely join preserving and X_{E_1} is join-dense in E_1^δ. The last two are equivalent because X_{D_2} is join-dense in D_2^δ. Now denoting the dual of h_i by f_i and the dual of $\Diamond_i$ by R_i for $i = 1$ and 2, we get

$$\begin{aligned} z \leq h_2^\delta(\Diamond_1^\sigma(x)) &\iff f_2(z) \leq \Diamond_1^\sigma(x) \\ &\iff f_2(z) R_1 x \end{aligned}$$

and

$$\begin{aligned} z \leq \Diamond_2^\sigma(h_1^\delta(x)) &\iff \exists z' \in X_{E_2}\ (z \leq \Diamond_2^\sigma(z') \text{ and } z' \leq h_1^\delta(x)) \\ &\iff \exists z' \in X_{E_2}\ (z R_2 z' \text{ and } f_1(z') \leq x) \end{aligned}$$

So the above diagram commutes if and only if

$$\forall z \in X_{D_2} \forall x \in X_{E_1} \left[f_2(z) R_1 x \iff \exists z' \in X_{E_2}\ (z R_2 z' \text{ and } f_1(z') \leq x) \right].$$

Note that the backward implication can be simplified as we can bring the quantifier outside to get

$$\forall z \in X_{D_2} \forall z' \in X_{E_2} \forall x \in X_{E_1} \left[(z R_2 z' \text{ and } f_1(z') \leq x) \implies f_2(z) R_1 x \right].$$

A special case of this condition is the one obtained by choosing $x = f_1(z')$:

$$\forall z \in X_{D_2} \forall z' \in X_{E_2} \left[z R_2 z' \implies f_2(z) R_1 f_1(z') \right].$$

On the other hand, since $\leq \circ R_1 \circ \leq\ = R_1$ the latter condition also implies the previous one. So the diagram commutes if and only if the following two conditions hold:

($\Diamond$back) $\forall z \in X_{D_2} \forall x \in X_{E_1} \left[f_2(z) R_1 x \implies \exists z' \in X_{E_2}\ (z R_2 z' \text{ and } f_1(z') \leq x) \right]$.
($\Diamond$forth) $\forall z \in X_{D_2} \forall z' \in X_{E_2} \left[z R_2 z' \implies f_2(z) R_1 f_1(z') \right]$.

In the case where $h_1 = h_2$ these are precisely the conditions dual to being a $\Diamond$-homomorphism between bounded distributive lattices. For box operations $\Box_1$ and $\Box_2$ with dual relations S_1 and S_2, respectively, we get order-dual dual conditions, namely:

($\Box$back) $\forall z \in X_{D_2} \forall x \in X_{E_1} \left[f_2(z) S_1 x \implies \exists z' \in X_{E_2}\ (z S_2 z' \text{ and } f_1(z') \geq x) \right]$.
($\Box$forth) $\forall z \in X_{D_2} \forall z' \in X_{E_2} \left[z S_2 z' \implies f_2(z) S_1 f_1(z') \right]$.

2.6 From Canonical Extensions to Esakia Spaces

As we have seen in Proposition 2 of Sect. 2.3, (π-)canonical extension provides a choice free approach to duality for Heyting algebras. In Sect. 2.5 we have seen how to obtain Stone duality from canonical extension. In this section we spell out how to move between the canonical extension of a Heyting algebra and its Esakia dual space. In order to witness the Heyting implication, we will make use of the results of Sect. 2.4 relating Heyting algebras to pairs of adjoint maps. It can also be done directly as we will indicate at the end of this section, however, we feel that the approach via adjoint pairs of maps is the most transparent and reflects most directly the spirit of the work of Leo Esakia.

Accordingly we need the following correspondence results:

1. Sublattices $D \hookrightarrow E$ of bounded distributive lattices E correspond dually to spectral quotients, but these can most simply be described not as certain equivalence relations but as certain quasi-orders on the dual Stone space [45]. A quasi-order gives rise to an equivalence relation and to an order on the quotient which will be the specialisation order of the spectral quotient space: Given a spectral space X, its spectral quotients are in one-to-one correspondence with the so-called compatible quasi-orders $\preceq \subseteq X \times X$ [17, Theorem 6]. A *compatible quasi-order* on a Stone space X is a quasi-order on X satisfying the following separation condition for all $x, y \in X$

$$x \not\preceq y \implies \exists U \subseteq X \ (U \text{compact open and a} \preceq\text{-downset}, y \in U \text{ and } x \notin U).$$

   Here U is a $\preceq$-downset provided for all $z, z' \in X$ we have $z \preceq z' \in U$ implies $z \in U$. Given a sublattice $D \hookrightarrow E$, the corresponding quasi-order is given by $\preceq_D = \{(x, y) \in X_E \times X_E \mid \forall a \in D \ (y \leq a \implies x \leq a)\}$ where the comparisons of x and y with a are made in E^δ (note $D \subseteq D^\delta \hookrightarrow E^\delta$). Given a compatible quasi-order, $\preceq$ on X_E, the dual space of the corresponding sublattice of E is the quotient space $X_D = (X_E/\approx, \tau_\preceq)$, where $\approx \, = \, \preceq \cap \succeq$ and $\tau_\preceq$ is given by those open subsets of the space X_E which are also $\preceq$-downsets. *We will denote the space $(X/\approx, \tau_\preceq)$ given by a given compatible order $\preceq$ on a Stone space X by* $X/\preceq$. The map dual to the embedding $D \hookrightarrow E$ is the quotient map $X_E \twoheadrightarrow X_E/\preceq$ and the relation R corresponding to $D \hookrightarrow E$ viewed as a 0 and $\vee$ preserving map from D to E is the relation $xR[x']_\approx$ iff $x \preceq x'$.
2. A pair of maps $\Diamond\colon D \leftrightarrows E \colon \Box$ is an adjoint pair with $\Diamond$ the lower adjoint and $\Box$ the upper adjoint if and only if the relation R dual to $\Diamond$ and the relation S dual to $\Box$ are converse to each other. That is, $S = R^{-1}$. The interesting direction of this fact is true because of the following string of equivalences:

$$\begin{aligned} xRx' &\iff x \leq \Diamond^{\delta}(x') \\ &\iff \kappa(x) \not\geq \Diamond^{\delta}(x') \\ &\iff \Box^{\delta}(\kappa(x)) \not\geq x' \\ &\iff \kappa(x') \geq \Box^{\delta}(\kappa(x)) \iff x'Sx. \end{aligned}$$

Now we just need one more correspondence result before we can get the Esakia duality for Heyting algebras. The following proposition is a generalisation of Theorem 4.5 of Chap. III in Esakia's book [14], which proves the same statement, but just for Heyting algebras. In addition (the hard direction) is the algebraic dual of the result needed in Priestley duality that each clopen downset comes from a lattice element.

Proposition 8. *Let B be a Boolean algebra with dual space X_B, and let D be a sublattice of B with corresponding compatible quasi-order $\preceq$ on X_B. Then D generates B as a Boolean algebra if and only if $\preceq$ is antisymmetric and thus a partial order.*

Proof. Suppose D generates B as a Boolean algebra, and let $x, x' \in X_B$ with $x \neq x'$. Since x and x' are filter elements of B^{δ}, there is $b \in B$ with $x \leq b$ but $x' \not\leq b$. Since D generates B as a Boolean algebra, $b = \bigvee_{i=1}^{n}(\bigwedge_{j=1}^{m_i} a_{ij})$, where each a_{ij} is either an element of D or the complement of one. Now $x \leq b$ and x an atom implies $x \leq \bigwedge_{j=1}^{m_i} a_{ij}$ for some i. Rewriting the latter conjunction in the form $(\bigwedge_{j=1}^{k} a_j) \wedge (\bigwedge_{j=k+1}^{m_i} \neg a_j)$, we obtain from $x' \not\leq b$ that $x' \not\leq (\bigwedge_{j=1}^{k} a_j) \wedge (\bigwedge_{j=k+1}^{m_i} \neg a_j)$ and thus there is a $j \in \{1, \ldots, k\}$ with $x' \not\leq a_j$ or there is $j \in \{k+1, \ldots, m_i\}$ with $x' \not\leq \neg a_j$. In the first case we obtain $x \leq a_j$ and $x' \not\leq a_j$ and in the second case we obtain $x \not\leq a_j$ and $x' \leq a_j$. Thus, by the definition of $\preceq$, either $x' \not\preceq x$ or $x \not\preceq x'$ and thus $\preceq$ is antisymmetric.

For the converse, fix first $x \in X_E$. For each $y \in X_E$ with $x \not\preceq y$, there is $a_y \in D$ with $y \leq a_y$ but $x \not\leq a_y$, and for each $y \in X_E$ with $y \not\preceq x$, there is $c_y \in D$ with $x \leq c_y$ but $y \not\leq c_y$. And thus the equivalence classes of $\approx\, = \,\preceq \cap \succeq$ are given by

$$\bigvee [x]_{\approx} = \left(\bigwedge_{x \not\preceq y \in X_E} \neg a_y\right) \wedge \left(\bigwedge_{y \not\preceq x \in X_E} c_y\right)$$

where the joins and meets are of course taken in E^{δ}.
Now suppose $\preceq$ is antisymmetric. Then $[x]_{\approx} = \{x\}$ and thus $\bigvee [x]_{\approx} = x$ for each $x \in X_E$. Thus, for $b \in B$ and for each $x \in X_E$ with $x \leq b$ we get

$$b \geq x = \left(\bigwedge_{x \not\preceq y \in X_E} \neg a_y\right) \wedge \left(\bigwedge_{y \not\preceq x \in X_E} c_y\right).$$

Let $<D>$ denote the Boolean subalgebra of B generated by D. Applying compactness of E^δ to the fact that $\left(\bigwedge_{x \not\leq y \in X_E} \neg a_y\right) \wedge \left(\bigwedge_{y \not\leq x \in X_E} c_y\right)$ is below b, we conclude that there is a finite submeet which gets below b. That is, there is $b_x \in <D>$ with $x \leq b_x \leq b$. Since X_E is join-dense in B it follows that

$$b = \bigvee_{b \geq x \in X_E} b_x.$$

Again by compactness of E^δ, there are $x_1, \ldots, x_n \in X_E$ with $x_i \leq b$ and $b \leq \bigvee_{i=1}^n b_{x_i}$. Since each $b_x \leq b$ we actually have equality and thus $b \in <D>$. □

Theorem 5. (Priestley duality [41]) *The category of bounded distributive lattices is dually equivalent to the category whose objects are Boolean spaces each equipped with a compatible partial order and whose morphisms are continuous and order preserving maps.*

Proof. This follows from the above results and the fact that the category of distributive lattices is equivalent to the category of lattice embeddings $D \hookrightarrow B$ such that D generates B with pairs of maps making commutative diagrams, e.g.

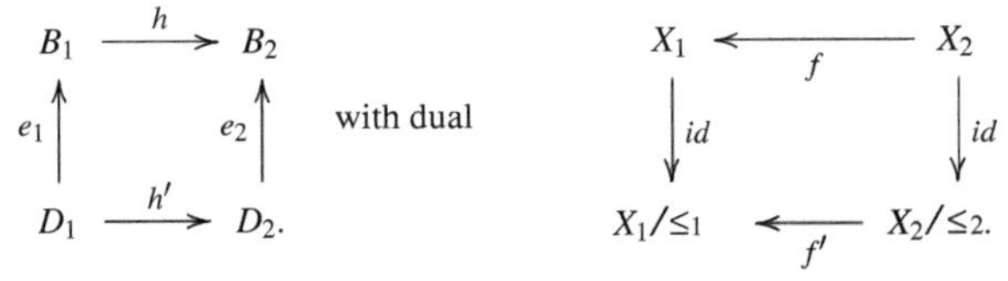

where the topology on the lower spaces are those that are downsets of the respective partial orders that are open in the respective topologies of the upper spaces. Saying that the diagram on the left commutes is equivalent to saying that the one on the right commutes, and this in turn is the same as saying that the maps f and f' are equal as set maps and that they are continuous both in the Boolean topology of the spaces on the top and in the spectral topology of the spaces on the bottom. This in turn is easily seen to be equivalent to saying that $f = f'$ is both continuous in the Boolean topology and order preserving. □

The ordered spaces $(X, \leq)$ where X is a Boolean space and $\leq$ is a compatible partial order on X are of course well known by now as *Priestley spaces* and the maps $f : (X, \leq) \to (Y, \leq)$ which are both continuous and order preserving are the Priestley maps. For the Esakia duality we need also the notion of a bounded morphism. A map $f : X \to Y$ between Priestley spaces is called a *bounded morphism* provided it is continuous, order preserving, and for each $x \in X$ and each $y \in Y$ with $y \leq f(x)$, there is $z \in X$ with $z \leq x$ and $y = f(z)$.

Theorem 6. (Esakia Duality [12]) *The category of Heyting algebras is dually equivalent to the category whose objects are Priestley spaces* $(X, \leq)$ *such that, for each* *clopen subset U of X, the set $\uparrow U$ is also clopen and whose morphisms are the bounded morphisms.*

Proof. For objects, we use the fact that a Heyting algebra corresponds to an adjoint pair $e\colon D \rightleftarrows B\colon g$ where e is an embedding whose image generates B. The dual of these embeddings e are precisely the Priestley spaces $(X, \leq)$ and the relation R corresponding to the quotient map $X \twoheadrightarrow (X, \leq)$ is, as remarked above, the relation

$$x R\{x'\}(= [x']_{\approx}) \iff x \leq x'.$$

Accordingly, we think of R simply as being $\leq$ (identifying $\{x\}$ with x for each $x \in X$). Note that, since the relation $R = \leq$ is the relation corresponding to the embedding e (as stated in item 1 in the beginning of this section), $\leq$ satisfies (D1) through (D3) (this is a consequence of the requirements for being a compatible quasi-order). The fact that e has an upper adjoint is, by the second item in the list at the beginning of this section, equivalent to the fact that the reverse relation $R^{-1} = \geq$ satisfies conditions (B1) through (B3). The condition (B1), $\geq \circ \geq \circ =$ being equal to $\geq$, is vacuously true. Condition (B2) states that $\downarrow x$ is closed in X and this also always holds because

$$\downarrow x = \bigcap \{U \mid x \in U, U \text{ a clopen downset}\}$$

is compact saturated by (D3). Finally (B3) requires that for each clopen subset U of X, the set $(\uparrow U)^c$ is a clopen downset, and this is equivalent to saying that for each clopen subset U of X, the set $\uparrow U$ itself is clopen. Therefore this last condition is the only additional condition on the Priestley space $(X, \leq)$.

Homomorphisms between Heyting algebras correspond to commutative diagrams

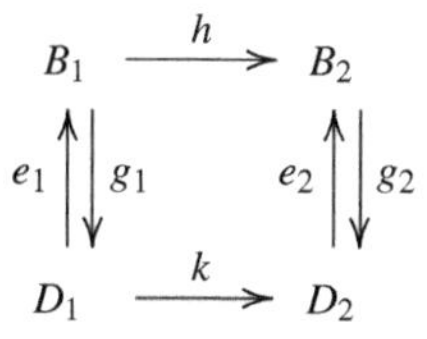

with dual

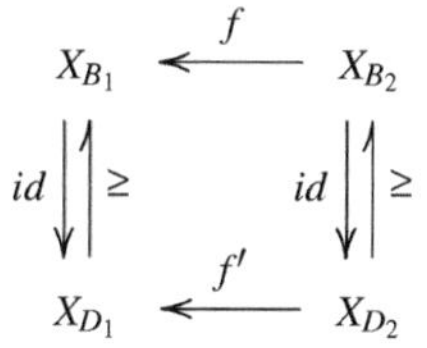

The commutativity on the algebraic side with respect to the maps e_i is just the Priestley map condition that $f = f'$ is continuous in the Boolean topologies and is order preserving. The commutativity of the diagram with respect to the maps g_i is precisely of the form treated at the very end of Sect. 2.5 for $\Box$ operations. Thus the additional requirements for this continuity are:

($\Box$back) $\forall z \in X_{D_2} \forall x \in X_{B_1} \left[f'(z) \geq_1 x \implies \exists z' \in X_{B_2} \ (z \geq_2 z' \text{ and } f(z') = x)\right]$.
($\Box$forth) $\forall z \in X_{D_2} \forall z' \in X_{B_2} \left[z \geq_2 z' \implies f'(z) \geq_1 f(z')\right]$.

Now replacing $id_i\colon X_{B_i} \to X_{D_i}$ by the corresponding Priestley space $(X_i, \leq_i)$ and using the fact that $f' = f$, we obtain the Esakia dual map

$$(X_1, \leq_1) \xleftarrow{\ f\ } (X_2, \leq_2)$$

which is order preserving and continuous and satisfies the conditions:

(□back) $\forall z \in X_2 \forall x \in X_1 \left[f(z) \geq_1 x \implies \exists z' \in X_2 \ (z \geq_2 z' \text{ and } f(z') = x) \right]$.
(□forth) f is order preserving.

However, since f is already required to be order preserving and continuous, all that remains is the (□back) condition, which is precisely the condition defining bounded morphisms. □

Remark 1. In the same manner as we have derived the Priestley and Esakia dualities from our descriptions of the corresponding categories of maps, we could derive the duality for S4 modal algebras as the duals of adjoint pairs $h\colon D \rightleftarrows B\colon e$, that is, Boolean spaces with a compatible quasi-order such that, for each clopen subset U, the set $\uparrow U$ is also clopen with continuous quasi-order preserving (quasi)bounded morphisms.

Remark 2. A more direct approach to duality for Heyting algebras is to take as dual space for a Heyting algebra A, the dual of the underlying lattice equipped with the ternary Kripke relation obtained from the binary implication operation with spectral maps that are bounded morphisms for this ternary relation. This is not as nice a presentation but is completely equivalent. We show here how to derive the ternary relation corresponding to implication on a Heyting algebra. Let A be a Heyting algebra and $\to\, : A^\partial \times A \to A$. Here A^∂ stands for the order dual of A. With this flip in the first coordinate, $\to$ is a dual operator and we can compute the corresponding relation $S = S_\to$ using the π-canonical extension of $\to$. As we've already seen in the unary case we get the same relation, with some switching of the order of coordinates as for its lower adjoint $\wedge$. For the following computation, one needs to know that the canonical extension of meet is the meet of the canonical extension [22, Lemma 4.4]. For all $x, y, z \in X_A$ we have

$$
\begin{aligned}
S(x, y, z) &\iff \kappa(x) \geq y \to^\pi \kappa(z) \\
&\iff x \not\leq y \to^\pi \kappa(z) \\
&\iff x \wedge^\sigma y \not\leq \kappa(z) \\
&\iff z \leq x \wedge y \\
&\iff z \leq x \text{ and } z \leq y
\end{aligned}
$$

which is of course interderivable with the binary relation $\leq$. We leave it as an exercise for the reader to check bounded morphisms for this ternary relation are precisely the same as those in the Esakia duality.

2.7 Esakia's Lemma and Sahlqvist Theory

Esakia formulated and proved Esakia's lemma in order to prove that the topological dual of a Heyting algebra homomorphism is a bounded morphism. Esakia's lemma has since played an important role in modal and intuitionistic model theory. In particular, it was used by Sambin and Vaccaro in their simplified proof of Sahlqvist's Theorem [44]. A generalisation of Esakia's lemma, formulated in algebraic terms [22, Lemma 3.8], was the key idea in our proof with Bjarni Jónsson of a fact regarding compositionality of canonical extensions of maps. We then used this result to prove the functoriality of canonical extension (this corresponds to Esakia's application of his lemma) as well as Sahlqvist-type theorems (corresponding to Sambin and Vaccaro's application of Esakia's Lemma).

We begin by stating Esakia's Lemma as Esakia stated it [12, Lemma 3].

Lemma 1. *If* (X, R) *is an Esakia space and* $\mathscr{C}$ *is a downward directed family of closed subsets of* X, *then*

$$R^{-1}\left[\bigcap \mathscr{C}\right] = \bigcap R^{-1}\left[\mathscr{C}\right].$$

In canonical extension language, this translates as follows. Let A be a Heyting algebra and let $B = A^-$ be its Booleanisation. Let $X = J^\infty(A^\delta)$, then $A^\delta = \mathscr{D}(X)$ and $B^\delta = \mathscr{P}(X)$. The subsets of X that are closed in topological terms are precisely the filter elements of B^δ. The relation R on the Esakia space of A is, as we have seen in the previous section, the relation $S_\Box$ dual to the box operation on B given by

$$\Box\colon B \xrightarrow{g} A \xrightarrow{e} B.$$

It is not hard to see that, on a Boolean algebra, a relation dual to a box operation is also dual to a diamond operation. Indeed the relation R on the Boolean space underlying the Esakia space of A is the relation $S_\Box$ for the above given box operation and it is also the relation $R_\Diamond$ for the conjugate diamond operation, $\Diamond = \neg\Box\neg$. For this operation, we saw in the previous section that $\Diamond^\delta$ on B^δ is just the operation $S \mapsto R^{-1}[S]$ on $\mathscr{P}(X)$. Thus Esakia's lemma says that for any down-directed family $\mathscr{C}$ of filter elements we have that

$$\Diamond^\delta\left(\bigwedge \mathscr{C}\right) = \bigwedge\{\Diamond^\delta(c) \mid c \in \mathscr{C}\},$$

where the infima are taken in B^δ. That is, while $\Diamond^\delta$ in general only preserves joins, it also preserves down-directed meets of filter elements. Order dually, of course this same statement also means that $\Box^\delta$ preserves directed joins of ideal elements.

This lemma, which holds for the canonical extension of the box operation associated with a Heyting algebra actually holds for canonical extensions of order preserving maps between distributive lattices in general. That is the content of Lemma 3.8 in [22], which was stated as follows.

Lemma 2. *Let* A and B *be bounded distributive lattices, and suppose* $f\colon A \to B$ *is isotone. For any down-directed set* $\mathscr{D}$ *of filter elements of* A^δ,

$$f^{\sigma}\left(\bigwedge \mathscr{D}\right) = \bigwedge\{f^{\sigma}(x) \mid x \in \mathscr{D}\}.$$

In [22], we used this lemma to prove that, for every $x \in X_B$ and for every element $u \in A^{\delta}$ that satisfies the inequality $x \leq f^{\sigma}(u)$, there is a minimal such solution below u. This is the crucial fact used in [22] to prove that if f is join preserving in each coordinate, then f^{σ} is Scott continuous. This, in conjunction with the following theorem, then leads to a proof of functoriality of canonical extension and to Sahlqvist-type results.

Theorem 7. [22, Theorem 4.3] *Let* $g: A \to B$ *be order preserving and* $f: B \to C$ *be such that* f^{σ} *is continuous, then* $(f \circ g)^{\sigma} = f^{\sigma} \circ g^{\sigma}$.

In order to prove functoriality of canonical extension one must show that $h \circ \Diamond_1 = \Diamond_2 \circ h$ implies that $h^{\delta} \circ \Diamond_1^{\sigma} = \Diamond_2^{\sigma} \circ h^{\delta}$. This is also crucial in the derivation of the description of bounded morphisms as we saw at the end of Sect. 2.5. The argument, on the basis of Theorem 7, goes as follows:

$$\begin{array}{rrl} & h \circ \Diamond_1 & = \Diamond_2 \circ h \\ \Longrightarrow & (h \circ \Diamond_1)^{\sigma} & = (\Diamond_2 \circ h)^{\sigma} \\ \stackrel{Th.7}{\Longrightarrow} & h^{\sigma} \circ (\Diamond_1)^{\sigma} & = (\Diamond_2)^{\sigma} \circ h^{\sigma} \\ \Longrightarrow & h^{\delta} \circ (\Diamond_1)^{\sigma} & = (\Diamond_2)^{\sigma} \circ h^{\delta}. \end{array}$$

The method for showing that equational properties holding on a lattice with additional operations lift to the canonical extension is similar, and this is the subject of Sahlqvist theory. An equation may be seen as the equality of two compositions of maps that are either basic operations or juxtapositions of such ($f: A \to B$ and $g: C \to D$ yield $[f, g]: A \times C \to B \times D$). Showing that the equation lifts is then a matter of showing that canonical extension commutes with composition and juxtaposition. In our paper [23] with Bjarni Jónsson, we no longer relied (directly) on Esakia's lemma for these kind of arguments, but developed a theory based on topology which allows a more transparent and uniform treatment of issues concerning the interaction of extending maps and composing them.

2.8 Esakia Spaces as Completions of Universal Models

One of the purposes of the dual space of a bounded distributive lattice is to supply a representation theorem: every bounded distributive lattice may be realised as a sublattice of a powerset lattice. For some lattices, one does not need something so complicated as the dual space to obtain such a representation. This is true, for example, in computer science in the study of classes of formal languages. By definition, these are given as sublattices of the powerset of the set of all words A^* in some alphabet A. This is also true for lattices that have "enough" join prime elements in the sense that each element is a join of join prime elements. This is true, e.g., for free bounded distributive lattices where the pure conjunctions of generators join generate and are join prime. This is also true for (finitely generated) free Heyting algebras. This latter fact is at the lattice theoretic origin of the so-called universal models of intuitionistic propositional logic.

Let $N \in \mathbb{N}$, $N \geq 1$, and let $A = F_{HA}(N)$ be the free Heyting algebra on N generators. Then A is infinite, but, as a lattice, it may be built incrementally as the direct limit (or colimit in categorical terms) of the finite sublattices A_n consisting of all elements of implicational rank less than or equal to n (meaning that there is a term describing such an element in terms of the generators in which the maximum number of nested implications is less than or equal to n). This direct limit, and its dual inverse limit of finite posets, have been studied extensively by Ghilardi [27–30] and others [3, 7, 8]. The dual inverse limit may be built up in a uniform way as follows.

Given a poset $(X, \leq)$, we say that a subset S of X is *rooted* if there exists $p \in S$ such that $q \leq p$ for each $q \in S$. In this case, we call p the *root* of S. It follows from the definition that a root of a rooted set is unique. We denote by $\mathscr{P}_r(X)$ the set of all rooted subsets of $(X, \leq)$. We also let $root : \mathscr{P}_r(X) \to X$ be the map sending each rooted subset S of X to its root. Now define the sequence $\{X_n\}_{n \in \mathbb{N}}$ of finite posets as follows:

$$X_0 = J(F_{DL}(N))(= \mathscr{P}(N)) \qquad X_1 = \mathscr{P}_r(X_0)$$

$$\text{For } n \geq 1 \quad X_{n+1} = \{\tau \in \mathscr{P}_r(X_n) \mid \forall T \in \tau \ \forall S \in X_n \ (S \leq T \implies \exists T' \in \tau \ (T' \leq T \text{ and } root(S) = root(T'))\}.$$

The condition defining X_{n+1}, which was first given in [27] might seem strange but it can be derived using correspondence theory in a straight forward way [3].

One can then show that the functions $root \colon \mathscr{P}_r(X_n) \to X_n$ remain surjective when one restricts the domains to X_{n+1}. We denote by ∇ the sequence

$$X_0 \xleftarrow{root} X_1 \xleftarrow{root} X_2 \ \cdots$$

The dual sequence

$$D_0 \xhookrightarrow{i_0} D_1 \xhookrightarrow{i_1} D_2 \xhookrightarrow{i_2} \cdots$$

has (the lattice underlying) $A = F_{HA}(N)$ as its direct limit, and the maps i_n are the duals of the root maps and are thus the upper adjoints of the forward image maps of the root maps. An observation, also dating back to Ghilardi, is that the maps in the sequence ∇ have upper adjoints, or in other words that the maps i_n send join primes to join primes and thus the join primes of the lattices D_n remain join prime all the way up the chain and thus also in A, which is then join-generated by the set $J(A)$ of join prime elements of A.

The upper adjoint maps are the maps $X_n \to X_{n+1}, x \mapsto {\downarrow}x$. Thus ∇ is both an inverse and a direct limit system

$$X_0 \underset{root}{\overset{\downarrow(\,)}{\rightleftarrows}} X_1 \underset{root}{\overset{\downarrow(\,)}{\rightleftarrows}} X_2 \;\ldots$$

From the above analysis of the sequence ∇, we see the following relationship:

$$\text{The Esakia space of } A \text{ is } \; X = \varprojlim X_n$$
$$\text{The set of join primes of } A \text{ is } \; J(A) = \varinjlim X_n.$$

Here the inverse limit may be taken in topological spaces and we obtian the Esakia space with its topology. The direct limit (or co-limit in category theoretic terms) may be taken in posets to obtain the collection of all join-irreducibles in A with the induced order. By the above analysis it is clear that A may be given a set representation in $J(A)$ via the embedding $A \hookrightarrow \mathscr{P}(J(A))$ where $a \in A$ is sent to $\{x \in J(A) \mid x \leq a\}$. Note that the sets $\{x \in J(A) \mid x \leq a\}$ for $a \in A$ are precisely the downsets of finite antichains in the poset $J(A)$. However, the poset $J(A)$ is fairly complicated and a smaller one suffices to obtain a set representation of A. This is the point of the so called *(N-)universal model.* The universal model, U, was already anticipated in [9] and originates with [43, 46]. See also [1, 33].The subject was revived in [2], where it was, among other, shown that the poset underlying the universal model consists of the finite height elements of the Esakia space of A (see [2, Theorem 3.2.9]). The interpretation that is part of the universal model is precisely the map sending $a \in A$ to $\{x \in U \mid x \leq a\}$. Also, the so-called *de Jongh formulas* show that the downset (in our order) (as well as the completement of the upset) of such a finite height element of the Esakia space X is (are) clopen (see [2, Theorem 3.3.2]). It is easy to see that the downset of a point in a Priestley space is clopen if and only if the point corresponds to a principal prime filter (and thus to a join prime element of the dual lattice). Thus the poset underlying the universal model U is contained in $J(A)$ and, as mentioned above, already the points in U are enough to obtain a set representation of A (see, e.g. [2, Theorem 3.2.20]). In this representation, an element $a \in A$ is of course sent to $\{x \in U \mid x \leq a\}$, the so-called *definable subsets* of the universal model U. While U is simpler than $J(A)$, no characterisation of the definable sets, $\{x \in U \mid x \leq a\}$ for $a \in A$, is known.

The fact that A is representable as a lattice of subsets of $J(A)$ and of U yields a connection between these posets and the Esakia space of A. This connection is a special case of results presented in [19] and this is the last topic in this survey of recent developments in duality theory as they relate to Esakia's work. Thus we will outline here those results of [19] that are concerned just with set representations of lattices. For further details see [19, Sect. 1] and the forthcoming journal paper [18]. The key initial observation of [19], relative to set representations of lattices, is that a set representation $D \hookrightarrow \mathscr{P}(X)$ may faithfully be seen as a special kind of quasi-uniform space.

A *quasi-uniform space* is a pair $(X, \mathscr{U})$, where X is a set, and $\mathscr{U}$ is a collection of subsets of $X \times X$ having the following properties:

1. $\mathscr{U}$ is a filter of subsets of $X \times X$ contained in the up-set of the diagonal $\Delta = \{(x, x) \mid x \in X\}$;
2. for each $U \in \mathscr{U}$, there exists $V \in \mathscr{U}$ such that $V \circ V \subseteq U$;

The collection $\mathscr{U}$ is called a quasi-uniformity and its elements are called *entourages* and should be thought of as the epsilon-neighboorhoods of the diagonal in a quasi-metric space, i.e., sets of the form $U \supseteq \{(x, y) \mid d(x, y) < \epsilon\}$ for some $\epsilon > 0$. The condition (2) corresponds to the triangle inequality. A quasi-uniform space is said to be a *uniform space* provided the converse, U^{-1}, of each entourage U is again an entourage of the space (this corresponds to the symmetry axiom for metrics).

A function $f\colon (X, \mathscr{U}) \to (Y, \mathscr{V})$ between quasi-uniform spaces is *uniformly continuous* provided $(f \times f)^{-1}(V) \in \mathscr{U}$ for each $V \in \mathscr{V}$. Sometimes we will write $f\colon X \to Y$ is $(\mathscr{U}, \mathscr{V})$-uniformly continuous instead to express this fact. A quasi-uniform space $(X, \mathscr{U})$ always gives rise to a topological space. This is the space X with the induced topology, which is given by $V \subseteq X$ is open provided, for each $x \in V$, there is $U \in \mathscr{U}$ such that $U(x) = \{y \in X \mid (x, y) \in U\} \subseteq V$. In general, several different quasi-uniformities on X give rise to the same topology. We will assume that all spaces are Komolgorov, that is, the induced topology is T_0. This requirement is equivalent to the intersection of all the entourages in $\mathscr{U}$ being a partial order rather than just a quasi-order on X. In case a quasi-uniform space $(X, \mathscr{U})$ is not separated, it may be mapped to its so-called Komolgorov quotient which is given by the equivalence relation obtained by intersecting all the $U \cap U^{-1}$ for $U \in \mathscr{U}$, or equivalently, given by the partial order reflection of the quasi-order corresponding to $\mathscr{U}$. For the basic theory of uniform spaces and quasi-uniform spaces, we refer to [6, 16].

We are now ready to explain how set representations may be viewed, up to isomorphism, as certain quasi-uniform spaces, which we will call Pervin spaces. First of all, instead of working with lattice representations $e\colon D \hookrightarrow \mathscr{P}(X)$, we will work with sublattices $\mathscr{L} \subseteq \mathscr{P}(X)$. Now, given a set X, we denote, for each subset $A \subseteq X$, by U_A the subset

$$(A^c \times X) \cup (X \times A) = \{(x, y) \mid x \in A \implies y \in A\}$$

of $X \times X$. Given a topology τ on X, the filter $\mathscr{U}_\tau$ generated by the sets U_A for $A \in \tau$ is a quasi-uniformity on X. The quasi-uniform spaces $(X, \mathscr{U}_\tau)$ were first introduced by Pervin [40] and are now known in the literature as Pervin spaces. Given a sublattice, $\mathscr{L} \subseteq \mathscr{P}(X)$, we define $(X, \mathscr{U}_{\mathscr{L}})$ to be the quasi-uniform space whose quasi-uniformity is the filter generated by the entourages U_L for $L \in \mathscr{L}$. Here we will call this more general class of quasi-uniform spaces *Pervin spaces*. The lattice $\mathscr{L}$ may be recovered from $(X, \mathscr{U}_{\mathscr{L}})$ as the blocks of the space. The *blocks* of a space $(X, \mathscr{U})$ are the subsets $A \subseteq X$ such that U_A is an entourage of the space, or equivalently, those for which the characteristic function $\chi_A\colon X \to 2$ is uniformly continuous with respect to the Sierpiński quasi-uniformity on 2, which is the one containing just 2^2 and $\{(0, 0), (1, 1), (1, 0)\}$.

The Pervin spaces are *transitive*, that is, they have a basis of transitive entourages. In addition, they are *totally bounded*: for every entourage U, there exists a finite cover $\mathscr{C}$ of the space X such that $C \times C \subseteq U$ for each $C \in \mathscr{C}$. It may also be shown that the Pervin spaces (as we define them here) are exactly the transitive and totally bounded quasi-uniform spaces. It is not hard to see that if $\mathscr{M} \subseteq \mathscr{P}(X)$ and $\mathscr{L} \subseteq \mathscr{P}(Y)$ are lattices of sets, then a map $f\colon (X, \mathscr{U}_{\mathscr{M}}) \to (Y, \mathscr{U}_{\mathscr{L}})$ is uniformly continuous if and only if f^{-1} induces a lattice homomorphism from $\mathscr{L}$ to $\mathscr{M}$ by restriction. Thus, the category of lattices of sets with morphisms that are commuting diagrams

$$\begin{array}{ccc} \mathscr{L} & \xrightarrow{\ h\ } & \mathscr{M} \\ \downarrow & & \downarrow \\ \mathscr{P}(Y) & \xrightarrow{\ \phi\ } & \mathscr{P}(X), \end{array}$$

where ϕ is a complete lattice homomorphism, is dually equivalent to the category of Pervin spaces with uniformly continuous maps.

Now we are ready to state the main result of Sect. 1 of [19]: The set representation of a lattice D given by Stone/Priestley duality is obtainable from *any* set representation $e\colon D \hookrightarrow \mathscr{P}(X)$ by taking the so-called bicompletion of the corresponding quasi-uniform Pervin space $(X, \mathscr{U}_{Im(e)})$.

To be more precise, we have:

Theorem 8. [19, Theorem 1.6] *Let D be a bounded distributive lattice, and let $e\colon D \hookrightarrow \mathscr{P}(X)$ be any embedding of D in a power set lattice and denote by $\mathscr{L}$ the image of the embedding e. Let $\widetilde{X}$ be the bicompletion of the Pervin space $(X, \mathscr{U}_{\mathscr{L}})$. Then $\widetilde{X}$ with the induced topology is the dual space of D.*

We give a few details on this theorem. For more details on bicompleteness see [16, Chap. 3]. The theory of completions of uniform spaces is well-understood, see e.g. [6, Chap. II.3]. However, for quasi-uniform spaces, the situation is much more delicate. Two of the most accepted and well behaved completions, namely the bicompletion [16] (which is equivalent to the pair completion and the strong completion) and the D-completion [10], actually agree for Pervin spaces. The bicompletion is particularly appropriate to the representation theory of distributive lattices since it relates the representations of the lattice, its order dual, and its booleanisation. In addition, the bicompletion is the simplest, as it mainly reduces to the completion theory of uniform spaces.

Let $(X, \mathscr{U})$ be a quasi-uniform space. The converse, $\mathscr{U}^{-1}$, of the quasi-uniformity $\mathscr{U}$, consisting of the converses U^{-1} of the entourages $U \in \mathscr{U}$, is again a quasi-uniformity on X. Further, the *symmetrisation* $\mathscr{U}^s$, which is the filter generated by the union of $\mathscr{U}$ and $\mathscr{U}^{-1}$, is a uniformity on X. It is easy to verify that if $(X, \mathscr{U}_{\mathscr{L}})$ is a Pervin space, then $(X, \mathscr{U}_{\mathscr{L}}^{-1})$ is the Pervin space corresponding to the order dual of $\mathscr{L}$ as embedded in $\mathscr{P}(X)$ via $\mathscr{L}^{\partial} \hookrightarrow \mathscr{P}(X)$ obtained by taking complements in $\mathscr{P}(X)$. Also $(X, \mathscr{U}^s_{\mathscr{L}})$ is the uniform Pervin space corresponding to the representation $\mathscr{L}^- \hookrightarrow \mathscr{P}(X)$ of the booleanisation $\mathscr{L}^-$ of $\mathscr{L}$.

Now, a quasi-uniform space $(X, \mathscr{U})$ is *bicomplete* if and only if $(X, \mathscr{U}^s)$ is a complete uniform space. It has been shown by Fletcher and Lindgren that the full category of bicomplete quasi-uniform spaces forms a reflective subcategory of the category of quasi-uniform spaces with uniformly continuous maps, and thus, for each quasi-uniform space $(X, \mathscr{U})$, there is a bicomplete quasi-uniform space $(\widetilde{X}, \widetilde{\mathscr{U}})$ and a uniformly continuous map $\eta_X : (X, \mathscr{U}) \to (\widetilde{X}, \widetilde{\mathscr{U}})$ with a universal property:

Theorem 9. [16, Chap. 3.3], *Let $(X, \mathscr{U})$ be a quasi-uniform space, $(Y, \mathscr{V})$ a bicomplete quasi-uniform space and let $f : X \to Y$ be a $(\mathscr{U}, \mathscr{V})$-uniformly continuous function. Then there exists a unique $\widetilde{f} : \widetilde{X} \to Y$ which is $(\widetilde{\mathscr{U}}, \mathscr{V})$-uniformly continuous such that $f = \widetilde{f} \circ \eta_X$.*

The bicompletion of a quasi-uniform space $(X, \mathscr{U})$ is closely related to that of its symmetrisation in that the symmetrisation of the bicompletion is equal to the (bi)completion of the symmetrisation. In the case of a quasi-uniform Pervin space $(X, \mathscr{U}_{\mathscr{L}})$ the symmetrisation is the uniform Pervin space $(X, \mathscr{U}_{\mathscr{L}^-})$ of the booleanisation $\mathscr{L}^-$ of $\mathscr{L}$. It is not too hard to show that the (uniform=bi) completion of $(X, \mathscr{U}_{\mathscr{L}^-})$ is the Boolean space dual to $\mathscr{L}^-$ given as a uniform space. That is, $\widetilde{X} = X_{\mathscr{L}^-}$ is the set of ultrafilters of $\mathscr{L}^-$ (or equivalently the set of prime filters of $\mathscr{L}$), and the uniformity corresponding to $\mathscr{L}^-$ is generated by the sets $U_B = (B^c \times \widetilde{X}) \cup (\widetilde{X} \times B)$ for $B \subseteq \widetilde{X}$ clopen (in the topology for $\mathscr{L}^-$, or equivalently, in the Priestley topology for $\mathscr{L}$). This is the unique uniformity inducing the Boolean topology on $\widetilde{X}$ since this space is compact Hausdorff, see [6, Chapter II.4, Theorem 1]. Thus this uniform space carries no more information than that of the topological dual space of $\mathscr{L}^-$.

The function $\eta_X : X \to \widetilde{X}$ underlying the embedding of $(X, \mathscr{U}_{\mathscr{L}^-})$ in $(\widetilde{X}, \widetilde{\mathscr{U}}_{\mathscr{L}^-})$ is the map which sends $x \in X$ to the point of $X_{\mathscr{L}^-}$ corresponding to the homomorphism $\chi_x : \mathscr{L}^- \to 2$ given by $\chi_x(L) = 1$ if and only if $x \in L$. Also, the $(\mathscr{U}_{\mathscr{L}^-}, \widetilde{\mathscr{U}}_{\mathscr{L}^-})$-uniform continuity of this map comes about in a particularly simple way as one can show that $\eta_X^{-1}(U_{\widehat{L}}) = U_L$ for each $L \in \mathscr{L}^-$.

Now we can formulate, what the bicompletion of $(X, \mathscr{U}_{\mathscr{L}})$ is: It is based on the map $\eta_X : X \to \widetilde{X}$ as given above, but the quasi-uniformity on $\widetilde{X}$ is generated by the sets $U_{\widehat{L}}$ for $L \in \mathscr{L}$ rather than by all the sets $\eta_X(U_L)$ for $L \in \mathscr{L}^-$. Thus the bicompletion of $(X, \mathscr{U}_{\mathscr{L}})$ is the Stone space of D 'in quasi-uniform form'. Alternatively, one can think of this space as an ordered uniform space and simply equip the Boolean space $(\widetilde{X}, \widetilde{\mathscr{U}}_{\mathscr{L}^-})$ with the order obtained by $\bigcap_{L \in \mathscr{L}} U_{\widehat{L}}$. This is then a uniform version of Priestley duality.

In closing, we record what this theory yields in the setting of finitely freely generated Heyting algebras. We have the following corollary to Theorem 8.

Corollary 3. *Let N be a positive natural number, and A the free N-generated Heyting algebra. Then the following statements hold:*

1. *The Esakia space* X *dual to* A *is homeomorphic to the bicompletion of the quasi-uniform Pervin space* $(J(A), \mathscr{U}_{\mathscr{L}})$ *where* $J(A)$ *is the set of join prime elements of* A *and* $\mathscr{L}$ *is the lattice of all downsets of finite antichains in* $J(A)$.
2. *The Esakia space* X *dual to* A *is homeomorphic to the bicompletion of the quasi-uniform Pervin space* $(U, \mathscr{U}_{\mathscr{D}})$ *where* U *is the frame underlying the* N*-universal model of intuitionistic logic and* $\mathscr{D}$ *is the lattice of all definable subsets of* U.

References

1. Bellissima F (1986) Finitely generated free heyting algebras. J Symb Logic 51(1):152–165
2. Bezhanishvili N (2006) Lattices of intermediate and cylindric modal logics. PhD thesis, University of Amsterdam.
3. Bezhanishvili N, Gehrke M (2011) Finitely generated free Heyting algebras via Birkhoff duality and coalgebra. Logical Methods Comput Sci 7(2:9):1–24.
4. Blok W, Dwinger P (1975) Equational classes of closure algebras. I. Indag Math 37:189–198
5. Blok W (1976) Varieties of interior algebras. PhD thesis, University of Amsterdam.
6. Bourbaki N (1998) General topology. Chapters 1–4, Elements of Mathematics. Springer, Berlin.
7. Butz C (1998) Finitely presented Heyting algebras. Technical Report, BRICS RS-98-30, Aarhus.
8. Darnière L, Junker M (2010) On Bellissima's construction of the finitely generated free Heyting algebras, and beyond. Arch Math Logic 49:743–771
9. de Jongh D (1968) Investigation on the intiutionistic propositional calculus. PhD thesis, University of Wisconsin.
10. Doitchinov D (1991) A concept of completeness of quasi-uniform spaces. Topol Appl 38(3):205–217
11. Dummett M, Lemmon E (1959) Modal logics between S4 and S5. Z Math Logik Grundlagen Math 5:250–264
12. Esakia L (1974) Topological Kripke models. Soviet Math Dokl 15:147–151
13. Esakia L (1984) On the variety of Grzegorczyk algebras. Sel Sov Math 3:343–366
14. Esakia L (1985) Heyting algebras. Duality theory. Metsniereba Press, Tbilisi (Russian)
15. Escardo M (2001) The regular-locally-compact coreflection of a stably locally compact locale. J Pure Appl Alg 157:41–55
16. Fletcher P, Lindgren W (1982) Quasi-uniform spaces, vol 77. Lecture notes in pure and applied mathematics, Marcel Dekker Inc., New York.
17. Gehrke M (2009) Stone duality and the recognisable languages over an algebra. In: Kurz A et al. (eds) CALCO 2009. Lecture notes in computer science, vol 5728. Springer, Berlin, pp 236–250.
18. Gehrke M, Stone duality, topological algebra, and recognition, preprint available at http://hal.archives-ouvertes.fr/hal-00859717
19. Gehrke M, Grigorieff S, Pin J (2010) Duality and equational theory of regular languages. In: Abramsky SEA (ed) ICALP 2010, part II. Lecture notes in computer science, vol 6199. Springer, Berlin, pp 151–162.
20. Gehrke M, Harding J (2001) Bounded lattice expansions. J. Algebra 238(1):345–371
21. Gehrke M, Jansana R, Palmigiano A, Δ_1-completions of a poset. Order (to appear). <!– Missing/Wrong Year –>
22. Gehrke M, Jónsson B (1994) Bounded distributive lattices with operators. Mathematica Japonica 40:207–215
23. Gehrke M, Jónsson B (2004) Bounded distributive lattices expansions. Mathematica Scandinavica 94(2):13–45

24. Gehrke M, Priestley H (2007) Canonical extensions of double quasioperator algebras: an algebraic perspective on duality for certain algebras with binary operations. J Pure Appl Alg 209(1):269–290
25. Gehrke M, Vosmaer J (2009) A view of canonical extension. In Bezhanishvili N, Löbner S, Schwabe K, Spada L (eds) TbiLLC. Lecture notes in computer science, vol 6618. Springer, Berlin, pp 77–100.
26. Gehrke M, Vosmaer J (2011) Canonical extensions and canonicity via dcpo presentations. Theor Comput Sci 412(25):2714–2723
27. Ghilardi S (1992) Free Heyting algebras as bi-Heyting algebras. Math Rep Acad Sci Can XV I (6):240–244
28. Ghilardi S (1995) An algebraic theory of normal forms. Ann Pure Appl Logic 71(3):189–245
29. Ghilardi S (2004) Unification, finite duality and projectivity in varieties of Heyting algebras. Ann Pure Appl Logic 127(1–3):99–115
30. Ghilardi S (2010) Continuity, freeness, and filtrations. J Appl Non-Class Logics 20(3):193–217
31. Ghilardi S, Meloni G (1997) Constructive canonicity in non-classical logics. Ann Pure Appl Logic 86(1):1–32
32. Goldblatt R (1989) Varieties of complex algebras. Ann Pure Appl Logic 44:173–242
33. Grigolia R (1987) Free algebras of non-classical logics. Metsniereba, Tbilisi (Russian)
34. Jónsson B, Tarski A (1951) Boolean algebras with operators I. Amer J Math 73:891–939
35. Jónsson B, Tarski A (1952) Boolean algebras with operators II. Amer J Math 74:127–162
36. Jung A (2004) Stably compact spaces and the probabilistic powerspace construction. In: Panangaden P, Desharnais J (eds) Bellairs workshop on domain-theoretic methods in probabilistic processes, electronic lecture motes in computer science, vol 87. Elsevier, p 15.
37. Makkai M, Reyes G (1995) Completeness results for intuitionistic and modal logic in a categorical setting. Ann Pure Appl Logic 72:25–101
38. McKinsey J, Tarski A (1946) On closed elements in closure algebras. Ann Math 47:122–162
39. Peremans W (1957) Embedding of a distributive lattice into a Boolean algebra. Indagationes Mathematicae 19:73–81
40. Pervin W (1962) Quasi-uniformization of topological spaces. Math Ann 147:316–317
41. Priestley H (1970) Representation of distributive lattices by means of ordered stone spaces. Bull Lond Math Soc 2:186–190
42. Rasiowa H, Sikorski R (1963) The mathematics of metamathematics, Monografie Matematyczne, vol 41. Państwowe Wydawnictwo Naukowe.
43. Rybakov V (1992) Rules of inference with parameters for intuitionistic logic. J Symb Logic 57:33–52
44. Sambin G, Vaccaro V (1989) A new proof of Sahlqvist's theorem on modal definability and completeness. J Symb Logic 54(3):992–999
45. Schmid J (2002) Quasiorders and sublattices of distributive lattices. Order 19:11–34
46. Shehtman V (1978) Rieger-Nishimura lattices. Soviet Math Doklady 19:1014–1018
47. Vosmaer J (2010) Logic, algebra and topology, Investigations into canonical extensions, duality theory and point-free topology. PhD thesis, University of Amsterdam.

Chapter 3
Free Modal Algebras Revisited: The Step-by-Step Method

Nick Bezhanishvili, Silvio Ghilardi and Mamuka Jibladze

In memory of Leo Esakia and Dito Pataraia

Abstract We review the step-by-step method of constructing finitely generated free modal algebras. First we discuss the global step-by-step method, which works well for rank one modal logics. Next we refine the global step-by-step method to obtain the local step-by-step method, which is applicable beyond rank one modal logics. In particular, we show that it works well for constructing the finitely generated free algebras for such well-known modal systems as **T**, **K4** and **S4**. This yields the notions of one-step algebras and of one-step frames, as well as of universal one-step extensions of one-step algebras and of one-step frames. We show that finitely generated free algebras for **T**, **K4** and **S4** and their dual spaces can be obtained by iterating the universal one-step extensions of one-step algebras and of one-step frames. In the final part of the chapter we compare our construction with recent literature, especially with [11] which undertakes a very similar approach.

Keywords Free modal algebras · One-step algebras · One-step frames · Duality · Step-by-step method

---

N. Bezhanishvili (✉)
Institute for Logic, Language and Computation, University of Amsterdam, Amsterdam, The Netherlands
e-mail: N.Bezhanishvili@uva.nl

S. Ghilardi
Department of Mathematics, Universita degli Studi di Milano, Milan, Italy
e-mail: silvio.ghilardi@unimi.it

M. Jibladze
Razmadze Mathematical Institute, Tbilisi State University, Tbilisi, Georgia
e-mail: jib@rmi.acnet.ge

G. Bezhanishvili (ed.), *Leo Esakia on Duality in Modal and Intuitionistic Logics*,
Outstanding Contributions to Logic 4, DOI: 10.1007/978-94-017-8860-1_3,

3.1 Introduction

Having a good description of finitely generated free algebras is an important tool for investigating propositional logics: free algebras give an insight on the shape of formulae and on the deduction mechanism that is independent of the particular syntactic methodology used for introducing a logical calculus. With a clear combinatorial and conceptual description of free algebras in mind, one can investigate better metatheoretical properties like admissibility of inference rules, solvability of equations, definability and interpretability matters, etc.

For modal logics (or some other non-classical logics such as intuitionistic logic), one way to characterize finitely generated free algebras is to use the relevant properties of their dual spaces: many of these algebras are atomic [3, 4, 9], thus restricting dual spaces to atoms still gives the possibility of having a representation theorem. The spaces of the atoms become the so-called 'universal models' and finitely generated free algebras can be described as the algebras of definable subsets of these models. Atoms, in turn, generate 'irreducible' or 'definable' finite models (along the suggestions of [21, 27]): such models can be described inductively, using for instance the height of the model. This line of research has been largely explored in a long series of papers in the 1970s and 1980s by the Georgian school (see, e.g., [15, 16, 24]), the Russian school (see, e.g., [30, 31]), and the Italian logician F. Bellissima (see, e.g., [3–5]). We refer to [6, 10] for an overview.

Still, duality can be used in another way to get descriptions of finitely generated free algebras: formulae naturally come equipped with a complexity measure, the measure counting maximum nested 'intensional' operators occurring in them. By 'intensional' operators we mean modal operators (or implication in the context of intuitionistic or relevance logic); non-intensional operators are the Boolean connectives (or a subset of them) and it is well known that only a finite number of combinations of such operators can be applied (up to logical equivalence) to a finite set of given formulae. Thus, finitely generated free algebras have a 'dual profinite' description as chain colimits of finite algebras defined by imposing complexity bounds. By finite duality, these finite algebras admit a description as finite discrete spaces and the intensional operators (which are only partially defined on them) induce a kind of combinatorial structure. The investigation of such combinatorial structure paves the way to a new, different description of finitely generated free algebras, a description that we call a *step-by-step* description: its essence is in fact the dual explanation of what it means to enrich a given set of formulae by one application of intensional operators followed by the finite closure with respect to the non-intensional operators. Of course, the whole construction should not destroy previously introduced intensional operators, that is why it applies to a 'one-step algebra' and results into an updated 'one-step algebra'.

The origin of these step-by-step constructions is two-fold: from the logical point of view they describe normal forms (in the sense of K. Fine [17]), and from the coalgebraic point of view they correspond to free coalgebra constructions ([1, 2, 8, 11, 20]). However, coalgebraic constructions work well for rank one logics (e.g.,

K and **D** in modal logic), but become unclear or quite involved when arbitrary subvarieties/logics are involved. Our plan is to exploit discrete dualities and the step-by-step combinatorics to illustrate potential applications of the method outside the rank one case.

We start the chapter by reviewing the general idea of the step-by-step method which, via duality, results in the dual description of free **K**-algebras and free algebras for rank one modal logics. We then extend this method to logics of higher rank using the new notion of a one-step algebra and its dual notion of a one-step frame. We adjust these ideas to particular non-rank one logics such as **T**, **K4** and **S4** obtaining (duals of) finitely generated free algebras of these logics in a transparent and modular way. Our construction follows closely the method developed in [11] for various non-rank one modal logics. We discuss the similarities and differences of our construction with that of [11], as well as with those of [20, 22]. We also list a number of (challenging) open problems.

In the chapter we discuss finitely generated free Heyting algebras only briefly, and refer the interested reader to [7, 19] for details. However, we believe that one-step algebras and one-step frames have a potential to play an important role in the theory of free algebras in various varieties of Heyting algebras.

We conclude our introduction by pointing out that another co-author of this paper was going to be Dito Pataraia. He developed interest towards the step-by-step method after Leo Esakia drew his attention to [19] where free Heyting algebras were described via this method. Both Dito and Leo were very interested in this method. Dito's interest towards this construction was mostly determined by its use in proving that every Heyting algebra can be realized as the subobject classifier of an elementary topos. Dito gave a few talks about this important theorem and his colleagues are now trying to reconstruct his very involved and original proof, a part of which essentially uses the step-by-step method. Dito had a number of deep observations on the step-by-step construction for free Heyting and modal algebras, and many of them were supposed to form part of this chapter. Leo was interested in this method as it gives an alternative and useful perspective on Esakia spaces of free Heyting algebras (see [18] for more details on this). In fact, Esakia duality for Heyting algebras plays a prominent role in this and nearly all other approaches that apply the ideas of duality to various constructions of Heyting algebras. With great sadness for their loss, but with a lot of admiration for their outstanding scientific achievements, their unique character and personality, we would like to dedicate this chapter to the memory of Leo Esakia and Dito Pataraia.

3.2 The Global Step-by-Step Method

In this section we recall the global step-by-step method and construct free **K**-algebras. As we will see, this method works nicely for rank one modal logics, but its extension to arbitrary modal logics, although possible, is quite laborious and involved [20]. We recall that a modal formula is of *rank one* if each occurrence of atomic formulas (i.e., propositional variables or constants) is under the scope of exactly one modal operator; moreover, the constants $\bot$, $\top$, may appear in rank one formulae without being under the scope of any modal operator. A modal logic is of rank one if it can be axiomatized by rank one modal formulae. In Sect. 3.3 we will refine the global step-by-step method to the local step-by-step method that works for modal logics of higher rank. This will result in a neat description of finitely generated free algebras for the well-known modal systems **T**, **K4** and **S4**.

Recall that a *modal algebra* is a pair $(B, \Diamond)$, where B is a Boolean algebra and $\Diamond : B \to B$ is a unary operation satisfying $\Diamond 0 = 0$ and $\Diamond(a \vee b) = \Diamond a \vee \Diamond b$ for each $a, b \in B$. A modal algebra $(B, \Diamond)$ is called a **T**-*algebra* if $a \leq \Diamond a$ and a **K4**-*algebra* if $\Diamond\Diamond a \leq \Diamond a$ for each $a \in B$. Finally, $(B, \Diamond)$ is an **S4**-*algebra* if it is both a **T**-algebra and a **K4**-algebra.

3.2.1 Algebraic View

The method we discuss in this section is taken essentially from [1, 20] (see also [8]). Thus, we only sketch the construction and refer the interested reader to any of [1, 8, 20] for details.

Given a Boolean algebra B, we let $V(B)$ denote the free Boolean algebra generated by the set $\{\Diamond a : a \in B\}$ and quotiented by the two axioms defining modal algebras. Alternatively $V(B)$ is the free Boolean algebra over the join-semilattice $\vee$, 0-reduct (equivalently $\wedge$, 1-reduct) of B. That is, the map $i_\Diamond^B : B \to V(B)$ (mapping $a \in B$ to $\Diamond a \in V(B)$) is such that it is a join-semilattice morphism (preserves $\vee$ and 0), and for any Boolean algebra A, any join-semilattice morphism $h : B \to A$ can be extended uniquely to a Boolean homomorphism $h^T : V(B) \to A$ so that $h^T \circ i_\Diamond^B = h$. Actually, V can be turned into an endofunctor on the category of Boolean algebras in a standard way: by letting $V(f)$ (for $f : B \to C$) be $(i_\Diamond^C \circ f)^T$. Note that the correspondence $h \mapsto h^T$ is bijective and, as a consequence, modal algebras can be equivalently defined as Boolean algebras B equipped with semilattice morphisms $\Diamond^T : V(B) \to B$ (we shall exploit this fact below).

Let B_0 be the free Boolean algebra on n-generators. For each $k \geq 0$ we let

$$B_{k+1} = B_0 + V(B_k),$$

where $+$ means the coproduct in the category of Boolean algebras. As Boolean algebras are locally finite, the coproduct of two finite Boolean algebras is again finite.

We define the maps $i_k : B_k \to B_{k+1}$ and $\Diamond_k^T : V(B_k) \to B_{k+1}$ as follows. Let $\Diamond_k^T$ be the second injection into the coproduct, and let i_k be defined recursively as follows: i_0 is the first coproduct injection and i_{k+1} is $id + V(i_k)$. Let B_∞ be the colimit of the following diagram in the category of Boolean algebras and Boolean homomorphisms

$$B_0 \xrightarrow{i_0} B_1 \to \cdots \to B_k \xrightarrow{i_k} B_{k+1} \to \cdots \tag{3.1}$$

Proposition 1 *For each* $k \geq 0$ *we have* $\Diamond^T_{k+1} \circ V(i_k) = i_{k+1} \circ \Diamond^T_k$. *Therefore,* $\{\Diamond^T_k : k \geq 0\}$ *can be extended to a map* $\Diamond^T_\infty : V(B_\infty) \to B_\infty$.

Proof (Sketch) That $\Diamond^T_{k+1} \circ V(i_k) = i_{k+1} \circ \Diamond^T_k$ holds can be easily seen by a direct computation (recall that the two diamonds are just the coproduct injections). To define $\Diamond^T_\infty$ one can then use the fact that V commutes with chain (more generally with filtered) colimits: thus, we can assume that the domain of $\Diamond^T_\infty$ is the colimit of

$$V(B_0) \xrightarrow{V(i_0)} V(B_1) \to \cdots \to V(B_k) \xrightarrow{V(i_k)} V(B_{k+1}) \to \cdots \tag{3.2}$$

in the category of Boolean algebras and Boolean homomorphisms. Now the maps $\Diamond^T_k : V(B_k) \to B_{k+1}$ form vertical maps from the chain (3.2) to the chain (3.1) commuting the related squares, hence it induces a colimit map which is our $\Diamond^T_\infty$. □

Let $\Diamond_k : B_k \to B_{k+1}$ be the map that corresponds to $\Diamond^T_k : V(B_k) \to B_{k+1}$ and let $\Diamond_\infty : B_\infty \to B_\infty$ be the map that corresponds to $\Diamond^T_\infty : V(B_\infty) \to B_\infty$. Then we have the following characterization of finitely generated free modal algebras.

Proposition 2 *The algebra* $\langle B_\infty, \Diamond_\infty \rangle$ *is the free modal algebra on* n *generators.*

Proof See [1, 8, 20]. □

Let L be a normal modal logic and $\mathbf{V}_L$ the corresponding variety of modal algebras. We also let $\mathrm{Ax}(L)$ be a (finite or infinite) equational axiomatization of $\mathbf{V}_L$. We will now briefly sketch how to extend the above method to obtain finitely generated free $\mathbf{V}_L$-algebras.

If we try to quotient (3.1) by the axioms of $\mathrm{Ax}(L)$, we need to interpret modal formulae into the steps of a chain colimit algebra and then take a quotient of the algebras in the chain. The definition of such interpretation must take into account the fact that the axioms have arbitrary modal rank, hence the interpretation involves many algebras at a time. If the axioms have modal rank one, the situation simplifies because we can modify uniformly the whole construction, by taking instead of V a suitable quotient of it [8, 20]. Examples of logics of rank one include $\mathbf{D} = \mathbf{K} + \Diamond\top = \mathbf{K} + \Box p \to \Diamond p$, $\mathbf{K} + \Diamond p \to \Box p$ and $\mathbf{K} + \Box p \leftrightarrow \Diamond p$.

3.2.2 Dual View

For the purposes of our paper it suffices to restrict ourselves to the discrete duality between finite modal algebras and finite relational structures. Almost all the results can be generalized to the infinite case by defining an appropriate Stone topology on relational structures (see, e.g., [18]). We chose to stick to the finite duality to keep the arguments simple.

We recall that there is a one-to-one correspondence between join-preserving maps between finite Boolean algebras and relations on their dual finite sets. In fact, the category of finite Boolean algebras and $\vee$, 0-preserving maps is dually equivalent to the category of finite sets and binary relations (see, e.g., [25, 28]).

Let X be a set and $\wp(X)$ its powerset. Then it is easy to see that for each Y, any relation $R \subseteq Y \times X$ uniquely corresponds to a map $f : Y \to \wp(X)$ defined by $f(y) = R(y) = \{x \in X : yRx\}$. Throughout this paper we will use twofold notations for binary relations interchangeably: as subsets of a cartesian product or as maps into the powerset.

Given a finite set X, let $\ni$ be the relation on $\wp(X) \times X$ defined by $U \ni x$ iff $x \in U$ for each $U \in \wp(X)$ and $x \in X$. Then X, $\wp(X)$ and $\ni$ have the following universal property: for each finite Y and $R \subseteq Y \times X$, there exists a unique map $f : Y \to \wp(X)$ (defined by $f(y) = R(y)$) such that we have $R = \ni \circ f$. The last equation refers to relational composition, i.e. it means that for each $x \in X$, $y \in Y$ we have yRx iff (there is $S \in \wp(X)$ such that $x \in S$ and $S = f(y)$) iff $x \in f(y)$.

Remark 1 In the general case we need to consider Stone spaces and continuous relations and maps. But the same correspondence holds in this case as well. That is, if Y is a Stone space (i.e., the dual of a Boolean algebra) and R is a continuous relation (i.e., the dual of a join-preserving map), then f is also a continuous map (i.e., the dual of a Boolean algebra homomorphism).

Translating this into algebraic terms gives us that the join-semilattice morphism dual to $\ni$ and the Boolean algebra dual to $\wp(X)$ satisfy the universal property of $V(B)$ and $i^B_\Diamond$ discussed in the previous section. Thus, as the universal property defines an object uniquely up to an isomorphism, we obtain the following theorem.

Proposition 3 ([20, Prop. 2.1], [32]) *Let B be a finite Boolean algebra and X its dual finite set. Then the algebra $V(B)$ is dual to $\wp(X)$. Moreover, the map $i^B_\Diamond : B \to V(B)$ is dual to the relation $\ni \subseteq \wp(X) \times X$.*

Remark 2 We note that this result can be generalized to the infinite case by considering Stone spaces and continuous maps and relations, and by taking the Vietoris space instead of the finite powerset. We also refer to [26, Chap. III.4] for generalizations of this result to compact regular frames.

Now we are ready to construct the duals of free modal algebras. Let X_0 be a 2^n-element set (the dual of B_0) and (because of the duality of $\wp$ and V (Proposition 3) and of $\times$ and $+$) let

$$X_{k+1} = X_0 \times \wp(X_k).$$

Proposition 4 *The sequence $(X_k)_{k<\omega}$ with the maps $\pi_k : X_0 \times \wp(X_k) \to X_k$ defined by*

$$\pi_0(x, U) = x, \quad \pi_k(x, U) = (x, \pi_{k-1}[U])$$

is dual to the sequence $(B_k)_{k<\omega}$ with the maps $i_k : B_k \to B_{k+1}$. In particular, the π_k are surjective. Moreover, the relation $R_k \subseteq (X_0 \times \wp(X_k)) \times X_k$ defined by

$$(x, U)R_k y \text{ iff } y \in U$$

is dual to $\Diamond_k : B_k \to B_{k+1}$.

Theorem 1 *Let* X_ω *be the limit in the category of Stone spaces and continuous maps of the diagram* $(X_k)_{k\in\omega}$ *with the maps* $\pi_{k+1} : X_{k+1} \to X_k$. *Let also* R_ω *be the limit of* $(R_k)_{k\in\omega}$ *in the category of Stone spaces and continuous relations defined by* $(x_i)_{i\in\omega} R_\omega (y_i)_{i\in\omega}$ *if* $x_{k+1} R_k y_k$ *for each* $k \in \omega$. *Then* (X_ω, R_ω) *is (isomorphic in a suitable category to) the dual of the free modal algebra* $(B_\omega, \Diamond_\omega)$.

Thus, via the global step-by-step method we described finitely generated free modal algebras and their dual spaces.

3.3 The Local Step-by-Step Method

The construction presented in the previous section is very useful for logics axiomatized by rank one equations. It, however, also has some drawbacks. For example, there is no manageable way to apply it to the well-known extensions of **K** such as **K4** and **S4** (it works for **T**, but for the other systems the adaptation is involved, see [20]). The point is that the definition of step $k + 1$ mentions not only step k but also step 0, which is rather unnatural. In this section we introduce a refinement of the construction. From an algebraic point of view, the new construction may be considered as just a trivial variant of the former one. Nevertheless it induces better constructions at the dual level. Its distinguishing feature is that the construction is local in that it relies on the universal property of the one-step construction (see [11, 19, 22] for similar ideas).

3.3.1 Algebraic View

As we pointed out in the introduction, the essence of the method we propose is to build free algebras in steps; a single step (taken *independently* on the whole chain of steps needed to build the free algebra as a colimit) applies the modal operators to the existing propositions and embeds the actual propositions into the new ones. This leads naturally to a *two-sorted* viewpoint: we have one algebra for the actual propositions and another one collecting actual and new propositions; moreover the two algebras are connected by an embedding and a diamond. All this is formally captured by the following definition.

Definition 1

1. A *one-step modal algebra* is a quadruple $(A_0, A_1, i_0, \Diamond_0)$, where A_0, A_1 are Boolean algebras, $i_0 : A_0 \to A_1$ is a Boolean morphism, and $\Diamond_0 : A_0 \to A_1$ is

a semilattice morphism. The algebras A_0, A_1 are called the *source* and the *target* Boolean algebras of the one-step modal algebra $(A_0, A_1, i_0, \Diamond_0)$.

2. A *one-step extension* of the one-step modal algebra $(A_0, A_1, i_0, \Diamond_0)$ is a one-step modal algebra $(A_1, A_2, i_1, \Diamond_1)$ (i.e., it is a one-step modal algebra whose source is the same as the target of $(A_0, A_1, i_0, \Diamond_0)$) satisfying $i_1 \Diamond_0 = \Diamond_1 i_0$.
3. The *universal one step-extension* of $(A_0, A_1, i_0, \Diamond_0)$ is a one-step extension $(A_1, A_2, i_1, \Diamond_1)$ such that for every other one-step extension $(A_1, A'_2, i'_1, \Diamond'_1)$, there is a unique Boolean morphism $\mu : A_2 \to A'_2$ such that $\mu \circ i_1 = i'_1$ and $\mu \circ \Diamond_1 = \Diamond'_1$. The meaning of the universal one-step extension is that it represents the general solution to the problem of adding $\Diamond_1 a_1$ for all $a_1 \in A_1$ while keeping (through i_1) the $\Diamond_0 a_0$ for $a \in A_0$.

Universal one-step extensions exist and are easily built through pushouts:

Proposition 5 *The universal one-step extension of* $(A_0, A_1, i_0, \Diamond_0)$ *is given by the following pushout taken in the category of Boolean algebras and Boolean homomorphisms.*

$$\begin{array}{ccc}
V(A_0) & \xrightarrow{V(i_0)} & V(A_1) \\
\Big\downarrow{\scriptstyle \Diamond_0^T} & & \Big\downarrow{\scriptstyle \Diamond_1^T} \\
A_1 & \xrightarrow{i_1} & A_2
\end{array}$$

where $A_2 = A_1 +_{V(A_0)} V(A_1)$.

Proof Immediate by the universal property of pushouts. □

Let B_0 be the free Boolean algebra on n-generators. We define a new sequence B_k by using pushouts. In the new sequence, the algebras B'_0, B'_1 and the morphisms i'_0, $\Diamond_0^T$ are as before; for $k \geq 1$, we have instead

$$B'_{k+1} := B'_k +_{V(B'_{k-1})} V(B'_k)$$

where i'_k, $\Diamond_k^T$ are the canonical maps into the pushout

$$\begin{array}{ccc}
V(B'_{k-1}) & \xrightarrow{\Diamond_{k-1}^T} & B'_k \\
\Big\downarrow{\scriptstyle V(i'_{k-1})} & & \Big\downarrow{\scriptstyle i'_k} \\
V(B'_k) & \xrightarrow{\Diamond_k^T} & B'_{k+1}
\end{array}$$

Let B'_∞ be the colimit of the diagram

$$B_0 \stackrel{i'_0}{\to} B'_1 \to \ldots \to B'_k \stackrel{i'_k}{\to} B'_{k+1} \to \ldots \tag{3.3}$$

Then we have

Proposition 6 *$\{\Diamond_k^T : k \geq 0\}$ can be extended to a map $\Diamond_\infty^T : V(B'_\infty) \to B'_\infty$ such that $(B'_\infty, \Diamond_\infty)$ is the n-generated free modal algebra.*

Proof (Sketch) That $\Diamond_{k+1}^T \circ V(i'_k) = i'_{k+1} \circ \Diamond_k^T$ holds now comes directly from the commutativity of the above pushout square. Thus, $\Diamond_\infty^T$ can be defined in the colimit like in the proof of Proposition 1. To show that the construction gives finitely generated free modal algebras, one can use Proposition 5, along the lines of e.g. [22] (alternatively, it is possible to show inductively that the construction is isomorphic to the global step-by-step construction of Proposition 2). □

3.3.2 Dual View

The dual construction is described through the notion of a one-step frame.

Definition 2 A *one-step frame* is a quadruple (X, Y, f, R), where X, Y are sets, $f : X \to Y$ is a map and $R \subseteq X \times \wp(Y)$ is a relation between X and Y.

The dual of a finite one-step frame (X, Y, f, R) is the one-step modal algebra $(\wp(Y), \wp(X), f^*, \Diamond_R)$, where f^* is the inverse image operation and $\Diamond_R$ is the semi-lattice morphism associated with R (for $A \subseteq Y$, we have $\Diamond_R(A) = \{x \in X \mid R(x) \cap A \neq \emptyset\}$). Every finite one-step modal algebra is the dual of a finite one-step frame (again, to extend this duality beyond the finite case, Stone spaces, continuous maps, continuous relations and Vietoris spaces are needed).

We can now dualize the local construction of Proposition 6 as follows: the dual of i'_k is given by a map $f_k : X_{k+1} \to X_k$ between finite sets; the dual of $\Diamond_k$ is a map $R_k : X_{k+1} \to \wp(X_k)$ (alternatively a relation $R_k \subseteq X_{k+1} \times X_k$). Then the dual of $(B'_k, B'_{k+1}, i'_k, \Diamond_k)$ is determined by the following diagram:

$$\begin{array}{ccc}
X_{k+1} & \xrightarrow{R_k} & \wp(X_k) \\
\downarrow{\scriptstyle f_k} & & \downarrow{\scriptstyle \wp(f_{k-1})} \\
X_k & \xrightarrow{R_{k-1}} & \wp(X_{k-1}) \\
\downarrow{\scriptstyle f_{k-1}} & & \\
X_{k-1} & &
\end{array}$$

where the square is a pullback, and for each map $f : X \to Y$, we assume that $\wp(f) = f[\cdot] : \wp(X) \to \wp(Y)$ is the direct image of f. Thus, we have

$$X_{k+1} = \{(x, S) \mid x \in X_k, S \in \wp(X_k), R_{k-1}(x) = f_{k-1}(S)\},$$

with

$$f_k(x, S) = x$$

and

$$R_k(x, S) = S.$$

We must also consider the dual of the first step of the chain leading to free modal algebras. This is more simple, being just a coproduct not a pushout. In short, if X_0 is the finite set dual to the Boolean algebra B_0, the dual of the one-step modal algebra $(B_0, B_0 + V(B_0), i_0, \Diamond_0)$ (where recall that $i_0, \Diamond_0$ are the two coproduct injections) is the one-step frame

$$(X_0 \times \wp(X_0), X_0, f_0, R_0) \tag{3.4}$$

where f_0 is the first projection and we have $R_0(x, S) = S$ for all $x \in X_0$, $S \subseteq X_0$.

3.3.3 Adding Equations

The main advantage of the second method is that, when building B'_{k+1}, it refers only to B'_k (and not also to B'_0): this makes descriptions of quotients modulo further equations easier. Suppose in fact that we are given some axioms $\mathrm{Ax}(L)$ for a logic L. Following a suggestion by Coumans and van Gool [11], we can rewrite an axiom in the form of a quasi-identity

$$t = 1 \to v = 1 \tag{3.5}$$

where the terms/formulae t, v have modal complexity less or equal to one (i.e., nested modal operators do not occur). To achieve this, one can repeat the following 'flattening' of quasi-equations $E \to v = 1$ sufficiently many times (we start with $E = \emptyset$): take a subterm $\Diamond v'$ of v, pick a fresh variable x and replace $E \to v = 1$ with $E \cup \{x = \Diamond v'\} \to v(x/\Diamond v') = 1$. Finally, quasi-identities having many premises can be turned into single-premise quasi-identities by taking conjunctions.

The quasi-equations of this kind can be interpreted in a one-step modal algebra $(A_0, A_1, i_0, \Diamond_0)$: once an assignment a of variables to members of A_0 is given, we can recursively define the element $t^{\mathsf{a}} \in A_1$ for every term t having modal complexity 0 or 1. An equation $t = 1$ of modal complexity at most 1 is valid in $(A_0, A_1, i_0, \Diamond_0)$ iff $t^{\mathsf{a}} = 1$ holds in A_1 for every a; similarly one can define validity of quasi-equations. Thus, an $\mathrm{Ax}(L)$-*one-step modal algebra* is a one-step modal algebra where all quasi-equations belonging to $\mathrm{Ax}(L)$ are valid; notice that this notion is relative not just to a logic L, but to a set $\mathrm{Ax}(L)$ of quasi-equations (i.e. of inference rules) chosen in order to axiomatize L.

Validity of conditions such as (3.5) can be forced by taking a quotient; since (3.5) is a quasi-equation (and not just an equation), we need a quotient which is iterated: one just quotients A_1 by the filter generated by all v^{a} such that $t^{\mathrm{a}} = 1$ holds in A_1 (varying a) and then repeats this procedure ω-times (or just sufficiently many finite times if A_1 is finite). In the end, one gets a one-step modal algebra $(A_0, A_1/F, q \circ i_0, q \circ \Diamond_0)$ (where F is the filter obtained in the end of the iteration and $q : A_1 \to A_1/F$ is the canonical map onto the quotient) that satisfies (3.5) and is universal with this property.

We define a chain

$$B_0 \xrightarrow{i''_0} B''_1 \to \cdots \to B''_k \xrightarrow{i''_k} B''_{k+1} \to \cdots \tag{3.6}$$

of Boolean algebras equipped with semilattice morphisms

$$B_0 \xrightarrow{\Diamond_0} B''_1 \to \cdots \to B''_k \xrightarrow{\Diamond_k} B''_{k+1} \to \cdots \tag{3.7}$$

satisfying the conditions $\Diamond_{k+1} \circ i''_k = i''_{k+1} \circ \Diamond_k$ (equivalently, $\Diamond^T_{k+1} \circ V(i''_k) = i''_{k+1} \circ \Diamond^T_k$) and such that for every $k \geq 0$, the one-step modal algebra $(B''_k, B''_{k+1}, i''_k, \Diamond_k)$ satisfies $\mathrm{Ax}(L)$. This is done by the same construction as in Proposition 6, with the only difference that we also apply the aforementioned quotient by $\mathrm{Ax}(L)$; that is, we define B''_{k+1} by taking a pushout

$$B''_k +_{V(B''_{k-1})} V(B''_k)$$

followed by a quotient by $\mathrm{Ax}(L)$. Let B''_∞ be the colimit of this diagram. Then:

Proposition 7 *For each $k \geq 0$, we have that $\Diamond^T_{k+1} \circ i''_k = i''_{k+1} \circ \Diamond^T_k$. Therefore, $\{\Diamond^T_k : k \geq 0\}$ can be extended to a map $\Diamond^T_\infty : V(B''_\infty) \to B''_\infty$, so that the algebra $(B''_\infty, \Diamond_\infty)$ is the n-generated free $\mathbf{V}_L$-algebra.*

Proof That $(B''_\infty, \Diamond_\infty)$ is free is proved in the same way as in Proposition 6. That we quotient each approximant by $\mathrm{Ax}(L)$ guarantees that $(B''_\infty, \Diamond_\infty)$ satisfies $\mathrm{Ax}(L)$, and hence is a $\mathbf{V}_L$-algebra. □

What is not guaranteed in general here is that the maps i''_k are injective; this is unavoidable, giving the fact that there are undecidable logics:

Proposition 8 *If for each $k \geq 0$ the maps $i''_k : B''_k \to B''_{k+1}$ are injective, then B''_k is isomorphic to a Boolean subalgebra of all terms of complexity k of the free $\mathbf{V}_L$-algebra, and moreover the logic L is decidable.*

Proof Given terms t, u whose complexity is less than, say k, we can define their canonical realizations $[t], [u] \in B''_\infty$ and $[t]_k, [u]_k \in B''_k$ (this is quite straightforward and intuitive, see, e.g., [20]); notice also that we have $\iota_k([t]_k) = [t]$ and $\iota_k([u]_k) = [u]$. Under the obvious indentification of terms and propositional formulae, it is evident that (since B''_∞ is the free V_L-algebra) $[t] = [u]$ holds iff $t \leftrightarrow u$ is provable in the logic L. Since ι_k is a function, we have that $[t]_k = [u]_k$ *implies* $[t] = [u]$; according to the standard algebraic colimit construction, the converse is true in case the maps $i''_{\tilde{k}}$ for $k \leq \tilde{k}$ are all injective, whence the claim of the proposition. The statement about decidability is clear: to check whether an identity $t = 1$ holds, it is sufficient to inspect whether $[t]_k = 1$ holds, where k is the complexity of t. □

The above relation $[t]_k = [u]_k$ is quite interesting from the proof-theoretic point of view: it means that it is possible to establish $t \leftrightarrow u$ via a proof involving formulae whose complexity does not exceed k. The existence of a proof whose complexity is bounded by the size of the formula to be proved is an evidence for a nice proof-theoretic behavior of the given axiomatization for a logic. It is also quite a desirable property sufficient to entail decidability. Thus the above one-step algebraic approach provides an intersting tool also from a purely proof-theoretic perspective.

From the dual point of view, one should try to understand in terms of dual one-step frames what it means for a one-step algebra $(A_0, A_1, i_0, \Diamond_0)$ to satisfy a set of quasi-equations (3.5). For this, one needs to develop the *one-step correspondence theory*. The goal of this one-step correspondence theory is to characterize in the two-sorted predicate language for one-step frames what it means for a one-step frame that the dual one-step modal algebra satisfies $\mathrm{Ax}(L)$ (ideally, the characterization should be manageable and possibly first-order).

Once this is understood, one has to understand further what it means from the dual point of view to build a quotient making a one-step algebra $(A_0, A_1, i_0, \Diamond_0)$ an algebra satisfying the quasi-equations (3.5) occurring in $\mathrm{Ax}(L)$. In view of the applications, it is sufficient to characterize the case in which A_0, A_1 are both finite. Armed by this characterization, if one is able to prove that the duals of the i''_k are surjective, one can conclude that the logic is decidable. If the duals of the i''_k are not surjective, one can try with a different axiomatization of the logic L. In conclusion, the duality task is threefold:

(dt1) develop one-step correspondence theory;
(dt2) have a nice characterization of the dual of the following operation: take a finite one-step algebra $(A_0, A_1, i_0, \Diamond_0)$ satisfying $\mathrm{Ax}(L)$, build the universal one-step extension of it and make it a one-step algebra satisfying $\mathrm{Ax}(L)$ again (we call this the *one step-*$\mathrm{Ax}(L)$*-extension* of $(A_0, A_1, i_0, \Diamond_0)$);
(dt3) have a nice characterization of the dual of the following operation: take a finite set X_0, build the one-step modal algebra dual to the one-step frame (3.4) and make it a one-step algebra satisfying $\mathrm{Ax}(L)$.

Usually (dt3) is quite easy, while (dt1)-(dt2) are different for each particular logic. We will discuss some cases in detail below.

3.4 Free K4-Algebras

We start by considering task (dt1) for **K4**. As we will see, it is accomplished by Proposition 9 below. We say that a one step-frame *validates* a quasi-equation if the corresponding one-step modal algebra $(\wp(Y), \wp(X), f^*, \Diamond_R)$ validates this quasi equation.

Definition 3 A one-step frame (X, Y, f, R) is *transitive* if it validates the **K4**-quasi-equation

$$a \leq \Diamond b \quad \Rightarrow \quad \Diamond a \leq \Diamond b \tag{3.8}$$

i.e., if

$$f^*(A) \subseteq \Diamond_R(B) \Rightarrow \Diamond_R(A) \subseteq \Diamond_R(B) \tag{3.9}$$

holds for all $A, B \subseteq Y$.

For $S \subseteq X$ and $x \in X$, define $S_x := \{\tilde{x} \in S \mid R(\tilde{x}) \subseteq R(x)\}$.

Proposition 9 *A one-step frame (X, Y, f, R) is transitive iff*

$$y \in R(x) \Rightarrow f^*(y) \cap X_x \neq \emptyset \tag{3.10}$$

holds for all $x \in X, y \in Y$.

Proof Assume (3.10) and pick, A, B, x such that $f^*(A) \subseteq \Diamond_R(B)$ and $x \in \Diamond_R(A)$. The goal is to show that $x \in \Diamond_R(B)$. From $x \in \Diamond_R(A)$, we obtain a $y \in R(x) \cap A$. Pick $z \in f^*(y) \cap X_x$. Since $f(z) = y \in A$, we have $z \in f^*(A) \subseteq \Diamond_R(B)$, hence $R(z) \cap B \neq \emptyset$, giving also $R(x) \cap B \neq \emptyset$ (because $z \in X_x$). Thus, $x \in \Diamond_R(B)$, as desired. Conversely, assume (3.9) and pick $y \in R(x)$. If $f^*(y) \cap X_x$ is empty, then for every $z \in f^*(y)$, there is $w \in R(z)$ such that $w \notin R(x)$. Let $A := \{y\}$ and let B be the complement of $R(x)$. Then $f^*(A) \subseteq \Diamond_R(B)$, hence $\Diamond_R(A) \subseteq \Diamond_R(B)$. Since $y \in R(x)$, it follows that $x \in \Diamond_R(B)$, i.e. that $R(x) \cap B$ is not empty, a contradiction because B is the complement of $R(x)$. □

If (X, Y, f, R) is a one-step frame, there is the largest $X^\sharp \subseteq X$ such that $(X^\sharp, Y, f, R)$ is transitive: in fact, the set of all $\tilde{X} \subseteq X$ such that for all $x \in \tilde{X}$

$$\forall y \in Y \ \ (y \in R(x) \Rightarrow f^*(y) \cap \tilde{X}_x \neq \emptyset)$$

(with f, R restricted to $\tilde{X}$ in the domain) is closed under unions and hence has the largest element $X^\sharp$. A *subframe* of a one-step frame is obtained by restricting functions and relations to some subset of a given frame. The largest subset $X^\sharp$ gives rise to the one-step subframe $(X^\sharp, Y, f_{|X^\sharp}, R_{|X^\sharp})$ (obtained by restricting f, R to $X^\sharp$ in the domain) that corresponds to the quotient modulo the quasi-equation (3.8). In general, one cannot say more than that: we just need to characterize the one-step subframe arising in tasks (dt2)–(dt3).

For (dt3) the situation is trivial: given any finite set X_0, the one-step frame (3.4) is already transitive, so the one-step transitive subframe we are looking for is the whole one-step frame in this case.

Task (dt2) requires to characterize the universal one-step **K4**-extension of a finite transitive one-step frame (the universal one-step extension of a finite transitive one-step frame is obviously defined to be the dual of the universal one-step **K4**-extension of the corresponding dual finite one-step **K4**-algebra).

In short, we get the following notion. Given a transitive finite one-step frame (X, Y, f, R), the *universal one-step* **K4**-*extension* of it is the largest transitive one-step frame $(X^\sharp, X, f', R') \subseteq (X', X, f', R')$, where X', f', R' are defined as follows

- $X' = \{(x, S) \in X \times \wp(X) \mid R(x) = f(S)\}$;
- $f'(x, S) = x$;
- $R'(x, S) = S$.

According to the above definitions, $X^\sharp$ is the largest $\widehat{X} \subseteq X'$ such that

$$(x, S) \in \widehat{X} \Rightarrow (\forall y \in S \, \exists S' \subseteq S \; (y, S') \in \widehat{X}). \tag{3.11}$$

To fully accomplish task (dt2), we need here a better explicit characterization of $X^\sharp$.

A subset $S \subseteq X$ of a one-step frame (X, Y, f, R) is said to be *transitive* (abbreviated $Tr(S)$) if $(S, Y, f_{|S}, R_{|S})$ is a transitive one-step frame (by $(-)_{|S}$ we denote the restriction of a relation or of a function to a subset S of its domain).

Proposition 10 $X^\sharp = \{(x, S) \mid R(x) = f(S) \;\&\; R(S) \subseteq R(x) \;\&\; Tr(S)\}$.

Proof We must show that (1) $X^\sharp$ satisfies condition (3.11); (2) if $\widehat{X}$ satisfies condition (3.11), then $\widehat{X} \subseteq X^\sharp$.

Ad (1): Take $(x, S) \in X^\sharp$ and $y \in S$. Define

$$S' = \{\widehat{y} \in S \mid f(\widehat{y}) \in R(y) \;\&\; R(\widehat{y}) \subseteq R(y)\}. \tag{3.12}$$

We show that $(y, S') \in X^\sharp$. First, $R(S') \subseteq R(y)$ and $f(S') \subseteq R(y)$ are immediate from the definition of S'. To show that $R(y) \subseteq f(S')$, notice that since S is transitive, for all $z \in R(y)$ there is $y_z \in S$ such that $f(y_z) = z$ and $R(y_z) \subseteq R(y)$. This shows that $z \in f(S')$. It remains to show that S' is transitive. Let $\widehat{y} \in S'$ and $w \in R(\widehat{y})$. Since $S' \subseteq S$ and S is transitive, there is $s \in S$ such that $f(s) = w$ and $R(s) \subseteq R(\widehat{y})$. We only need to prove that $s \in S'$, i.e. that (a) $f(s) \in R(y)$ and (b) $R(s) \subseteq R(y)$. Since $\widehat{y} \in S'$, we have $R(\widehat{y}) \subseteq R(y)$ and this implies $f(s) \in R(y)$ (because $f(s) = w \in R(\widehat{y})$). Thus, (a) holds. For (b), observe that $\widehat{y} \in S'$ implies $R(\widehat{y}) \subseteq R(y)$. Since we also have $R(s) \subseteq R(\widehat{y})$, we obtain $R(s) \subseteq R(y)$, i.e. (b) holds.

Ad (2): Let $\widehat{X}$ satisfy (3.11) and let $(x, S) \in \widehat{X}$. We show that $(x, S) \in X^\sharp$. To show that $R(S) \subseteq R(x)$, take $y \in S$. Then, according to (3.11), there is $S' \subseteq S$ such that $(y, S') \in \widehat{X}$. Thus, $R(y) = f(S') \subseteq f(S) = R(x)$, and consequently, $R(y) \subseteq R(x)$. So $R(S) \subseteq R(x)$ ($y \in S$ is arbitrary), as required. It remains to verify that S is transitive. Consider $y \in S$ and $z \in R(y)$. Then there is S' such that

$(y, S') \in \widehat{X}$, which implies $R(y) = f(S')$. Therefore, there is $s \in S'$ such that $f(s) = z$. However, $R(S') \subseteq R(y)$ follows from $(y, S') \in \widehat{X}$ (we just proved that this applies to all members of $\widehat{X}$), hence $R(s) \subseteq R(y)$. □

The following proposition says that we can also apply Proposition 8 in this case.

Proposition 11 *If (X, Y, f, R) is a transitive finite one-step frame, then f' restricted to $X^\sharp$ is surjective.*

Proof It is sufficient to show that for every $x \in X$, we have that the pair (x, S) belongs to $X^\sharp$, where S is given by

$$S = \{\widehat{y} \in X \mid f(\widehat{y}) \in R(x) \,\&\, R(\widehat{y}) \subseteq R(x)\}.$$

This follows from the fact that X is transitive (in the same way as case (i) above).□

Let X_0 be a 2^n-element set and let

$$X_1 = X_0 \times \wp(X_0), \qquad X_{k+1} = X_k^\sharp \;\; (k \geq 1).$$

We also let $f_k : X_{k+1} \to X_k$ and $R_k : X_{k+1} \to \wp(X_k)$ be defined by $f_k(x, S) = x$ and $R_k(x, S) = S$. Then using duality and Propositions 7, 8, 10, and 11 we arrive at the following result (we refer to the statement of Theorem 1 for the indication of the appropriate categories where limits below are taken in):

Theorem 2 *The limit (X_ω, R_ω) of the sequence $\{(X_k, X_{k+1}, f_k, R_k) : k < \omega\}$ is (isomorphic to) the dual of the free n-generated* **K4***-algebra. Moreover, each X_k is dual to the algebra of all* **K4***-equivalent terms of complexity k.*

3.5 Free S4-Algebras

We first deal with the **T**-case.

Definition 4 A one-step frame (X, Y, f, R) is *reflexive* if it validates the **T**-equation

$$a \leq \Diamond a$$

i.e., if

$$f^*(A) \subseteq \Diamond_R(A) \tag{3.13}$$

holds for all $A \subseteq Y$.

Task (dt1) is accomplished by the following easy proposition:

Proposition 12 *A one-step frame* (X, Y, f, R) *is reflexive iff*

$$f(x) \in R(x) \tag{3.14}$$

holds for all $x \in X$.

Proof Suppose $x \in f^*(A)$ for some $A \subseteq Y$. By (3.14), $f(x) \in R(x)$ and so $R(x) \cap A \neq \emptyset$. Thus, $x \in \Diamond_R(A)$, satisfying (3.13). Conversely, suppose $x \in X$ is such that $f(x) \notin R(x)$. Let $A = X \setminus R(x)$. Then $f(x) \in A$ and $R(x) \cap A = \emptyset$. So $x \in f^*(A)$ and $x \notin \Diamond_R(A)$. Therefore, $f^*(A) \nsubseteq \Diamond_R(A)$, refuting (3.13). □

Clearly the largest reflexive one-step subframe of a one-step frame (X, Y, f, R) is obtained by taking the subset of X formed by those x such that $f(x) \in R(x)$ and by restricting f and R to it. This observation accomplishes also task (dt3). For (dt2), we have an obvious notion of the *universal one-step* **T**-*extension* of a reflexive one-step frame (X, Y, f, R). This is the largest reflexive one-step frame $(X^\sharp, X, f', R') \subseteq (X', X, f', R')$, where X', f', R' are defined as follows

- $X' = \{(x, S) \in X \times \wp(X) \mid R(x) = f(S)\}$;
- $f'(x, S) = x$;
- $R'(x, S) = S$.

We immediately have that

Proposition 13 $X^\sharp = \{(x, S) \mid R(x) = f(S) \;\&\; x \in S\}$.

Proof The proof is similar to the proof of Proposition 10. We must show that $X^\sharp$ is the largest subset $\widehat{X} \subseteq X'$ satisfying the condition

$$(x, S) \in \widehat{X} \Rightarrow x \in S. \tag{3.15}$$

But this immediately follows from the definition of $X^\sharp$. □

As a consequence, we obtain:

Proposition 14 *If* (X, Y, f, R) *is a reflexive finite one-step frame such that* f *is surjective, then* f' *restricted to* $X^\sharp$ *is surjective.*

Proof Let $x \in X$. We need to find $S \subseteq X$ such that $R(x) = f(S)$ and $x \in S$. Let $S = f^*(R(x))$. Then, as f is surjective, we have $f(S) = f(f^*(R(x))) = R(x)$. As X is reflexive, by Proposition 12, $f(x) \in R(x)$. So $x \in f^*(R(x)) = S$. Therefore, $(x, S) \in X^\sharp$ and $f'(x, S) = x$. Thus, f' is surjective. □

Then if X_k's, f_k's and R_k's are as in the previous section (of course $X^\sharp$ is as in Proposition 13), we have the following:

Theorem 3 *The limit* (X_ω, R_ω) *of the sequence* $\{(X_k, X_{k+1}, f_k, R_k) : k < \omega\}$ *is (isomorphic to) the dual of the free* n*-generated* **T**-*algebra. Moreover, each* X_k *is dual to the algebra of all* **T**-*equivalent terms of complexity* k.

Finally, the relevant tasks for **S4** are obtained by combining the above results for **T** and **K4**. The combination is not entirely straightforward because we need to carefully revisit all proofs. More precisely, the *universal one-step* **S4**-*extension* of a reflexive and transitive one-step frame (X, Y, f, R) is the largest reflexive and transitive one-step frame $(X^\sharp, X, f', R') \subseteq (X', X, f', R')$, where X', f', R' are defined as follows

- $X' = \{(x, S) \in X \times \wp(X) \mid R(x) = f(S)\}$;
- $f'(x, S) = x$;
- $R'(x, S) = S$.

We have:

Proposition 15 $X^\sharp = \{(x, S) \mid R(x) = f(S) \,\&\, R(S) \subseteq R(x) \,\&\, Tr(S) \,\&\, x \in S\}$.

Proof We can repeat word by word the proof of Proposition 10 (we only need to observe that if (X, Y, f, R) is reflexive, then S' as defined in (3.12) is such that $y \in S'$). □

Next we let the X_k's, f_k's and R_k's be the same as in the previous section with $X^\sharp$ as in Proposition 15. Then, since we can prove surjectivity by the same argument used in Proposition 11, we obtain the following:

Theorem 4 *The limit* (X_ω, R_ω) *of the sequence* $\{(X_k, X_{k+1}, f_k, R_k) : k < \omega\}$ *is (isomorphic to) the dual of the free* n*-generated* **S4**-*algebra. Moreover, each* X_k *is dual to the algebra of all* **S4**-*equivalent terms of complexity* k.

3.6 Comparison to Other Approaches

In the recent literature, various approaches have been proposed in order to build free modal algebras as chain colimits iterating some one-step construction. In the end, all constructions must be isomorphic, however the differences they induce in the dual combinatorics of finite frames might be significant, especially from the point of view of concrete manageability and ease of manipulation. This is why it is worth trying to make a closer comparison. The origins of the global step-by-step method for the basic modal logic **K** go back to Abramsky's 1988 British Logic Colloquium talk (the paper [1] based on this talk appeared a decade later). In [20] a detailed account of the global step-by-step method for **K** is given and, moreover, it is extended to other modal logics. In [8] the same construction is put in the context of coalgebra, underlying direct applications to rank one modal logics and is generalized to the basic positive modal logic.

In [22] the step-by-step approach to the construction of free algebras is applied to the **S4** case. The solution adopted there is to build a chain of finite **S4**-algebras

$$(B_0, \Diamond_0) \xrightarrow{i_0} (B_1, \Diamond_1) \to \cdots \to (B_k, \Diamond_k) \xrightarrow{i_k} (B_{k+1}, \Diamond_{k+1}) \to \cdots \quad (3.16)$$

where the morphisms i_k are *continuous* and *relatively open*, which means that they fully preserve 'old' diamonds (i.e. diamonds of the kind $\Diamond_k i_{k-1}(a)$) and just semi-preserve 'last' (i.e. the other) diamonds. In essence, this means that in each algebra there are 'core' diamonds that are defined once and for all and some other diamonds that are defined 'by completion' in a temporary fashion. The merit of this construction is that the duals of the $(B_k, \Diamond_k)$ are relatively nicely defined preordered sets. However, it is not clear how far this technique (implicitly relying on the existence of a kind of '**S4**-completion by continuity') can be pushed because its adaptation to general logics looks unclear.

To overcome this potential problem, in [11] free algebras are built up from a chain such as (3.16), where the algebras $(B_k, \Diamond_k)$ are *partial*, i.e. diamonds are only partially defined there. In addition, it is required that the image of i_k is in the domain of the next diamond $\Diamond_{k+1}$. If we view a partial algebra $(B, \Diamond)$ as a pair

$$B_0 \subseteq B, \qquad B_0 \stackrel{\Diamond}{\to} B$$

(where B_0 is the Boolean subalgebra of B which is the domain of $\Diamond$), then it is clear that

$$(B_0, B, i, \Diamond)$$

is a one-step modal algebra in our sense (here i is the inclusion morphism). This is essentially the alternative description of partial algebras as described in [11, Remark 2.2]. The only difference is that our definition of a one-step modal algebra *does not require injectivity* of i. This difference might be considered quite immaterial – and in fact it is whenever one is able to prove that in the end the duals of the i_k are surjective maps (see Propositions 11,14). However, taking the injectivity requirement out is essential for *our recipe based on the tasks* (dt1)-(dt2)-(dt3) to be formulated and applied. Thus, this approach supplies an algebraic tool to investigate proof-theoretic aspects, as explained in Proposition 8.

The step-by-step method for free Heyting algebras was first introduced in [19]. As an application of this method, it was shown that free finitely generated Heyting algebras are in fact bi-Heyting algebras. In [7] a modular approach to this construction was developed. It is similar to the local step-by-step approach of [11] and of the current paper. It is our understanding that the technique of [7] can be rephrased in terms of one-step Heyting algebras and the corresponding one-step frames.

3.7 Open Problems

We conclude by listing some open problems. The basic question is whether the local step-by-step method applies (or can be adjusted) to other important modal logics. As possible candidates we suggest the logic $\mathbf{wK4} = \mathbf{K} + (\Diamond\Diamond p \to p \vee \Diamond p)$ of weakly transitive frames [13] or, more generally, the logics of n-transitive frames [29].

Another challenging test case is the provability logic **GL**. It is known that **GL** is axiomatized over **K4** by a one-step rule

$$(\Box p \leq p) \rightarrow (p = 1),$$

see [12, 23]. Thus, the technique developed in Sect. 3.3 can be applied to **GL**. Also [12, 14] give a similar axiomatization of the logic **Grz** over **K4** by the rule

$$(\Box(p \rightarrow \Box p) \leq p) \rightarrow (p = 1).$$

Strictly speaking this is not a one-step rule, but we can up to equivalence flatten the rule to

$$(q = p \rightarrow \Box p) \;\&\; (\Box q \leq p) \rightarrow (p = 1)$$

so that it fits our purposes. We leave it as an open problem whether free **GL** and **Grz**-algebras could be described using the technique developed in this paper.

Finally, our results (Proposition 9 and 12) suggest that there might exist some Sahlqvist-like correspondence for one-step frames. An investigation of this correspondence is an interesting topic on its own, but it will also undoubtedly shed a new light on the problem of obtaining free algebras for modal logics via the step-by-step method.

Acknowledgments The authors are very grateful to the referee for many useful suggestions that improved the presentation of the paper. They would also like to express special thanks to Guram Bezhanishvili for his invaluable help and encouragement in writing this paper. The first author would like to acknowledge the support of the Netherlands Organisation for Scientific Research under grant 639.032.918. The first and third authors would also like to acknowledge the support of the Rustaveli Science Foundation of Georgia under grant FR/489/5-105/11. The second author acknowledges the support of the PRIN 2010-2011 project "Logical Methods for Information Management" funded by the Italian Ministry of Education, University and Research (MIUR).

References

1. Abramsky S (2005) A Cook's tour of the finitary non-well-founded sets. In: Artemov S et al (eds) We will show them: essays in honour of Dov Gabbay. College Publications, pp 1–18
2. Adámek J, Trnková V (1990) Automata and algebras in categories. Mathematics and its applications (East European Series), vol 37. Kluwer Academic Publishers Group, Dordrecht
3. Bellissima F (1984) Atoms in modal algebras. Z Math Logik Grundlag Math 30(4):303–312
4. Bellissima F (1985) An effective representation for finitely generated free interior algebras. Algebra Univers 20(3):302–317
5. Bellissima F (1986) Finitely generated free Heyting algebras. J Symbolic Logic 51:152–165
6. Bezhanishvili N (2006) Lattices of intermediate and cylindric modal logics. PhD thesis, University of Amsterdam
7. Bezhanishvili N, Gehrke M (2011) Finitely generated free Heyting algebras via Birkhoff duality and coalgebra. Log Methods Comput Sci 7(2:9):1–24
8. Bezhanishvili N, Kurz A (2007) Free modal algebras: a coalgebraic perspective. In: CALCO 2007. LNCS, vol 4624, pp 143–157. Springer, Berlin

9. Blok W (1977) The free closure algebra on finitely many generators. Indag Math 39(5):362–379
10. Chagrov A, Zakharyaschev M (1997) Modal logic. The Clarendon Press, Oxford
11. Coumans D, Van Gool S (2012) On generalizing free algebras for a functor. J Logic Comput 23(3):645–672
12. Esakia L (1981) Diagonal constructions, Löb's formula and Cantor's scattered spaces (Russian). In: Studies in logic and semantics, Metsniereba, Tbilisi, pp 128–143
13. Esakia L (2001) Weak transitivity—a restitution. In: Logical investigations, No. 8 (Russian) (Moscow, 2001). Nauka, Moscow, pp 244–255
14. Esakia L (2011) A modal transcription of the Hausdorff residue. In: Logic, language, and computation. Lecture Notes in Computer Science vol 6618. Springer, Berlin, pp 46–52
15. Esakia L, Grigolia R (1975) Christmas trees. On free cyclic algebras in some varieties of closure algebras. Bull Sect Logic 4:95–102
16. Esakia L, Grigolia R (1977) The criterion of Brouwerian and closure algebras to be finitely generated. Bull Sect Logic 6:46–52
17. Fine K (1975) Normal forms in modal logic. Notre Dame J Formal Logic 16:229–237
18. Gehrke M (2014) Canonical extensions, Esakia spaces, and universal models (This volume)
19. Ghilardi S (1992) Free Heyting algebras as bi-Heyting algebras. Math Rep Acad Sci Canada XVI 6:240–244
20. Ghilardi S (1995) An algebraic theory of normal forms. Ann Pure Appl Logic 71:189–245
21. Ghilardi S (1995) Irreducible models and definable embeddings. In: Logic colloquium '92 (Veszprém, 1992). Studies in Logic Language and Information. pp 95–113. CSLI Publ, Stanford
22. Ghilardi S (2010) Continuity, freeness, and filtrations. J Appl Non-Classical Logics 20(3):193–217
23. Goldblatt R (1985) An algebraic study of well-foundedness. Studia Logica 44(4):423–437
24. Grigolia R (1987) Free algebras of non-classical logics. Metsniereba, Tbilisi (Russian)
25. Halmos P (1962) Algebraic logic. Chelsea Publishing Co., New York
26. Johnstone P (1982) Stone spaces. Cambridge University Press, Cambridge
27. de Jongh D (1968) Investigations on the intuitionistic propositional calculus. PhD thesis, University of Wisconsin
28. Jónsson B, Tarski A (1951) Boolean algebras with operators. I Amer J Math 73:891–939
29. Kracht M (1999) Tools and techniques in modal logic. Studies in logic and the foundations of mathematics, vol 142. North-Holland Publishing Co., Amsterdam
30. Rybakov V (1997) Admissibility of logical inference rules. Studies in logic and the foundations of mathematics, vol 136. North-Holland Publishing Co., Amsterdam
31. Shehtman V (1978) Rieger—Nishimura lattices. Soviet Math Dokl 19:1014–1018
32. Venema Y, Vosmaer J (2014) Modal logic and the vietoris functor (This volume)

Chapter 4
Easkia Duality and Its Extensions

Sergio A. Celani and Ramon Jansana

In memory of Leo Esakia

Abstract In recent years Esakia duality for Heyting algebras has been extended in two directions. First to weak Heyting algebras, namely distributive lattices with an implication with weaker properties than that of the implication of a Heyting algebra, and secondly to implicative semilattices. The first algebras correspond to subintuitionistic logics, the second ones to the conjunction and implication fragment of intuitionistic logic. Esakia duality has also been complemented with dualities for categories whose objects are Heyting algebras and whose morphisms are maps that preserve less structure than homomorphisms of Heyting algebras. In this chapter we survey these developments.

Keywords Weak Heyting algebras, Distributive semilattices, Implicative semilattices, Priestley duality, Esakia duality

4.1 Introduction

In 1937 Marshall H. Stone published what in the mathematical language of our days can be described as the first duality between the category BDL of bounded distributive lattices and the category of certain topological spaces [52]. The dual space of a

S. A. Celani
Facultad de Ciencias Exactas, Universidad Nacional del Centro,
Pinto, 399, 7000 Tandil, Argentina
e-mail: scelani@exa.unicen.edu.ar

R. Jansana (✉)
Department of Logic, History and Philosophy of Science, University of Barcelona,
Montalegre 6, 08001 Barcelona, Spain
e-mail: jansana@ub.edu

G. Bezhanishvili (ed.), *Leo Esakia on Duality in Modal and Intuitionistic Logics*, Outstanding Contributions to Logic 4, DOI: 10.1007/978-94-017-8860-1_4,

bounded distributive lattice is nowadays known as a spectral space [19, 38] or as a coherent space [39]. Later, in 1970 Hilary Priestley developed a different duality between BDL and the category of certain ordered topological spaces whose objects are ordered Boolean spaces satisfying a separation condition relating the order with the clopen sets of the space. Since then these spaces are known as Priestley spaces and this duality is known as Priestley duality [47, 48]. In [19] William H. Cornish showed that the category of spectral spaces and the category of Priestley spaces are isomorphic. For a book exposition of Priestley duality the reader is referred to [21] and for a book exposition of Stone duality for bounded distributive lattices to [4, 39].

In 1974 Leo Esakia published the paper [25] where first a duality for S4-algebras is presented and then a duality for Heyting algebras is derived from it. When Esakia published [25] he was unaware of Priestley's work and only later, after becoming familiar with it, he realized that his duality for Heyting algebras is a restricted version of Priestley duality. For a nice exposition of Esakia duality for Heyting algebras and how it is related to the duality for S4-algebras we address the reader to Mai Gehrke's paper [28] in the present volume.

In recent years Esakia duality for Heyting algebras has been generalized in two directions. In [16] Esakia duality has been generalized to algebras that, like Heyting algebras, are bounded distributive lattices with an implication but the implication has weaker properties than the implication of a Heyting algebra. Because of this fact, these algebras are called weak Heyting algebras. In [8, 11] Esakia duality has been generalized to implicative semilattices and bounded implicative semilattices; that is, to subalgebras of the $(\wedge, \rightarrow, 1)$-reducts and of the $(\wedge, \rightarrow, 0, 1)$-reducts of Heyting algebras, respectively. From both dualities, for weak Heyting algebras and for implicative semilattices, Esakia duality can be obtained as a special case.

Both generalizations are interesting from the logical point of view. As it is well-known, Heyting algebras provide the algebraic semantics for intuitionistic logic and this logic has also a well-understood relational semantics. Esakia duality establishes the link between these two semantics. In an analogous way, implicative semilattices [45], also known as Brouwerian semilattices [40], provide the algebraic semantics for the $(\wedge, \rightarrow, 1)$-fragment of intuitionistic logic. Therefore, the generalization of Esakia duality to implicative semilattices provides a relational semantics for this fragment. Implicative semilatices also play an important role in universal algebra, as shown by Köhler and Pigozzi [41].

To show the logical interest in weak Heyting algebras and the topological duality for them, let $\{\wedge, \vee, \rightarrow, 0, 1\}$ be the algebraic similarity type of Heyting algebras, which is also the set of connectives for the propositional language $\mathscr{L}_I$ of intuitionistic logic. In this language some fragments of modal logic can be formalized by interpreting $\rightarrow$ as strict implication and defining $\Box\varphi$ as $1 \rightarrow \varphi$. From this perspective it is a sensible query to find a propositional logic that has as natural extensions both the intuitionistic logic and different strict implication fragments of normal modal logics. This logic was introduced as a consequence relation in [15], under the name wK_σ. Previously K. Dǒsen studied it as a set of theorems in [22]. In [15] it was provided with a complete relational semantics. The logic wK_σ belongs to the family of logics called subintuitionistic; see [15] and the references therein for information on the logics of this family. To give a precise (and useful) definition of wK_σ we can proceed by first defining the translation $(.)^\dagger$ from the formulas of $\mathscr{L}_I$ to the formulas of the usual language of modal logic with connectives $\wedge, \vee, \supset, \Box, 0, 1$, where $\supset$ is classical implication. The definition of $(.)^\dagger$ is as follows: $(.)^\dagger$ commutes with conjunction and disjunction, maps 0 to 0, 1 to 1 and $(\varphi \to \psi)^\dagger = \Box(\varphi \supset \psi)$. Then the logic wK_σ can be defined by:

$$\Gamma \vdash_{wK_\sigma} \varphi \ \text{ iff } \ (\Gamma)^\dagger \models (\varphi)^\dagger,$$

where $\models$ denotes the local consequence relation associated with the class of all Kripke frames and $\Gamma \cup \{\phi\}$ is a set of formulas of $\mathscr{L}_I$. The class of algebras that corresponds to wK_σ is the variety of weak Heyting algebras introduced in [16] and the topological duality for these algebras presented in [16] establishes the link between the algebraic and relational semantics for wK_σ. From the duality for weak Heyting algebras one can obtain the dualities for modal algebras and for Heyting algebras as particular cases. It is interesting to point out that there are varieties of weak Heyting algebras that can not be seen neither as varieties of Heyting algebras nor as varieties of modal algebras, for example the variety of Basic algebras introduced in [3].

In recent years there also has been an interest in dualities for categories whose objects are Heyting algebras but whose morphisms are maps that only preserve some of, instead of all, the operations of a Heyting algebra, see [5, 8, 11]. The dual categories also have as objects Esakia spaces but the morphisms are partial maps. These dualities have been applied in [5] to provide an algebraic proof of M. Zacharyashev's result [60, 62] that all superintuitionistic (or intermediate) logics are axiomatizable by canonical formulas (in Zacharyashev's sense).

In this chapter we survey both generalizations of Esakia duality discussed above, as well as the dualities for categories of Heyting algebras whose morphisms are weaker than Heyting algebra homomorphisms. There is some work related to the material presented in this chapter that we do not survey but is worth pointing out. We do not survey the specialization to Heyting algebras of Stone duality for bounded distributive lattices and the Stone type duality for implicative semilattices obtained in [13]. We also do not survey the bitopological duality for bounded distributive lattices and for Heyting algebras obtained in [6], strongly related to Esakia duality.

In order to make the chapter as self contained as possible, we start in Sect. 4.2 by presenting first, and very concisely, Priestley duality and then, in Sect. 4.3, Esakia duality. We assume the reader is familiar with the algebraic notions of distributive lattice, relative pseudo-complement and Heyting algebra (see [4]), as well as modal algebras and their dual spaces (see [30, 51, 54]). The structure of the rest of the chapter is as follows. In Sect. 4.4 we introduce weak Heyting algebras and in Sect. 4.5 we present the Priestley style duality for these algebras developed in [16]. Section 4.6 describes the duality for implicative semilattices obtained in [8, 11]. Finally, Sect. 4.7 expounds the categories of Esakia spaces with partial maps as morphisms.

4.2 Preliminaries on Priestley Duality

We use the following notational conventions. Let X be a set. If $U \subseteq X$, then U^c denotes the complement of U w.r.t. X, namely $X - U$. The power set of X will be denoted by $\mathscr{P}(X)$. If R is a binary relation on X and $U \subseteq X$ we let

$$R^{-1}[U] := \{x \in X : (\exists y \in U)(xRy)\},$$
$$R[U] := \{y \in X : (\exists x \in U)(xRy)\}.$$

If U is a singleton, say $\{x\}$, we will write $R[x]$ and $R^{-1}[x]$ instead. A set $U \subseteq X$ is *R-closed*, or an *R-up-set*, if for every $x \in U$, $R[x] \subseteq U$.

Let $\langle X, \leq\rangle$ be a poset. A set $Y \subseteq X$ is an *up-set* if it is $\leq$-closed; that is, if for every $x \in Y$ and every $y \in X$ such that $x \leq y$, we have $y \in Y$. Dually, we have the notion of *down-set*. Given a set $U \subseteq X$, let $\downarrow U$ denote the down-set generated by U; that is, $\downarrow U = \{x \in X : (\exists y \in U)\ x \leq y\}$. Dually, $\uparrow U$ is the up-set generated by U. Given $x \in X$, we abbreviate $\downarrow\{x\}$ and $\uparrow\{x\}$ by $\downarrow x$ and $\uparrow x$, respectively. The set of all up-sets of X will be denoted by $\mathscr{P}^{\uparrow}(X)$. This set together with set-theoretic union, $\bigcup$, and intersection, $\bigcap$, is a complete distributive lattice which is relatively pseudo-complemented; that is, for every $U, V \in \mathscr{P}^{\uparrow}(X)$ there exists a unique $U \rightarrow_X V \in \mathscr{P}^{\uparrow}(X)$ such that for every $W \in \mathscr{P}^{\uparrow}(X)$,

$$U \cap W \subseteq V \quad \text{iff} \quad W \subseteq U \rightarrow_X V.$$

Thus, the binary operation $\rightarrow_X$ on $\mathscr{P}^{\uparrow}(X)$ when added to the lattice $\mathscr{P}^{\uparrow}(X)$ turns it into a Heyting algebra. The operation $\rightarrow_X$ satisfies

$$U \rightarrow_X V = \{x \in X : \uparrow x \cap U \subseteq V)\} = (\downarrow(U - V))^c.$$

Definition 1 A *Priestley space* is a triple $X = \langle X, \leq, \tau\rangle$ where $\langle X, \leq\rangle$ is a poset, $\langle X, \tau\rangle$ is a compact topological space and the following separation condition is satisfied: for every $x, y \in X$, if $x \not\leq y$, then there exists a clopen up-set U such that $x \in U$ and $y \notin U$.

It follows that for every Priestley space $\langle X, \leq, \tau\rangle$, the space $\langle X, \tau\rangle$ is a Boolean space (compact, Hausdorff and 0-dimensional). If Y is a subset of a Priestley space X, the topological closure of Y in X will be denoted by $Cl_X(Y)$; the subscript will be omitted if no confusion is likely to arise. If $\langle X_1, \leq_1, \tau_1\rangle$ and $\langle X_2, \leq_2, \tau_2\rangle$ are Priestley spaces, a map $f : X_1 \to X_2$ is *order preserving* when for every $x, y \in X_1$, if $x \leq_1 y$ then $f(x) \leq_2 f(y)$, and it is *continuous* when for every $U \in \tau_2$, $f^{-1}[U] \in \tau_1$. Note that $f : X_1 \to X_2$ is order preserving if and only if for every $x \in X_1$, $f[\uparrow x] \subseteq \uparrow f(x)$. The Priestley spaces, taken as objects, together with the continuous and order preserving maps between them, taken as morphisms, constitute the *category of Priestley spaces* that we denote by PriSp. We call the morphisms of this category *Priestley morphisms*.

The contravariant functors D : PriSp $\to$ BDL and X : BDL $\to$ PriSp that establish Priestley duality between the categories BDL and PriSp are defined as follows.

If X is a Priestley space, let $D(X)$ be the set of all clopen up-sets of X. This set is a ring of sets and $D(X) = \langle D(X), \cup, \cap, \emptyset, X\rangle$ is a bounded distributive lattice. This distributive lattice is the dual of X and it is, by definition, the image of X by the functor D.

If $f : X \to Y$ is a continuous and order preserving map between Priestley spaces X and Y, then the map $D(f) : D(Y) \to D(X)$ defined by

$$D(f) = f^{-1}(U),$$

for every $U \in D(Y)$, is a bounded lattice homomorphism. This map is the image of f by the functor D and has the following properties: $D(f)$ is onto $D(X)$ if and only if f is one-to-one, and $D(f)$ is one-to-one if and only if f is onto Y. Moreover, D is indeed a functor. If X, Y and Z are Priestley spaces and $f : X \to Y$ and $g : Y \to Z$ are continuous and order preserving maps, then $D(g \circ f) = D(f) \circ D(g)$. Also, if $id_X : X \to X$ is the identity morphism from a Priestley space X to itself, then $D(id_X)$ is the identity homomorphism from $D(X)$ to $D(X)$.

Let $L = \langle L, \vee, \wedge, 0, 1\rangle$ be a bounded distributive lattice and let $X(L)$ be the set of prime filters of L. For every $a \in L$, let

$$\sigma(a) := \{x \in X(L) : a \in x\}.$$

The topology τ_L on $X(L)$ is the topology generated by the subbasis

$$\{\sigma(a) : a \in L\} \cup \{\sigma(a)^c : a \in L\}.$$

The space $X(L) = \langle X(L), \subseteq, \tau_L\rangle$ is a Priestley space and is the dual space of L; that is, it is by definition the image of L by the functor X.

If L_1 and L_2 are bounded distributive lattices and $h : L_1 \to L_2$ is a bounded lattice homomorphism, then the map $X(h) : X(L_2) \to X(L_1)$ defined by

$$X(h)(x) = h^{-1}(x),$$

for every $x \in X(L_2)$, is continuous and order preserving. It is the image of h by the functor X. Moreover, h is one-to-one if and only if $X(h)$ is onto $X(L_1)$, and h is onto L_2 if and only if $X(h)$ is one-to-one. X is indeed a functor because if L_1, L_2 and L_3 are bounded distributive lattices and $h : L_1 \to L_2$ and $h' : L_2 \to L_3$ are bounded lattice homomorphisms, then $X(h' \circ h) = X(h) \circ X(h')$; and obviously, if $id_L : L \to L$ is the identity homomorphism from a bounded distributive lattice to itself, then $X(id_L)$ is the identity Priestley morphism from $X(L)$ to $X(L)$.

The natural transformations that justify that the functors X : BDL $\to$ PriSp and D : PriSp $\to$ BDL establish a dual equivalence between BDL and PriSp are given by the maps σ_L and ε_X, defined as follows. Let L be a bounded distributive lattice. The map $\sigma_L : L \to D(X(L))$ is defined by $\sigma_L(a) := \sigma(a)$, for every $a \in L$. This map is a bounded lattice isomorphism. Moreover, for every Priestley space $X = \langle X, \leq, \tau \rangle$, the map $\varepsilon_X : X \to X(D(X))$ is defined by

$$\varepsilon_X(x) := \{U \in D(X) : x \in U\},$$

for every $x \in X$. This map is a homeomorphism and an order-isomorphism.

Stone duality for Boolean algebras follows easily from Priestley duality. The full subcategory of **BDL** whose objects are the Boolean lattices has as dual category the full subcategory of **PriSp** whose objects are the Priestley spaces $\langle X, \leq, \tau \rangle$ where $\leq$ is the identity relation; in this situation $D(X)$ is the Boolean lattice of the clopen subsets of the space. Since the category of Boolean lattices is isomorphic to the category of Boolean algebras, this last category is dually equivalent to the subcategory of **PriSp** that is isomorphic to the category of Boolean spaces under the identification of a Priestley space $\langle X, \leq, \tau \rangle$, where $\leq$ is the identity relation, with the Boolean space $\langle X, \tau \rangle$. From this, Stone duality for Boolean algebras follows at once.

There exists a correspondence between algebraic concepts for bounded distributive lattices and order-topological concepts for Priestley spaces. We mention the correspondence between the most basic ones.

Let L be a bounded distributive lattice and $X(L)$ be its dual space. The lattice of filters of L ordered by inclusion is dually isomorphic to the lattice of closed up-sets of $X(L)$, also order by inclusion. The dual isomorphism is given by the map Φ defined by

$$\Phi(F) := \bigcap \{\sigma(a) : a \in F\}$$

for every filter F of L. The inverse of Φ is such that $\Phi^{-1}(U) = \{a \in L : U \subseteq \sigma(a)\}$ for every closed up-set U of $X(L)$. The image by Φ of a prime filter x of L is the principal up-set $\uparrow x$ of $X(L)$. In a similar manner, the map Ψ that sends an ideal I of L to the set $\bigcup\{\sigma(a) : a \in I\}$ establishes an isomorphism between the lattice of ideals of L ordered by inclusion and the lattice of open up-sets of $X(L)$ also order by inclusion. A prime ideal I of L is mapped to the up-set $(\downarrow x)^c$ of $X(L)$, where x is the prime filter $L - I$. If L is a Boolean lattice, we have the well-known correspondence between filters of L and closed sets of its Boolean space and between ideals of L and open sets.

The lattice of congruences of L ordered by inclusion is dually isomorphic to the lattice of closed subsets of $X(L)$ also ordered by inclusion. The isomorphism is given by the map Θ from the set of closed subsets of $X(L)$ to the set of congruences of L given by

$$\Theta(Y) := \{(a, b) \in L \times L : \sigma(a) \cap Y = \sigma(b) \cap Y\}$$

for every closed set Y of $X(L)$. The inverse of Θ can be defined as follows. Let us first say that a congruence θ of L is *compatible* with $x \in X(A)$ if for all $a, b \in L$, if $a\theta b$ and $a \in x$, then $b \in x$. Then,

$$\Theta^{-1}(\theta) = \{x \in X(L) : \theta \text{ is compatible with } x\}.$$

Finally, the subalgebras of L are in one-to-one correspondence with the Priestley quasi-orders of $X(L)$ [1, 18, 53]. Let $X = \langle X, \leq, \tau \rangle$ be a Priestley space. A *Priestley quasi-order* on X is a quasi-order R on X that extends $\leq$ and such that for every $x, y \in X$, if $x \not R y$, then there exists a clopen R-up-set $U \subseteq X$ such that $x \in U$ and $y \notin U$.[1] The map H from the lattice of universes of subalgebras of L to the lattice of Priestley quasi-orders defined by

$$H(M) := \{(x, y) \in X(L) \times X(L) : x \cap M \subseteq y\}$$

for every universe M of a subalgebra of L, establishes a dual isomorphism between these two lattices. Its inverse is the map that sends a Priestley quasi-order R on $X(L)$ to the set

$$M_R := \{a \in L : R[\sigma(a)] \subseteq \sigma(a)\}.$$

4.3 Esakia Duality

The duality between bounded distributive lattices and Priestley spaces specializes to Heyting algebras giving what is now with full justice known as Esakia duality. Although, as was pointed out in the Introduction, L. Esakia derived in [25] his duality for Heyting algebras from a duality for S4-algebras, and only later realized that it is a restricted version of Priestley duality, it is better for our purposes to describe Esakia duality as such a restriction. We do this in this section. It should be mention that M. Adams independently obtained the duality in an unpublished paper, as it is pointed out in [20].

Recall that a *Heyting algebra* is an algebra $A = \langle A, \vee, \wedge, \rightarrow, 0, 1 \rangle$ where $\langle A, \vee, \wedge, 0, 1 \rangle$ is a bounded and relatively pseudo-complemented distributive lattice and $\rightarrow$ is the operation of relative pseudo-complementation (or residuation); that is,

$$a \wedge b \leq c \quad \text{iff} \quad a \leq b \rightarrow c,$$

for every $a, b, c \in A$.

The category HA of Heyting algebras has as objects Heyting algebras and as morphisms the homomorphisms between them. This category is isomorphic to the category of bounded and relatively pseudo-complemented distributive lattices with morphisms the bounded lattice homomorphisms that in addition preserve the relative

---

1 The term 'Priestley quasi-order' is introduced in [6, 10].

pseudo-complementation, or residuation.[2] This category is not a full subcategory of BDL. So, to obtain a duality for Heyting algebras using Priestley spaces as the dual objects, one not only needs to characterize the dual Priestley space of a bounded relatively pseudo-complemented distributive lattice, but also the dual of a homomorphism between bounded relatively pseudo-complemented distributive lattices that in addition preserves relative pseudo-complementation.

Definition 2 An *Esakia space* is a Priestley space $\langle X, \leq, \tau \rangle$ with the additional property that for every clopen set $U \subseteq X$, $\downarrow U$ is clopen.

An *Esakia morphism* between two Esakia spaces $\langle X_1, \leq_1, \tau_1 \rangle$ and $\langle X_2, \leq_2, \tau_2 \rangle$ is an order preserving and continuous map f from X_1 to X_2 such that for every $x \in X_1$ and every $z \in X_2$, $f(x) \leq_2 z$ implies that there exists $y \in X_1$ such that $x \leq_1 y$ and $f(y) = z$; in other words, it is an order preserving and continuous map such that for every $x \in X_1$, $\uparrow f(x) \subseteq f[\uparrow x]$.[3]

The Esakia spaces with Esakia morphisms between them constitute a subcategory of PriSp which is not a full subcategory.[4] It is denoted in this chapter by EsSp. The functors that establish the duality between HA and EsSp are in essence the functors D and X of Priestley duality suitably restricted and modified.

Recall that the operation $\rightarrow_X$ defined in the lattice $\mathcal{P}^{\uparrow}(X)$ of up-sets of X by

$$U \rightarrow_X V := \{x \in X : \uparrow x \cap U \subseteq V)\}$$

is the residual of the intersection operation between up-sets; that is,

$$U \cap V \subseteq W \quad \text{iff} \quad U \subseteq V \rightarrow_X W,$$

for every $U, V, W \in \mathcal{P}^{\uparrow}(X)$; and the bounded distributive lattice $\mathcal{P}^{\uparrow}(X)$ endowed with the operation $\rightarrow_X$ is a Heyting algebra.

The fundamental fact about Esakia spaces is the following:

Proposition 1 *A Priestley space $X = \langle X, \leq, \tau \rangle$ is an Esakia space if and only if $D(X)$ is closed under the operation $\rightarrow_X$.*

From this proposition Esakia duality follows easily. The dual of an Esakia space X is the Heyting algebra $D(X) = \langle D(X), \cup, \cap, \rightarrow_X, \emptyset, X \rangle$ and the dual of a Heyting algebra $A = \langle A, \vee, \wedge, \rightarrow, 0, 1 \rangle$ is the Priestley space $X(A)$ of the lattice reduct $\langle A, \vee, \wedge, 0, 1 \rangle$. This space is an Esakia space and the map σ_A is an isomorphism

---

[2] Note that if L and L' are bounded relatively pseudo-complemented distributive lattices (the residuation operation, or implication, is not part of the signature) and $h : L \to L'$ is a bounded lattice homomorphism, then we only have that for $a, b \in L$, $h(a \to b) \leq h(a) \to' h(b)$.

[3] In [25] the maps between Esakia spaces that are order preserving and satisfy that for every $x \in X_1$, $\uparrow f(x) \subseteq f[\uparrow x]$ are called *strongly isotone*. That is, they are the maps such that for every $x \in X_1$, $\uparrow f(x) = f[\uparrow x]$.

[4] The full subcategory with Esakia spaces as objects has as morphisms the continuous and order preserving maps between them, but not all of them satisfy that for every $x \in X_1$, $\uparrow f(x) \subseteq f[\uparrow x]$.

between A and the Heyting algebra $\langle D(X(A)), \cup, \cap, \rightarrow_X (A), \emptyset, X\rangle$. Moreover, if $h : A_1 \rightarrow A_2$ is a homomorphism from a Heyting algebra A_1 to a Heyting algebra A_2, then the map $X(h)$ from $X(A_2)$ to $X(A_1)$ is an Esakia morphism and if f is an Esakia morphism from an Esakia space X_1 to an Esakia space X_2, then the map $D(f)$ is a Heyting algebra homomorphism from $D(X_2)$ to $D(X_1)$.

It is worth pointing out that L. Esakia proved the lemma that came to be known as Esakia's lemma precisely to establish the correspondence between Heyting algebra homomorphisms and Esakia morphisms. This lemma has proved to be very useful in proving Shalqvist style correspondence results for modal logic using topological arguments.

The correspondence between the algebraic concepts for bounded distributive lattices and order-topological concepts for Priestley spaces presented in the previous section specializes to Heyting algebras and Esakia spaces as follows.

Let A be a Heyting algebra and $X(A)$ its dual Esakia space. Filters of A correspond to closed up-sets of $X(A)$ as in the case of bounded distributive lattices; here the presence of the implication does not play any role. Similarly, ideals of A correspond to open up-sets of $X(A)$.

The dual isomorphism Θ between the lattice of congruences of the bounded distributive lattice reduct of A and the lattice of closed sets of its Priestley space, when restricted to the congruences of A provides an isomorphism between that lattice of congruences of A and the lattice of closed up-sets of the dual Esakia space $X(A)$. One way to prove this result is to use the well-known fact that in any Heyting algebra A the lattice of congruences of A is isomorphic to the lattice of filters of A.

The isomorphism H between the lattice of the universes of bounded sublattices of A and the lattice of Priestley quasi-orders of $X(A)$, when restricted to the universes of subalgebras of A, provides an isomorphism between the lattice of these universes and the lattice of the Priestley quasi-orders of $X(A)$ that satisfy an extra condition. These Priestley quasi-orders are characterized in [6, 10] where they are called Esakia quasi-orders. They are defined as follows. A binary relation R on the set of points of an Esakia space $\langle X, \leq, \tau\rangle$ is an *Esakia quasi-order* if it is a Priestley quasi-order that in addition satisfies that for all $x, y \in X$, if xRy, then there exists $z \in X$ such that $x \leq z$ and $\langle z, y\rangle \in R \cap R^{-1}$.

Another way to characterize subalgebras of a Heyting algebra is by means of Esakia equivalence relations of the dual space. Let X be an Esakia space and let $\sim$ be an equivalence relation on X. For $x \in X$ let $[x] := \{y \in X : x \sim y\}$, and for $Y \subseteq X$ let $[Y] := \bigcup\{[y] : y \in Y\}$. We call $Y \subseteq X$ *saturated* if $Y = [Y]$. An equivalence relation $\sim$ is an *Esakia equivalence relation* if $\sim$ satisfies the following two conditions:

1. $x \not\sim y$ implies there exists a saturated clopen set U of X such that $x \in U$ and $y \notin U$.
2. $(\forall x, y, z \in X)[(x \sim y \;\&\; y \leq z) \Rightarrow (\exists z' \in X)(x \leq z' \;\&\; z' \sim z)]$.

If E satisfies only condition (1), then E is an equivalence relation which is a Priestley quasi-order. We call such equivalence relations *Priestley equivalence relations.*

Thus, an Esakia equivalence relation is a Priestley equivalence relation satisfying condition (2).

There is an order-isomorphism between the set of Esakia quasi-orders of an Esakia space X, ordered by inclusion, and the set of Esakia equivalence relations on X also ordered by inclusion. The correspondence is obtained as follows [6, 10]. If R is an Esakia quasi-order on X, then $\sim_R := R \cap R^{-1}$ is an Esakia equivalence relation on X; conversely, if $\sim$ is an Esakia equivalence relation on X, then $R_\sim := (\leq \circ \sim)$ is an Esakia quasi-order on X. Moreover, $R_{\sim_R} = R$ and $\sim_{R_\sim} = \sim$. It follows that for a Heyting algebra A, the complete lattice of universes of subalgebras of A is dually order isomorphic to the poset of Esakia equivalence relations on $X(A)$. In particular, if A is a Boolean algebra, since in the Priestley space of A the relation $\leq$ becomes $=$, Esakia equivalence relations become Priestley equivalence relations, and so we obtain the following well-known characterization of subalgebras of a Boolean algebra. For a Boolean algebra A, the complete lattice of subalgebras of A is dually isomorphic to the poset of Priestley equivalence relations on $X(A)$ (ordered by $\subseteq$).

The lattices of subalgebras of distributive lattices and of Heyting algebras using duality have been studied in [2, 6, 10, 25, 34, 37, 53, 57, 58].

4.4 Weak Heyting Algebras

The notion of weak Heyting algebra introduced in [16] is a weakening of the concept of Heyting algebra.[5] As we mentioned in the Introduction, weak Heyting algebras are the variety of algebras that corresponds to the subintuitionistic logic wK_σ introduced in [15].

The algebraic similarity type of weak Heyting algebras is the same as the algebraic similarity type of Heyting algebras, namely $\{\wedge, \vee, \rightarrow, 0, 1\}$. As we mentioned in the Introduction, one can interpret the symbol $\rightarrow$ as the strict implication of modal logic and define $\Box\varphi$ as $1 \rightarrow \varphi$. From this perspective, weak Heyting algebras encompass (in a sense that we make precise shortly) both modal algebras and Heyting algebras. From the duality for weak Heyting algebras one can obtain the dualities for modal algebras and Heyting algebras as particular cases.

In this section we present the definition and basic algebraic facts about weak Heyting algebras, as well as examples of varieties of weak Heyting algebras; some of them already appeared in the literature before the notion of weak Heyting algebra was introduced. In the next section we present a Priestley style duality for weak Heyting algebras. The proofs of all the results we mention in this and next section can be found in [16].

Definition 3 An algebra $\langle A, \vee, \wedge, \rightarrow, 0, 1\rangle$ is a *weak Heyting algebra*, or a WH-algebra, if its $\{\wedge, \vee, 0, 1\}$-reduct $\langle A, \vee, \wedge, 0, 1\rangle$ is a bounded distributive lattice

[5] In [16] they are called weakly Heyting, but weak Heyting appears to be a better terminology. It comes from [12].

and $\rightarrow$ is a binary operation on A satisfying the following conditions for every $a, b, c \in A$:

C1 $(a \rightarrow b) \wedge (a \rightarrow c) = a \rightarrow (b \wedge c)$.
C2 $(a \rightarrow c) \wedge (b \rightarrow c) = (a \vee b) \rightarrow c$.
C3 $(a \rightarrow b) \wedge (b \rightarrow c) \leq a \rightarrow c$.
C4 $a \rightarrow a = 1$.

From the definition it is immediate that the class of WH-algebras is a variety. We denote by WH this variety as well as the category with objets WH-algebras and morphisms the homomorphisms between them.

The following elementary facts about WH-algebras are useful.

Proposition 2 *Let A be a WH-algebra. For every $a, b, c \in A$,*

1. *if $a \leq b$, then $c \rightarrow a \leq c \rightarrow b$ and $b \rightarrow c \leq a \rightarrow c$;*
2. *if $a \leq b$, then $a \rightarrow b = 1$;*
3. *$(a \rightarrow b) \wedge (a \rightarrow c) \leq a \rightarrow (c \vee b)$.*

It is also useful to point out that the converses of conditions 2 and 3 of the proposition above, which hold in Heyting algebras, do not necessarily hold in WH-algebras.

Heyting algebras are obviously WH-algebras. Other examples of WH-algebras arise in a natural way from modal algebras. Let $\mathbf{B} = \langle B, \wedge, \vee, ^{-}, \Diamond, 0, 1\rangle$ be a modal algebra, that is, a Boolean algebra $\langle B, \wedge, \vee, ^{-}, 0, 1\rangle$ together with an operator $\Diamond$ on B, namely, a unary operation $\Diamond : B \rightarrow B$ that distributes over nonempty finite joins and such that $\Diamond 0 = 0$. As usual, one defines $\Box a = \overline{\Diamond \overline{a}}$. The strict implication $\rightarrow$ on B is the binary operation defined by

$$a \rightarrow b = \overline{\Diamond(a \wedge \overline{b})},$$

for all $a, b \in B$. Thus,

$$a \rightarrow b = \Box(\overline{a} \vee b) \quad \text{and} \quad \Box a = 1 \rightarrow a.$$

It is easy to see that the algebra $\langle B, \wedge, \vee, \rightarrow, 0, 1\rangle$ is a WH-algebra. In fact, all the WH-algebras whose lattice reduct is a Boolean lattice are obtainable in this way. Let $A = \langle A, \wedge, \vee, \rightarrow, 0, 1\rangle$ be a WH-algebra such that $\langle A, \wedge, \vee, 0, 1\rangle$ is a Boolean lattice (i.e. a complemented bounded distributive lattice). The algebra $\langle A, \wedge, \vee, ^{-}, \Diamond, 0, 1\rangle$, where $^{-}$ is the complement operation and $\Diamond$ is defined by $\Diamond a = \overline{1 \rightarrow \overline{a}}$, is a modal algebra and its strict implication operation is the original $\rightarrow$. We will say that a WH-algebra is a WH-*modal algebra* if its lattice reduct is a Boolean lattice.

Other examples of WH-algebras that appeared in the literature before the introduction of the concept of weak Heyting algebra are the basic algebras introduced by Ardeshir and Ruitenburg [3] and the subresiduated lattices of Epstein and Horn [24].

Basic algebras correspond to the Basic Propositional Logic introduced by Visser [55, 56] as a logic in the language of intuitionistic logic that has the same relation to the modal logic $K4$ as intuitionistic logic has to $S4$. Subresiduated lattices were introduced in [24] as the algebras that correspond to the strict implication fragment (in the sense made precise in the Introduction) of the modal logic $S4$.

Definition 4 A *basic algebra* is a WH-algebra that in addition satisfies the inequality

(I) $x \leq 1 \rightarrow x$

and a *subresiduated lattice* is a WH-algebra that in addition satisfies the inequalities

(T) $x \rightarrow y \leq z \rightarrow (x \rightarrow y)$,
(R) $x \wedge (x \rightarrow y) \leq y$.

Remark 1 Theorem 1 of [24] shows that a subresiduated lattice is an algebra $A = \langle A, \wedge, \vee, \rightarrow, 0, 1\rangle$ where $\langle A, \wedge, \vee, 0, 1\rangle$ is a bounded distributive lattice and the following four inequalities hold: $1 \leq (x \wedge y) \rightarrow y$, $x \rightarrow y \leq z \rightarrow (x \rightarrow y)$, (R) and (T). From this theorem and some properties of subresiduated lattices proved in [24] it is easy to see that an algebra $A = \langle A, \wedge, \vee, \rightarrow, 0, 1\rangle$ is a subresiduated lattice if and only if it is a WH-algebra satisfying the inequalities (R) and (T). This fact justifies our definition.

In every WH-algebra the inequality (I) implies (T). Indeed, on the one hand (I) implies $x \rightarrow y \leq 1 \rightarrow (x \rightarrow y)$ and, on the other hand, $1 \rightarrow (x \rightarrow y) \leq z \rightarrow (x \rightarrow y)$ holds in every WH-algebra. Therefore, if we combine conditions (I), (T) and (R) to obtain subvarieties of WH we end up with at most five subvarieties. In fact there are exactly five obtainable in this way: the variety of subresiduated lattices, denoted SRL, the variety of basic algebras, denoted B, the variety of WH-algebras that satisfy (R), these will be called RWH-*algebras*, the variety of WH-algebras that satisfy (T), these will be called TWH-*algebras*, and finally the variety of Heyting algebras (which are the WH-algebras that satisfy the three inequalities (R), (T) and (I)). The variety of RWH-algebras will be denoted by RWH and the variety of TWH-algebras by TWH. If in addition we denote by HA the variety of Heyting algebras, the relation between all these varieties is depicted in Fig. 4.1.

Other varieties of weak Heyting algebras have been considered in the literature; for example, [14] studies n-linear weak Heyting algebras and [12] introduces the notion of pre-Heyting algebra.

4.5 Priestley Style Duality for WH-Algebras

A fundamental difference between Heyting algebras and WH-algebras is that WH-algebras cannot be defined as special distributive lattices in which the implication operation is uniquely determined. For example, if L is a bounded distributive lattice, the map $f : L \times L \rightarrow L$ such that $f(a, b) = 1$ for every $a, b \in L$ satisfies conditions

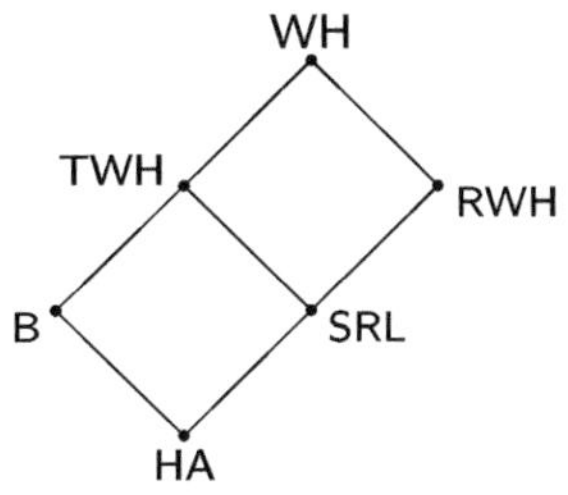

Fig. 4.1 The subvarieties of the variety of WH-algebras

C1-C4 in the definition of a weak Heyting algebra. Because of this, when developing a Priestley style duality for WH-algebras, in order to account for the implication, we have to add an extra structure to the dual Priestley space of the distributive lattice reduct of a WH-algebra. Unlike the case of Heyting algebras, simply adding additional conditions that the dual space of the distributive lattice reduct has to satisfy is not enough.

In this section we introduce a category dual to the category of WH-algebras whose objects are Priestley spaces with an additional binary relation that dualizes the implication. This section is based on [16] where all the results we present here can be found with their proofs.

Let X be a set and let R be a binary relation on X. The binary operation $\to_R$ on $\mathscr{P}(X)$ is defined as follows:

$$U \to_R V = \{x \in X : R[x] \cap U \subseteq V\},$$

for every $U, V \subseteq X$. It is easy to see that

$$U \to_R V = X - R^{-1}[U - V].$$

Note that if R is the partial order $\leq$ of a Priestley space X, then the operation $\to_R$ is exactly the residuation operation $\to_X$ considered before to obtain the Heyting algebra of the up-sets of X.

It is not difficult to see that if $X = \langle X, \leq, \tau\rangle$ is a Priestley space and R is a binary relation on X, then the set $D(X)$ of clopen up-sets is closed under the operation $\to_R$ if and only if for all $U, V \in D(X)$, $R^{-1}[U - V]$ is a clopen down-set.

Moreover, it is also not difficult to see that in any structure $X = \langle X, \leq, R, \tau\rangle$ where $\langle X, \leq, \tau\rangle$ is a Priestley space and R is a binary relation such that (i) $R[x]$ is a closed subset of X for every $x \in X$ and (ii) $R^{-1}[U - V]$ is a clopen down-set for all $U, V \in D(X)$, it holds that $(\leq \circ R) \subseteq R$.

Let $\langle X, \leq, R\rangle$ be a poset endowed with a binary relation R satisfying the condition $(\leq \circ R) \subseteq R$. This implies that for every $U, V \in \mathscr{P}^{\uparrow}(X)$, $U \to_R V \in \mathscr{P}^{\uparrow}(X)$. Therefore, the bounded distributive lattice of up-sets of $\langle X, \leq\rangle$ can be expanded to an algebra of the type of a WH-algebra by adding the operation $\to_R$. It is easy to show that the obtained algebra is a WH-algebra.

Let $X = \langle X, \leq, R, \tau \rangle$ be a Priestley space with a binary relation R satisfying $(\leq \circ R) \subseteq R$. That the dual bounded distributive lattice $D(X)$ of $\langle X, \leq, \tau \rangle$ is closed under the operation $\rightarrow_R$ can be characterized by a topological condition that does not involve $D(X)$.

Lemma 1 *Let $X = \langle X, \leq, R, \tau \rangle$ be a Priestley space endowed with a binary relation R on X such that $(\leq \circ R) \subseteq R$. Then $D(X)$ is closed under the operation $\rightarrow_R$ on $\mathscr{P}^{\uparrow}(X)$ if and only if for every clopen set Y, $R^{-1}[Y]$ is clopen.*

The considerations above justify the following definition of WH-spaces. These spaces are the objects of the category dual to the category of WH-algebras, which will be defined shortly.

Definition 5 A *WH-space* is a structure $\langle X, \leq, R, \tau \rangle$ such that

1. $\langle X, \leq, \tau \rangle$ is a Priestley space,
2. $(\leq \circ R) \subseteq R$,
3. $R[x]$ is a closed subset of X for all $x \in X$,
4. for every clopen subset Y of X, $R^{-1}[Y]$ is clopen.

In [32] Paul Halmos introduced *Boolean relations* on a Boolean space as the binary relations on the set of points of the space that satisfy properties (3) and (4) above. These relations are also studied in [59]. The spaces $\langle X, R, \tau \rangle$ where $\langle X, \tau \rangle$ is a Boolean space and R is a Boolean relation are the duals of modal algebras according to the duality theory for these algebras. We call them *modal spaces*. Similarly, a WH-space is a structure that combines in a very natural way being a Priestley space and a modal space; that is, it is a structure $\langle X, \leq, R, \tau \rangle$ such that (i) $\langle X, R, \tau \rangle$ is a modal space, (ii) $\langle X, \leq, \tau \rangle$ is a Priestley space and (iii) $(\leq \circ R) \subseteq R$. For information on the topological duality for modal algebras we address the reader to [51]. A brief description can be found in [7]. It is also worth to consult [54].

Definition 6 Let $\langle X, \leq, \tau \rangle$ be a Priestley space. We say that a binary relation R on X is a WH-*relation* if it is a Boolean relation on the Stone space $\langle X, \tau \rangle$ that satisfies $(\leq \circ R) \subseteq R$. Note that the empty relation and $X \times X$ are WH-relations.

From Lemma 1 and the considerations that precede it, it immediately follows that WH-spaces can be defined equivalently as the structures $\langle X, \leq, R, \tau \rangle$ where $\langle X, \leq, \tau \rangle$ is a Priestley space, $R[x]$ is a closed subset of X for every $x \in X$, and for all $U, V \in D(X)$, $R^{-1}[U - V]$ is a clopen down-set. Therefore, if $X = \langle X, \leq, R, \tau \rangle$ is a WH-space, then the lattice $D(X)$ is closed under the operation $\rightarrow_R$, and augmented with the restriction of this operation to the domain of $D(X)$, it is a subalgebra of the WH-algebra $\mathscr{P}^{\uparrow}(X)$. Thus, it is a WH-algebra. This algebra is, by definition, the *dual WH-algebra* of the WH-space X. Slightly abusing the notation, we also denote it by $D(X)$.

We proceed to obtain the dual of a WH-algebra. Let A be a WH-algebra. We define the binary relation R_A on the set $X(A)$ of prime filters of A as follows:

$$(x, y) \in R_A \quad \text{iff} \quad (\forall a, b \in A)(a \to b \in x \,\&\, a \in y \Longrightarrow b \in y).$$

Let us consider the structure $X(A) = \langle X(A), \subseteq, \tau_A, R_A \rangle$, where $\langle X(A), \subseteq, \tau_A \rangle$ is the Priestley space dual of the lattice reduct of A.

The following extension of the prime filter lemma for distributive lattices to WH-algebras is fundamental to obtain the duality for WH-algebras. Let A be a WH-algebra. To state the lemma, we extend the relation R_A between prime filters defined above to arbitrary filters by using the same defining condition, but now applied to all filters. Moreover, if F is a filter of A and X is a nonempty subset of A, we define the set $D_F(X)$ as follows:

$$D_F(X) := \{b \in A : \text{ there is a nonempty and finite } Y \subseteq X \text{ with } \bigwedge Y \to b \in F\}.$$

Lemma 2 (Existence of Prime Filters) *Let A be a WH-algebra, F a filter and I an ideal of A and let $X \subseteq A$ be nonempty. If $D_F(X) \cap I = \emptyset$, then there exists a prime filter x such that*

$$D_F(X) \subseteq x, \quad (F, x) \in R_A \quad \textit{and} \quad x \cap I = \emptyset.$$

The next lemma is crucial in establishing the desired duality.

Lemma 3 *For every WH-algebra A and every $a, b \in A$,*

$$\sigma(a \to b) = \sigma(a) \to_{R_A} \sigma(b).$$

Hence, $\sigma(A)$ is closed under the operation $\to_{R_A}$.

Proof Suppose that $a \to b \notin x$. We will prove that there is $y \in X(A)$ such that $(x, y) \in R_A$, $a \in y$ and $b \notin y$. Consider the set $D_x(\{a\}) = \{c \in A : a \to c \in x\}$. This set is disjoint from the ideal $\downarrow b$ because otherwise there would exist $c \leq b$ such that $a \to c \in x$. But then $a \to b \in x$, which contradicts the assumption. By Lemma 2, there exists $y \in X(A)$ such that $(x, y) \in R_A$, $a \in y$ and $b \notin y$. Thus, $y \notin \sigma(a) \to_{R_A} \sigma(b)$. The other inclusion is immediate. □

Proposition 3 *For every WH-algebra A, $X(A) = \langle X(A), \subseteq, \tau_A, R_A \rangle$ is a WH-space.*

Proof The proof of condition (2) of Definition 5 is easy. Condition (4) follows from the fact that $D(X(A))$ is $\sigma(A)$ using lemmas 3 and 1. It remains to prove condition (3). Let $x, y \in X(A)$ be such that $y \notin R_A[x]$. Then, by the definition of R_A, there are $a, b \in A$ with $a \to b \in x$, $a \in y$ and $b \notin y$. Therefore, $x \in \sigma(a) \to_{R_A} \sigma(b)$ and $y \in \sigma(a) - \sigma(b)$. So, since $R_A[x] \cap (\sigma(a) - \sigma(b)) = \emptyset$, y is not in the closure of $R_A[x]$. Thus, $R_A[x]$ is closed. □

A consequence of Lemma 3 is that for every WH-algebra A, the map σ_A is a WH-algebra isomorphism between A and $D(X(A))$. Note that the domain of $D(X(A))$ is the set $\sigma[A]$.

Recall from Priestley duality that for every Priestley space $\langle X, \leq, \tau \rangle$, the map ε_X from X to the set $X(D(X))$, which is defined by the condition

$$\varepsilon_X(x) = \{U \in D(X) : x \in U\},$$

for every $x \in X$, is a homeomorphism and an order isomorphism between the space $\langle X, \leq, \tau \rangle$ and the Priestley dual of $D(X)$. For every WH-space $\langle X, \leq, R, \tau \rangle$, the map ε_X, in addition to being a homeomorphism and an order isomorphism between $\langle X, \leq, \tau \rangle$ and $\langle X(D(X)), \subseteq, \tau_{D(X)} \rangle$, is also an isomorphism between the relational structures $\langle X, R \rangle$ and $\langle X(D(X)), R_{D(X)} \rangle$. This follows from the proposition below.

Proposition 4 *Let $\langle X, \leq, \tau \rangle$ be a Priestley space endowed with a binary relation R such that for all $U, V \in D$, $R^{-1}[U - V]$ is a clopen down-set. Then the following conditions are equivalent:*

1. *For all $x \in X$, $R[x]$ is a closed subset of X.*
2. *For all $x, y \in X$, if $(\varepsilon_X(x), \varepsilon_X(y)) \in R_D$, then $(x, y) \in R$, where R_D is the relation associated with the WH-algebra $D(X)$.*

Proof To prove that (1) implies (2), let $x, y \in X$ be such that $(\varepsilon_X(x), \varepsilon_X(y)) \in R_D$ and suppose that $(x, y) \notin R$. Since $R[x]$ is closed, there exist $U, V \in D$ such that $R[x] \cap (U - V) = \emptyset$ and $y \in U - V$. Then $x \in U \rightarrow_R V$ and $y \in U$. As $(\varepsilon_X(x), \varepsilon_X(y)) \in R_D$, it follows that $y \in V$, which is impossible. To prove the other implication, let $x, y \in X$ be such that $y \in Cl(R[x])$ and suppose that $y \notin R[x]$. Then $(\varepsilon_X(x), \varepsilon_X(y)) \notin R_D$. It follows that there exist $U, V \in D$ such that $x \in U \rightarrow_R V$ and $y \in U - V$. Since $y \in Cl(R[x])$, we have $R[x] \cap (U - V) \neq \emptyset$, which is a contradiction. □

Consequently, we obtain that if $X = \langle X, \leq, R, \tau \rangle$ is a WH-space, the lattice $D(X)$ with the operation $\rightarrow_R$ is a WH-algebra whose associated WH-space $X(D(X))$ is isomorphic to $\langle X, \leq, R, \tau \rangle$.

The morphisms between WH-spaces are the maps defined as follows.

Definition 7 Let X_1 and X_2 be WH-spaces. A map $f : X_1 \longrightarrow X_2$ is a *WH-morphism* if it is a morphism between Priestley spaces (i.e., it is continuous and order preserving), and it is in addition a *bounded morphism* (or p-morphism) between $\langle X_1, R_1 \rangle$ and $\langle X_2, R_2 \rangle$, which means that it satisfies the following two conditions:

1. if $(x, y) \in R_1$, then $(f(x), f(y)) \in R_2$,
2. if $(f(x), z) \in R_2$, then there is $y \in X_1$ such that $(x, y) \in R_1$ and $f(y) = z$.

More concisely stated, f is a bounded morphism if it satisfies $f[R_1[x]] = R_2[f(x)]$ for every $x \in X_1$.

Notice that if the WH-spaces are modal spaces, then the morphisms between them are the usual bounded morphisms.

Theorem 1 *Let X_1, X_2 be WH-spaces and A_1, A_2 be WH-algebras.*

1. *If $f : X_1 \to X_2$ is a WH-morphism, the dual map $D(f) : D(X_1) \to D(X_2)$, defined by $D(f)(U) = f^{-1}(U)$, for every $U \in D(X_2)$, is a WH-homomorphism.*
2. *If $h : A_1 \to A_2$ is a WH-homomorphism, the dual function $X(h) : X(A_2) \to X(A_1)$, defined by $X(h)(x) = h^{-1}(x)$, for every $x \in X(A_2)$, is a WH-morphism.*

The next theorem follows from the above results and Priestley duality for bounded distributive lattices.

Theorem 2 *The category* WH *of WH-algebras is dually equivalent to the category* WHS *whose objects are WH-spaces and whose morphisms are WH-morphisms.*

Remark 2 If we add the trivial partial ordering, namely the identity relation, to a modal space $\langle X, R, \tau\rangle$, we obtain the WH-space $\langle X, =, R, \tau\rangle$. In this way modal spaces can be seen as the WH-spaces whose partial order is the identity relation, and the well-known duality between modal algebras and modal spaces can be obtained by restricting the duality between WH-algebras and WH-spaces to these objects. Let us call the WH-spaces of the form $\langle X, =, R, \tau\rangle$ *WH-modal spaces*. The bounded distributive lattice reduct of the dual WH-algebra $D(X)$ of a WH-modal space $\langle X, =, R, \tau\rangle$ is the Boolean lattice $B(X) = \langle D(X), \cup, \cap, \emptyset, X\rangle$ of the clopen subsets of $\langle X, \tau\rangle$. Augmented with the complement operation, the WH-algebra $D(X)$ of $\langle X, =, R, \tau\rangle$ is term-wise definitionally equivalent[6] to the modal algebra $\langle B(X), \Diamond_R\rangle$, where $\Diamond_R$ is defined by $\Diamond_R(U) = R^{-1}[U]$ for every $U \in D(X)$. Accordingly we can say that the WH-algebras which are the duals of the WH-modal spaces are modal algebras.

When we restrict Theorem 2 to the full subcategory of WH-modal algebras, we obtain the dual equivalence between the category of WH-modal algebras and the category of WH-modal spaces. By the above considerations, these categories are isomorphic to the category of modal algebras and to the category of modal spaces, respectively. Therefore, we obtain the well-known duality between these two categories as a corollary.

Moreover, as we already mentioned, for a Heyting algebra A, R_A is the inclusion relation on $\mathscr{P}(A)$ and, as we will see below, the dual WH-spaces of Heyting algebras are the WH-spaces where $\leq = R$. Thus, these spaces can be identified with Esakia spaces and the morphisms between them are exactly the Esakia morphisms between Esakia spaces. Consequently, the above theorem also implies Esakia duality for Heyting algebras.

It is worth mentioning that the WH-relations on the dual Priestley space of a bounded distributive lattice L correspond to the binary operations on L that expand L to a WH-algebra. This correspondence is obtained as follows.

Let L be a bounded distributive lattice. We refer to its partial ordering by $\leq_L$. Let $\preceq_L$ be the relation between binary operations on L defined by

---

[6] Two algebras of different similarity type are term-wise definitionally equivalent if every principal operation of one is definable by a term of the other.

$$f \preceq_L g \iff \forall a, b \in L,\ f(a,b) \leq_L g(a,b).$$

Let WHI(L) be the set of binary operations f on L such that the expansion $\langle L, f\rangle$ is a WH-algebra and let WHR($X(L)$) be the set of WH-relations on the dual Priestley space $X(L)$ of L.

Theorem 3 *Let L be a bounded distributive lattice and let $\langle X(L), \subseteq, \tau_L\rangle$ be its dual Priestley space. There is a dual isomorphism Φ between the posets $\langle \mathrm{WHI}(L), \preceq_L\rangle$ and $\langle \mathrm{WHR}(X(L)), \subseteq\rangle$, given by $\Phi(f) := R_{\langle L, f\rangle}$.*

The Priestley spaces that correspond to finite bounded distributive lattices are customarily identified with their partial order reducts because their topology is the discrete topology. In this way one obtains Birkhoff's duality between finite distributive lattices and finite posets. A similar situation holds for finite WH-algebras.

Theorem 4 *Let L be a finite bounded distributive lattice and let $\langle X, \leq\rangle$ be its associated dual partial ordering. There is a dual isomorphism between the ordered set $\langle \mathrm{WHI}(L), \preceq_L\rangle$ and the set, ordered by inclusion, of the binary relations R on X such that $\leq \circ R \subseteq R$. Moreover, this last ordered set is a lattice whose infimum operation is intersection and whose supremum operation is union.*

Before discussing the dual categories of some subvarieties of WH-algebras, we mention that [12] describes finitely generated free weak Heyting algebras and provides a coalgebraic perspective on WH-spaces. Namely, it is shown that the category of WH-spaces is isomorphic to the category of Vietoris coalgebras on the category of Priestley spaces, a result analogous to the well-known isomorphism between the category of modal algebras and the category of Vietoris coalgebras on Stone spaces [42].

The duality for WH-algebras just described specializes to the five subviarieties of WH considered in Sect. 4.4:

- The category of subresiduated lattices is dually equivalent to the category of WH-spaces $\langle X, \leq, R, \tau\rangle$ with R reflexive and transitive.
- The category of RWH-algebras is dually equivalent to the category of WH-spaces $\langle X, \leq, R, \tau\rangle$ with R reflexive.
- The category of TWH-algebras is dually equivalent to the category of WH-spaces $\langle X, \leq, R, \tau\rangle$ with R transitive.
- The category of basic algebras is dually equivalent to the category of WH-spaces $\langle X, \leq, R, \tau\rangle$ with $R \subseteq \leq$.

Basic algebras and RWH-algebras have interesting characterizations, related to the condition

$$a \wedge b \leq c \quad \text{iff} \quad a \leq b \to c, \tag{4.1}$$

expressing that the implication is the residual of the meet. This condition characterizes the WH-algebras that are Heyting algebras. A WH-algebra A is a basic algebra if and only if the implication from left to right of (4.1) holds, and it is a RWH-algebra if and only if the other implication of (4.1) holds. Therefore, the implication from left to right is satisfied in A if and only if in the dual WH-space $X(A)$ of A it holds that R_A is contained in the inclusion relation. It is easily shown that the other implication holds if and only if the inclusion relation on $X(A)$ is contained in R_A. As a corollary, it follows from the duality between the categories of WH-algebras and of WH-spaces that for every WH-space $\langle X, \leq, R, \tau\rangle$, $D(X)$ is a RWH-algebra if and only if the relation $\leq$ is contained in the relation R, and that the category of RWH-algebras is dually equivalent to the category of WH-spaces $\langle X, \leq, R, \tau\rangle$ such that $\leq \subseteq R$.

The above considerations imply that Heyting algebras, as WH-algebras, can be characterized in terms of their dual WH-spaces in an easy way. Since a WH-algebra A is a Heyting algebra if and only if condition (4.1) holds, a WH-algebra A is a Heyting algebra if and only if the WH-relation R_A on the set $X(A)$ is the inclusion relation. Therefore, a WH-space $\langle X, \leq, R, \tau\rangle$ is (isomorphic to) the WH-space of a Heyting algebra if and only if $\leq = R$. Thus, as we already mentioned, Esakia duality follows from the duality between the category of WH-algebras and the category of WH-spaces.

The correspondence between the algebraic concepts for WH-algebras and order-topological concepts for WH-spaces goes in parallel with the correspondence between the algebraic concepts for bounded distributive lattices and order-topological concepts for Priestley spaces we presented in Sect. 4.2.

Let A be a WH-algebra. The map Θ defined in Sect. 4.2, whose domain is the set of closed subsets of $X(A)$, establishes a dual isomorphism between the lattice of closed R_A-up-sets of $X(A)$ and the lattice $\mathrm{Con}(A)$ of congruences of A.

The duals of filters of A are the duals of filters of its lattice reduct, namely the closed up-sets of $X(A)$. In modal algebras, the open filters are the filters that are closed under the dual operator $\Box$. In every modal algebra, the lattice of open filters is isomorphic to the lattice of congruences. A similar notion of filter can be defined for a WH-algebra A. A filter F of A is an *open filter* if for every $a \in F$, $1 \to a \in F$. The closed up-sets that correspond to open filters are the closed up-sets of $X(A)$ which are also R_A-up-sets. Moreover, these closed sets correspond to the increasing congruences of A, where a congruence θ of A is *increasing* if for every $x, y \in X(A)$, if θ is compatible with x and $x \subseteq y$, then θ is compatible with y.

Let A be a WH-algebra and M the universe of a bounded sublattice of A. Then M is the universe of a subalgebra of A if and only if $H(M) \circ R_A \subseteq R_A \circ \left(\subseteq_M^{-1} \cap \subseteq_M\right)$, where H is the map defined in Sect. 4.2. Moreover, the map H establishes a dual isomorphism between the lattice of universes of subalgebras of A and the lattice of the Priestley quasi-orders R on $X(A)$ with the property that $R \circ R_A \subseteq R_A \circ \left(R^{-1} \cap R\right)$. Applying this result to Heyting algebras, we obtain the well-known theorem that says that for every Heyting algebra A, the map H establishes a dual isomorphism between the lattice of universes of subalgebras of A and the lattice of Esakia quasi-orders on $X(A)$. The reason is that R_A is the inclusion relation and for a Priestley quasi-order R, we have $R \circ \subseteq = R$.

4.6 Duality for Bounded Implicative Semilattices

In this section we expound the duality for bounded implicative semilattices developed in [8, 11], where it is called generalized Esakia duality. Bounded implicative semilattices can be described as the subalgebras of the $(\wedge, \rightarrow, 0, 1)$-reducts of Heyting algebras. As we mentioned in the Introduction, bounded implicative semilattices provide an algebraic semantics for the fragment of intuitionistic logic with the connectives $\wedge, \rightarrow, 0, 1$. As we will see, the dual space of a bounded implicative semilattice is a Priestley space augmented with a dense subset that satisfies certain conditions. When the bounded implicative semilattice happens to be a Heyting algebra, that is, when every two elements have a supremum in the semilattice order, then its dual space is an Esakia space. Thus, Esakia duality easily follows from the duality we present in this section.

The structure of the section is as follows. First we recall the definition of a (bounded) implicative semilattice and then we proceed to expound the Priestley style duality for these algebras. To do so, we first need to describe the Priestley style duality for distributive meet-semilattices developed in [8, 9] and for their order duals in [35, 36]. The reader can find proofs of all the results we present in this section in [8, 9, 11].

Definition 8 An *implicative semilattice* is an algebra $A = \langle A, \wedge, \rightarrow, 1\rangle$ where $\langle A, \wedge, 1\rangle$ is a relatively pseudo-complemented meet-semilattice with greatest, or top, element 1 and $\rightarrow$ is the binary operation on A of relative pseudo-complementation (or residuation); that is, for all a, b, c,

$$a \wedge c \leq b \text{ iff } c \leq a \rightarrow b.$$

A *bounded implicative semilattice* is an algebra $A = \langle A, \wedge, \rightarrow, 0, 1\rangle$ where $\langle A, \wedge, \rightarrow, 1\rangle$ is an implicative semilattice and 0 is a least, or bottom, element.

The meet-semilattice order $\leq$ of an implicative semilattice $\langle A, \wedge, \rightarrow, 1\rangle$ satisfies

$$a \leq b \text{ iff } a \rightarrow b = 1,$$

for all $a, b \in A$. The class of implicative semilattices forms a variety and so does the class of bounded implicative semilattices. An axiomatization can be found in [44].

The meet-semilattice reduct of a bounded implicative semilattice is a bounded distributive meet-semilattice, where we recall that a bounded meet-semilattice $L = \langle L, \wedge, 0, 1\rangle$ is *distributive* if for every $a, b_1, b_2 \in L$ with $b_1 \wedge b_2 \leq a$ there exist $c_1, c_2 \in L$ such that $b_1 \leq c_1, b_2 \leq c_2$ and $a = c_1 \wedge c_2$.

Generalized Esakia duality for bounded implicative semilattices is built on generalized Priestley duality for bounded distributive meet-semilattices in a similar way Esakia duality builds on Priestley duality. Therefore, before expounding it we need to recall briefly the generalized Priestley duality for bounded distributive meet-semilattices. This duality was developed for bounded join-semilattices and homomorphisms between bounded join-semilattices that in addition preserve existing finite meets in [35, 36], and for bounded distributive meet-semilattices and all homomorphisms between them in [8, 9]. The main difference between the duality for the category whose morphisms are the homomorphism that also preserve existing finite joins (or dually, existing finite meets, if we consider join-semilattices) and the duality for the category with all homomorphisms is that in the first case the dual of a morphism is a function and in the second case it is a relation. The idea of considering relations as duals of homomorphisms in the context of topological dualities for implicative semilattices originates in [13]. But it can be traced back to the work of Halmos [32, 33], where hemimorphisms between Boolean algebras (i.e. maps that preserve join and 0) are described dually by means of ceratin relations between their dual spaces, an idea also exploited in [51] and [18].[7] This is closely related to the duality for modal algebras because in a modal algebra A the operator $\Diamond$ is a hemimorphism of the Booelan reduct of A to itself, and so in the dual Boolean space of A the operator $\Diamond$ corresponds to a binary relation. All this is in accordance with recent discoveries of the need of using relations as duals of morphisms in the general theory of dualities and canonical extensions, for example in the discrete dualities presented in [23], and also in the theory of RS-frames given in [27].

4.6.1 Generalized Priestley Duality

Let $L = \langle L, \wedge, 0, 1\rangle$ be a bounded meet-semilattice. A *filter* of L is a nonempty set $F \subseteq L$ which is an up-set w.r.t. $\leq$ and is closed under finite meets. The set of all filters of L ordered by set-theoretic inclusion is a complete lattice where the infimum of a set of filters is their intersection. A filter is proper if it is not L. A filter is *prime* if it is a prime element of the lattice of filters; that is, if it is proper and for all filters F_1, F_2 of L, whenever $F_1 \cap F_2 \subseteq F$, then $F_1 \subseteq F$ or $F_2 \subseteq F$. It is well known that a bounded meet-semilattice is distributive if and only if the lattice of its filters is a distributive lattice.

The notions of filter, proper filter and prime filter for bounded meet-semilattices extend to bounded implicative semilattices; they are just filters, proper filters and prime filters of the bounded meet-semilattice reduct. In any bounded implicative semilattice A it holds that a set $F \subseteq A$ is a filter if and only if $1 \in F$ and for every $a, b \in A$, if $a, a \to b \in F$, then $b \in F$.

For a bounded distributive meet-semilattice L, if we define a topology on the set $\mathrm{Pr}(L)$ of prime filters of L as we do in the case of bounded distributive lattices to obtain the dual Priestley space, that is, by considering the topology generated by the subbasis

$$\{\sigma(a) : a \in L\} \cup \{\sigma(a)^c : a \in L\},$$

[7] Halmos introduced the term 'hemimorphism' in the above sense, but in the literature we find 'hemimorphism' applied to the meet and top preserving maps as well, see e.g. [51].

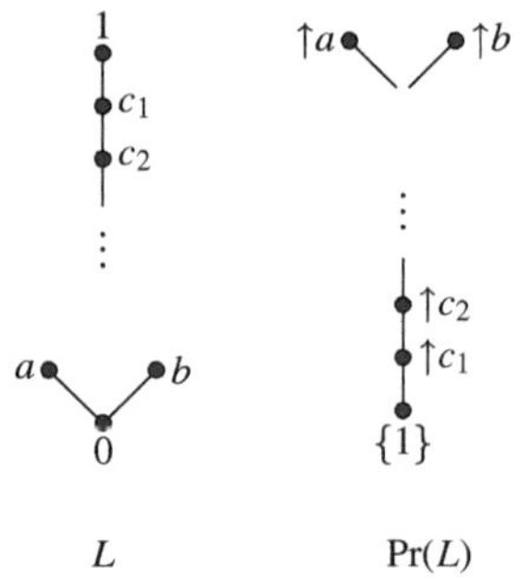

Fig. 4.2 L and $\mathrm{Pr}(L)$

where $\sigma(a) := \{P \in \mathrm{Pr}(L) : a \in P\}$ for every $a \in L$, we may end up with a non compact space, and the same holds for bounded implicative semilattices, as the following example shows.

Example 1 Let L be the bounded distributive meet-semilattice shown in Fig. 4.2, which is an implicative semilattice because the meet operation is residuated. The prime filters of L are the principal up-sets $\uparrow 1$, $\uparrow a$, $\uparrow b$ and $\uparrow c_n$ for each n. Thus, the set $\mathrm{Pr}(L)$ of prime filters of L, ordered by set-inclusion, looks as shown in Fig. 4.2. Clearly $\mathrm{Pr}(L)$ with the topology just considered is not compact because the sequence $\{\uparrow c_n : n \in \omega\}$ has no limit point. In order to make $\mathrm{Pr}(L)$ compact, we need to add to $\mathrm{Pr}(L)$ the limit of this sequence. Therefore, we need to add to $\mathrm{Pr}(L)$ the filter $\{1, c_1, c_2, \dots\}$, which is not a prime filter.

This indicates that to obtain a Priestley style duality for bounded distributive meet-semilattices we need to work with a collection of filters that in many cases includes properly the collection of prime filters. The filters of that collection are called optimal filters in [9]. They can be introduced through the distributive envelope of a bounded distributive meet-semilattice that we proceed to define.

Let L be a bounded distributive meet-semilattice. Let $X = \mathrm{Pr}(L)$ be the set of prime filters of L and let $\leq$ be set-theoretic inclusion. Then $\langle X, \leq \rangle$ is a poset. Moreover, let $\sigma[L] = \{\sigma(a) : a \in L\}$, where $\sigma(a) = \{F \in \mathrm{Pr}(L) : a \in F\}$ for every $a \in L$. The map σ is a bounded meet-semilattice homomorphism from L to $\mathcal{P}^{\uparrow}(X)$; that is, it preserves the bounds and meet. Let $\mathrm{De}(L)$ denote the sublattice of $\mathcal{P}^{\uparrow}(X)$ generated by $\sigma[L]$. This lattice is the *distributive envelope* of L. For several equivalent characterizations of $\mathrm{De}(L)$ and more information on the distributive envelope, we refer to [8, 9, 36]. Here we need to know that $\sigma : L \to \mathrm{De}(L)$ is in addition one-to-one and preserves all existing finite joins in L.

A filter F of L is *optimal* if there exists a prime filter P of $\mathrm{De}(L)$ such that $F = \sigma^{-1}[P]$. As was shown in [8, Lem. 4.20], each prime filter of L is optimal, but there are optimal filters of L which may not be prime. Indeed, the distributive envelope of the meet-semilattice L shown in Fig. 4.2 is shown in Fig. 4.3. Obviously $P = \uparrow(\sigma(a) \cup \sigma(b))$ is a prime filter of $\mathrm{De}(L)$, so $F = \sigma^{-1}[P] = \{1, c_1, c_2, \dots\}$ is an optimal filter of L, which is not a prime filter of L.

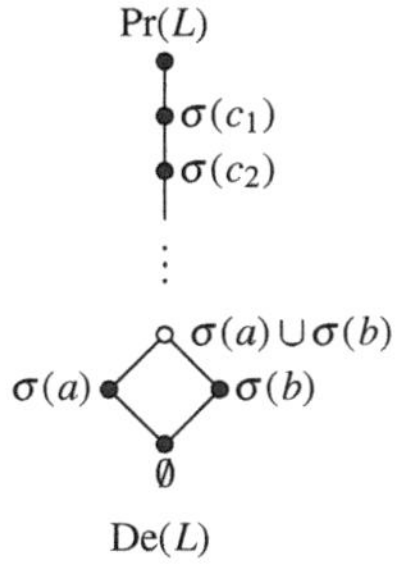

Fig. 4.3 De(L)

The optimal filters of a bounded distributive meet-semilattice L correspond to the weakly prime ideals of the dual join-semilattice L^{∂} of L; this ideals are the key ingredients of the duality developed in [35, 36]. The optimal filters of L are exactly the pseudoprime elements (joins of prime ideals) [29, Def. I-3.24] of the lattice of filters of L. Moreover, the optimal filters turn out to be the filters that are complements of prime Frink ideals. Recall that a Frink ideal [26] is a set $I \subseteq L$ such that for every finite $A \subseteq I$ the set of all upper bounds of the set of all lower bounds of A is included in I, and that a Frink ideal I is prime if $a \in I$ or $b \in I$ whenever $a \wedge b \in I$. If L is a bounded distributive lattice, then the Frink ideals are the ideals of L and so the optimal filters of L are exactly the prime filters of L.

The basic category of bounded distributive meet-semilattices for which we present a duality is the category **BDM** whose objects are bounded distributive meet-semilattices and whose morphisms are meet-semilattice homomorphisms between them that preserve the top element. If L_1, L_2 are bounded distributive meet-semilattices, a map $h : L_1 \to L_2$ is a *meet-semilattice homomorphism preserving the top element* if for every $a, b \in L_1$, $h(a \wedge_1 b) = h(a) \wedge_2 h(b)$ and $h(1) = 1$.

Some facts on optimal and prime filters of a bounded distributive meet-semilattice that are important to obtain the duality are listed in the two propositions below.

Let L be a distributive meet-semilattice. We denote by L_* the set of optimal filters of L and by L_+ the set of prime filters. Moreover, for every $a \in L$, we set $\phi_L(a) := \{x \in L_* : a \in x\}$.

Proposition 5 *1. $L_+ \subseteq L_*$, and the equality holds whenever L is a lattice,*
2. for every $x \in L_*$ there exists $y \in L_+$ such that $x \subseteq y$; i.e. in the poset $\langle L_*, \subseteq\rangle$, $\downarrow L_+ = L_*$,
3. for every $x \in L_*$, $x \in L_+$ if and only if the set $\{\varphi_L(a) : a \notin x\}$ is updirected (under inclusion).

The topology τ on L_* that we consider is the topology generated by the subbasis

$$\{\varphi_L(a) : a \in L\} \cup \{\varphi_L(b)^c : b \in L\}.$$

Then:

Proposition 6 *4.* $\langle L_*, \subseteq, \tau\rangle$ *is a Priestley space and is homeomorphic and order isomorphic to the Priestley space of the distributive envelope* $\mathrm{De}(L)$ *of* L*,*

5. L_+ *is a dense subset of* L_**,*
6. *the maximal elements of* $\varphi_L(a)^c$ *are prime filters,*
7. *if* U *is a clopen up-set with the property that the maximal elements of* U^c *are prime filters, then there is* $a \in L$ *such that* $U = \varphi_L(a)$*.*

The seven facts just mentioned in Propositions 5 and 6 motivate the definition of the objects of the dual category of BDM we are going to introduce. To define these objects we need some notions and notation. Let X be a Priestley space and let X_0 be a dense subset of X. A clopen up-set U of X is said to be X_0-*admissible* if $\max(U^c) \subseteq X_0$.[8] This condition holds if and only if U is a clopen up-set such that $U^c = \downarrow(X_0 - U)$. Let X^* denote the set of X_0-admissible clopen up-sets of X. For $x \in X$ we let $\mathscr{I}_x = \{U \in X^* : x \notin U\}$.

Definition 9 A quadruple $X = \langle X, \tau, \leq, X_0\rangle$ is a *generalized Priestley space* if:

1. $\langle X, \tau, \leq\rangle$ is a Priestley space,
2. X_0 is a dense subset of X,
3. $X = \downarrow X_0$,
4. for every $x \in X$, $x \in X_0$ iff $\mathscr{I}_x$ is updirected,
5. for every $x, y \in X$, we have $x \leq y$ iff $(\forall U \in X^*)(x \in U \Rightarrow y \in U)$.

If X is a generalized Priestley space, then X^* is a bounded distributive meet-semilattice, where meet is intersection and $\emptyset$ and X are the bounds. Conversely, for a bounded distributive meet-semilattice L, facts (1)–(7) in Propositions 5 and 6 above show the quadruple $L_* = \langle L_*, \tau, \subseteq, L_+\rangle$ is a generalized Priestley space and the map $\varphi_L : L \to {L_*}^*$ is an order-isomorphism. Moreover, if X is a generalized Priestley space, then the map $\psi : X \to {X^*}_*$ defined by

$$\psi_X(x) = \{U \in X^* : x \in U\},$$

for every $x \in X$, is a homeomorphism and an order isomorphism such that $\psi_X[X_0] = {X^*}_+$.

Now we define the morphisms of our category, which are relations between generalized Priestley spaces.

Definition 10 Let X and Y be generalized Priestley spaces and let $R \subseteq X \times Y$. For every $Z \subseteq Y$ we define

$$R^*[Z] = \{x \in X : R[x] \subseteq Z\}.$$

---

[8] If $X = \langle X, \tau, \leq\rangle$ is a Priestley space and $Y \subseteq X$, $\max(Y)$ denotes the set of $\leq$-maximal elements of Y.

A relation $R \subseteq X \times Y$ is a *generalized Priestley morphism* if:

1. $R[x] = \bigcap\{U \in Y^* : R[x] \subseteq U\}$ for every $x \in X$,
2. if $U \in Y^*$, then $R^*[U] \in X^*$.

Thus, if $R \subseteq X \times Y$ is a generalized Priestley morphism, then R^* can be turned into a meet-semilattice homomorphism $R^* : Y^* \to X^*$ preserving the top element by letting $R^*(U) = R^*[U]$ [9, Lem. 6.5]; this map is the dual of R.

The dual of a meet-semilattice homomorphism $h : L_1 \to L_2$ preserving the top element is the relation $h_* \subseteq (L_2)_* \times (L_1)_*$ defined by

$$y\, h_* x \quad \text{iff} \quad h^{-1}[y] \subseteq x$$

for every $y \in (L_2)_*$ and every $x \in (L_1)_*$. This relation is a generalized Priestley morphism.

The usual relational composition of two generalized Priestley morphisms may not be a generalized Priestley morphism. To obtain the category **GPS** of generalized Priestley spaces and generalized Priestley morphisms, a composition operation of two generalized Priestley morphisms has to be defined. This is done in [8, 9]. Let X, Y and Z be generalized Priestley spaces, and let $R \subseteq X \times Y$ and $S \subseteq Y \times Z$ be generalized Priestley morphisms. The composition is the relation $S{*}R \subseteq X \times Z$ defined by

$$x(S{*}R)z \quad \text{iff} \quad (\forall U \in Z^*)((R \circ S)[x] \subseteq U \Rightarrow z \in U),$$

where $R \circ S$ is the usual relational composition of R and S. Then

$$x(S{*}R)z \quad \text{iff} \quad (\forall U \in Z^*)(x \in R^*[S^*[U]] \Rightarrow z \in U).$$

It holds [9, Lem. 6.8] that if R and S are generalized Priestley morphisms, then so is $S * R$ and $(S * R)^* = R^* \circ S^*$. Moreover, the operation $*$ is associative and for every generalized Priestley space X, the relation $\leq_X$ is the identity morphism of **GPS**. Also, if L_1, L_2 and L_3 are bounded distributive meet-semilattices and $h : L_1 \to L_2$, $k : L_2 \to L_3$ are meet-semilattice homomorphisms preserving top, then $(k \circ h)_* = h_* * k_*$.

In [8, 9] it is shown that the maps $(.)_* :$ **BDM** $\to$ **GPS** and $(.)^* :$ **GPS** $\to$ **BDM** that arise from the definitions above are indeed functors that establish a dual equivalence between **BDM** and **GPS** with natural transformations the maps φ_L and the generalized Priestley relations R_X defined as follows. Let X be a generalized Priestley space. The relation $R_X \subseteq X \times X^*{}_*$ is defined from the map ψ_X by

$$x R_X P \quad \text{iff} \quad \psi_X(x) \subseteq P$$

for every $x \in X$ and every $P \in X^*{}_*$.

Theorem 5 *The categories* BDM *and* GPS *are dually equivalent.*

Recall that in Priestley duality for bounded distributive lattices the dual of a homomorphism h is onto if and only if h is one-to-one, and it is one-to-one if and only if h is onto. There are nice characterizations of the duals in GPS of onto and of one-to-one morphisms in BDM. Let X and Y be generalized Priestley spaces. We say that a generalized Priestley morphism $R \subseteq X \times Y$ is *1-1* if for every $x \in X$ and every $U \in X^*$ with $x \notin U$ there exists $V \in Y^*$ such that $R[U] \subseteq V$ and $R[x] \nsubseteq V$, and we say that it is *onto* if for every $y \in Y$ there exists $x \in X$ such that $R[x] = \uparrow y$. It holds that R is 1-1 if and only if its dual $h_R : Y^* \to X^*$ is an onto meet-homomorphism preserving top; and also that R is onto if and only if $h_R : Y^* \to X^*$ is a one-to-one meet-homomorphism preserving top. Dually, if L_1 and L_2 are bounded distributive meet-semilattices and $h : L_1 \to L_2$ is a meet-homomorphism preserving top, then h is one-to-one if and only if its dual h_* is onto; and also h is onto if and only if h_* is 1-1.

In [8, 9] other categories of generalized Priestley spaces are considered by restricting the class of morphisms. Let X and Y be generalized Priestley spaces. A generalized Priestley morphism $R \subseteq X \times Y$ is *total* if $R^{-1}[Y] = X$ and it is *functional* if for every $x \in X$ there exists $y \in X$ such that $R[x] = \uparrow y$. So, every functional generalized Priestley morphism is total. It is easy to see that the composition $*$ of two total generalized Priestley morphisms is total and the composition $*$ of two generalized functional Priestley morphisms is functional.

If L_1 and L_2 are distributive meet-semilattices and $h : L_1 \to L_2$ is a meet-semilattice homomorphism preserving top, then h preserves bottom if and only if h_* is total, and h preserves all existing finite joins in L_1 if and only if h_* is functional. It follows from these facts that the category $\mathsf{BDM}^{\mathsf{B}}$ of all bounded meet-semilattices with the meet-semilattice homomorphisms preserving the bounds is dually equivalent to the category of generalized Priestley spaces with the total generalized Priestley morphisms, and the category $\mathsf{BDM}^{\mathsf{J}}$ of all bounded meet-semilattices with the meet-semilattice homomorphisms preserving the bounds and the existing finite joins is dually equivalent to the category of generalized Priestley spaces and functional generalized Priestley morphisms.

4.6.2 Generalized Esakia Duality

Before specializing generalized Priestley duality to bounded implicative semilattices to obtain generalized Esakia duality, let us see how the map φ_L behaves with respect to the relative pseudo-complements that may exist in a bounded distributive meet-semilattice L.

Let L be a bounded distributive meet-semilattice. We consider the poset $\langle L_*, \subseteq \rangle$ and the relative pseudo-complement, or residuation, operation $\to_{L_*}$ on the lattice of up-sets of L_*. Recall that this operation satisfies

$$U \to_{L_*} V := \{x \in L_* : \uparrow x \cap U \subseteq V\}.$$

Lemma 4 *Let L be a bounded distributive meet-semilattice. Then $\varphi_L : L \to {L_*}^*$ preserves existing relative pseudo-complements; that is, if the relative pseudo-complement of $a, b \in L$ exists in L and we denote it by $a \to b$, then*

$$\varphi_L(a \to b) = \varphi_L(a) \to_{L_*} \varphi_L(b).$$

Proof Let $a, b \in L$. Since $\varphi_L(a \wedge b) = \varphi_L(a) \cap \varphi_L(b)$, we have $\varphi_L(a) \cap \varphi_L(a \to b) = \varphi_L(a \wedge (a \to b)) \subseteq \varphi_L(b)$, and so $\varphi_L(a \to b) \subseteq \varphi_L(a) \to_{L_*} \varphi_L(b)$. If $x \notin \varphi_L(a \to b)$, then $a \to b \notin x$. Let F be the filter of L generated by $\{a\} \cup x$. If there is $c \in F \cap \downarrow b$, then there is $d \in x$ such that $a \wedge d \leq c \leq b$. Therefore, $d \leq a \to b$, and so $a \to b \in x$, a contradiction. Thus, $F \cap \downarrow b = \emptyset$, and by the prime filter lemma (see, e.g., [31, Sect. II.5, Lem. 2]), there is $y \in L_+ \subseteq L_*$ such that $F \subseteq y$ and $b \notin y$. It follows that $x \subseteq y$, $a \in y$ and $b \notin y$. Therefore, $y \in \uparrow x \cap \varphi_L(a)$ and $y \notin \varphi_L(b)$. Thus, $\uparrow x \cap \varphi_L(a) \nsubseteq \varphi_L(b)$, and so $x \notin \varphi_L(a) \to_{L_*} \varphi_L(b)$. Consequently, $\varphi_L(a \to b) = \varphi_L(a) \to_{L_*} \varphi_L(b)$. □

Let X be a generalized Priestley space. By [8, Cor. 6.16], $X^* \cup \{U^c : U \in X^*\}$ is a subbasis of X, and as X^* is closed under finite intersections, we can write each clopen subset U of X in the form $U = \bigcup_{i=1}^n (U_i \cap \bigcap_{j=1}^m V_j^c) = \bigcup_{i=1}^n (U_i - \bigcup_{j=1}^m V_j)$ for some $U_i, V_j \in X^*$.

Definition 11 Let X be a generalized Priestley space. A clopen subset U of X is called *Esakia clopen* if $U = \bigcup_{i=1}^n (U_i \cap V_i^c) = \bigcup_{i=1}^n (U_i - V_i)$ for some $U_1, \ldots, U_n, V_1, \ldots, V_n \in X^*$.

It is not difficult to see [11, Lem. 3.5] that if U is an Esakia clopen set of a generalized Priestley space X, then $\max(U) \subseteq X_0$. And it is worth pointing out that the converse of this implication is not true in general; see Example 2 below.

Definition 12 We call a generalized Priestley space X a *generalized Esakia space* if $\downarrow U$ is clopen for each Esakia clopen U of X.

Proposition 7 *1. If A is a bounded implicative semilattice, then $A_* = \langle A_*, \tau, \subseteq, A_+\rangle$ is a generalized Esakia space.*
2. If X is a generalized Esakia space, then X^* is closed under the operation $\to_X$ and $X^* = \langle X^*, \cap, \to_X, \emptyset, X\rangle$ is a bounded implicative semilattice.

Proof (1) Suppose that A is a bounded implicative semilattice. Then A is a bounded distributive meet-semilattice, and so A_* is a generalized Priestley space. Let U be Esakia clopen in A_*. Then $U = \bigcup_{i=1}^n (\varphi_A(a_i) - \varphi_A(b_i))$ for some $a_i, b_j \in A$. By Lemma 4, $\varphi_A(a_i) \to \varphi_A(b_i) = [\downarrow(\varphi_A(a_i) - \varphi_A(b_i))]^c \in {A_*}^*$. Therefore, $\downarrow(\varphi_A(a_i) - \varphi_A(b_i))$ is clopen in A_* for each $i \leq n$. Thus, $\downarrow U = \bigcup_{i=1}^n \downarrow(\varphi_A(a_i) - \varphi_A(b_i))$ is clopen in A_*, and so A_* is a generalized Esakia space.

(2) Suppose that X is a generalized Esakia space. Then X is a generalized Priestley space, and so $\langle X^*, \cap, X, \emptyset\rangle$ is a bounded distributive meet-semilattice. Let

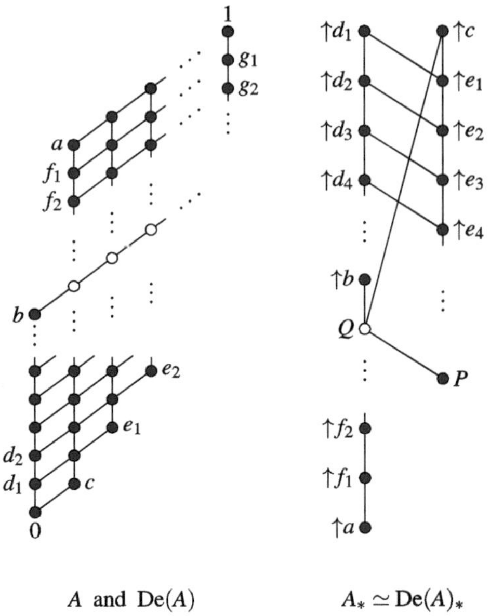

Fig. 4.4 A, $\mathrm{De}(A)$, A_*, $\mathrm{De}(A)_*$

$U, V \in X^*$. Then $U - V$ is Esakia clopen. Therefore, $\downarrow(U - V)$ is clopen in X and $\max\downarrow(U - V) = \max(U - V) \subseteq \max(V^c) \subseteq X_0$. As $(U \to V)^c = \downarrow(U - V)$, it follows that $U \to_X V \in X^*$. Consequently, $\langle X^*, \cap, \to, X, \emptyset\rangle$ is a subalgebra of the $(\cap, \to, X, \emptyset)$-reduct of the Heyting algebra $\mathscr{P}^{\uparrow}(X)$. So it is a bounded implicative semilattice. □

If $X = \langle X, \tau, \leq, X_0\rangle$ is a generalized Esakia space, one may expect that $\langle X, \tau, \leq\rangle$ is not only a Priestley space but an Esakia space. The next example shows that this may not be the case in general. Algebraically it means that the distributive envelope $\mathrm{De}(A)$ of a bounded implicative semilattice A may *not* be a Heyting algebra.

Example 2 Consider the implicative semilattice A shown in Fig. 4.4. The distributive envelope $\mathrm{De}(A)$ of A and the dual space A_* of A are also shown in Fig. 4.4. Note that the black circles indicate the elements of A and the white circles indicate the elements of $\mathrm{De}(A) - A$. Then A_* is a generalized Esakia space order-homeomorphic to the dual space $\mathrm{De}(A)_*$ of $\mathrm{De}(A)$. We denote by P the filter $\{1, g_1, g_2, \ldots\}$ and by Q the filter $\uparrow a \cup \bigcup\{\uparrow f_n : n \in \omega\}$. It is easy to see that Q is the only optimal filter of A which is not prime. Thus, $A_+ = A_* - \{Q\}$. It is also easy to calculate that $a \to b = b$ and $a \to c = c$ in A, but that $\sigma(a) \to (\sigma(b) \cup \sigma(c))$ does not exist in $\mathrm{De}(A)$. Consequently, $\mathrm{De}(A)$ is not a Heyting algebra. Stated dually, $U = \{\uparrow a, \uparrow f_1, \uparrow f_2, \ldots, Q\}$ is clopen in A_*, but $\downarrow U = U \cup \{P\}$ is not clopen in A_*. Thus, A_* is not an Esakia space. Of course, U is not Esakia clopen because A_* is a generalized Esakia space. Note that $V = U \cup \{\uparrow d_1\}$ is an example of non-Esakia clopen in A_* such that $\max(V) \subseteq X_0$. showing that to be Esakia clopen is not equivalent to having the maximal elements in X_0.

The morphisms between generalized Esakia spaces are defined as follows.

Definition 13 Let X and Y be generalized Esakia spaces. We call a generalized Priestley morphism $R \subseteq X \times Y$ a *generalized Esakia morphism* if

$$(\forall x \in X)(\forall y \in Y_0)(xRy \Rightarrow (\exists z \in X_0)(x \leq z \ \&\ R[z] = \uparrow y)).$$

Equivalently, $R \subseteq X \times Y$ is a generalized Esakia morphism if and only if $R[x] \cap Y_0 = R[\uparrow x \cap X_0] \cap Y_0$ for every $x \in X$. A generalized Esakia morphism is *total* if it is total as a generalized Priestley morphism and it is *functional* if it is functional as a generalized Priestley morphism.

Similar to the generalized Priestley morphism case, we have that if R and S are generalized Esakia morphisms, then so is the composition $S * R$.

The duals of implicative semilattice homomorphisms are generalized Esakia morphisms and the duals of generalized Esakia morphisms are implicative semilattice homomorphisms.

Proposition 8 *1. Let A and B be bounded implicative semilattices and let $h : A \to B$ be an implicative semilattice homomorphism. Then the relation $h_* \subseteq B_* \times A_*$ is a generalized Esakia morphism. Moreover, if h is a bounded homomorphism, then h_* is total, and if h preserve all existing finite joins, then h_* is functional.*
2. Let X and Y be generalized Esakia spaces and let $R \subseteq X \times Y$ be a generalized Esakia morphism. Then $R^* : Y^* \to X^*$ is an implicative semilattice homomorphism. Moreover, if R is total, then R^* is a bounded homomorphism, and if R is functional, then R^* preserves all existing finite joins.

It is important to notice that the condition in the definition of generalized Esakia morphism can not be strengthened to

$$(\forall x \in X)(\forall y \in Y)(xRy \Rightarrow (\exists z \in X)(x \leq z \ \&\ R[z] = \uparrow y))$$

as Example 4.5 in [11] shows.

Let GES denote the category of generalized Esakia spaces and generalized Esakia morphisms, in which $*$ is the composition of two morphisms and $\leq_X$ is the identity morphism for each object X. Let also GES^T denote the subcategory of GES whose objects are the generalized Esakia spaces and whose morphisms are the total generalized Esakia morphisms, and let GES^F denote the subcategory of GES^T whose objects are the generalized Esakia spaces and whose morphisms are the functional generalized Esakia morphisms. The category GES^F is a non-full subcategory of GES^T and GES^T is a non-full subcategory of GES.

Let BIM denote the category of bounded implicative semilattices and implicative semilattice homomorphisms, BIM^B the category of bounded implicative semilattices and bounded implicative semilattice homomorphisms, and BIM^J the category of bounded implicative semilattices and implicative semilattice homomorphisms preserving all existing finite joins. We have that BIM^J is a non-full subcategory of BIM^B and BIM^B is a non-full subcategory of BIM.

Theorem 6 *The categories* BIM, $\mathsf{BIM}^{\mathsf{B}}$ *and* $\mathsf{BIM}^{\mathsf{J}}$ *are dually equivalent to the categories* GES, $\mathsf{GES}^{\mathsf{T}}$ *and* $\mathsf{GES}^{\mathsf{F}}$, *respectively.*

There is a category of generalized Esakia spaces dual to $\mathsf{BIM}^{\mathsf{J}}$ whose morphisms are functions instead of relations. To describe it we first deal with the bounded distributive meet-semilattice case.

Let X and Y be generalized Priestley spaces. A map $f : X \to Y$ is a *strong Priestley morphism* if it is order preserving and for every $U \in Y^*$, $f^{-1}[U] \in X^*$. Note that any strong Priestley morphism is a Priestley morphism between the corresponding Priestley spaces, but the converse may not be true. Now let X and Y be generalized Esakia spaces. A map $f : X \to Y$ is a *strong Esakia morphism* if it is a strong Priestley morphism that in addition satisfies:

$$(\forall x \in X)(\forall y \in Y_0)(f(x) \leq y \Rightarrow (\exists z \in X_0)(x \leq z \;\&\; f(z) = y)).$$

Let X and Y be generalized Priestley spaces and $R \subseteq X \times Y$ a functional generalized Priestley morphism. The map $f^R : X \to Y$ defined by

$$f^R(x) = \text{ the least element of } R[x]$$

is a strong Priestley morphism. And if X, Y, Z are generalized Priestley spaces and $R \subseteq X \times Y$ and $S \subseteq Y \times Z$ are functional Priestley morphisms, then $f^{S*R} = f^S \circ f^R$. In case X and Y are generalized Esakia spaces and $R \subseteq X \times Y$ is a functional generalized Esakia morphism, f^R is a strong Esakia morphism.

Let X and Y be generalized Priestley spaces and $f : X \to Y$ a strong Priestley morphism. The relation $R^f \subseteq X \times Y$ defined by

$$xR^f y \quad \text{iff} \quad f(x) \leq y,$$

for every $x \in X$ and every $y \in Y$, is a functional generalized Priestley morphism. Moreover, if X, Y, Z are generalized Priestley spaces and $f : X \to Y$ and $g : Y \to Z$ are strong Priestley morphisms, then $R^{g \circ f} = R^g * R^f$. If X and Y are generalized Esakia spaces and $f : X \to Y$ is a strong Esakia morphism, then R^f is a functional generalized Esakia morphism.

It is easy to see that if X, Y are generalized Priestley spaces, $R \subseteq X \times Y$ is a functional generalized Priestley morphism and $f : X \to Y$ is a strong Priestley morphism, then $R^{f^R} = R$ and $f^{R^f} = f$.

Let $\mathsf{GPS}^{\mathsf{S}}$ be the category of generalized Priestley spaces and strong Priestley morphisms and let $\mathsf{GES}^{\mathsf{S}}$ be the category of generalized Esakia spaces and strong Esakia morphisms. From the comments above, it is easy to see that the categories $\mathsf{GPS}^{\mathsf{S}}$ and $\mathsf{GPS}^{\mathsf{F}}$ are isomorphic and so are the categories $\mathsf{GES}^{\mathsf{S}}$ and $\mathsf{GPS}^{\mathsf{F}}$. Thus:

Theorem 7 *The categories* $\mathsf{GES}^{\mathsf{S}}$ *and* $\mathsf{BIM}^{\mathsf{J}}$ *are dually equivalent.*

This theorem implies Esakia duality for Heyting algebras. Esakia spaces can be identified with the generalized Esakia spaces X where $X_0 = X$. Under this identification, Esakia morphisms are strong Esakia morphisms. So, the category of Esakia spaces is isomorphic to the full subcategory of GES^S whose objects are the generalized Esakia spaces X where $X_0 = X$. The dual category of this category is a subcategory of BIM^J whose objects are the bounded implicative semilattices that are the $(\wedge, \to, 0, 1)$-reducts of Heyting algebras. Therefore, it is isomorphic to the category of Heyting algebras. Thus, this last category is dually equivalent to the category of Esakia spaces.

Moreover, in the present setting it is very natural to consider categories whose objects are Heyting algebras but whose morphisms preserve less structure than homomorphisms of Heyting algebras. Let A and B be Heyting algebras. We say that a map $h : A \to B$ is a *$(\wedge, \to)$-homomorphism* if $h(a \wedge b) = h(a) \wedge h(b)$ and $h(a \to b) = h(a) \to h(b)$ for all $a, b \in A$; this implies that $h(1) = 1$. We say that it is a *$(\wedge, \to, 0)$-homomorphism* if in addition it preserves the bottom element, and we say that it is a *$(\wedge, \vee, \to)$-homomorphism* if it is a $(\wedge, \to)$-homomorphism that preserves nonempty finite joins. Note that if $h : A \to B$ is an onto $(\wedge, \to)$-homomorphism, then it is an onto Heyting algebra homomorphism.

Let $\mathsf{HA}^{\wedge,\to}$, $\mathsf{HA}^{\wedge,\to,0}$ and $\mathsf{HA}^{\wedge,\vee,\to,0}$ be the categories with objects Heyting algebras and with morphisms $(\wedge, \to)$-homomorphisms, $(\wedge, \to, 0)$-homomorphisms and $(\wedge, \vee, \to, 0)$-homomorphisms, respectively. Then identifying again Esakia spaces with the generalized Esakia spaces X where $X_0 = X$, Theorem 6 allows us to obtain categories with Esakia spaces as objects and generalized Esakia morphisms, total generalized Esakia morphisms and functional generalized Esakia morphisms as morphisms that are dual to the categories $\mathsf{HA}^{\wedge,\to}$, $\mathsf{HA}^{\wedge,\to,0}$ and $\mathsf{HA}^{\wedge,\vee,\to,0}$, respectively.

Although we only dealt with implicative semilattices with bottom, the duality for bounded implicative semilattices we presented can be extended easily to the category of all implicative semilattices in the same way it is done in [9, Sect. 9] for distributive meet-semilattices possibly without a bottom element.

We conclude this section by a brief discussion of the dual concepts of filters, ideals, Frink ideals and congruences.

The filters, ideals and Frink ideals of a bounded implicative semilattice are the filters, ideals and Frink ideals of its meet-semilattice reduct. So, to obtain their duals it is enough to describe the duals of this notions for bounded distributive meet-semilattices. The maps Φ and Ψ that we defined in Sect. 4.2 from the lattice of filters and ideals of a bounded distributive lattice have their analogues for bounded distributive meet-semilattices.

Let L be a bounded distributive meet-semilattice. We define the map Φ with domain the lattice of filters of L and the map Ψ with domain the set of Frink ideals of L as follows:

$$\Phi(F) := \bigcap\{\varphi_L(a) : a \in F\} \qquad \Psi(I) := \bigcup\{\varphi_L(a) : a \in I\}$$

for every filter F of L and every Frink ideal I of L. The map Ψ sets an isomorphism between the lattice of Frink ideals of L and the lattice of open up-sets of L_*, and an isomorphism of the ordered set of ideals of L with the ordered set of open up-sets U of L_* such that $L_* - U = \downarrow(L_+ - U)$. And the the map Φ sets a dual isomorphism between the lattice of filters of L and the lattice of closed up-sets C of L_* satisfying $L_* - C = \downarrow(L_+ - C)$. In particular, prime Frink ideals of L correspond to the open upsets of L_* of the form $(\downarrow x)^c$ for some $x \in L_*$, and prime ideals of L correspond to the open upsets of L_* of the form $(\downarrow x)^c$ for some $x \in L_+$. Similarly, since there are 1-1 correspondences between prime filters and prime ideals of L and between optimal filters and prime Frink ideals of L, optimal filters of L correspond to the closed upsets of L_* of the form $\uparrow x$ for some $x \in L_*$, and prime filters of L correspond to the closed upsets of L_* of the form $\uparrow x$ for some $x \in L_+$.

In each implicative semilattice, the lattice of its filters is dually isomorphic to the lattice of its congruences. So, it follows from the just stated correspondence that if A is an implicative semilattice, then there is an isomorphism between the lattice of congruences of A and the lattice of closed up-sets C of the dual Esakia space A_* such that $A_* - C = \downarrow(A_+ - C)$.

The dual characterization of subalgebras of bounded distributive meet-semilattices and of bounded implicative semilattices is rather involved. We address the interested reader to [10] where the dual characterization of homomorphic images of bounded distributive meet-semilattices and of bounded implicative semilattices can also be found.

4.7 Categories of Esakia Spaces with Partial Maps as Morphisms

In [11] a different duality for the categories $\mathsf{HA}^{\wedge,\rightarrow}$, $\mathsf{HA}^{\wedge,\rightarrow,0}$ and $\mathsf{HA}^{\wedge,\vee,\rightarrow,0}$ is also proved, where objects of the dual categories are Esakia spaces but morphisms are partial maps instead of relations. Slightly different dualities for these categories are obtained in [5], where morphisms are also partial maps. These last dualities are used in [5] to obtain an enlightening algebraic approach to Zakhariaschev's results on canonical formulas for intuitionistic logic. M. Zakharyaschev [60, 62] introduced canonical formulas and proved that every superintuitionistic (or intermediate) logic can be axiomatized by canonical formulas. He later generalized the result to cover all extensions of the modal logic K4. Zakharyaschev's theorem has many useful consequences. For instance, it was used by him to prove that the disjunction-free fragment of a superintuitionistic logic with the disjunction property coincides with the disjunction-free fragment of intuitionistic propositional logic [61] (a result proved independently and by a different technique by Minari [43]). Zakharyaschev's proof is rather complicated and uses model-theoretic techniques, but [5] provides a more simple proof using algebraic techniques and duality in a way that also provides an algebraic explanation of some of the concepts used in Zakharyaschev's proof.

We expound in this section the dualities in [5]. The reader can find the results and their proofs there.

Recall that a partial map f from a set X to a set Y is a map from a subset of X to Y. Let X and Y be Esakia spaces. A partial map $f : X \to Y$ is a *partial Esakia morphism* if

1. for all $x, y \in \mathrm{dom}(f)$, if $x \le y$, then $f(x) \le f(y)$,
2. if $x \in \mathrm{dom}(f)$, $y \in Y$ and $f(x) \le y$, then there exists $z \in \mathrm{dom}(f)$ such that $x \le z$ and $f(z) = y$,
3. $f[\uparrow x]$ is closed for every $x \in X$,
4. if U is a clopen up-set of Y, then $X - \downarrow f^{-1}[Y - U]$ is a clopen up-set of X,
5. for every $x \in X$, $x \in \mathrm{dom}(f)$ if and only if there exists $y \in Y$ such that $f[\uparrow x] = \uparrow y$.

The following fact is important to mention. If $f : X \to Y$ is a partial Esakia morphism from an Esakia space X to an Esakia space Y, then $\mathrm{dom}(f)$ is a closed subset of X. Thus, $\mathrm{dom}(f)$ with the subspace topology is a Boolean space. It then follows that $f : \mathrm{dom}(f) \to Y$ is a continuous (total) function. It also follows that if U is a closed subset of X, then $f(U)$ is a closed subset of Y.

Let A and B be Heyting algebras and let $h : A \to B$ be a $(\wedge, \to)$-homomorphism. The partial map $h_\bullet : X(A) \to X(B)$ with

$$\mathrm{dom}(h_\bullet) := \{x \in X(B) : h^{-1}[x] \in X(A)\}$$

and such that for every $x \in \mathrm{dom}(h_\bullet)$,

$$h_\bullet(x) = h^{-1}[x]$$

is a partial Esakia morphism from $X(B)$ to $X(A)$.

Now let X and Y be Esakia spaces and let $f : X \to Y$ be a partial Esakia morphism. The map $f^\bullet : D(Y) \to D(X)$ defined by

$$f^\bullet(U) := X - \downarrow f^{-1}[Y - U],$$

for every $U \in D(Y)$, is a $(\wedge, \to)$-homomorphism from $D(Y)$ to $D(X)$.

The set-theoretic composition of composable partial Esakia morphisms need not be an Esakia morphism as shown in Example 3.18 of [5]. So to obtain a category with morphisms partial Esakia morphism between Esakia spaces, a composition operation has to be defined.

Let X, Y, Z be Esakia spaces and $f : X \to Y, g : Y \to Z$ be partial Esakia morphisms. The map $g \star f : X \to Z$ is defined as follows. First we set

$$\mathrm{dom}(g \star f) := \{x \in X : g(f[\uparrow x]) = \uparrow z \text{ for some } z \in Z\}$$

and then we set for each $x \in \mathrm{dom}(g \star f)$,

$$(g \star f)(x) := \text{ the unique } z \text{ such that } \uparrow z = g(f[\uparrow x]).$$

This map is a partial Esakia morphism and $(g \star f)^{\bullet} = f^{\bullet} \circ g^{\bullet}$. Moreover, the operation $\star$ between composable partial Esakia morphisms is associative, and the identity map on an Esakia space is a partial Esakia morphism. Thus, we have that Esakia spaces and partial Esakia morphisms with the composition operation $\star$ form a category, denoted ES^{p}. Furthermore, if we assign to every Esakia space X its dual Heyting algebra $D(X)$ and to every partial Esakia morphism $f : X \to Y$ the $(\wedge, \to)$-homomorphism $f^{\bullet} : D(Y) \to D(X)$, we obtain a contravariant functor from ES^{p} to $\mathsf{HA}^{\wedge,\to}$. To obtain a contravariant functor in the reverse direction that assigns to every Heyting algebra A its dual Esakia space and to every $(\wedge, \to)$-homomorphism $h : A \to B$ between Heyting algebras the partial Esakia morphism $h_{\bullet} : X(B) \to X(A)$, we need to know that if A, B, C are Heyting algebras and $h : A \to B, k : B \to C$ are $(\wedge, \to)$-homomorphisms, then $(k \circ h)_{\bullet} = h_{\bullet} \star k_{\bullet}$. This is indeed the case as shown in [5]. These two contravariant functors establish an equivalence between $\mathsf{HA}^{\wedge,\to}$ and ES^{p}. The natural transformations are the same as for Esakia duality.

A partial Esakia morphism $f : X \to Y$ from an Esakia space X to an Esakia space Y is *well* if $X = {\downarrow}\mathrm{dom}(f)$. The dual $f_{\bullet} : D(Y) \to D(X)$ of a well partial Esakia morphism $f : X \to Y$ is a $(\wedge, \to, 0)$-homomorphism. Moreover, if A and B are Heyting algebras and $h : A \to B$ is a $(\wedge, \to, 0)$-homomorphism, then the dual partial Esakia morphism $h_{\bullet} : X(B) \to X(A)$ is well. Also, the composition $\star$ of two composable well partial Esakia morphisms is well and the identity map from an Esakia space to itself is also well. These facts imply that the category $\mathsf{HA}^{\wedge,\to,0}$ is dually equivalent to the subcategory of ES^{p} whose objects are Esakia spaces and whose morphisms are well partial Esakia morphisms.

Let X, Y be Esakia spaces. A partial Esakia morphism $f : X \to Y$ is *strong* if for every $x \in X$ such that $f[{\uparrow}x] \neq \emptyset$, we have $x \in \mathrm{dom}(f)$. The dual $f_{\bullet} : D(Y) \to D(X)$ of a strong partial Esakia morphism $f : X \to Y$ is a $(\wedge, \vee, \to)$-homomorphism. Moreover, if A and B are Heyting algebras and $h : A \to B$ is a $(\wedge, \vee, \to)$-homomorphism, then the dual partial Esakia morphism $h_{\bullet} : X(B) \to X(A)$ is strong. It follows that the category $\mathsf{HA}^{\wedge,\vee,\to}$ of Heyting algebras and $(\wedge, \vee, \to)$-homomorphisms is dually equivalent to the subcategory of ES^{p} whose objects are Esakia spaces and whose morphisms are strong partial Esakia morphisms.

Acknowledgments We would like to thank an anonymous referee for the useful comments on the presentation of the chapter and also Guram Bezhanishvili for his useful comments and his careful reading of the chapter that lead to the improvement of English and style. The work on the chapter has been possible thanks to the Marie Curie Actions-International Research Staff Exchange Scheme (IRSES) MaToMUVI-FP7-PEOPLE-2009-IRSES from the European Union. The second author has also been partially supported by the grants 2009SGR-1433 of the AGAUR of the Generalitat de Catalunya and MTM2011-25747 of the Spanish Ministerio de Ciencia e Innovación, which includes eu feder funds.

References

1. Adams M (1973) The Frattini sublattice of a distributive lattice. Algebra Universalis 3:216–228
2. Adams M (1986) Maximal subalgebras of Heyting algebras. Proc Edinburgh Math Soc 29:359–365
3. Ardeshir M, Ruitenburg W (1998) Basic propositional calculus I. Math Logic Q 44:317–343
4. Balbes R, Dwinger P (1974) Distributive lattices. University of Missouri Press, Columbia
5. Bezhanishvili G, Bezhanishvili N (2009) An algebraic approach to canonical formulas: intuitionistic case. Rev Symbolic Logic 2:517–549
6. Bezhanishvili G, Bezhanishvili N, Gabelaia D, Kurz A (2010) Bitopological duality for distributive lattices and Heyting algebras. Math Struct Comput Sci 20:359–393
7. Bezhanishvili G, Ghilardi S, Jibladze M (2011) An algebraic approach to subframe logics. Modal case. Notre Dame J Formal Logic 52:187–202
8. Bezhanishvili G, Jansana R (2008) Duality for distributive and implicative semilattices. Preprints of University of Barcelona Research Group in Non-Classical Logics
9. Bezhanishvili G, Jansana R (2011) Priestley style duality for distributive meet-semilattices. Studia Logica 98:83–123
10. Bezhanishvili G, Jansana R (2011) Generalized Priestley quasi-orders. Order 28:201–220
11. Bezhanishvili G, Jansana R (2013) Esakia style duality for implicative semilattices. Appl Categorical Struct 21:181–208
12. Bezhanishvili N, Gehrke M (2011) Finitely generated free Heyting algebras via Birkhoff duality and coalgebra. Logical Methods Comput Sci 2(9):1–24
13. Celani S (2003) Representation of Hilbert algebras and implicative semilattices. Cent Eur J Math 1:561–572
14. Celani S (2006) *n*-linear weakly Heyting algebras. Math Logic Q 52:404–416
15. Celani S, Jansana R (2003) A closer look at some subintuitionistic logics. Notre Dame J Formal Logic 42:225–255
16. Celani S, Jansana R (2005) Bounded distributive lattices with a strict implication. Math Logic Q 51:219–246
17. Cignoli R (1991) Distributive lattice congruences and Priestley spaces. In: Actas del primer congreso Antonio Monteiro, Universidad Nacional del Sur, Bahía Blanca, pp 81–84
18. Cignoli R, Lafalce S, Petrovich A (1991) Remarks on Priestley duality for distributive lattices. Order 8:299–315
19. Cornish W (1975) On H. Priestley's dual of the category of bounded distributive lattices. Matematički Vesnik 12:329–332
20. Davey B, Galati J (2003) A coalgebraic view of Heyting duality. Studia Logica 75:259–270
21. Davey B, Priestley H (2002) Introduction to lattices and order, 2nd edn. Cambridge University Press, Cambridge
22. Dŏsen K (1993) Modal translations in *K* and *D*. In: de Rijke M (ed) Diamonds and defaults. Kluwer Academic Publishers, Dordrecht, pp 103–127
23. Dunn M, Gehrke M, Palmigiano A (2005) Canonical extensions and relational completeness of some substructural logics. J Symbolic Logic 70:713–740
24. Epstein G, Horn A (1976) Logics which are characterized by subresiduated lattices. Zeitschrift fur Mathematiche Logik und Grundlagen der Mathematik 22:199–210
25. Esakia L (1974) Topological Kripke models. Soviet Math Dokl 15:147–151
26. Frink O (1954) Ideals in partially ordered sets. Am Math Mon 61:223–234
27. Gehrke M (2006) Generalized Kripke frames. Studia Logica 84:241–275
28. Gehrke M (2014) Canonical extensions, Esakia spaces, and universal models (this volume)
29. Gierz G, Hofmann K, Keimel K, Lawson J, Mislove M, Scott D (2003) Continuous lattices and domains, Encyclopedia of Mathematics and its Applications. Cambridge University Press, Cambridge
30. Goldblatt R (1989) Varieties of complex algebras. Ann Pure Appl Logic 44:173–242

31. Grätzer G (1998) General lattice theory, 2nd edn. Birkhauser, Bassel (New appendices by the author with Davey B, Freese R, Ganter B, Greferath M, Jipsen P, Priestley H, Rose H, Schmidt E, Schmidt S, Wehrung F, Wille R)
32. Halmos P (1955) Algebraic logic, I. Monadic Boolean algebras. Compositio Mathematica 12:217–249
33. Halmos P (1962) Algebraic logic. Chelsea Pub. Co., New York
34. Hansoul G (1996) Priestley duality for some subalgebra lattices. Studia Logica 56:133–149
35. Hansoul G (2003) Priestley duality for distributive semilattices. Institut de Mathématique, Université de Liège, Preprint 97.011
36. Hansoul G, Poussart C (2008) Priestley duality for distributive semilattices. Bull Soc Roy Sci Liège 77:104–119
37. Hansoul G, Vrancken-Mawet L (1987) The subalgebra lattice of a Heyting algebra. Czech Math J 37:34–41
38. Hochster M (1969) Prime ideal structure in commutative rings. Trans Am Math Soc 142:43–60
39. Johnstone P (1982) Stone spaces. Cambridge University Press, Cambridge
40. Köhler P (1981) Brouwerian semilattices. Trans Am Math Soc 268:126
41. Köhler P, Pigozzi D (1980) Varieties with equationally definable principal congruences. Algebra Universalis 11:213–219
42. Kupke C, Kurz A, Venema Y (2004) Stone coalgebras. Theor Comput Sci 327:109–134
43. Minari P (1986) Intermediate logics with the same disjunctionless fragment as intuitionistic logic. Studia Logica 45:207–222
44. Monteiro L (1963) Axiomes indépendants pour les algèbres de Łukasiewicz trivalentes. Bulletin de la Societé des Sciences Mathématiques et Physiques de la R. P. Roumanie, Nouvelle Série 7, pp 199–202
45. Nemitz W (1965) Implicative semilattices. Trans Am Math Soc 117:128–142
46. Petrovich A (1996) Distributive lattices with an operator. Studia Logica 56:205–224
47. Priestley H (1970) Representation of distributive lattices by means of ordered Stone spaces. Bull Lond Math Soc 2:186–190
48. Priestley H (1972) Ordered topological spaces and the representation of distributive lattices. Proc Lond Math Soc 24(3):507–530
49. Priestley H (1974) Stone lattices: a topological approach. Fundamenta Mathematicae 84:127–143
50. Rautenberg W (1979) Klassiche und nichtklassiche Aussagenlogik. Vieweg & Sohn, Braunschweig, Fried
51. Sambin G, Vaccaro V (1988) Topology and duality in modal logic. Ann Pure Appl Logic 37:249–296
52. Stone M (1937) Topological representations of distributive lattices and Brouwerian logics. Časopis pro pestǒvvaný matematiky a fysiky 67:1–25
53. Schmid J (2002) Quasiorders and sublattices of distributive lattices. Order 19:11–13
54. Venema Y (2007) Algebras and coalgebras. In: Blackburn P, van Benthem J, Wolter F (eds) Handbook of Modal Logic, vol 3 of Studies in Logic and Practical Reasoning. Elsevier, Amsterdam, pp 331–426
55. Visser A (1981) Aspects of diagonalization and provability. PhD thesis, University of Utrecht, Utrecht
56. Visser A (1981) A propositional logic with explicit fixed points. Studia Logica 40:155–175
57. Vrancken-Mawet L (1982) Le lattis des sous-algèbres d'un algèbre de Heyting finie. Bull Soc Rot Sci Liège 51:82–94
58. Vrancken-Mawet L (1984) The lattice of R-subalgebras of a bounded distributive lattice. Comment Math Univ Carolinae 25:1–17
59. Wright F (1957) Some remarks on Boolean duality. Portugaliae Mathematicae 16:109–117
60. Zakharyaschev M (1983) On intermediate logics. Soviet Mathematics Doklady 27(2):274–277
61. Zakharyaschev M (1987) The disjunction property of superintuitionistic and modal logics. Matematicheskie Zametki 42:729–738, 763
62. Zakharyaschev M (1989) Syntax and semantics of superintuitionistic logics. Algebra Logic 28:262–282

Chapter 5
On the Blok-Esakia Theorem

Frank Wolter and Michael Zakharyaschev

In memory of Leo Esakia

Abstract We discuss the celebrated Blok-Esakia theorem on the isomorphism between the lattices of extensions of intuitionistic propositional logic and the Grzegorczyk modal system. In particular, we present the original algebraic proof of this theorem found by Blok, and give a brief survey of generalisations of the Blok-Esakia theorem to extensions of intuitionistic logic with modal operators and coimplication.

Keywords Modal logic · Intuitionistic logic · Modal algebra · Heyting algebra · Intermediate logics

5.1 Introduction

The Blok-Esakia theorem, which was proved independently by the Dutch logician Wim Blok [6] and the Georgian logician Leo Esakia [13] in 1976, is a jewel of non-classical mathematical logic. It can be regarded as a culmination of a long sequence of results, which started in the 1920–1930s with attempts to understand the logical aspects of Brouwer's intuitionism by means of classical modal logic and involved such big names in mathematics and logic as K. Gödel, A.N. Kolmogorov, P.S. Novikov and A. Tarski. Arguably, it was this direction of research that attracted mathematical logicians to modal logic rather than the philosophical analysis of

F. Wolter (✉)
Department of Computer Science, University of Liverpool, Liverpool, UK
e-mail: wolter@liverpool.ac.uk

M. Zakharyaschev
Department of Computer Science and Information Systems,
Birkbeck, University of London, London, UK
e-mail: michael@dcs.bbk.ac.uk

G. Bezhanishvili (ed.), *Leo Esakia on Duality in Modal and Intuitionistic Logics*,
Outstanding Contributions to Logic 4, DOI: 10.1007/978-94-017-8860-1_5,

modalities by Lewis and Langford [43]. Moreover, it contributed to establishing close connections between logic, algebra and topology. (It may be of interest to note that Blok and Esakia were rather an algebraist and, respectively, a topologist who applied their results in logic.)

Blok's and Esakia's aims were to understand and describe the structure of the extremely complex lattices of modal and superintuitionistic (aka intermediate) logics—or, in algebraic terms, the lattices of varieties of topological Boolean and Heyting algebras.[1] Their theorem provided means for a comparative study of these lattices and gave a 'superintuitionistic classification' of the lattice of modal logics containing **S4**. Esakia [16] believed that one could give a complete description of the structure of all modal companions of an arbitrary superintuitionistic logic. In particular, he aimed to describe the structure of all modal companions of intuitionistic propositional logic **Int**, discovered that the McKinsey system **S4.1** was one of them and that the Grzegorczyk [33] system **Grz** was the largest one. It is to be noted that the first to observe and investigate the close relationship between the lattices of extensions of **Int** and **S4** were Dummett and Lemmon [11], who—in 1959—used the relational representations of topological Boolean and Heyting algebras that are known to us as Kripke frames. Maksimova and Rybakov [47] in 1974 laid a solid algebraic foundation to the area.

This chapter is a brief overview of results related to the Blok-Esakia theorem, which supplements the earlier survey [10]. In Sect. 5.2, we discuss the role and place of the Blok-Esakia theorem in the theory of modal and superintuitionistic logics. In Sect. 5.3, we give Blok's original algebraic proof of this theorem, which has never been properly published. Section 5.4 surveys generalisations of the Blok-Esakia theorem to intuitionistic modal logics, and, in Sect. 5.5, we discuss its extension to intuitionistic logic with coimplication.

5.2 Modal Companions of Superintuitionistic Logics

According to the (informal) Brouwer-Heyting-Kolmogorov semantics of intuitionistic logic, a statement is true if it has a proof. Orlov [57] and Gödel [25] formalised this semantics by means of a modal logic where the formula $\Box\varphi$ stands for 'φ is provable.' (Novikov [55] read $\Box\varphi$ as 'φ is establishable.') Their modal logic contained classical propositional logic,[2] **Cl**, three properly modal axioms

$$\Box(p \to q) \to (\Box p \to \Box q), \qquad \Box p \to p, \qquad \Box p \to \Box\Box p,$$

[1] Topological Boolean algebras [60] are also known as closure algebras [48], interior algebras [6] and **S4**-algebras. Heyting algebras are called pseudo-Boolean algebras in [60].

[2] Actually, Orlov [57] considered a somewhat weaker logic, which can be regarded as the first relevant system.

and the inference rules $\varphi/\Box\varphi$ (if we have derived φ, then φ is provable), *modus ponens* and substitution. Gödel [25] observed that the resulting logic is equivalent to one of the systems in the Lewis and Langford [43] nomenclature, namely **S4**, and conjectured that propositional intuitionistic logic **Int**, as axiomatised by Heyting [35], can be defined by taking

$$\varphi \in \mathbf{Int} \quad \text{iff} \quad T(\varphi) \in \mathbf{S4}, \tag{5.1}$$

where $T(\varphi)$ is the modal formula obtained by prefixing $\Box$ to every subformula[3] of the intuitionistic formula φ. This conjecture was proved by McKinsey and Tarski [49] in 1948; many other proofs of this fundamental result were given later by Maehara [44], Hacking [34], Schütte [67], Novikov [55], et al.

It has been known since Gödel's [24] that there are infinitely many (more precisely, continuum-many [36]) logics between **Int** and **Cl**. Moreover, some of them are 'constructive' in the same way as **Int**, for instance, the Kleene realisability logic [38, 54, 65] or the Medvedev logic of finite problems [50]. The logics sitting between **Int** and **Cl** were called *intermediate logics* by Umezawa [72, 73]; in the 1960s, Kuznetsov suggested the name *superintuitionistic logics* (*si-logics,* for short) for all extensions of **Int**. We denote the class of si-logics by Ext**Int**. The class of normal (that is, closed under the necessitation rule $\varphi/\Box\varphi$) extensions of **S4** will be denoted by NExt**S4**. Thus,

$$\mathrm{Ext}\mathbf{Int} = \{\mathbf{Int} + \Gamma \mid \Gamma \subseteq \mathscr{L}_I\},$$
$$\mathrm{NExt}\mathbf{S4} = \{\mathbf{S4} \oplus \Sigma \mid \Sigma \subseteq \mathscr{L}_M\},$$

where $\mathscr{L}_I$ is the set of propositional (intuitionistic) formulas, $\mathscr{L}_M$ is the set of modal formulas, $+$ stands for 'add the formulas in Γ and take the closure under *modus ponens* and substitution,' while $\oplus$ also requires the closure under necessitation.

Dummett and Lemmon [11] extended the translation T to the whole class of si-logics. More precisely, with every si-logic $L = \mathbf{Int} + \Gamma$ they associated the modal logic $\tau L = \mathbf{S4} \oplus \{T(\varphi) \mid \varphi \in \Gamma\}$ and showed that L is embedded in τL by T: for every $\varphi \in \mathscr{L}_I$, we have

$$\varphi \in L \quad \text{iff} \quad T(\varphi) \in \tau L. \tag{5.2}$$

It turned out, in particular, that $\tau\mathbf{Cl} = \mathbf{S5}$, $\tau\mathbf{KC} = \mathbf{S4.2}$, $\tau\mathbf{LC} = \mathbf{S4.3}$, where

$$\mathbf{Cl} = \mathbf{Int} + p \vee \neg p, \qquad \mathbf{S5} = \mathbf{S4} \oplus p \to \Box\Diamond p,$$
$$\mathbf{KC} = \mathbf{Int} + \neg p \vee \neg\neg p, \qquad \mathbf{S4.2} = \mathbf{S4} \oplus \Diamond\Box p \to \Box\Diamond p,$$
$$\mathbf{LC} = \mathbf{Int} + (p \to q) \vee (q \to p), \quad \mathbf{S4.3} = \mathbf{S4} \oplus \Box(\Box p \to q) \vee \Box(\Box q \to p).$$

[3] There are different variants of the translation T; in fact, it is enough to prefix $\Box$ to propositional variables, implications and negations only.

One of the questions considered in [11] was to identify those properties of logics that were preserved under the map τ.

Grzegorczyk [33] found a proper extension of **S4** into which **Int** can also be embedded by T. His logic is known now as the *Grzegorczyk logic*

$$\mathbf{Grz} = \mathbf{S4} \oplus \Box(\Box(p \to \Box p) \to p) \to p.$$

Thus, we have, for every $\varphi \in \mathscr{L}_I$:

$$\varphi \in \mathbf{Int} \quad \text{iff} \quad T(\varphi) \in \mathbf{Grz}. \tag{5.3}$$

In fact, according to the Blok-Esakia theorem, **Grz** is the *largest* extension of **S4** into which **Int** is embeddable by T. Esakia [13] observed that **Int** was also embeddable into the *McKinsey logic* $\mathbf{S4.1} = \mathbf{S4} \oplus \Box\Diamond p \to \Diamond\Box p$.

A systematic study of the embeddings of si-logics into modal logics was launched by Maksimova and Rybakov [47], Blok [6] and Esakia [13, 15, 16]. Maksimova and Rybakov introduced two more maps:

$$\rho\colon \mathrm{NExt}\mathbf{S4} \to \mathrm{Ext}\mathbf{Int} \qquad \text{and} \qquad \sigma\colon \mathrm{Ext}\mathbf{Int} \to \mathrm{NExt}\mathbf{S4}$$

where

- $\rho M = \{\varphi \in \mathscr{L}_I \mid T(\varphi) \in M\}$, for every $M \in \mathrm{NExt}\mathbf{S4}$; Esakia called ρM the *superintuitionistic fragment of* M, and M a *modal companion of* ρM;
- $\sigma L = \tau L \oplus \mathbf{Grz}$, for every $L \in \mathrm{Ext}\mathbf{Int}$ (Maksimova and Rybakov [47] used a somewhat different map, which was later shown to be equivalent to σ by Blok and Esakia).

Thus, for example, $\rho\mathbf{Grz} = \rho\mathbf{S4.1} = \mathbf{Int}$, $\tau\mathbf{Int} = \mathbf{S4}$, and $\sigma\mathbf{Int} = \mathbf{Grz}$.

The results of Maksimova and Rybakov [47], Blok [6] and Esakia [13, 15, 16] on the relationship between Ext**Int** and NExt**S4** can be summarised as follows:

1. The set of all modal companions of any si-logic L forms the interval

$$\rho^{-1}(L) = \{M \in \mathrm{NExt}\mathbf{S4} \mid \tau L \subseteq M \subseteq \sigma L\},$$

   with τL being the smallest and σL the greatest modal companions of L in NExt**S4**.[4] Note that this interval always contains an infinite descending chain of logics; for some si-logics, it may contain continuum-many modal logics.
2. The map ρ is a lattice homomorphism from NExt**S4** onto Ext**Int**, τ is a lattice isomorphism from Ext**Int** into NExt**S4**, and all the three maps ρ, τ and σ preserve infinite sums and intersections of logics [47].

[4] That every si-logic L has *a* greatest modal companion was first established by Maksimova and Rybakov [47], who gave an answer to an open question by R. Bull; however, they did not observe that greatest modal companion is actually $\tau L \oplus \mathbf{Grz}$.

3. (**The Blok-Esakia Theorem**) The map σ is a lattice isomorphism from Ext**Int** onto NExt**Grz**.
4. Rybakov [66] also observed that, for any $L \in$ Ext**Int**, the lattice Ext$L$ is isomorphically embeddable into $\rho^{-1}L$. It follows, for example, that there are a continuum of modal companions of **Int**.

The emerging relationship between the lattices Ext**Int** and NExt**S4** can be described semantically. Recall (see, e.g., [9, 27] for details and further references) that *general frames* for **Int** are structures of the form $\mathfrak{F} = (W, R, P)$, where W is a non-empty set, R a partial order on W and P is a collection of upward closed subsets of W (with respect to R) that contains $\emptyset$ and is closed under $\cap$, $\cup$ and the operation $\rightarrow$ defined by taking

$$X \rightarrow Y = \{x \in W \mid \forall y\,(xRy \wedge y \in X \rightarrow y \in Y)\}.$$

If P contains *all* upward closed subsets in W, then $\mathfrak{F}$ is called a *Kripke frame* and denoted by $\mathfrak{F} = (W, R)$. Every si-logic L is characterised by the class $\mathsf{Fr}L$ of general frames validating L. *General frames* for **S4** are triples of the form $\mathfrak{F} = (W, R, P)$, where R a *quasi-order* on $W \neq \emptyset$ and $P \subseteq 2^W$ is a Boolean algebra of subsets of W closed under the operation $\Box$ defined by taking

$$\Box X = \{x \in W \mid \forall y\,(xRy \rightarrow y \in X)\}.$$

General frames of the form $\mathfrak{F} = (W, R, 2^W)$ are called *Kripke frames* and denoted by $\mathfrak{F} = (W, R)$. Every logic $M \in$ NExt**S4** is characterised by the class $\mathsf{Fr}M$ of general frames validating M. For example, a Kripke frame $\mathfrak{F} = (W, R)$ is in Fr**Grz** iff $\mathfrak{F}$ does not contain an infinite ascending chain of the form $x_1Rx_2Rx_3\ldots$ with $x_i \neq x_{i+1}$, $i \geq 1$. We call such frames *Noetherian*. The smallest non-Noetherian frame contains two distinct points accessible from each other; we denote this frame by $\mathfrak{C}_2$.

Given a frame $\mathfrak{F} = (W, R, P)$ for **S4** and a point $x \in W$, we denote by $C(x)$ the *cluster* generated by x in $\mathfrak{F}$, that is, the set

$$C(x) = \{y \in W \mid xRy \text{ and } yRx\}.$$

(Thus, the frame $\mathfrak{C}_2$ above is just a two-point cluster.) The *skeleton* of $\mathfrak{F}$ is the general frame $\rho\mathfrak{F} = (\rho W, \rho R, \rho P)$ for **Int** defined by taking $\rho X = \{C(x) \mid x \in X\}$, for $X \in P$, $C(x)\rho RC(y)$ iff xRy, and

$$\rho P = \{\rho X \mid X \in P \text{ and } X = \Box X\}.$$

Conversely, given a frame $\mathfrak{F} = (W, R, P)$ for **Int**, denote by $\sigma\mathfrak{F}$ the frame $(W, R, \sigma P)$ for **S4**, where σP is the Boolean closure of P in 2^W. Note that the operator σ does not preserve Kripke frames as, for example, $\sigma(\omega, \leq)$ is not a Kripke frame. Another way of converting an intuitionistic frame $\mathfrak{F} = (W, R, P)$ into a modal one is by expanding its points into clusters. Given a cardinal κ, $0 < \kappa \leq \omega$, define $\tau_\kappa \mathfrak{F} = (\kappa W, \kappa R, \kappa P)$ by replacing every $x \in W$ with a κ-cluster with the points x_i, for $i \in \kappa$, and taking κP to be the Boolean closure of $\{X_I \mid I \subseteq \kappa \text{ and } X \in \sigma P\}$, where $X_I = \{x_i \mid i \in I \text{ and } x \in X\}$ [79]. One can show that both $\rho\sigma\mathfrak{F}$ and $\rho\tau_\kappa\mathfrak{F}$ are isomorphic to $\mathfrak{F}$.

Given a class $\mathscr{K}$ of frames, we set $\rho\mathscr{K} = \{\rho\mathfrak{F} \mid \mathfrak{F} \in \mathscr{K}\}$; a similar notation will be used for the operators σ and τ_κ. The logic determined by $\mathscr{K}$ is denoted by $\mathsf{Log}\mathscr{K}$ (it will always be clear from the context whether it is a si- or modal logic). Now, we have:

(ρ) for any $M \in$ NExt**S4** and $\mathscr{K}$, $M = \mathsf{Log}\mathscr{K}$ iff $\rho M = \mathsf{Log}\rho\mathscr{K}$,
(τ) for any $L \in$ Ext**Int** and $\mathscr{K}$, $L = \mathsf{Log}\mathscr{K}$ iff $\tau L = \mathsf{Log}\{\tau_\kappa\mathscr{K} \mid \kappa < \omega\}$,
(σ) for any $L \in$ Ext**Int** and $\mathscr{K}$, $L = \mathsf{Log}\mathscr{K}$ iff $\sigma L = \mathsf{Log}\sigma\mathscr{K}$.

Thus, we can think of NExt**S4** as a two-dimensional structure: in one dimension, we can change the skeleton of frames and thereby change the si-fragment ρM of a modal logic M; in the other, we can change the size of clusters in frames, which keeps the same si-fragment ρM but varies the logic between $\tau\rho M$ and $\sigma\rho M$.

A little bit different perspective can be obtained by employing the machinery of canonical formulas (see [4, 9, 80] for details and further references). For simplicity, let us imagine that all logics in Ext**Int** and NExt**S4** are *subframe logics*, that is, their classes of frames are closed under taking (not necessarily generated) subframes. All such logics are Kripke complete [17, 78], so we can only deal with Kripke frames. Given a finite rooted quasi-order $\mathfrak{F}$, one can construct a modal formula, $\alpha(\mathfrak{F})$, such that, for any frame $\mathfrak{G}$, we have $\mathfrak{G} \not\models \alpha(\mathfrak{F})$ iff $\mathfrak{F}$ is a p-morphic image of a subframe of $\mathfrak{G}$; in this case we also say that $\mathfrak{G}$ is *sub-reducible* to $\mathfrak{F}$. A similar intuitionistic formula, $\beta(\mathfrak{F})$, can be associated with any finite rooted partial order $\mathfrak{F}$. The formulas of the form $\alpha(\mathfrak{F})$ and $\beta(\mathfrak{F})$ are called *subframe formulas*. As shown in [17, 78], all subframe modal and si-logics can be axiomatised by the respective subframe formulas. (We note in passing that the subframe si-logics are precisely those logics in Ext**Int** that can be axiomatised by purely implicative formulas [78, 81].)

Given a si-logic $L = \mathbf{Int} + \{\beta(\mathfrak{F}_i) \mid i \in I\}$, every logic $M \in \rho^{-1}L$ can be represented in the form

$$M = \mathbf{S4} \oplus \{\alpha(\mathfrak{F}_i) \mid i \in I\} \oplus \{\alpha(\mathfrak{F}_j) \mid j \in J\}, \tag{5.4}$$

where every frame $\mathfrak{F}_j$, $j \in J$, contains a cluster with at least two points. The logic $\mathbf{S4} \oplus \{\alpha(\mathfrak{F}_i) \mid i \in I\}$ is obviously τL, while $\sigma L = \tau L \oplus \alpha(\mathfrak{C}_2)$. The lattice $\rho^{-1}\mathbf{Cl}$ of modal companions of classical logic **Cl** looks as follows:

$$\tau\mathbf{Cl} = \mathbf{S5} \subset \cdots \subset \mathbf{S5} \oplus \alpha(\mathfrak{C}_n) \subset \cdots \subset \mathbf{S5} \oplus \alpha(\mathfrak{C}_2) = \mathsf{Log}\{\mathfrak{C}_1\},$$

where $\mathfrak{C}_n$ is a cluster with n points. However, for other si-logics L, the lattice $\rho^{-1}L$ may be very complex.

Every $M \in \mathrm{NExt}\mathbf{S4}$ can be represented as

$$M = M^* \oplus \tau\rho M, \quad \text{with} \quad M^* \subseteq \mathbf{Grz}.$$

Muravitsky [53] called the logic M^* a *modal component* of M and observed that the modal components of M form a dense sublattice of $\mathrm{NExt}\mathbf{S4}$ with $M \cap \mathbf{Grz}$ as its greatest element. The problem whether this sublattice always has a least element was left open in [53]. We only note here that a least element does exist if M is a subframe logic.

The semantic characterisations given above can be used to investigate whether this or that property of logics is preserved under the maps ρ, τ and σ. For example, all the three maps preserve decidability, the finite model property and the disjunction property [47, 79]; Kripke completeness is preserved by ρ, τ but not by σ [47, 68, 79]; interpolation is preserved only under ρ [46]. (For more preservation results and further references consult [9, 10].)

In this chapter, we do not consider embeddings of **Int** and its extensions into the logic of formal provability (in Peano Arthmetic) **GL**, found by Boolos [7], Goldblatt [26] and Kuznetsov and Muravitskij [42]. A discussion of these results can be found in [10]; see also the chapters in this volume written by T. Litak and A. Muravitsky. Artemov [1] analyses the Brouwer-Heyting-Kolmogorov interpretation of intuitionistic logic in the context of his logics of proofs **LP** closely related to **S4**. Relationships between first-order si- and modal logics are investigated in [23].

5.3 An Algebraic Proof of the Blok-Esakia Theorem

In this section, we give a sketch of the algebraic proof of the Blok-Esakia theorem that was found by Blok in his PhD thesis [6] but never published in a journal. (A proof using the machinery of canonical formulas was given in [9]; Jerabek [37] considered modal companions of si-logics from the point of view of inference rules and also gave a proof of the Blok-Esakia theorem.)

We remind the reader that si- and modal logics are determined by varieties of Heyting and, respectively, topological Boolean algebras. A *Heyting algebra* $\mathfrak{A} = (A, \wedge, \vee, \to, \bot, \top)$ extends a bounded distributive lattice $(A, \wedge, \vee, \bot, \top)$ with a binary operator $a \to b$ for the relative pseudo-complement of a with respect to b; that is, for all $c \in A$, we have $a \wedge c \leq b$ iff $c \leq a \to b$. The class of all Heyting algebras is a variety (equationally definable); we denote it by H. Subvarieties V of H are in 1–1 correspondence to si-logics: for any class $\mathcal{V}$ of Heyting algebras, the set

$$L(\mathcal{V}) = \{\varphi \in \mathcal{L}_I \mid \forall \mathfrak{A} \in \mathcal{V}\ \mathfrak{A} \models (\varphi = \top)\}$$

is a si-logic and, conversely, for every si-logic L,

$$\mathscr{V}(L) = \{\mathfrak{A} \mid \forall\varphi \in L\ \mathfrak{A} \models (\varphi = \top)\}$$

is a variety of Heyting algebras. Moreover, $L(\mathscr{V}(L)) = L$ and $\mathscr{V}(L(\mathscr{V})) = \mathscr{V}$ for any si-logic L and any variety $\mathscr{V}$ of Heyting algebras. These results can be proved directly or using duality between Heyting algebras and general frames for **Int**: for any such general frame $\mathfrak{F} = (W, R, P)$, the set P with operations $\cap$, $\cup$, and $\rightarrow$ defined above forms a Heyting algebra denoted by $\mathfrak{F}^+$. Conversely, for every Heyting algebra $\mathfrak{A}$, one can construct a general frame $\mathfrak{A}_+ = (W, R, P)$ whose domain W consists of all prime filters X in $\mathfrak{A}$ with XRY iff $X \subseteq Y$, and $V \in P$ iff there exists $a \in A$ with $V = \{X \in W \mid a \in X\}$. Moreover, $\mathfrak{A}$ is isomorphic to $(\mathfrak{A}_+)^+$.

A *topological Boolean algebra*, or an **S4**-*algebra*, $\mathfrak{A} = (A, \wedge, \vee, \neg, \bot, \top, \Box)$ extends a Boolean algebra $(A, \wedge, \vee, \neg, \bot, \top)$ with a unary operator $\Box$ satisfying the following equations, for all $a, b \in A$:

$$\Box\top = \top, \qquad \Box(a \wedge b) = \Box a \wedge \Box b, \qquad \Box a \leq a, \qquad \Box a \leq \Box\Box a.$$

The class of all **S4**-algebras is a variety; we denote it by $\mathscr{V}(\mathbf{S4})$. Subvarieties V of $\mathscr{V}(\mathbf{S4})$ are in 1–1 correspondence to normal extensions of **S4**: for any class $\mathscr{V}$ of **S4**-algebras, the set

$$L(\mathscr{V}) = \{\varphi \in \mathscr{L}_M \mid \forall\mathfrak{A} \in \mathscr{V}\ \mathfrak{A} \models (\varphi = \top)\}$$

is a logic in NExt**S4** and, conversely, for every logic $L \in$ NExt**S4**,

$$\mathscr{V}(L) = \{\mathfrak{A} \mid \forall\varphi \in L\ \mathfrak{A} \models (\varphi = \top)\}$$

is a variety of **S4**-algebras. Moreover, $L(\mathscr{V}(L)) = L$ and $\mathscr{V}(L(\mathscr{V})) = \mathscr{V}$ for any $L \in$ NExt**S4** and any variety $\mathscr{V}$ of **S4**-algebras. Similarly to the representation of Heyting algebras by frames for **Int** above, one can represent **S4**-algebras by general frames for **S4**. For any such general frame $\mathfrak{F} = (W, R, P)$ for **S4**, the set P with the operations intersection, union, complement, and $\Box$ defined above forms an **S4**-algebra denoted by $\mathfrak{F}^+$. Conversely, for every **S4**-algebra $\mathfrak{A}$, one can construct a general frame $\mathfrak{A}_+ = (W, R, P)$ whose domain W consists of all ultrafilters X in $\mathfrak{A}$ with XRY iff $\{a \mid \Box a \in X\} \subseteq Y$, and $V \in P$ iff there exists $a \in A$ with $V = \{X \in W \mid a \in X\}$. And again, $\mathfrak{A}$ is isomorphic to $(\mathfrak{A}_+)^+$.

We are in the position now to describe the relationship between si-logics and normal extensions of **S4** at the level of Heyting and **S4**-algebras.

From **S4**-*algebras to Heyting algebras*. For any **S4**-algebra $\mathfrak{A} = (A, \wedge, \vee, \neg, \bot, \top, \Box)$, we define a Heyting algebra $\rho\mathfrak{A}$ by taking

$$\rho\mathfrak{A} = (\rho A, \wedge, \vee, \rightarrow, \bot, \top),$$

where $\rho A = \{\Box a \mid a \in A\}$ and $a \to b = \Box(\neg a \vee b)$. Alternatively, one can obtain (an isomorphic copy of) $\rho\mathfrak{A}$ by applying the operation ρ defined for general frames to $\mathfrak{A}_+$ and then taking the induced algebra; that is, $\rho\mathfrak{A}$ is isomorphic to $(\rho(\mathfrak{A}_+))^+$.

From Heyting algebras to **S4**-*algebras*. Conversely, with every Heyting algebra $\mathfrak{A}$ one can associate an **S4**-algebra $\sigma\mathfrak{A}$ in the following way. First, given a bounded distributive lattice $\mathfrak{D} = (D, \wedge, \vee, \bot, \top)$, we construct the *free Boolean extension* $\mathfrak{B}$ of $\mathfrak{D}$ with domain $B = \sigma D \supseteq D$, which is the (uniquely determined) Boolean algebra generated by D such that, for any bounded lattice homomorphism $f : \mathfrak{D} \to \mathfrak{C}$ into a Boolean algebra $\mathfrak{C}$, there exists a unique Boolean homomorphism $h : \mathfrak{B} \to \mathfrak{C}$ with $h\restriction D = f$. Now, given a Heyting algebra $\mathfrak{A} = (A, \wedge, \vee, \to, \bot, \top)$, we obtain the **S4**-algebra $\sigma\mathfrak{A}$ by setting in the free Boolean extension of its underlying bounded distributive lattice

$$\Box a = \bigwedge_{i=1}^{n} (a_i \to b_i), \quad \text{for } a = \bigwedge_{i=1}^{n} (\neg a_i \vee b_i).$$

One can show that $\sigma\mathfrak{A} \in \mathscr{V}(\mathbf{Grz})$ and that $\mathfrak{A} \models (\varphi = \top)$ iff $\sigma\mathfrak{A} \models (T(\varphi) = \top)$. $\sigma\mathfrak{A}$ can also be obtained by first forming $\mathfrak{A}_+ = (W, R, P)$ and then taking the **S4**-algebra $(W, R, \sigma P)^+$ induced by $(W, R, \sigma P)$, where σP has been defined above.

Given classes $\mathscr{K}$ and $\mathscr{H}$ of **S4**- and Heyting algebras, respectively, we set

$$\rho\mathscr{K} = \{\rho\mathfrak{A} \mid \mathfrak{A} \in \mathscr{K}\} \quad \text{and} \quad \sigma\mathscr{H} = \{\sigma\mathfrak{A} \mid \mathfrak{A} \in \mathscr{H}\}.$$

We denote by $\mathsf{H}\mathscr{K}$, $\mathsf{S}\mathscr{K}$, $\mathsf{P}\mathscr{K}$, and $\mathsf{P_U}\mathscr{K}$ the classes of subalgebras, homomorphic images, products, and ultraproducts of algebras in $\mathscr{K}$, respectively. Recall that a class $\mathscr{K}$ of algebras (of the same signature) is a variety if, and only if, it is closed under subalgebras, homomorphic images, and products. Every first-order definable class (and, hence, every variety) is closed under ultraproducts. The following lemma can be proved by showing that $\rho\mathscr{V}$ is closed under subalgebras, homomorphic images, and products [5, 6]:

Lemma 1. *For any variety $\mathscr{V}$ of* **S4**-*algebras, $\rho\mathscr{V}$ is a variety of Heyting algebras.*

For a variety $\mathscr{V}$ of Heyting algebras, $\sigma\mathscr{V}$ is not always a variety. We denote by $\sigma^*\mathscr{V}$ the variety of **S4**-algebras *generated by $\sigma\mathscr{V}$*. The following result implies the Blok-Esakia Theorem:

Theorem 1. (i) *For every variety $\mathscr{V}$ of Heyting algebras, $\rho\sigma^*\mathscr{V} = \mathscr{V}$.*
(ii) *For every variety $\mathscr{V}$ of* **Grz**-*algebras, $\sigma^*\rho\mathscr{V} = \mathscr{V}$.*

For a detailed and instructive exposition of the main steps of the proof of Theorem 1, we refer the reader to [3]. Here we focus on (ii) and, in particular, the following technical lemma from Blok's PhD thesis, which is the key to the algebraic proof of the Blok-Esakia theorem.

Lemma 2. *Let $\mathfrak{A} \in \mathcal{V}(\mathbf{Grz})$ be a countable algebra and let $\mathfrak{B}$ be a subalgebra of $\mathfrak{A}$ such that*

- *$\rho A \subseteq B$;*
- *there exists $c \in A$ such that A is the Boolean closure of $B \cup \{c\}$ in $\mathfrak{A}$ (denoted, slightly abusing notation, $\mathfrak{A} = [\mathfrak{B} \cup \{c\}]_{BA}$).*

Then $\mathfrak{A} \in \mathsf{SP_U}\mathfrak{B}$.

Proof (*sketch*). We follow the proof given in Blok's PhD thesis [6]. Suppose that $B = \{b_i \mid i < \omega\}$ and let U be a non-principal ultrafilter on ω. We remind the reader of the definition of the ultraproduct $\prod_{i<\omega} \mathfrak{B}/U$. First, we define an equivalence relation $\sim_U$ by taking $g \sim_U g'$ iff $\{i < \omega \mid g(i) = g'(i)\} \in U$, for any $g, g' \in \prod_{i<\omega} B$, and set $[g] = \{g' \mid g \sim_U g'\}$. The domain of $\prod_{i<\omega} \mathfrak{B}/U$ is $\{[g] \mid g \in \prod_{i<\omega} B\}$. For $b \in B$, let $\widehat{b} = (b, b, \dots) \in \prod_{i<\omega} B$. The map $f\colon \mathfrak{B} \to \prod_{i<\omega} \mathfrak{B}/U$ defined by taking $f(b) = [\widehat{b}]$ is an embedding of the **S4**-algebra $\mathfrak{B}$ into the **S4**-algebra $\prod_{i<\omega} \mathfrak{B}/U$. We show that f extends to an embedding $\widehat{f}$ of the **S4**-algebra $\mathfrak{A}$ into the **S4**-algebra $\prod_{i<\omega} \mathfrak{B}/U$.

For $n \geq 0$, let

$$C_n = \{b_i \in B \mid b_i \leq c,\ i \leq n\}, \qquad c_n = \bigvee_{b \in C_n} b, \qquad \widehat{c} = (c_n)_{n<\omega}.$$

First, using a Lemma on the existence of *Boolean* embeddings from [31, p. 84] one can show that f can be extended to a Boolean embedding $\widehat{f}\colon \mathfrak{A} \to \prod_{i<\omega} \mathfrak{B}/U$ with $\widehat{f}(c) = [\widehat{c}]$. The next, and crucial, part of the proof is to show that $\widehat{f}$ commutes with the $\Box$-operator. Then $\mathfrak{A} \in \mathsf{SP_U}\mathfrak{B}$, as required. To show that $\widehat{f}$ commutes with $\Box$, let $a \in A$. Then

$$a = (c \vee d_1) \wedge (\neg c \vee d_2) \wedge d_3,$$

for some $d_1, d_2, d_3 \in B$. It suffices to show that

(a) $\widehat{f}(\Box(c \vee d_1)) = \Box\widehat{f}(c \vee d_1)$,
(b) $\widehat{f}(\Box(\neg c \vee d_2)) = \Box\widehat{f}(\neg c \vee d_2)$,
(c) $\widehat{f}(\Box d_3) = \Box\widehat{f} d_3$,

since then we shall have:

$$\begin{aligned}
\widehat{f}(\Box a) &= \widehat{f}(\Box((c \vee d_1) \wedge (\neg c \vee d_2) \wedge d_3)) \\
&= \widehat{f}(\Box(c \vee d_1) \wedge \Box(\neg c \vee d_2) \wedge \Box d_3) \\
&= \widehat{f}(\Box(c \vee d_1)) \wedge \widehat{f}(\Box(\neg c \vee d_2)) \wedge \widehat{f}(\Box d_3) \\
&= \Box\widehat{f}(c \vee d_1) \wedge \Box\widehat{f}(\neg c \vee d_2) \wedge \Box\widehat{f} d_3 \\
&= \Box\widehat{f}(a).
\end{aligned}$$

Now, (c) follows from $d_3 \in B$ and the condition that f is a homomorphism. For (a), let $b = d_1$. We observe that

$$\Box(c \vee b) = \Box((\Box(c \vee b) \wedge \neg b) \vee b)$$

because $\Box(c \vee b) \wedge \neg b \leq c$. We have $\Box(c \vee b) \wedge \neg b \in B$ since $\Box(c \vee b) \in \rho A \subseteq B$ and $b \in B$. Hence $\Box(c \vee b) \wedge \neg b = b_n$ for some $n < \omega$. We obtain $c_n \geq b_n$ and, for all $m \geq n$,

$$\Box(c \vee b) = \Box((\Box(c \vee b) \wedge \neg b) \vee b) \leq \Box(c_m \vee b) \leq \Box(c \vee b).$$

Thus, $\Box(c \vee b) = \Box(c_m \vee b)$ for all $m \geq n$. The equation $\widehat{f}(\Box(c \vee b)) = \Box\widehat{f}(c \vee b)$ follows.

To show (b), let $b = d_2$, $p = \Box(\neg c \vee b)$, and $q = \Box((c \wedge \neg b) \vee p)$. We note that $q = \Box(\neg(\neg c \vee b) \vee \Box(\neg c \vee b))$. We obtain $\neg p \wedge q \leq c \wedge \neg b$. Since $\mathfrak{A} \in \mathscr{V}(\mathbf{Grz})$, we obtain, for all x,

$$\mathfrak{A} \models \Box(\neg\Box(\neg x \vee \Box x) \vee \Box x) = \Box x$$

and, therefore,

$$\begin{aligned}\Box(\neg q \vee p) &= \Box(\neg\Box(\neg(\neg c \vee b) \vee \Box(\neg c \vee b)) \vee \Box(\neg c \vee b)) \\ &= \Box(\neg c \vee b).\end{aligned}$$

We have $\neg p \wedge q \in B$ since $\rho A \subseteq B$, and so $\neg p \wedge q = b_n$ for some $n < \omega$. From $\neg p \wedge q \leq c \wedge \neg b$ we obtain $b_n \leq c_m \wedge \neg b$ for all $m \geq n$, and therefore $\neg b_n \geq \neg c_m \vee b$, for all $m \geq n$. Hence

$$\Box(\neg c \vee b) = \Box(\neg q \vee p) = \Box\neg b_n \geq \Box(\neg c_m \vee b) \geq \Box(\neg c \vee b).$$

Thus, we obtain $\Box(\neg c \vee b) = \Box(\neg c_m \vee b)$ for all $m \geq n$. The required equation $\widehat{f}(\Box(\neg c \vee b)) = \Box\widehat{f}(\neg c \vee b)$ follows. □

We are now in the position to show that $\sigma^*\rho\mathscr{V} = \mathscr{V}$, for any variety $\mathscr{V}$ of **Grz**-algebras. The inclusion $\sigma^*\rho\mathscr{V} \subseteq \mathscr{V}$ is clear. Since any variety is generated by its finitely generated members, to prove $\mathscr{V} \subseteq \sigma^*\rho\mathscr{V}$ it is sufficient to show that all finitely generated $\mathfrak{A} \in \mathscr{V}$ are in the variety generated by $\sigma\rho\mathscr{V}$. Let $\mathfrak{A} \in \mathscr{V}$ be generated by $\{a_1, \ldots, a_n\}$. $\sigma\rho\mathfrak{A}$ is (isomorphic to) a subalgebra of $\mathfrak{A}$. Consider the sequence

$$[\sigma\rho\mathfrak{A} \cup \{a_1\}]_{BA}, \ldots, [\sigma\rho\mathfrak{A} \cup \{a_1, \ldots, a_n\}]_{BA} = \mathfrak{A}.$$

By Lemma 2, it follows by induction that

$$[\sigma\rho\mathfrak{A} \cup \{a_1, \ldots, a_i\}]_{BA} \in \mathscr{V}(\sigma\rho\mathfrak{A}) \subseteq \sigma^*\rho\mathscr{V},$$

for $1 \leq i \leq n$. Thus, $\mathfrak{A} \in \sigma^*\rho\mathscr{V}$, as required.

Intuitionistic logic and its extensions can be embedded in modal logics different from normal extensions of **S4** using different translations; for details and references, the reader can consult [10]. In the remainder of this chapter, we briefly consider extensions of **Int** with extra operators.

5.4 Blok-Esakia Theorems for Intuitionistic Modal Logics

Modal extensions of intuitionistic propositional logic are notoriously much harder to investigate than si-logics and standard (uni)modal logics. In fact, it is already non-trivial to define what an intuitionistic analogue of a given modal logic should be—for intuitionistic $\Box$ and $\Diamond$ are not supposed to be dual. Servi [18, 20], for instance, used a generalisation of the translation T to argue that her systems were 'true' intuitionistic analogues of classical modal logics. In this section, we briefly discuss two extensions of the Blok-Esakia theorem to intuitionistic modal logics.

We begin by considering the most obvious basic system $\mathbf{IntK}_\Box$, which is obtained by adding to **Int** the standard axiom $\Box(p \wedge q) \leftrightarrow (\Box p \wedge \Box q)$ and the necessitation inference rule $\varphi/\Box\varphi$ of the minimal modal logic **K** ($\Diamond\varphi$ can be defined as $\neg\Box\neg\varphi$; note, however, that this $\Diamond$ does not distribute over disjunction). As before, NExt$\mathbf{IntK}_\Box$ denotes the family of logics of the form $\mathbf{IntK}_\Box \oplus \Gamma$, where Γ is a set of modal formulas. An example of a logic in this family is Kuznetsov's [41] intuitionistic provability logic

$$\mathbf{I}^\Delta = \mathbf{IntK}_\Box \oplus p \to \Box p \oplus (\Box p \to p) \to p \oplus ((p \to q) \to p) \to (\Box q \to p),$$

an intuitionistic analogue of the provability logic **GL**. (Esakia suggested the name **KM** for this logic; see Muravitsky's chapter in this volume for a detailed account.) Muravitskij [51, 52] actually proved that the lattices NExt$\mathbf{I}^\Delta$ and NExt**GL** are isomorphic (this result and some generalisations are discussed in Litak's chapter).

A *Kripke frame* for $\mathbf{IntK}_\Box$ is a structure of the form $\mathfrak{F} = (W, R, R_\Box)$, where R is a partial order and $R_\Box$ a binary relation on W such that $R \circ R_\Box \circ R = R_\Box$. The intuitionistic connectives are interpreted in $\mathfrak{F}$ by means of R, while $\Box$ is interpreted via $R_\Box$. Algebraically, every logic $L \in$ NExt$\mathbf{IntK}_\Box$ corresponds to the variety of Heyting algebras with modal operators validating L. For more details on algebraic and relational semantics of these logics and their duality, the reader is referred to [71, 76].

We embed logics in NExt$\mathbf{IntK}_\Box$ into extensions of the fusion (aka independent join) $\mathbf{S4} \otimes \mathbf{K}$ of the modal logics **S4** and **K**. Assuming that the necessity operators in **S4** and **K** are denoted by $\Box_I$ and $\Box$, respectively, we consider the translation T which prefixes $\Box_I$ to every subformula of a given formula in the language of $\mathbf{IntK}_\Box$. As before, we say that T *embeds* $L \in$ NExt$\mathbf{IntK}_\Box$ into $M \in$ NExt($\mathbf{S4} \otimes \mathbf{K}$) if, for every (unimodal) formula φ,

$$\varphi \in L \quad \text{iff} \quad T(\varphi) \in M.$$

In this case M is called a *bimodal companion* of L.

For every logic $M \in \mathrm{NExt}(\mathbf{S4} \otimes \mathbf{K})$, let

$$\rho M = \{\varphi \mid T(\varphi) \in M\},$$

and let σ be the map from NExt$\mathbf{IntK}_\Box$ into $\mathrm{NExt}(\mathbf{S4} \otimes \mathbf{K})$ defined by taking

$$\sigma(\mathbf{IntK}_\Box \oplus \Gamma) = (\mathbf{Grz} \otimes \mathbf{K}) \oplus mix \oplus T(\Gamma), \quad \text{where} \quad mix = \Box_I \Box \Box_I p \leftrightarrow \Box p.$$

Here, the axiom mix reflects the condition $R \circ R_\Box \circ R = R_\Box$ on frames for $\mathbf{IntK}_\Box$. The following extension of the embedding results discussed in Sect. 5.2 was proved in [76, 77]:

Theorem 2. (i) *The map ρ is a lattice homomorphism from NExt(*$\mathbf{S4} \otimes \mathbf{K}$*) onto NExt*$\mathbf{IntK}_\Box$*, which preservs decidability, Kripke completeness, tabularity and the finite model property.*

(ii) *Each logic* $\mathbf{IntK}_\Box \oplus \Gamma$ *is embedded by T into any logic M in the interval*

$$(\mathbf{S4} \otimes \mathbf{K}) \oplus T(\Gamma) \subseteq M \subseteq (\mathbf{Grz} \otimes \mathbf{K}) \oplus mix \oplus T(\Gamma).$$

(iii) *The map σ is an isomorphism from* NExt$\mathbf{IntK}_\Box$ *onto* $\mathrm{NExt}((\mathbf{Grz} \otimes \mathbf{K}) \oplus mix)$ *preserving the finite model property and tabularity.*

Very few general completeness and decidability results are known for intuitionistic modal logics. The theorem above provides means for obtaining such results for logics in NExt$\mathbf{IntK}_\Box$. For example, one can show that if a si-logic $\mathbf{Int} + \Gamma$ is decidable (Kripke complete or has the finite model property) then the logic $\mathbf{IntK}_\Box \oplus \Gamma$ enjoys the same property (for details and more results, the reader is referred to [76, 77]).

Intuitionistic modal logics with independent $\Box$ and $\Diamond$ can be defined as extensions of the basic system $\mathbf{IntK}_{\Box\Diamond}$, which contains the axioms and rules of $\mathbf{IntK}_\Box$ as well as the following axioms for $\Diamond$:

$$\Diamond(p \vee q) \leftrightarrow \Diamond p \vee \Diamond q \qquad \text{and} \qquad \neg\Diamond\bot.$$

Kripke frames for $\mathbf{IntK}_{\Box\Diamond}$ are of the form $(W, R, R_\Box, R_\Diamond)$, where R is a partial order (interpreting the intuitionistic connectives), while $R_\Box$ and $R_\Diamond$ are binary relations on W (interpreting, respectively, $\Box$ and $\Diamond$) such that the following conditions hold: $R \circ R_\Box \circ R = R_\Box$ and $R^{-1} \circ R_\Diamond \circ R^{-1} = R_\Diamond$.

Perhaps the most prominent logics in NExt$\mathbf{IntK}_{\Box\Diamond}$ were constructed by Prior [59] and Fischer Servi [19, 20]. Fischer Servi introduced a weak connection between the necessity and possibility operators in her system

$$\mathbf{FS} = \mathbf{IntK}_{\Box\Diamond} \oplus \Diamond(p \to q) \to (\Box p \to \Diamond q) \oplus (\Diamond p \to \Box q) \to \Box(p \to q).$$

Frames for **FS** satisfy the following conditions:

$$xR_{\Diamond}y \;\to\; \exists z\,(yRz \wedge xR_{\Box}z \wedge xR_{\Diamond}z),$$
$$xR_{\Box}y \;\to\; \exists z\,(xRz \wedge xR_{\Box}y \wedge zR_{\Diamond}y).$$

A remarkable feature of **FS** is that the standard first-order translation not only embeds **K** into classical first-order logic but also **FS** into intuitionistic first-order logic; for details, consult [32, 70]. Another important extension of $\mathbf{IntK}_{\Box\Diamond}$ is the logic

$$\begin{aligned}\mathbf{MIPC} = \mathbf{FS} \oplus\ & \Box p \to p \oplus \Box p \to \Box\Box p \oplus \Diamond p \to \Box\Diamond p \oplus \\ & p \to \Diamond p \oplus \Diamond\Diamond p \to \Diamond p \oplus \Diamond\Box p \to \Box p\end{aligned}$$

introduced by Prior [59]. **MIPC** is an intuitionistic analogue of the modal logic **S5** in the sense that it is equivalent to the one-variable fragment of intuitionistic first-order logic in the same way as **S5** is equivalent to the one-variable fragment of classical first-order logic. (Note, by the way, that the two-variable intuitionistic logic is undecidable [40], unlike the corresponding classical logic, which is NExpTime-complete [30].) **MIPC** is determined by the class of its Kripke frames $(W, R, R_{\Box}, R_{\Diamond})$, where $R_{\Box}$ is a quasi-order, $R_{\Diamond} = R_{\Box}^{-1}$ and $R_{\Box} = R \circ (R_{\Box} \cap R_{\Diamond})$.

The extension of **MIPC** with the duality axiom $\neg\Box\neg p \to \Diamond p$ [21, 56, 64] is known as *weak* **S5** and denoted by **WS5**. Bezhanishvili [2] showed that, for every formula φ, we have $\varphi \in \mathbf{WS5}$ iff $\neg\neg\varphi \in \mathbf{MIPC}$ (remember that, according to Glivenko's theorem, $\varphi \in \mathbf{Cl}$ iff $\neg\neg\varphi \in \mathbf{Int}$). Kripke frames $(W, R, R_{\Box}, R_{\Diamond})$, characterising **WS5**, are frames for **MIPC** such that $R_{\Box}$ is an equivalence relation on W.

Bezhanishvili [3] proved an analogue of the Blok-Esakia theorem for **WS5** and the extension of **Grz** (in the language with $\Box_I$) with *universal modalities*. Modal logics with universal modalities were introduced by Goranko and Passy [28] who, for any (classical) modal logic L with $\Box_I$, defined the (classical) bimodal logic L_u with two boxes, $\Box_I$ and $\forall$, by taking

$$L_u \;= L \oplus \{\text{axioms of } \mathbf{S5} \text{ for } \forall\} \;\oplus\; \forall p \to \Box_I p.$$

For example, the logic $\mathbf{S4}_u$ can be interpreted in topological spaces by regarding $\Box_I$ as the interior operator and $\forall$ as 'for all points in the space.' Because of this, $\mathbf{S4}_u$ plays a prominent role in spatial representation and reasoning; see [22] and references therein. By adding to $\mathbf{S4}_u$ the axiom $\forall(\Diamond_I p \to \Box_I p) \to \forall p \vee \forall\neg p$, we obtain the logic $\mathbf{S4}_u\mathbf{C}$ which is characterised by connected topological spaces [69].

Bezhanishvili [3] defined a translation T from the language of **WS5** to the language of $\mathbf{S4}_u$ by extending the standard Gödel translation of **Int** into **S4** with two more clauses $T(\Box\varphi) = \forall T(\varphi)$ and $T(\Diamond\varphi) = \exists T(\varphi)$, and showed that this translation is an embedding of **WS5** into both $\mathbf{S4}_u$ and $\mathbf{Grz}_u$. It also embeds the logic

$$\mathbf{WS5C} = \mathbf{WS5} \oplus \Box(p \vee \neg p) \rightarrow (p \rightarrow \Box p)$$

into both $\mathbf{S4}_u\mathbf{C}$ and $\mathbf{Grz}_u\mathbf{C} = \mathbf{Grz}_u \oplus \forall(\Diamond_I p \rightarrow \Box_I p) \rightarrow \forall p \vee \forall \neg p$. Moreover, the following extension of the Blok-Esakia theorem holds for T:

- the lattice NExt**WS5** is isomorphic to the lattice NExt$\mathbf{Grz}_u$, and
- the lattice NExt**WS5C** is isomorphic to the lattice NExt$\mathbf{Grz}_u\mathbf{C}$.

A Blok-Esakia theorem for the lattice of *all extensions* of $\mathbf{IntK}_{\Box\Diamond}$ is obtained in [76]. In contrast to the target classical modal logics considered above, the modal logic constructed in [76] has, in addition to the **S4/Grz**-modality, a modal operator that is not normal (but still has a natural Kripke semantics).

5.5 The Blok-Esakia Theorem for the Heyting-Brouwer Logic

In the 1970s, Cecylia Rauszer suggested the extension of the signature of intuitionistic logic by means of a new binary operator for *coimplication*, which we denote here by $\check{\rightarrow}$. Algebraically, $\check{\rightarrow}$ is defined in terms of intuitionistic disjunction in the same way as the intuitionistic implication is defined in terms of intuitionistic conjunction and thus re-establishes, in an extension of intuitionistic logic, the symmetry between classical disjunction and conjunction that is given up in the signature of intuitionistic logic. The translation T of intuitionistic formulas to modal formulas can be extended by setting

$$T(\varphi \check{\rightarrow} \psi) = \Diamond_P(T(\psi) \wedge \neg T(\varphi)),$$

where $\Diamond_P$ is the basic Priorean tense operator for 'at some time in the past' that is interpreted by the inverse of the accessibility relation for the modal $\Box$. To emphasise symmetry, in this section, we denote the modal operator $\Box$ by $\Box_F$ for 'always in the future.' It turns out that many properties of the translation T still hold for this translation of coimplication in Priorean tense logic. In particular, a natural Blok-Esakia theorem holds. Interestingly, Leo Esakia [12, 14] considered both logics and made significant contributions to the study of algebras and their dual Kripke frames for both tense logics and intuitionistic logic extended by coimplication.

The basic logic in the extended language is called *Heyting-Brouwer logic*, **HB**, and is axiomatised by adding to any standard Hilbert-style axiomatisation of **Int** the axioms (we set $\check{\neg} = p \check{\rightarrow} \top$)

$$p \rightarrow (q \vee (q \check{\rightarrow} p)), \qquad (q \check{\rightarrow} p) \rightarrow \check{\neg}(p \rightarrow q),$$
$$(r \check{\rightarrow} (q \check{\rightarrow} p)) \rightarrow ((p \vee q) \check{\rightarrow} p), \qquad \neg(q \check{\rightarrow} p) \rightarrow (p \rightarrow q), \qquad \neg(p \check{\rightarrow} p),$$

and the rule (RN): $p/\neg\check{\neg} p$. **HB** and its first-order extensions have been investigated in [61–63].

In the same way as intuitionistic logic, **HB** is determined by Kripke frames that are partial orders and in which

– $w \models \varphi \check{\rightarrow} \psi$ iff there exists v with vRw, $v \models \psi$, and $v \not\models \varphi$.

An algebraic semantics for **HB** is given by Heyting-Brouwer algebras (aka double Heyting algebras, biHeyting-algebras, and Semi-Boolean algebras) which have been investigated in, for example, [39, 45, 62]. For recent progress in proof theory for **HB** we refer the reader to [8, 29, 58] (where, mostly, **HB** is called bi-intuitionistic logic).

The basic tense logic into which **HB** is embedded by T is called **S4.t**. It is the normal bimodal logic with operators $\Box_F$ and $\Box_P$ (and their duals $\Diamond_F$ and $\Diamond_P$) which both satisfy the axioms for **S4** and the Priorean tense axioms

$$p \to \Box_P \Diamond_F p \quad \text{and} \quad p \to \Box_F \Diamond_P p.$$

In the same way as **S4**, the tense logic **S4.t** is determined by Kripke frames that are quasi-orders. The following equivalence follows directly from completeness with respect to Kripke semantics: for all φ,

$$\varphi \in \mathbf{HB} \quad \text{iff} \quad T(\varphi) \in \mathbf{S4.t}.$$

We now extend the mappings τ, ρ, and σ between si-logics and normal extensions of **S4** to normal extensions of **HB** and **S4.t**. A normal super-Heyting-Brouwer logic (shb-logic) is an extension of **HB** that is closed under modus ponens, substitution, and (RN). By NExt$L$ we denote the lattice of shb-logics containing a shb-logic L. For a set Γ of intuitionistic formulas with coimplication, we denote by $\mathbf{HB} \oplus \Gamma$ the smallest shb-logic containing Γ. Similarly, a normal extension of **S4.t** is an extension of **S4.t** closed under substitution, modus ponens, $p/\Box_P p$, and $p/\Box_F p$. By NExt$L$ we denote the lattice of normal tense logics containing a normal tense logic L and by $L \oplus \Gamma$ we denote the smallest normal extension of L containing Γ.

The analogue of **Grz** in tense logic is given by **Grz.t**, which is obtained from **S4.t** by setting

$$\mathbf{Grz.t} = \mathbf{S4.t} \oplus \{\Box_F(\Box_F(p \to \Box_F p) \to p) \to p, \Box_P(\Box_P(p \to \Box_P p) \to p) \to p\}.$$

Note that we use the axiom for **Grz** for the future and the past. Using it for the future only would give a weaker logic without the finite model property [74] which is a tense companion of **HB** but not sufficiently strong for a Blok-Esakia theorem. We set

– for $L = \mathbf{HB} \oplus \Gamma$, $\tau L = \mathbf{S4.t} \oplus \{T(\varphi) \mid \varphi \in \Gamma\}$,
– for $M \in \text{NExt}\mathbf{S4.t}$, $\rho M = \{\varphi \mid T(\varphi) \in M\}$,
– for $L \in \text{NExt}\mathbf{HB}$, $\sigma L = \mathbf{Grz.t} \oplus \tau L$.

Now, using an extension of the algebraic methods used in Blok's thesis, the following is shown in [75]:

1. The map ρ is a lattice homomorphism from NExt**S4.t** onto NExt**HB**; τ is a lattice isomorphism from NExt**HB** into NExt**S4.t**. The three maps ρ, τ and σ preserve infinite sums and intersections of logics.
2. The map σ is a lattice isomorphism from NExt**HB** onto NExt**S4.t**.

Wolter [75] also considers extensions of those mappings and the Blok-Esakia theorem to non-normal super Heyting-Brouwer logics [logics that are not closed under (RN)] and modal extensions of super Heyting-Brouwer logic. However, in contrast to the situation for si-logics, the preservation properties of those mappings have not yet been investigated in any detail.

Acknowledgments We are grateful to Guram Bezhanishvili, Alexander Chagrov, Alex Citkin, Tadeusz Litak, Larisa Maksimova, Aleksei Muravitsky and Vladimir Rybakov for their help and comments.

References

1. Artemov S (2001) Explicit provability and constructive semantics. Bull Symbolic Logic 7:1–36
2. Bezhanishvili G (2001) Glivenko type theorems for intuitionistic modal logics. Studia Logica 67(1):89–109
3. Bezhanishvili G (2009) The universal modality, the center of a Heyting algebra, and the Blok-Esakia theorem. Ann Pure Appl Logic 161(3):253–267
4. Bezhanishvili G, Bezhanishvili N (2011) An algebraic approach to canonical formulas: modal case. Studia Logica 99(1–3):93–125
5. Blok W, Dwinger P (1975) Equational classes of closure algebras. Indagationes Math 37:189–198
6. Blok W (1976) Varieties of interior algebras. PhD thesis, University of Amsterdam
7. Boolos G (1980) On systems of modal logic with provability interpretations. Theoria 46:7–18
8. Buisman L, Goré R (2007) A cut-free sequent calculus for bi-intuitionistic logic. In: TABLEAUX, Springer, Berlin, pp 90–106
9. Chagrov A, Zakharyaschev M (1997) Modal logic, volume 35 of Oxford logic guides. Clarendon Press, Oxford
10. Chagrov A, Zakharyaschev M (1992) Modal companions of intermediate propositional logics. Studia Logica 51:49–82
11. Dummett M, Lemmon E (1959) Modal logics between S4 and S5. Zeitschrift für Math Logik und Grundlagen der Mathematik 5:250–264
12. Esakia L (1975) The problem of dualism in the intuitionistic logic and Brouwerian lattices. In: 5th international congress of logic, methodology and philosophy of science, Canada, pp 7–8
13. Esakia L (1976) On modal companions of superintuitionistic logics. In: VII Soviet Symposium on logic. Kiev (Russian)
14. Esakia L (1978) Semantical analysis of bimodal (tense) systems. In: Logic, semantics and methodology, Metsniereba Press, Tbilisi, pp 87–99 (Russian)
15. Esakia L (1979) On varieties of Grzegorczyk algebras. In: Mikhailov A (ed) Studies in non-classical logics and set theory, Moscow, Nauka, pp 257–287 (Russian)
16. Esakia L (1979) To the theory of modal and superintuitionistic systems. In: Smirnov V (ed) Proceedings of the USSR symposium on the theory of logical inference, Nauka, Moscow, pp 147–172 (Russian)
17. Fine K (1985) Logics containing K4. Part II. J Symbolic Logic 50:619–651

18. Fischer Servi G (1977) On modal logics with an intuitionistic base. Studia Logica 36:141–149
19. Fischer Servi G (1980) Semantics for a class of intuitionistic modal calculi. In: M Dalla Chiara (ed) Italian studies in the philosophy of science, Reidel, Dordrecht, pp 59–72
20. Fischer Servi G (1984) Axiomatizations for some intuitionistic modal logics. Rendiconti di Matematica di Torino 42:179–194
21. Font J (1986) Modality and possibility in some intuitionistic modal logics. Notre Dame J Formal Logic 27:533–546
22. Gabbay D, Kurucz A, Wolter F, Zakharyaschev M (2003) Many-dimensional modal logics: theory and applications. Elsevier, Amsterdam
23. Gabbay D, Shehtman V, Skvortsov D (2009) Quantification in nonclassical logic. Elsevier, Amsterdam
24. Gödel K (1932) Zum intuitionistischen Aussagenkalkül. Anzeiger der Akademie der Wissenschaften in Wien 69:65–66
25. Gödel K (1933) Eine Interpretation des intuitionistischen Aussagenkalküls. Ergebnisse eines mathematischen Kolloquiums 4:39–40
26. Goldblatt R (1978) Arithmetical necessity, provability and intuitionistic logic. Theoria 44:38–46
27. Goldblatt R (2003) Mathematical modal logic: a view of its evolution. J. Appl Logic 1(5–6):309–392
28. Goranko V, Passy S (1992) Using the universal modality: gains and questions. J Logic Comput 2:5–30
29. Goré R, Postniece L, Tiu A (2010) Cut-elimination and proof search for bi-intuitionistic tense logic. In: Advances in modal logic, pp 156–177
30. Grädel E, Kolaitis P, Vardi M (1997) On the decision problem for two-variable first order logic. Bull Symbolic Logic 3:53–69
31. Grätzer G (1971) Lattice theory: first concepts and distributive lattices. Freeman Co, San Francisco
32. Grefe C (1998) Fischer Servi's intuitionistic modal logic has the finite model property. In: Kracht M, de Rijke M, Wansing H, Zakharyaschev M (eds) Advances in modal logic. CSLI Publications, Stanford, pp 85–98
33. Grzegorczyk A (1967) Some relational systems and the associated topological spaces. Fundamenta Math 60:223–231
34. Hacking I (1963) What is strict implication? J Symbolic Logic 28:51–71
35. Heyting A (1930) Die formalen Regeln der intuitionistischen Logik. Sitzungsberichte der Preussischen Akademie der Wissenschaften, pp 42–56
36. Jankov V (1968) The construction of a sequence of strongly independent superintuitionistic propositional calculi. Sov Math Doklady 9:806–807
37. Jerábek E (2009) Canonical rules. J Symbolic Logic 74(4):1171–1205
38. Kleene S (1945) On the interpretation of intuitionistic number theory. J Symbolic Logic 10:109–124
39. Köhler R (1980) A subdirectly irreducible double Heyting algebra which is not simple. Algebra Univers 10:189–194
40. Kontchakov R, Kurucz A, Zakharyaschev M (2005) Undecidability of first-order intuitionistic and modal logics with two variables. Bull Symbolic Logic 11(3):428–438
41. Kuznetsov A (1985) Proof-intuitionistic propositional calculus. Doklady Academii Nauk SSSR 283:27–30 (Russian)
42. Kuznetsov A, Muravitskij A (1980) Provability as modality. In: Actual problems of logic and methodology of science, Naukova Dumka, Kiev, pp 193–230 (Russian)
43. Lewis C, Langford C (1932) Symbolic logic. Appleton-Century-Crofts, New York
44. Maehara S (1954) Eine Darstellung der intuitionistischen Logik in der klassischen. Nagoya Math J 7:45–64
45. Makkai M, Reyes G (1995) Completeness results for intuitionistic and modal logic in a categorical setting. Ann Pure Appl Logic 72(1):25–101

46. Maksimova L (1982) Failure of the interpolation property in modal companions of Dummett's logic. Algebra Logic 21:690–694
47. Maksimova L, Rybakov V (1974) Lattices of modal logics. Algebra Logic 13:105–122
48. McKinsey J, Tarski A (1944) The algebra of topology. Ann Math 45:141–191
49. McKinsey J, Tarski A (1948) Some theorems about the sentential calculi of Lewis and Heyting. J Symbolic Logic 13:1–15
50. Medvedev Y (1966) Interpretation of logical formulas by means of finite problems. Sov Math Doklady 7:857–860
51. Muravitskij A (1985) Correspondence of proof-intuitionistic logic extensions to provability logic extensions. Sov Math Doklady 31:345–348
52. Muravitsky A (1989) The correspondence of proof-intuitionistic logic extensions to proof-logic extensions. Mat Logika i Algoritm Probl Trudy Inst Mat 12:104–120
53. Muravitsky A (2006) The embedding theorem: its further developments and consequences. Part 1. Notre Dame J Formal Logic 47(4):525–540
54. Nelson D (1947) Recursive functions and intuitionistic number theory. Trans Am Math Soc 61:307–368
55. Novikov P (1977) Constructive mathematical logic from the point of view of classical logic. Nauka, Moscow (Russian)
56. Ono H (1977) On some intuitionistic modal logics. Publ Res Inst Math Sci Kyoto Univ 13:55–67
57. Orlov I (1928) The calculus of compatibility of propositions. Math USSR Sbornik 35:263–286 (Russian)
58. Pinto L, Uustalu T (2010) Relating sequent calculi for bi-intuitionistic propositional logic. In: CL&C, pp 57–72
59. Prior A (1957) Time and modality. Clarendon Press, Oxford
60. Rasiowa H, Sikorski R (1963) The mathematics of metamathematics. Polish Scientific Publishers, Warsaw
61. Rauszer C (1974) A formalization of the propositional calculus of H-B logic. Studia Logica 33:23–34
62. Rauszer C (1974) Semi-boolean algebras and their applications to intuitionistic logic with dual operators. Fundam Math 83:219–249
63. Rauszer C (1977) Applications of Kripke models to Heyting-Brouwer logic. Studia Logica 36:61–71
64. Reyes G, Zawadowski M (1993) Formal systems for modal operators on locales. Studia Logica 52(4):595–614
65. Rose G (1953) Propositional calculus and realizability. Trans Am Math Soc 75:1–19
66. Rybakov V (1976) Hereditarily finitely axiomatizable extensions of the logic *S*4. Algebra Logic 15:115–128
67. Schütte K (1968) Vollständige systeme modaler und intuitionistischer logik. Springer, Berlin
68. Shehtman V (1980) Topological models of propositional logics. Semiot Inf Sci 15:74–98 (Russian)
69. Shehtman V (1999) "Everywhere" and "here". J Appl Non-Classical Logics 9(2–3)
70. Simpson A (1994) The proof theory and semantics of intuitionistic modal logic. PhD thesis, University of Edinburgh
71. Sotirov V (1984) Modal theories with intuitionistic logic. In: Proceedings of the conference on mathematical logic, Sofia, Bulgarian Academy of Sciences, pp 139–171
72. Umezawa T (1955) Über die Zwischensysteme der Aussagenlogik. Nagoya Math J 9:181–189
73. Umezawa T (1959) On intermediate propositional logics. J Symbolic Logic 24:20–36
74. Wolter F (1995) The finite model property in tense logic. J Symbolic Logic 60:757–774
75. Wolter F (1998) On logics with coimplication. J Philos Logic 27:387
76. Wolter F, Zakharyaschev M (1997) On the relation between intuitionistic and classical modal logics. Algebra Logic 36:121–155
77. Wolter F, Zakharyaschev M (1999) Intuitionistic modal logics as fragments of classical bimodal logics. In: Orlowska E (ed) Logic at work. Springer, Berlin, pp 168–186

78. Zakharyaschev M (1989) Syntax and semantics of intermediate logics. Algebra Logic 28:262–282
79. Zakharyaschev M (1991) Modal companions of superintuitionistic logics: syntax, semantics and preservation theorems. Math USSR Sbornik 68:277–289
80. Zakharyaschev M (1992) Canonical formulas for $K4$. Part I: basic results. J Symbolic Logic 57:1377–1402
81. Zakharyaschev M (1996) Canonical formulas for $K4$. Part II cofinal subframe logics. J Symbolic Logic 61:421–449

Chapter 6
Modal Logic and the Vietoris Functor

Yde Venema and Jacob Vosmaer

Dedicated to the memory of Leo Esakia, who was and will remain a great source of inspiration, both as a logician and as a person

Abstract In [16], Esakia uses the Vietoris topology to give a coalgebra-flavored definition of topological Kripke frames, thus relating the *Vietoris topology*, *modal logic* and *coalgebra*. In this chapter, we sketch some of the thematically related mathematical developments that followed. Specifically, we look at Stone duality for the Vietoris hyperspace and the Vietoris powerlocale, and at recent work combining coalgebraic modal logic and the Vietoris functor.

Keywords Modal logic · Vietoris topology · Stone duality · Coalgebra

6.1 Introduction

The *Vietoris hyperspace* is a topological construction on compact Hausdorff spaces, which was introduced in 1922 by Leopold Vietoris [41] as a generalization of the Hausdorff metric. Given a compact Hausdorff space X, one can obtain the Vietoris topology on $\mathrm{K}\,X$, the set of compact subsets of X, by generating a topology from a basis, consisting of all sets of the form

$$\nabla\{U_1, \ldots, U_n\} := \{F \in \mathrm{K}\,X \mid F \subseteq \bigcup_{i=1}^{n} U_i \text{ and } \forall i \leq n,\ F \cap U_i \neq \emptyset\},$$

Y. Venema (✉) · J. Vosmaer
Institute for Logic, Language and Computation, University of Amsterdam,
Amsterdam, The Netherlands
e-mail: Y.Venema@uva.nl

J. Vosmaer
e-mail: contact@jacobvosmaer.nl

G. Bezhanishvili (ed.), *Leo Esakia on Duality in Modal and Intuitionistic Logics*,
Outstanding Contributions to Logic 4, DOI: 10.1007/978-94-017-8860-1_6,

where $\{U_1, \ldots, U_n\}$ ranges over the collection of finite sets of opens in X. Alternatively, one can generate the Vietoris topology from a subbasis, consisting of open sets

$$[U] := \{F \in \mathrm{K}\,X \mid F \subseteq U\} \text{ and } \langle U \rangle := \{F \in \mathrm{K}\,X \mid F \cap U \neq \emptyset\},$$

where U ranges over the open subsets of X. This construction can be seen as a functor on the category of compact Hausdorff spaces and continuous functions: if $f\colon X \to Y$ is continuous, then so is $\mathrm{V}\,f\colon \mathrm{V}\,X \to \mathrm{V}\,Y$, where $\mathrm{V}\,f\colon F \mapsto f[F]$ is taking forward images.

With his 1974 paper [16], Leo Esakia was the first to point out that there is a connection between the Vietoris topology and *modal logic*: he defines his *topological Kripke frames* using the Vietoris topology, and links these structures to modal algebras via a Stone-type duality. In fact, from a modern viewpoint, Esakia's topological Kripke frames are *coalgebras*, and his duality is a key example of a nontrivial algebra/coalgebra duality. This chapter will explore some of the further connections within this picture—comprising the Vietoris topology, modal logic and coalgebra—that have since been discovered in the mathematical landscape. In particular, we will look at how modal logic can help one to understand the Vietoris construction.

Generally, modal logicians think of topological structures and Stone-type dualities as *tools* for understanding modal logics; tools that are of interest primarily or at least partly because the standard Kripke semantics is too coarse a tool for bringing out subtle differences between modal logics. In this chapter, we take an opposite view, namely of modal logic, and *coalgebraic logic*, as a tool for understanding the Vietoris topology. In Sect. 6.2 we consider the basic case. We discuss the use of Boolean modal logic for describing the Stone dual of the Vietoris functor on Stone spaces, and the relation of this idea to coalgebra and Esakia's work [16]. In Sect. 6.3 we see that the relation between the Vietoris construction and modal logic generalizes from Stone spaces to compact Hausdorff spaces. This takes us into locale theory, where the modal logic approach has been used to generalize the Vietoris construction even beyond compact Hausdorff spaces, to stably locally compact spaces. As examples of situations where we find spaces which are not compact Hausdorff, we consider distributive lattices and algebraic domains. Finally, in Sect. 6.4, we investigate a recent perspective on the Vietoris construction, namely, via the *nabla* modality and *Moss' coalgebraic logic*. This leads to a new presentation of the Vietoris construction in locale theory, as well as a new direction of generalization.

This chapter can serve as a first guide through the mathematical landscape that we just sketched, by providing a tour along some well-known results, and relating these to new work. Throughout, we have assumed that the reader has at least some basic familiarity with the following subjects: propositional modal logic and its Kripke semantics, basic general topology and category theory, Stone duality for Boolean algebras, and frames and locales as used in point-free topology. At the end of each (sub)section we provide some historical notes and pointers to the literature (in particular we provide references for facts that are mentioned without proof in the main text).

6.2 The Main Ideas in the Boolean Case

In his 1974 paper [16], Esakia presented duality results for *topological Kripke frames* and modal algebras by building on Stone duality for Boolean algebras. Topological Kripke frames, more commonly known as descriptive general frames, play an important role in the model theory of modal logic, because unlike "ordinary" (discrete) Kripke frames, they provide a complete semantics for modal logic. In his definition of a topological Kripke frame, Esakia interestingly uses the Vietoris topology and the idea that Kripke frames can be seen as what we nowadays call coalgebras. These choices together foreshadow two influential ideas, which can be seen as red threads running through the research we discuss in this chapter:

1. Modal logic can be used to present the *Stone dual of the Vietoris functor*;
2. Certain "modal variants" of Stone duality can be *categorically separated* into dualities for their base logics and their modalities by stating them as *algebra/coalgebra duality* results.

In this section we will discuss the above two ideas in the "basic" case of Boolean algebras and Stone spaces. Our givens are the contravariant functors $\mathrm{K}\Omega\colon \mathbf{Stone} \to \mathbf{BA}$ and $\mathrm{spec}\colon \mathbf{BA} \to \mathbf{Stone}$, which constitute the dual equivalence $\mathbf{BA} \simeq \mathbf{Stone}^{op}$, and the covariant endofunctor $\mathrm{V}\colon \mathbf{Stone} \to \mathbf{Stone}$. We can present these three functors in one picture as follows:

$$\mathbf{BA} \underset{\mathrm{spec}}{\overset{\mathrm{K}\Omega}{\underset{\longrightarrow}{\overset{\longleftarrow}{\simeq}}}} \mathbf{Stone} \circlearrowleft \mathrm{V} \tag{6.1}$$

Can we do something about the asymmetry in this picture? Can we define a functor on Boolean algebras, in "algebraic" terms, which is dual to V? In Sect. 6.2.1, we will see that this is indeed the case. Specifically, we can use modal logic to describe a functor $\mathrm{M_f}\colon \mathbf{BA} \to \mathbf{BA}$, which can be seen as the Stone dual of $\mathrm{V}\colon \mathbf{Stone} \to \mathbf{Stone}$.

$$\mathrm{M_f} \circlearrowright \mathbf{BA} \underset{\mathrm{spec}}{\overset{\mathrm{K}\Omega}{\underset{\longrightarrow}{\overset{\longleftarrow}{\simeq}}}} \mathbf{Stone} \circlearrowleft \mathrm{V} \tag{6.2}$$

We can do two things with the resulting picture: we can use it to frame Esakia's duality as an algebra/coalgebra duality, which is what we will do in Sect. 6.2.2, but we can also view it as an archetype, and ask ourselves: can we generalize this picture? In Sect. 6.3, we will see that $\mathrm{M_f}$ is essentially a restriction of the Vietoris powerlocale, a more general construction on locales, and that one can also prove various duality results for $\mathrm{M_f}$.

6.2.1 The Stone Dual of the Vietoris Functor

Our goal in this subsection is to present the fact that the functor $\mathrm{M_f}$, which is presented using *modal logic*, is the Stone dual of $\mathrm{V}\colon \mathbf{Stone} \to \mathbf{Stone}$. To visualize this we pull apart diagram (6.2), which gives us the following:

$$\begin{array}{ccc} \mathbf{BA} & \underset{\text{spec}}{\overset{\mathrm{K}\Omega}{\leftrightarrows}}\ \simeq & \mathbf{Stone} \\ {\scriptstyle \mathrm{M_f}}\downarrow & & \downarrow{\scriptstyle \mathrm{V}} \\ \mathbf{BA} & \underset{\text{spec}}{\overset{\mathrm{K}\Omega}{\leftrightarrows}}\ \simeq & \mathbf{Stone} \end{array} \tag{6.3}$$

What we mean by saying that $\mathrm{M_f}$ is the Stone dual of V is that the above diagram commutes up to isomorphism. We will make this claim more precise shortly. The subscript "f" on $\mathrm{M_f}$ denotes that this is a construction which constructs *finitary* algebras; we will see an infinitary version of $\mathrm{M_f}$ in Sect. 6.3.

Starting from a Boolean algebra $\mathbb{A} = \langle A; \wedge, \vee, \neg, 0, 1\rangle$, we can define a new Boolean algebra "based on" $\mathbb{A}$ using the following presentation by generators and relations:

Definition 1. Let $\mathbb{A}$ be a Boolean algebra. We define $\mathrm{M_f}\,\mathbb{A}$ to be the Boolean algebra generated by the set $\{\Box a \mid a \in A\} \cup \{\Diamond a \mid a \in A\}$, subject to the following relations:

$$\begin{array}{rr} \Box 1 = 1; & \Diamond 0 = 0; \\ \Box(a \wedge b) = \Box a \wedge \Box b; & \Diamond(a \vee b) = \Diamond a \vee \Diamond b; \\ \Box(a \vee b) \leq \Box a \vee \Diamond b; & \Box a \wedge \Diamond b \leq \Diamond(a \wedge b). \end{array}$$

One may obtain $\mathrm{M_f}\,\mathbb{A}$ by taking the quotient, over the relations listed, of the free Boolean algebra generated by the set $\{\Box a \mid a \in A\}$ and $\{\Diamond a \mid a \in A\}$. In this definition, the sets $\{\Box a \mid a \in A\}$ and $\{\Diamond a \mid a \in A\}$ represent two distinct copies of A; we use boxes and diamonds to denote the respective elements of these sets in order to underline the connection with modal logic. Observe that the relations are nothing more than an algebraic axiomatization of the Boolean modal logic **K**; the last two relations (the *interaction axioms*) imply that $\Diamond \neg a$ is the Boolean complement of $\Box a$: simply substitute $\neg a$ for b (also see Remark 1).

The action of $\mathrm{M_f}$ on Boolean algebra homomorphisms is defined as follows: given a Boolean algebra homomorphism $f\colon \mathbb{A} \to \mathbb{B}$, we can map the generators of $\mathrm{M_f}\,\mathbb{A}$ into $\mathrm{M_f}\,\mathbb{B}$ in the straightforward way, namely by sending

$$\Box a \mapsto \Box f(a) \text{ and } \Diamond a \mapsto \Diamond f(a).$$

Since this mapping respects the relations on $\mathrm{M_f}\,\mathbb{A}$, we obtain a unique Boolean algebra homomorphism $\mathrm{M_f}\, f\colon\ \mathrm{M_f}\,\mathbb{A} \to \mathrm{M_f}\,\mathbb{B}$. This completes our description of the functor $\mathrm{M_f}\colon \mathbf{BA} \to \mathbf{BA}$.

We will now state more precisely what we mean by saying that diagram (6.3) commutes.

Fact 1. *There exist natural isomorphisms such that for any Boolean algebra $\mathbb{A}$ and any Stone space X, we have*

1. $\mathrm{M_f}(\mathrm{K}\Omega\, X) \simeq \mathrm{K}\Omega(\mathrm{V}\, X)$, *and*
2. $\mathrm{spec}(\mathrm{M_f}\,\mathbb{A}) \simeq \mathrm{V}(\mathrm{spec}\,\mathbb{A})$.

Proof Sketch It follows from the fact that $\mathrm{K}\Omega$ and spec form a dual equivalence of categories that statements (1) and (2) are in fact equivalent. Below we will sketch a proof of the fact that $\mathrm{spec}(\mathrm{M_f}\,\mathbb{A}) \simeq \mathrm{V}(\mathrm{spec}\,\mathbb{A})$ for any Boolean algebra $\mathbb{A}$. We leave the proof of the naturality of this isomorphism to the reader, and use without warning the fact that in this setting, the compact sets coincide with the closed ones.

1. The elements of $\mathrm{V}(\mathrm{spec}\,\mathbb{A})$, i.e., the closed subsets of $\mathrm{spec}\,\mathbb{A}$, are in a 1-1 correspondence with $\mathrm{Filt}\,\mathbb{A}$, the filters of $\mathbb{A}$. We can topologize $\mathrm{Filt}\,\mathbb{A}$ by generating a topology from
$$[a] := \{F \in \mathrm{Filt}\,\mathbb{A} \mid a \in F\}, \text{ and}$$
$$\langle a\rangle := \{F \in \mathrm{Filt}\,\mathbb{A} \mid \forall b \in F,\ a \wedge b > 0\},$$
where a ranges over the elements of $\mathbb{A}$. Using this topology, $\mathrm{Filt}\,\mathbb{A}$ is homeomorphic to $\mathrm{V}(\mathrm{spec}\,\mathbb{A})$.
2. We view the elements of $\mathrm{spec}(\mathrm{M_f}\,\mathbb{A})$, i.e., the ultrafilters of $\mathrm{M_f}\,\mathbb{A}$, as Boolean homomorphisms $p\colon \mathrm{M_f}\,\mathbb{A} \to \mathbf{2}$, where $\mathbf{2}$ is the two-element Boolean algebra. Given a homomorphism $p\colon \mathrm{M_f}\,\mathbb{A} \to \mathbf{2}$, we define
$$F_p := \{a \in A \mid p(\Box a) = 1\}.$$
This gives us a map from $\mathrm{spec}(\mathrm{M_f}\,\mathbb{A})$ to $\mathrm{Filt}\,\mathbb{A}$.
3. Conversely, given a filter $F \in \mathrm{Filt}\,\mathbb{A}$, we define a map p_F from the generators of $\mathrm{M_f}\,\mathbb{A}$ to $\mathbf{2}$ by specifying
$$p_F(\Box a) = \begin{cases} 1 \text{ if } a \in F; \\ 0 \text{ otherwise,} \end{cases}$$
for the $\Box$ -generators, and
$$p_F(\Diamond a) = \begin{cases} 1 \text{ if } \forall b \in F, a \wedge b > 0; \\ 0 \text{ otherwise,} \end{cases}$$
for the $\Diamond$-generators. One can verify that this mapping extends to a Boolean homomorphism $p_F\colon \mathrm{M_f}\,\mathbb{A} \to \mathbf{2}$ by checking the relations from Definition 1. Thus, we have defined a map from $\mathrm{Filt}\,\mathbb{A}$ to $\mathrm{spec}(\mathrm{M_f}\,\mathbb{A})$.
4. Finally, we must show that the assignments $p \mapsto F_p$ and $F \mapsto p_F$ are both continuous, and that for all $F \in \mathrm{Filt}\,\mathbb{A}$, $F = F_{p_F}$ and for all $p\colon \mathrm{M_f}\,\mathbb{A} \to \mathbf{2}$, $p = p_{F_p}$. □

Remark 1. In Boolean modal logic, the modalities $\Box$ and $\Diamond$ are interdefinable. For the functor $\mathrm{M_f}\colon \mathbf{BA} \to \mathbf{BA}$, this is reflected by the following fact. Given a Boolean algebra $\mathbb{A}$, we define $\mathrm{M}_{\Box}\,\mathbb{A}$ to be the Boolean algebra generated by the set $\{\Box a \mid a \in A\}$, subject to the relations $\Box 1 = 1$ and $\Box(a \wedge b) = \Box a \wedge \Box b$. One can easily show that $\mathrm{M}_{\Box}$ is a functor on the category of Boolean algebras; moreover, there exists a natural isomorphism such that for any Boolean algebra $\mathbb{A}$, $\mathrm{M_f}\,\mathbb{A} \simeq \mathrm{M}_{\Box}\,\mathbb{A}$; this isomorphism can be obtained by sending each $\Box a$-generator of $\mathrm{M_f}\,\mathbb{A}$ to the corresponding $\Box a$ in $\mathrm{M}_{\Box}\,\mathbb{A}$, and each $\Diamond a$ of $\mathrm{M_f}\,\mathbb{A}$ to $\neg\Box\neg a$ in $\mathrm{M}_{\Box}\,\mathbb{A}$. Indeed, all of the narrative in Sect. 6.2.1 above could have been stated in terms of the functor $\mathrm{M}_{\Box}$ rather than $\mathrm{M_f}$.

Notes

The Boolean case of Stone duality for the Vietoris functor, as discussed above, is discussed in more detail by Kupke et al. [26]. See the notes for Sect. 6.3.1 for more sources.

6.2.2 Algebra/Coalgebra Duality

In this subsection we will use our new knowledge of the functor $\mathrm{M_f}\colon \mathbf{BA} \to \mathbf{BA}$ to state an archetypical algebra/coalgebra duality result: the duality between *Vietoris coalgebras over Stone spaces* and $\mathrm{M_f}$*-algebras over Boolean algebras*. We then discuss the relation of this duality with the original results of Esakia, and its impact on the completeness theory of modal logic.

6.2.2.1 Algebras and Coalgebras

First, we recall the categorical notions of F-algebras and coalgebras. Let $\mathrm{F}\colon \mathbf{C} \to \mathbf{C}$ be an endofunctor on a category $\mathbf{C}$. The category $\mathrm{Alg_C(F)}$, of F-*algebras over* $\mathbf{C}$ has as its *objects* all $\mathbf{C}$-morphisms of the shape $h\colon \mathrm{F}X \to X$, where X, the 'carrier set' of the algebra, ranges over the objects of $\mathbf{C}$. A *morphism* between F-algebras $h\colon \mathrm{F}X \to X$ and $h'\colon \mathrm{F}X' \to X'$ is a $\mathbf{C}$-morphism $f\colon X \to X'$ such that $f \circ h = h' \circ \mathrm{F} f$, i.e., such that the following square commutes:

$$\begin{array}{ccc} \mathrm{F}X & \xrightarrow{\;h\;} & X \\ {\scriptstyle \mathrm{F}f}\downarrow & & \downarrow{\scriptstyle f} \\ \mathrm{F}X' & \xrightarrow[\;h'\;]{} & X' \end{array}$$

Below we will see that the category of modal algebras and modal algebra homomorphisms can be presented as a category of F-algebras over **BA**, the category of Boolean algebras.

The category $\mathrm{Coalg}_{\mathbf{C}}(F)$, of F-*coalgebras over* **C**, is defined dually: F-coalgebras are morphisms $h\colon X \to \mathrm{F}\,X$, and morphisms of F-coalgebras must make a similar square commute:

$$\begin{array}{ccc} X & \xrightarrow{\;h\;} & \mathrm{F}\,X \\ {\scriptstyle f}\downarrow & & \downarrow{\scriptstyle \mathrm{F}f} \\ X' & \xrightarrow[\;h'\;]{} & \mathrm{F}\,X' \end{array}$$

An important example of F-coalgebras is given by *Kripke frames*. If $\mathrm{P}\colon \mathbf{Set} \to \mathbf{Set}$ is the covariant powerset functor, then the category of Kripke frames and bounded morphisms can be presented as $\mathrm{Coalg}_{\mathbf{Set}}(\mathrm{P})$, the category of P-coalgebras over **Set**. If $\langle X, R\rangle$ is a Kripke frame, then we can equivalently present the accessibility relation $R \subseteq X \times X$ as the *successor map* $\rho_R\colon X \to \mathrm{P}\,X$, where $\rho_R\colon x \mapsto \{y \in X \mid Rxy\}$. Moreover, one can easily verify that coalgebra morphisms between P-coalgebras are precisely bounded morphisms.

In [16], Esakia defined *topological* Kripke frames in a similar way: a topological Kripke frame consists of a Stone space X and a binary relation $R \subseteq X \times X$ such that $\rho_R\colon X \to \mathrm{V}\,X$ is continuous as a map into the Vietoris hyperspace of X. This is noteworthy because the idea to view Kripke frames as P-coalgebras only started to gain popularity through the work of Aczel in the late 1980s [5].

6.2.2.2 Duality for Vietoris Coalgebras

Using Stone duality and Fact 1, it is now an elementary exercise in category theory to see that the category of $\mathrm{M_f}$-algebras over **BA** is dually equivalent to the category of V-coalgebras over **Stone**.

Fact 2. $\mathrm{Alg}_{\mathbf{BA}}(\mathrm{M_f}) \simeq \big(\mathrm{Coalg}_{\mathbf{Stone}}(\mathrm{V})\big)^{op}$.

In order to relate this fact to Esakia's results, we need to do a little more work. Particularly, on the algebraic side, Esakia is not working with algebras for the functor $\mathrm{M_f}$, but with the category **MA** of *modal algebras* and modal algebra homomorphisms. Interestingly, the categories $\mathrm{Alg}_{\mathbf{BA}}(\mathrm{M_f})$ and **MA** are *isomorphic*:

Fact 3. $\mathrm{Alg}_{\mathbf{BA}}(\mathrm{M_f}) \cong \mathbf{MA}$.

Proof Sketch Let $\mathbb{A}$ be a Boolean algebra with underlying set A, and let (f, g) be a pair of functions $f, g\colon A \to A$. We call (f, g) a *modal expansion of* $\mathbb{A}$ if the algebraic structure $\langle \mathbb{A}, f, g\rangle$ is a modal algebra, i.e. f preserves $\wedge$ and 1, g preserves $\vee$ and 0, and $\neg \circ f = g \circ \neg$. The key insight underlying the proof of Fact 3 concerns the existence, for a given Boolean algebra $\mathbb{A}$, of a 1-1 correspondence between the modal expansions of $\mathbb{A}$ and the set $\mathrm{Hom}_{\mathbf{BA}}(\mathrm{M_f}\,\mathbb{A}, \mathbb{A})$ of Boolean algebra homomorphisms from $\mathrm{M_f}\,\mathbb{A}$ to $\mathbb{A}$: if (f, g) is a modal expansion of $\mathbb{A}$, then the assignment $\Box a \mapsto f(a)$ and $\Diamond a \mapsto g(a)$ uniquely determines a Boolean homomorphism from $\mathrm{M_f}\,\mathbb{A}$ to $\mathbb{A}$, and conversely, if $h\colon \mathrm{M_f}\,\mathbb{A} \to \mathbb{A}$ is a Boolean homomorphism, then the maps $a \mapsto h(\Box a)$ and $a \mapsto h(\Diamond a)$ define a modal expansion of $\mathbb{A}$. $\Box$

From Facts 1 and 2, we can now deduce the following modern version of Esakia's duality result, which states that modal algebras are dually equivalent to Vietoris coalgebras over Stone spaces:

Fact 4. $\mathbf{MA} \simeq \left(\mathrm{Coalg}_{\mathbf{Stone}}(\mathrm{V})\right)^{op}$.

To conclude this subsection, we briefly indicate how the duality between modal algebras and Vietoris coalgebras is used in the completeness theory of modal logic. Again, the key insight here is that Vietoris coalgebras can be seen as topological Kripke frames; in particular, by forgetting the topology of this structure, we obtain an ordinary Kripke frame. This 'forgetting' can be formalized as a functor U from the category $\mathrm{Coalg}_{\mathbf{Stone}}(\mathrm{V})$ to the category $\mathrm{Coalg}_{\mathbf{Set}}(\mathrm{P})$ of P-coalgebras over **Set**, which as we know is isomorphic to the category of Kripke frames and bounded morphisms. The completeness of modal logic can then be proved by showing that every modal algebra $\mathbb{A}$ can be embedded into the full complex algebra of the underlying Kripke frame of the dual Vietoris coalgebra of $\mathbb{A}$. We will briefly revisit the relation between modal logic and coalgebra in Sect. 6.4.2.1.

Notes

There are many good introductions to Stone duality; our notation stems from [22]. More detailed discussions of duality for modal algebras and Vietoris coalgebras can be found in the work of Abramsky [3] and Kupke et al. [26].

Regarding Esakia's duality for topological Kripke frames, it should be noted that in his paper [16], Esakia is mainly interested in the duality between closure algebras and reflexive, transitive topological Kripke frames, and the duality between Heyting algebras and (what are now called) *Esakia spaces*: reflexive, transitive and anti-symmetric topological Kripke frames. The coalgebraic view of Esakia spaces, already present in Esakia's original paper, has also been discussed by Davey and Galati [15].

6.3 Varying the Base Categories

In Sect. 6.2.1, we have seen that the functor $\mathrm{V}\colon \mathbf{Stone} \to \mathbf{Stone}$, the Vietoris hyperspace construction restricted to Stone spaces, is dual to the functor $\mathrm{M_f}\colon \mathbf{BA} \to \mathbf{BA}$, which is presented using Boolean modal logic. In this section we will see that in the compact Hausdorff case, the Vietoris hyperspace is dual to a construction on locales which uses geometric modal logic: the *Vietoris powerlocale*.

In Sect. 6.3.1, we will see how the duality from Sect. 6.2.1 can be extended to compact regular locales and compact Hausdorff spaces, and how this locale-theoretic approach suggests a generalization of the Vietoris hyperspace from compact

Hausdorff spaces to a hyperspace construction on *stable locally compact spaces*. In Sect. 6.3.2, we look at an important example which is not covered by the compact Hausdorff case, namely Stone duality for distributive lattices and both coherent and Priestley spaces. Finally, in Sect. 6.3.3, we will see how the locale-theoretic Vietoris construction also is the Stone dual of the *Plotkin powerdomain* construction on algebraic domains.

6.3.1 The Vietoris Powerlocale

Vietoris introduced his hyperspace construction to topologize the set of all closed subsets of a compact Hausdorff space. To extend the duality result from Sect. 6.2.1 beyond Stone spaces, we can use Stone duality as it is used in locale theory: as the categorical *equivalence* between spatial locales and sober spaces.

6.3.1.1 Compact Hausdorff Spaces and Compact Regular Locales

Using the Axiom of Choice, the equivalence between spatial locales and sober spaces restricts to an equivalence between *compact regular locales* and compact Hausdorff spaces. Recall that if $\mathbb{A}$ is a locale and $a, b \in \mathbb{A}$, we say a is *well inside* b ($a \lessdot b$) if there is a c such that $a \wedge c = 0$ and $b \vee c = 1$. Equivalently, $a \lessdot b$ iff $a^* \vee b = 1$, where a^* is the pseudo-complement of a. If $U, V \in \Omega X$ are open subsets of a topological space X, then $U \lessdot V$ iff $\mathrm{cl}(U) \subseteq V$. We say $\mathbb{A}$ is *regular* if for every $a \in \mathbb{A}$, $a = \bigvee\{b \mid b \lessdot a\}$. Furthermore, $\mathbb{A}$ is *compact* if for every non-empty directed set S, $1 \leq \bigvee S$ implies $1 \in S$.

Knowing that Ω, the functor sending a space to its locale of opens, and pt, the functor sending a locale to its space of points, constitute an equivalence between **KRegLoc**, the category of compact regular locales, and **KHaus**, the category of compact Hausdorff spaces and continuous maps, we can now draw the following picture:

$$\mathbf{KRegLoc} \underset{\mathrm{pt}}{\overset{\Omega}{\underset{\longrightarrow}{\overset{\longleftarrow}{\simeq}}}} \mathbf{KHaus} \circlearrowleft \mathrm{V} \qquad (6.4)$$

Again, the question is: can we find an endofunctor on **KRegLoc**, defined in "algebraic" terms, corresponding to $\mathrm{V}\colon \mathbf{KHaus} \to \mathbf{KHaus}$? Indeed we can, using the following modification of Definition 1.

Definition 2. Let $\mathbb{A}$ be a locale. We define $\mathrm{M}\,\mathbb{A}$, the *Vietoris powerlocale of* $\mathbb{A}$, to be the locale generated by the set $\{\Box a \mid a \in A\} \cup \{\Diamond a \mid a \in A\}$, subject to the following relations:

$$\begin{array}{rcl} \Box 1 = 1; & & \Diamond 0 = 0; \\ \Box(a \wedge b) = \Box a \wedge \Box b; & & \Diamond(a \vee b) = \Diamond a \vee \Diamond b; \\ \Box \text{ preserves directed joins;} & & \Diamond \text{ preserves directed joins;} \\ \Box(a \vee b) \leq \Box a \vee \Diamond b; & & \Box a \wedge \Diamond b \leq \Diamond(a \wedge b). \end{array}$$

The action of M on frame homomorphisms is defined as in the case of **BA**.

Readers who raise their eyebrows at the above definition, worrying about the fact that we are using generators and relations to define an algebra with an *infinitary* signature, can rest assured: for locales, this is not a problem; see [22, Sect. II.1] or [38]. Observe that the only difference between Definitions 1 and 2, apart from the shift from Boolean algebras to locales, is the additional stipulation that $\Box$ and $\Diamond$ preserve directed joins. From a logical viewpoint, this amounts to a shift from Boolean propositional logic to *geometric* propositional logic, i.e., the logic of finite conjunctions and infinite disjunctions which is preeminent in locale theory and topos theory. Also note that although we are currently interested in the restriction of M to compact regular locales, Definition 2 is stated for arbitrary locales.

We can draw the following diagram now that we have our functor M on locales; as before, we will see that the diagram commutes up to natural isomorphism.

$$\begin{array}{ccc} \mathbf{KRegLoc} & \underset{\text{pt}}{\overset{\Omega}{\underset{\longrightarrow}{\overset{\longleftarrow}{\simeq}}}} & \mathbf{KHaus} \\ {\scriptstyle M}\downarrow & & \downarrow{\scriptstyle V} \\ \mathbf{KRegLoc} & \underset{\text{pt}}{\overset{\Omega}{\underset{\longrightarrow}{\overset{\longleftarrow}{\simeq}}}} & \mathbf{KHaus} \end{array} \tag{6.5}$$

In other words, M restricted to compact regular locales is the Stone dual of V on compact Hausdorff spaces:

Fact 5. *The functor* $\mathrm{M}\colon \mathbf{Loc} \to \mathbf{Loc}$ *restricts to an endofunctor on compact regular locales. Moreover, there exist natural isomorphisms such that for any compact Hausdorff space X and for any compact regular locale $\mathbb{A}$, we have*

1. $\mathrm{M}(\Omega X) \simeq \Omega(\mathrm{V} X)$*, and*
2. $\mathrm{pt}(\mathrm{M}\,\mathbb{A}) \simeq \mathrm{V}(\mathrm{pt}\,\mathbb{A})$.

6.3.1.2 Beyond Compact Hausdorff: Stably Locally Compact Spaces

Recall that in Sect. 6.3.1.1 we asked ourselves what the Stone dual of the Vietoris hyperspace construction on compact Hausdorff spaces is. Now that we have defined the Vietoris powerlocale construction for arbitrary locales, we can ask ourselves: what is the Stone dual of the Vietoris powerlocale, beyond the compact Hausdorff case? In other words, if $\mathbb{A}$ is a spatial locale and $X = \mathrm{pt}\,\mathbb{A}$ is its equivalent sober space of points, can we define a hyperspace $\mathrm{V}\,X$ based on X, such that $\mathrm{M}\,\mathbb{A}$ is equivalent to $\mathrm{V}\,X$?

In its full generality, this question is ill posed. For example, if we take $\mathbb{A}$ to be the open-set lattice of $\mathbb{Q}$, the set of rational numbers equipped with their usual topology, then $\mathrm{M}\,\mathbb{A}$ does not have a Stone dual because it is not *spatial*, i.e., $\mathrm{M}\,\mathbb{A} \not\simeq \Omega \circ \mathrm{pt}(\mathrm{M}\,\mathbb{A})$ (see p. 177 of [23]). Below we will see, however, that we can ask *and* affirmatively answer this question in the case of stably locally compact spaces.

Recall that a topological space is *sober* if it is T_0 and if every irreducibly closed set is the closure of a singleton. A subset U of a topological space X is *saturated* if it is an intersection of opens; equivalently, U is saturated if it is an upper set in the specialization order of X. A topological space is *stably locally compact* if X is sober, locally compact, and binary intersections of compact saturated sets are compact.

Definition 3. Let X be a stably locally compact space. A *lens* is an intersection of a saturated set with a closed set. We define $\mathrm{V}\,X$, the Vietoris hyperspace of X, to be the collection of compact lenses of X with the topology generated by the usual subbasic opens,

$$[U] = \{L \in \mathrm{V}\,X \mid L \subseteq U\} \text{ and}$$
$$\langle U\rangle = \{L \in \mathrm{V}\,X \mid L \cap U \neq \emptyset\},$$

where U ranges over the opens of X.

The choice of compact *lenses*, rather than arbitrary compact subsets of X, is dictated by the desideratum that $\mathrm{V}\,X$ is again T_0: the original space X may have too many compact subsets.

What are the localic analogs of stably locally compact spaces? Recall that the *way-below* relation on a dcpo (directed complete partial order) $\mathbb{D}$ is defined as follows: we say that a is way below b ($a \ll b$) if for every directed set S with $b \leq \bigvee S$, there is a $c \in S$ such that $a \leq c$. A dcpo $\mathbb{D}$ is *continuous* if for every $a \in \mathbb{D}$, the set $\{b \in \mathbb{D} \mid b \ll a\}$ is directed and $a = \bigvee\{b \in L \mid b \ll a\}$. Now let $\mathbb{A}$ be a locale. We say $\mathbb{A}$ is *stably locally compact* if the dcpo reduct of $\mathbb{A}$ is continuous and for all $a, b, c \in \mathbb{A}$, if $a \ll b$ and $a \ll c$ then $a \ll b \wedge c$.

Fact 6. 1. *Both* M *and* V *preserve stable local compactness;*
2. *If $\mathbb{A}$ is a stably locally compact locale and X is a stably locally compact space, then both* $\mathrm{M}(\Omega\,X) \simeq \Omega(\mathrm{V}\,X)$ *and* $\mathrm{V}(\mathrm{pt}\,\mathbb{A}) \simeq \mathrm{pt}(\mathrm{M}\,\mathbb{A})$.

Notes

The equivalence between the categories **KRegLoc** and **KHaus** was established by Isbell [20], see also [6]. The Vietoris powerlocale was first introduced by Johnstone [22, Chap. III, Sect. 4], where he also proves the results contained in Fact 5. We also recommend [22] as an introduction to locale theory and the duality between compact regular locales and compact Hausdorff spaces. For an introduction to stably locally compact spaces, we refer to Gierz et al. [18].

The results contained in Fact 6 are also due to Johnstone [23]. For a discussion of the equivalence between stably locally compact locales and stably locally compact spaces, we suggest reading [25, Sects. 1.2 and 1.3]. An alternative account of the Vietoris construction in both localic and spatial form is given by Simmons [35].

Finally, we would like to point out two alternative approaches to the question "What is the Stone dual of the Vietoris powerlocale?". Firstly, this question has often been approached by (more) *constructive* means, diverging from the "classical" perspective we take in this chapter. This is the case in the work of Johnstone [23] we referred to in Sect. 6.3.1 and of Vickers [39]. Secondly we would like to point out the work of Palmigiano and Venema [32], who use *Chu spaces* to find the Stone dual of the Vietoris powerlocale, taking inspiration from the success of *relation lifting* (see Sect. 6.4) in coalgebraic logic. Yet another approach uses so-called de Vries algebras [9].

6.3.2 Distributive Lattices and the Vietoris Construction

We will now look at the Vietoris functor in relation to an important example of stably locally compact spaces which are not necessarily Hausdorff, namely, the Stone duals of distributive lattices: coherent spaces and Priestley spaces. In this subsection we will look at four different versions of the Vietoris functor, each of which acts on a category (dually) equivalent to **DL**, the category of bounded distributive lattices and (bounded) lattice homomorphisms (throughout this chapter, lattices are assumed to be bounded). The final aim is to show that the three squares in diagram (6.6) commute up to isomorphism.

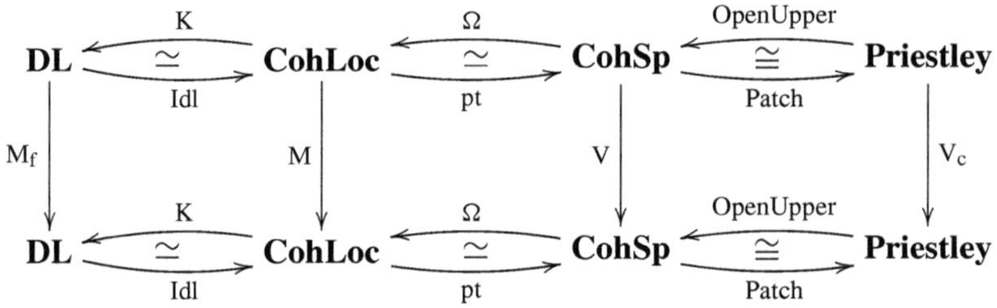

(6.6)

In Sect. 6.3.2.1, we look at a distributive lattice version of the functor M_f and its relation to M. In Sect. 6.3.2.2, we will see how M restricted to coherent locales corresponds to the compact lens hyperspace of Definition 3. Finally, in Sect. 6.3.2.3, we will see how to construct the *convex* Vietoris hyperspace of a Priestley space.

6.3.2.1 Distributive Lattices and Coherent Locales

We start by looking closer at the left square in diagram (6.6).

$$
\begin{array}{ccc}
\mathbf{DL} & \underset{\mathrm{Idl}}{\overset{\mathrm{K}}{\underset{\longrightarrow}{\overset{\longleftarrow}{\simeq}}}} & \mathbf{CohLoc} \\
{\scriptstyle \mathrm{M_f}}\downarrow & & \downarrow{\scriptstyle \mathrm{M}} \\
\mathbf{DL} & \underset{\mathrm{Idl}}{\overset{\mathrm{K}}{\underset{\longrightarrow}{\overset{\longleftarrow}{\simeq}}}} & \mathbf{CohLoc}
\end{array}
$$

By **CohLoc** we denote the category of coherent locales and coherent maps. A locale $\mathbb{A}$ is *coherent* if $\mathbb{A}$ is algebraic, meaning that every $a \in \mathbb{A}$ is a directed join of finite (also called compact) elements, and if additionally $\mathrm{K}\,\mathbb{A}$, the poset of finite elements of $\mathbb{A}$, forms a (distributive) lattice. Equivalently, $\mathbb{A}$ has to be the ideal completion of a distributive lattice. In fact, the ideal completion functor Idl is one half of a dual equivalence between the category **DL** of distributive lattices and lattice homomorphisms, and **CohLoc** of coherent locales and coherent maps; the other half is the functor K which sends a coherent locale to its distributive lattice of finite elements.

To understand the vertical arrows in the left square of diagram (6.6) we need to introduce the functor $\mathrm{M_f}$ on distributive lattices.

Definition 4. Let $\mathbb{A}$ be a distributive lattice. We define $\mathrm{M_f}\,\mathbb{A}$ to be the distributive lattice generated by the set $\{\Box a \mid a \in A\} \cup \{\Diamond a \mid a \in A\}$, subject to the following relations:

$$
\begin{aligned}
\Box 1 &= 1; & \Diamond 0 &= 0; \\
\Box(a \wedge b) &= \Box a \wedge \Box b; & \Diamond(a \vee b) &= \Diamond a \vee \Diamond b; \\
\Box(a \vee b) &\leq \Box a \vee \Diamond b; & \Box a \wedge \Diamond b &\leq \Diamond(a \wedge b).
\end{aligned}
$$

The action of M on lattice homomorphisms is defined as before: given $f : \mathbb{A} \to \mathbb{B}$, we let $\mathrm{M}\,f$ be the extension of $\Box a \mapsto \Box f(a)$ and $\Diamond a \mapsto \Diamond f(a)$.

Note that Definition 4 differs from Definition 1 only because we are generating a distributive lattice rather than a Boolean algebra. This difference is quite subtle due to the following fact.

Fact 7. *Let* $\mathrm{U} : \mathbf{BA} \to \mathbf{DL}$ *denote the forgetful functor that sends a Boolean algebra to its underlying (distributive) lattice. Then there exists a natural isomorphism such that for any Boolean algebra* $\mathbb{A}$, *we have* $\mathrm{U}(\mathrm{M_f}\,\mathbb{A}) \simeq \mathrm{M_f}(\mathrm{U}\,\mathbb{A})$.

The above fact corresponds to the well-known fact in Boolean modal logic that any modal formula containing arbitrary negations is equivalent to a modal formula in which negations are only applied to proposition letters—this observation can also be used in a proof of Fact 7.

We can now explicitly state the content of the first square of diagram (6.6), namely, that the functor $\mathrm{M_f}$ on distributive lattices is equivalent to the Vietoris powerlocale M restricted to coherent locales.

Fact 8. *The Vietoris powerlocale functor* M: **Loc** → **Loc** *restricts to an endofunctor on coherent locales. Moreover, there exist natural isomorphisms such that for any coherent locale* $\mathbb{A}$ *and for any distributive lattice* $\mathbb{L}$*, we have*

1. $\mathrm{M}(\mathrm{Idl}\,\mathbb{L}) \simeq \mathrm{Idl}(\mathrm{M_f}\,\mathbb{L})$*, and*
2. $\mathrm{M_f}(\mathrm{K}\,\mathbb{A}) \simeq \mathrm{K}(\mathrm{M}\,\mathbb{A})$.

6.3.2.2 Coherent Locales and Coherent Spaces

We move on to the middle square of diagram (6.6), in which we encounter the Vietoris functor on coherent spaces.

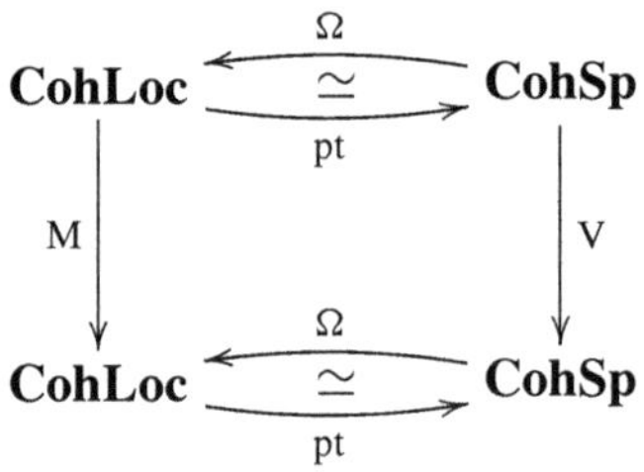

By **CohSp** we denote the category of *coherent spaces* and *coherent maps*. Recall that a coherent space is a (compact) sober space with a basis of compact opens, with the additional property that any finite intersection of compact opens is compact. (Coherent spaces/maps are also known as *spectral* spaces/maps.) A continuous map between coherent spaces is called coherent if the inverse image of a compact open set is compact.

Definition 5. Let X be a coherent space. We define $\mathrm{V}\,X$ to be the Vietoris hyperspace of compact lenses introduced in Definition 3. Moreover, if $f: X \to Y$ is a coherent map between coherent spaces, we define $\mathrm{V}\,f: \mathrm{V}\,X \to \mathrm{V}\,Y$ as follows:

$$\mathrm{V}\,f : L \mapsto \uparrow f[L] \cap \mathrm{cl}(f[L]),$$

where $\uparrow f[L]$ is the saturation of $f[L]$, i.e., its upward closure in the specialization order, and $\mathrm{cl}(f[L])$ is the closure of $f[L]$.

The reason we need to take a "lens closure" in the definition of $\mathrm{V}\,f$ above is that unlike compactness, the property of being a lens is not stable under forward images of continuous functions.

Fact 9. *The construction* V *described above is well-defined, and it is an endofunctor on the category of coherent spaces and coherent maps. Moreover, there exist natural isomorphisms such that for any coherent locale* $\mathbb{A}$ *and for any coherent space* X*, we have*

1. $\mathrm{V}(\mathrm{pt}\,\mathbb{A}) \simeq \mathrm{pt}(\mathrm{M}\,\mathbb{A})$, *and*
2. $\Omega(\mathrm{V}\,X) \simeq \mathrm{M}(\Omega\,X)$.

We can refine Definition 5 by exploiting the special role of compact open sets in coherent spaces:

Fact 10. *Let X be a coherent space. If $U \subseteq X$ is compact open in X, then so are $[U]$ and $\langle U \rangle$ in $\mathrm{V}\,X$. In fact, the sets of the form $[U]$ and $\langle U \rangle$, with U ranging over the compact opens of X, form a sub-base for the topology on $\mathrm{V}\,X$.*

Here we are essentially using the fact that $\mathrm{M_f}\colon \mathbf{DL} \to \mathbf{DL}$ is the Stone dual of $\mathrm{V}\colon \mathbf{CohSp} \to \mathbf{CohSp}$.

6.3.2.3 Coherent Spaces and Priestley Spaces

We will now discuss the final square of diagram (6.6), and learn about the Vietoris construction for Priestley spaces.

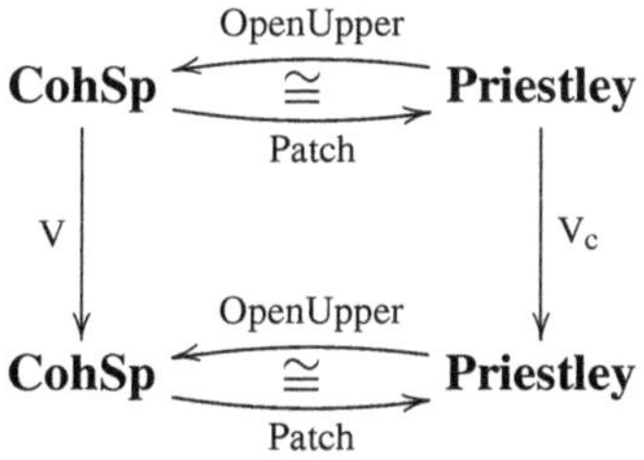

(6.7)

By **Priestley** we denote the category of *Priestley spaces* and order-preserving continuous maps. A Priestley space is a partially ordered compact space $\langle X, \leq, \tau \rangle$, with the additional property that if $x, y \in X$ such that $x \not\leq y$, then there exists a clopen upper set $U \subseteq X$ such that $x \in U \not\ni y$. As a consequence, Priestley spaces are Hausdorff. The categories **CohSp** and **Priestley** are *isomorphic*: we can transform coherent spaces into Priestley spaces and vice versa, and these transformations are mutually inverse. If $\langle X, \tau \rangle$ is a coherent space, then $\langle X, \leq_\tau, \mathrm{patch}(\tau) \rangle$ is a Priestley space, where $\leq_\tau$ is the specialization order of τ and $\mathrm{patch}(\tau)$ is the *patch topology* of τ, i.e., the topology generated by the open sets of τ and the complements of compact saturated sets. This allows one to define a functor $\mathrm{Patch}\colon \mathbf{CohSp} \to \mathbf{Priestley}$, which leaves the set-theoretic functions underlying coherent maps unchanged. We can also go from Priestley spaces to coherent spaces: if $\langle X, \leq, \sigma \rangle$ is a Priestley space, then $\langle X, \sigma^{\uparrow} \rangle$ is a coherent space, where $\sigma^{\uparrow}$ is the collection of open upper sets of $\langle X, \leq, \sigma \rangle$. This gives us a functor $\mathrm{OpenUpper}\colon \mathbf{Priestley} \to \mathbf{CohSp}$, which again leaves the functions underlying the morphisms unchanged. The functors OpenUpper and Patch form an isomorphism of categories: if X is a coherent space and if Y is a Priestley space, then

$$\mathrm{OpenUpper}\,\big(\mathrm{Patch}\,X\big) = X \text{ and } \mathrm{Patch}\,\big(\mathrm{OpenUpper}\,Y\big) = Y.$$

For a detailed account of this connection, see Cornish [14].

Before we introduce the Vietoris construction on Priestley spaces, we will take a closer look at the patch topology, and in particular the patch topology of $\mathrm{V}\,X$ when X is a coherent space.

Fact 11. *Let $\langle X, \tau\rangle$ be a coherent space. The patch topology* patch(τ) *of τ is generated by the following base:*

$$\{U \setminus V \mid U, V \textit{compact open in } \tau\}.$$

Topological properties with respect to τ often correspond to *order*-topological properties with respect to patch(τ).

Fact 12. *Let $\langle X, \tau\rangle$ be a coherent space, and let $\leq$ be its specialization order.*

1. *The* open *subsets of X are precisely the* patch-open upper *subsets of X.*
2. *The* closed *subsets of X are precisely the* patch-closed lower *subsets of X.*
3. *The* compact saturated *subsets of X are precisely the* patch-closed upper *subsets of X.*
4. *The* compact open *subsets of X are precisely the* patch-clopen upper *subsets of X.*

Lemma 1. *Let $\langle X, \tau\rangle$ be a coherent space, let $\leq$ be its specialization order, and let L be a compact lens. Then (1) L is patch-compact; and (2) $\downarrow L$ is closed.*

Proof. Let L be a compact lens. Since L is a lens, $L = \uparrow L \cap \mathrm{cl}(L)$. Because all opens are upper sets, a subset $\mathscr{C} \subseteq \tau$ covers L iff it covers $\uparrow L$; it follows that L is compact iff $\uparrow L$ is compact. Since we assumed that L is compact, so is $\uparrow L$, whence by Fact 12(3) above, $\uparrow L$ must be patch-closed. By Fact 12(2), $\mathrm{cl}(L)$ is also patch-closed. It follows that $L = \uparrow L \cap \mathrm{cl}(L)$ is patch-closed, and because patch(τ) is a compact Hausdorff topology, L is also patch-compact. This proves statement (1); as for the second statement, since $\langle X, \leq, \mathrm{patch}(\tau)\rangle$ is a Priestley space, it follows from e.g. [22, Chap. 7, Sect. 1] that $\downarrow L$ is patch-closed. By Fact 12(2), $\downarrow L$ is also closed w.r.t. τ. □

A subset U of a poset $\mathbb{P}$ is called *convex* if $U = \uparrow U \cap \downarrow U$. If U, V are subsets of $\mathbb{P}$, we say that U is below V in the *Egli-Milner order* ($U \leq_{EM} V$) if both $U \subseteq \downarrow V$ and $\uparrow U \subseteq V$. In other words, $U \leq_{EM} V$ iff

$$\forall x \in U, \exists y \in V \text{ such that } x \leq y, \text{ and } \forall y \in V, \exists x \in U \text{ such that } x \leq y. \quad (6.8)$$

Proposition 1. *Let $\langle X, \tau\rangle$ be a coherent space and let $\leq$ be its specialization order.*

1. *The compact lenses of X are precisely the patch-compact convex subsets of X.*
2. *The specialization order of* $\mathrm{V}\,X$ *is* $\leq_{EM}$.

3. *The patch topology of* $\mathrm{V}\,X$ *is generated by sets of the form*

$$[U], \langle U\rangle, [X \setminus U], \langle X \setminus U\rangle,$$

where U *ranges over the compact opens of* X.

Proof. 1. Suppose $L \subseteq X$ is a compact lens. Then by Lemma 1(1), L is patch-compact. Since $L = \uparrow L \cap \mathrm{cl}(L)$ and $\mathrm{cl}(L)$ is always a lower set, it is easy to see that L is convex. For the converse, suppose that L is a patch-compact convex set. Because $\tau \subseteq \mathrm{patch}(\tau)$, L must also be compact w.r.t. τ. Moreover, since $L = \uparrow L \cap \downarrow L$, and $\downarrow L$ is closed by Lemma 1(2), we see that L is a lens.

2. Let L and M be points of $\mathrm{V}\,X$, i.e., compact lenses of X. Observe that L is below M in the specialization order of $\mathrm{V}\,X$ iff

$$\forall U \in \tau, L \in [U] \Rightarrow M \in [U], \text{ and } \forall U \in \tau, L \in \langle U\rangle \Rightarrow M \in \langle U\rangle. \tag{6.9}$$

Suppose that (6.9) holds for L and M. Then since $\uparrow L = \bigcap\{U \in \tau \mid L \subseteq U\}$, it follows from the left half of (6.9) that $M \subseteq \uparrow L$. Moreover, if we take $U = X \setminus \downarrow M$, then by Lemma 1(2), U is open. Now $M \notin \langle U\rangle$, so by the right side of (6.9), $L \notin \langle U\rangle$, i.e., $L \cap (X \setminus \downarrow M) = \emptyset$, so that $L \subseteq \downarrow M$. We conclude that $L \leq_{EM} M$.

Conversely, suppose that $L \leq_{EM} M$, so that $M \subseteq \uparrow L$ and $L \subseteq \downarrow M$. If U is an open set such that $L \in [U]$, i.e., such that $L \subseteq U$, then $\uparrow L \subseteq U$ so since we assumed $M \subseteq \uparrow L$, $M \in [U]$. And if U is an open set such that $M \notin \langle U\rangle$, i.e., such that $M \cap U = \emptyset$, then since U is an upper set, it is also the case that $\downarrow M \cap U = \emptyset$. But then since we assumed that $L \subseteq \downarrow M$, we see that $L \cap U = \emptyset$, so that $L \notin \langle U\rangle$. It follows that (6.9) holds.

3. Observe that if U is a compact open set, then since

$$[X \setminus U] = \mathrm{V}\,X \setminus \langle U\rangle \text{ and } \langle X \setminus U\rangle = \mathrm{V}\,X \setminus [U], \tag{6.10}$$

it follows from Fact 11 that $[X \setminus U]$ and $\langle X \setminus U\rangle$ are patch-open sets in $\mathrm{V}\,X$.

It follows from Fact 10 that every compact open of $\mathrm{V}\,X$ can be expressed as a finite union of finite intersections of sets of the form $[U]$ and $\langle U\rangle$, where U ranges over compact opens of X. Using De Morgan's laws and the distributive laws, one can see that the *complement* of a compact open set in $\mathrm{V}\,X$ can therefore be expressed as a finite union of finite intersections of sets $\mathrm{V}\,X \setminus [U]$ and $\mathrm{V}\,X \setminus \langle U\rangle$, with U still ranging over compact opens. Using (6.10), we see therefore that the complements of compact opens of $\mathrm{V}\,X$ can be obtained as finite unions of finite intersections of sets $[X \setminus U]$ and $\langle X \setminus U\rangle$. It now follows by Fact 11 that the patch topology of $\mathrm{V}\,X$ is generated by sets of the form $[U]$, $\langle U\rangle$, $[X \setminus U]$, $\langle X \setminus U\rangle$, with U ranging over the compact opens of X. □

We will now define the Vietoris construction on Priestley spaces.

Definition 6. Let X be a Priestley space. We define $\mathrm{V_c}\,X$, the *Vietoris convex hyperspace* of X, to be the collection of compact convex subsets of X, ordered by the

Egli-Milner order $\leq_{EM}$ and topologized by the usual subbasic opens $[U]$ and $\langle U \rangle$, with U ranging over the clopen upper and clopen lower sets. If $f: X \to Y$ is a morphism of Priestley spaces, i.e., if f is a continuous order-preserving map, then we define

$$\mathrm{V_c} f : F \mapsto \uparrow f[F] \cap \downarrow f[F].$$

In other words, $\mathrm{V_c} f$ sends each compact set F to the "convex closure" of its forward image $f[F]$.

In light of Proposition 1, the following should come as no surprise:

Theorem 13. *The construction* $\mathrm{V_c}$ *described above is an endofunctor on the category of Priestley spaces and continuous order-preserving maps.*

In fact, the Vietoris convex hyperspace on Priestley spaces coincides with the Vietoris hyperspace of compact lenses on coherent spaces, i.e., diagram (6.7) *commutes:*

$$\mathrm{V_c} \circ \mathrm{Patch} = \mathrm{Patch} \circ \mathrm{V} \textit{ and } \mathrm{V} \circ \mathrm{OpenUpper} = \mathrm{OpenUpper} \circ \mathrm{V_c}.$$

Notes

Facts 8 and 9 can be found (implicitly) in Johnstone's [23]; we do not know a reference for Fact 7, which corresponds to a well-known fact in modal logic.

The origins of Definition 6 and Theorem 13 are not entirely clear to us. Definition 6 is mentioned by Palmigiano in a paper [31] which focuses on a different kind of Vietoris construction for Priestley spaces. Theorem 13 is stated by Bezhanishvili and Kurz [10], who then refer to [23] and [31]. None of these sources spells out a proof however, so we decided to include one here.

A detailed discussion of Facts 11 and 12, and the isomorphism between the categories of coherent spaces and Priestley spaces, both in relation to *bitopological spaces*, can be found in [8]. An earlier discussion of the patch topology can be found in [19].

6.3.3 Algebraic Domains and the Plotkin Powerdomain

In this final subsection of Sect. 6.3, we will look at Stone duality for the Vietoris powerlocale from an opposite perspective. Namely, we will look at algebraic domains and the Plotkin powerdomain, and we will see that the Stone dual of the Plotkin powerdomain is the Vietoris powerlocale.

Domains, the structures which are studied in domain theory for applications such as semantics for programming languages, are ordered structures which one can simultaneously regard as topological spaces. Crucially, the topology of a domain is uniquely determined by its order (namely, it is the *Scott topology*), and conversely, the order on a domain is uniquely determined by its topology (namely, it is the *specialization order*). From a topological viewpoint, one could say that domains are classes of T_0 spaces which are defined using order-theoretic properties of their specialization orders.

In several important cases, the natural topology of a domain (the Scott topology) can be understood via Stone duality. In this subsection we will consider algebraic domains, a class of directed complete partial orders (dcpo's) which happen to have the property that they are sober in their Scott topologies. Consequently, algebraic domains can be understood in *three* different ways: (1) as dcpo's, (2) as topological spaces, and (3) dually, as locales.

First, we recall the definition of algebraic domains. An element p of a dcpo $\mathbb{D}$ is called *finite* if for all directed S such that $p \leq \bigvee S$, there is a $c \in S$ such that $p \leq c$. We denote the poset of finite elements of $\mathbb{D}$ by $\mathrm{K}\,\mathbb{D}$. We say $\mathbb{D}$ is an *algebraic domain* if $\mathbb{D}$ is a dcpo such that for all $a \in D$, the set $\{b \in \mathrm{K}\,\mathbb{D} \mid b \leq a\}$ is directed and $a = \bigvee\{b \in \mathrm{K}\,\mathbb{D} \mid b \leq a\}$. Every algebraic domain $\mathbb{D}$ is completely determined by its finite elements; specifically, $\mathbb{D} \simeq \mathrm{Idl}(\mathrm{K}\,\mathbb{D})$, where Idl stands for taking the ideal completion. (Note that $\mathrm{K}\,\mathbb{D}$ is a join semilattice, so that ideals can be defined as usual.)

The *Scott topology* on a domain $\mathbb{D}$ is defined as the collection of all upper sets which are inaccessible by directed joins; we denote this topology (and also the locale it induces) by $\Sigma\mathbb{D}$. This allows us to transform domains into locales. Moreover, if we convert the locale $\Sigma\mathbb{D}$ back into a space of points using Stone duality, we find that $\mathrm{pt}(\Sigma\mathbb{D})$, viewed as a dcpo, is isomorphic to $\mathbb{D}$, *assuming* $\mathbb{D}$ *is algebraic*. (The order on $\mathrm{pt}(\Sigma\mathbb{D})$ is the specialization order.)

Powerdomain constructions were introduced in domain theory to model *branching* of computational processes. One particular powerdomain construction is the so-called *Plotkin powerdomain*, which is defined as a free dcpo semi-lattice construction. For algebraic domains, the following surprising characterization of the Plotkin powerdomains is known: if $\mathbb{D}$ is an algebraic domain, then its Plotkin powerdomain can be presented as the ideal completion of the convex subsets of $\mathrm{K}\,\mathbb{D}$, ordered by the Egli-Milner order (see 6.8).

Given the Plotkin powerdomain construction on algebraic domains, and the fact that algebraic domains can be seen as the dual spaces of locales, we can now ask ourselves the question: what is the Stone dual of the Plotkin powerdomain? The answer is that the formation of the Plotkin powerdomain corresponds exactly to the formation of the Vietoris powerlocale.

Fact 14. *Let* $\mathbb{D}$ *be an algebraic domain and let* $\mathrm{Pl}\,\mathbb{D}$ *be its Plotkin powerdomain. Then* $\mathrm{M}(\Sigma\mathbb{D}) \simeq \Sigma(\mathrm{Pl}\,\mathbb{D})$.

Notes

For a general introduction to domain theory we refer to [18], or to [4] in connection with power constructions. Fact 14 is due to Robinson [33]. A natural generalization of it would be to consider continuous rather than algebraic domains. Vickers [40] discusses powerdomains and powerlocales in the context of continuous lattices, but he does not address the specific problem of generalizing Fact 14.

Above, we have left out a discussion of Abramsky's *Domain theory in logical form* [2], for lack of space. In a nutshell, Abramsky exploits Stone duality for the intersection of algebraic domains

(Sect. 6.3.3) and coherent spaces. Within this context, the Plotkin powerdomain is Stone dual to the functor $\mathrm{M_f}$ on distributive lattices, a fact which is used to study *bisimulation* in [1]. For an introduction to the very powerful framework of "Domain theory in logical form" we refer the reader to [2] or [4].

6.4 The Vietoris Construction and the Nabla Modality

If we look at our discussion of Stone duality for the Vietoris functor in Sect. 6.2.1, we see an asymmetry in the presentations of the hyperspace topology on the one hand and the logical/algebraic powerlocale constructions on the other hand. The hyperspace of a compact Hausdorff space X can be topologized in two equivalent ways, namely using *basic opens* of the shape

$$\nabla\{U_1, \ldots, U_n\} := \{F \in \mathrm{K}\,X \mid F \subseteq \bigcup_{i=1}^{n} U_i \text{ and } \forall i \leq n,\ F \cap U_i \neq \emptyset\},$$

(∇ is pronounced "nabla") versus using *subbasic opens* of the shape

$$[U] := \{F \in \mathrm{K}\,X \mid F \subseteq U\} \text{ and}$$
$$\langle U \rangle := \{F \in \mathrm{K}\,X \mid F \cap U \neq \emptyset\},$$

where U and the U_i range over the opens of X. The powerlocales $\mathrm{M}\,\mathbb{A}$ and $\mathrm{M_f}\,\mathbb{A}$, on the other hand, we only presented using box ($\Box$) and diamond ($\Diamond$) in combination with positive modal logic.

The co-existence of these two distinct definitions of the Vietoris construction on topological spaces naturally raises the question, how to give a presentation of the Vietoris powerlocale directly in terms of nabla (∇); similarly, it is an interesting problem how to axiomatize modal logic in terms of the nabla modality. In this section we will see how ideas from the theory of *coalgebra*, and more specifically, *coalgebraic modal logic* may be used to address and solve these problems. As a by-product of this coalgebraic approach, we will see that the Vietoris construction V can be seen as an *instance* of a more general construction which is parametrized by a 'coalgebra functor' on the category **Set**: Given such a functor T we will define the notion of a T-powerlocale functor on the category of locales, in such a way that the Vietoris construction corresponds to the case where T is the power set functor P.

We will first have a brief look at the nabla modality as a derived connective in Sect. 6.4.1. In Sect. 6.4.2, we will introduce the syntax and semantics of Moss' coalgebraic modal logic. In Sect. 6.4.3, we introduce the Carioca axiom system, which is sound and complete with respect to Moss' coalgebraic logic. In Sect. 6.4.4 we will then show how these axioms can be applied to the Vietoris powerlocale, and how they even lead to a notion of generalized powerlocale.

6.4.1 Nabla-Expressions

In this subsection, we will look at the nabla modality as a derived connective. From this point of view, the nabla modality is simply an *expression* consisting of $\Box$ and $\Diamond$ modalities. The main result we discuss is the fact that every element of a powerlocale can be expressed as a disjunction of nabla expressions.

Definition 7. Let $\mathbb{A}$ be a locale, distributive lattice or Boolean algebra. A *nabla-expression over* $\mathbb{A}$ is a term of the shape

$$\Box\left(\bigvee\alpha\right) \wedge \bigwedge_{a\in\alpha} \Diamond a,$$

where $\alpha \subseteq \mathbb{A}$ is a finite subset of $\mathbb{A}$.

It is not hard to see that if $\mathbb{A} = \Omega X$, then the nabla-expressions over $\mathbb{A}$ correspond precisely to the basic open subsets $\nabla\{U_1, \ldots, U_n\}$ of the Vietoris hyperspace $\mathrm{V}\,X$. The fact that the sets $\nabla\{U_1, \ldots, U_n\}$ form a basis for the Vietoris topology, rather than a subbasis, can also be expressed algebraically:

Fact 15. 1. *If* $\mathbb{A}$ *is a locale then every element of* $\mathrm{M}\,\mathbb{A}$ *can be expressed as a join of nabla-expressions over* $\mathbb{A}$*;*
2. *If* $\mathbb{A}$ *is a distributive lattice or a Boolean algebra, then every element of* $\mathrm{M_f}\,\mathbb{A}$ *can be expressed as a* finite *join of nabla-expressions over* $\mathbb{A}$.

Proof Sketch We will briefly discuss the case where $\mathbb{A}$ is a distributive lattice. Suppose $x \in \mathrm{M_f}\,\mathbb{A}$. Because $\mathrm{M_f}\,\mathbb{A}$ is generated by (equivalence classes of) elements of the shape $\Box\, a$ and $\Diamond\, b$, we may assume that x is a disjunction of terms of the shape

$$\bigwedge_I \Box a_i \wedge \bigwedge_J \Diamond b_j,$$

where I, J are finite index sets and the a_i, b_j come from $\mathbb{A}$. It will suffice to show that such conjunctions can be obtained as disjunctions of nabla-expressions.

Because $\Box$ preserves finite meets, we will assume we have a single $\Box$-conjunct $\Box\, a$ (if $I = \emptyset$, this will be the term $\Box\, 1$). We will now show that the following term can be obtained as a disjunction of at most two nabla-expressions:

$$\Box a \wedge \bigwedge_J \Diamond b_j.$$

For the case that $|J| = 0$, we leave it as an exercise for the reader to show that

$$\Box a = (\Box a \wedge \Diamond a) \vee \left(\Box(\bigvee\emptyset) \wedge \bigwedge\emptyset\right),$$

which is a binary disjunction of nabla-expressions.

We will now assume that $|J| > 0$, and we will show that in this case we get just one nabla-expression. To do this, we will use the following equations, which can be easily derived from the axioms in Definition 4:

$$\Box c \wedge \Diamond d = \Box c \wedge \Diamond(c \wedge d); \tag{6.11}$$
$$\Box c \wedge \Diamond d = \Box c \wedge \Diamond c \wedge \Diamond d. \tag{6.12}$$

We now see that

$$\begin{aligned}
&\Box a \wedge \textstyle\bigwedge_J \Diamond b_j \\
&= \Box a \wedge \textstyle\bigwedge_J \Diamond(b_j \wedge a) && \text{by (11)}(|J|\text{times}), \\
&= \Box a \wedge \Diamond a \wedge \textstyle\bigwedge_J \Diamond(b_j \wedge a) && \text{by (12) since} |J| > 0, \\
&= \Box \big(a \vee \textstyle\bigvee_J (b_j \wedge a)\big) \wedge \Diamond a \wedge \textstyle\bigwedge_J \Diamond(b_j \wedge a) && \text{by order theory.}
\end{aligned}$$

The final expression above is now indeed a nabla-expression, for

$$\alpha = \{a\} \cup \{b_j \wedge a \mid j \in J\}.$$

Since we assumed $x \in \mathrm{M_f}\,\mathbb{A}$ to be a finite disjunction of conjunctions of $\Box a$'s and $\Diamond b_j$'s, and since each such conjunction is the disjunction of at most two nabla-expressions, it follows that x itself is also a finite disjunction of nabla-expressions. The same argument can be applied in the locale case. □

Notes

What we call "nabla expressions" above have been used, in one form or another, both in modal logic and in locale/domain-theoretic investigations of the powerlocale. For modal logic, see e.g. the normal forms used by Fine [17]; for locale theory, see e.g. Johnstone [22, 23] and Robinson [33]. None of these sources, however, explicitly state or prove Fact 15.

6.4.2 Moss' Coalgebraic Logic

In this subsection we introduce the syntax and semantics of Moss' coalgebraic logic. We start with an observation about the semantics of nabla-expressions in Kripke frames. We will then very briefly review some of the background of coalgebra and coalgebraic logic in Sect. 6.4.2.1. In Sect. 6.4.2.2, we introduce relation lifting, a technique which sits at the heart of Moss' coalgebraic logic. In Sect. 6.4.2.3, we then introduce the syntax and semantics of Moss' coalgebraic logic.

Suppose that $\mathfrak{F} = \langle X, R\rangle$ is a Kripke frame, where $R \subseteq X \times X$, and suppose we have a nabla-expression

$$\Box \left(\textstyle\bigvee_{i=1}^{n} \varphi_i\right) \wedge \textstyle\bigwedge_{i=1}^{n} \Diamond \varphi_i .$$

For simplicity, we assume $\varphi_1, \ldots, \varphi_n$ are closed formulas, i.e., they contain no proposition letters. What is the semantics of our nabla-expression? If $x \in X$, then

$$x \Vdash_{\mathfrak{F}} \Box \left(\bigvee_{i=1}^{n} \varphi_i\right) \wedge \bigwedge_{i=1}^{n} \Diamond \varphi_i$$

if and only if

$$\forall y \in R[x],\ \exists i \leq n,\ y \Vdash_{\mathfrak{F}} \varphi_i \text{ and } \forall i \leq n,\ \exists y \in R[x],\ y \Vdash_{\mathfrak{F}} \varphi_i, \tag{6.13}$$

where $R[x]$ is the set of R-successors of x. If we view $\Vdash_{\mathfrak{F}}$ as a binary relation between X and the set of all closed modal formulas, then we can abbreviate (6.13) as follows:

$$R[x]\ (\Vdash_{\mathfrak{F}})_{EM}\ \{\varphi_1, \ldots, \varphi_n\},$$

where $(\cdot)_{EM}$ stands for taking the *Egli-Milner* lifting of a binary relation (see 6.8 in Sect. 6.3.2.3).

Guided by this observation, we now consider a variant of the standard modal language in which we take the ∇ modality to be a *primitive* modality, with the following semantics on a given Kripke frame $\mathfrak{F}$:

$$x \Vdash_{\mathfrak{F}} \nabla\{\varphi_1, \ldots, \varphi_n\} \text{ iff } R[x]\ (\Vdash_{\mathfrak{F}})_{EM}\ \{\varphi_1, \ldots, \varphi_n\}. \tag{6.14}$$

It is not hard to verify that using (6.14),

$$x \Vdash_{\mathfrak{F}} \Box \varphi \text{ iff } x \Vdash_{\mathfrak{F}} \nabla\{\varphi\} \vee \nabla\emptyset,$$

and that

$$x \Vdash_{\mathfrak{F}} \Diamond \varphi \text{ iff } x \Vdash_{\mathfrak{F}} \nabla\{\varphi, \top\}.$$

What makes the reformulation interesting is that the semantics (6.14) allows for *coalgebraic generalizations*. As we will see, the key for turning the above observation about Kripke frames into a logical language and semantics for more general coalgebras is to use *relation lifting*.

6.4.2.1 Coalgebra and Coalgebraic Modal Logic

The theory of *coalgebra* aims to provide a general mathematical framework for the study of state-based evolving systems. Given an endofunctor T on the category **Set** of sets with functions, we already saw the definition of a coalgebra of type T, or briefly: a T-coalgebra is a pair (S, σ) where S is some set and $\sigma : S \to \mathrm{T}S$. The set S is called the *carrier* of the coalgebra, elements of which are called *states*; σ is called the *transition map* of the coalgebra. A T-coalgebra morphism between coalgebras $\sigma : S \to \mathrm{T}S$ and $\sigma' : S' \to \mathrm{T}S'$ is simply a function $f : S \to S'$ such that $\mathrm{T}f \circ \sigma = \sigma' \circ f$.

$$\begin{array}{ccc} S & \xrightarrow{\sigma} & \mathrm{T}\,S \\ {\scriptstyle f}\downarrow & & \downarrow{\scriptstyle \mathrm{T}\,f} \\ S' & \xrightarrow[\sigma']{} & \mathrm{T}\,S' \end{array}$$

The coalgebraic approach to state-based systems combines mathematical simplicity with wide applicability: many features of computation, such as input, output, non-determinism, probability or interaction between agents, can be encoded in the functor T. Examples of coalgebras are Kripke frames, Kripke models, deterministic automata, topologies (with continuous open maps), and Markov chains.

The key notion of equivalence in coalgebra is that of two states s and s' in coalgebras (S, σ) and (S', σ') being *behaviorally equivalent*, notation: $(S, \sigma), s \simeq (S', \sigma'), s'$; this relation holds if there are coalgebra morphisms f, f' with a common codomain such that $f(s) = f'(s')$. As the name suggests, behaviorally equivalent states are considered to display the same behavior, and hence, to be essentially the same.

Coalgebraic logics are designed and studied in order to reason formally about coalgebras and their behavior; one of the main applications of this approach is the design of specification and verification languages for coalgebras. An (abstract) coalgebraic logic is a pair $(\mathscr{L}, \Vdash^{\mathscr{L}})$ such that $\mathscr{L}$ is a set of *formulas* and $\Vdash$ is a collection of relations associating with each T-coalgebra (S, σ) a binary relation $\Vdash^{\mathscr{L}}_{(S,\sigma)} \subseteq S \times \mathscr{L}$. If $s \Vdash^{\mathscr{L}}_{(S,\sigma)} \varphi$ we say that the formula φ is *true* or *satisfied* at s in (S, σ), and we will often write $(S, \sigma), s \Vdash \varphi$.

A natural criterion for a coalgebraic logic is that it cannot make a distinction between behaviorally equivalent states. A formula φ is *behaviorally invariant* if for all pairs of behaviorally equivalent pointed coalgebras $(S, \sigma), s \simeq (S', \sigma'), s'$ it holds that $(S, \sigma), s \Vdash \varphi \iff (S', \sigma'), s' \Vdash \varphi$. A coalgebraic language is *adequate* if all of its formulas are behaviorally invariant. An example of an adequate language is the classical modal logic interpreted on P-coalgebras, i.e., on Kripke frames.

Given the prominence of Kripke frames and models as examples of coalgebras, it is not surprising to see that standard modal logic can be suitably generalized to provide adequate coalgebraic logics for coalgebras of arbitrary type. There are in fact distinct ways to do this; here we will focus on the approach based on the notion of relation lifting.

6.4.2.2 Relation Lifting

Relation lifting is nothing more than a particular way of extending a coalgebra type functor $\mathrm{T}\colon \mathbf{Set} \to \mathbf{Set}$ to a functor $\overline{\mathrm{T}}\colon \mathbf{Rel} \to \mathbf{Rel}$ on the category of sets and binary relations. For our purposes, we restrict attention to transition types that preserve *weak pullbacks*.

A *weak pullback* of two morphisms $f\colon X \to Z$ and $g\colon Y \to Z$ with a shared codomain Z is a pair of morphisms $p_X\colon P \to X$ and $p_Y\colon P \to Y$ with a shared domain P, such that (1) $f \circ p_X = g \circ p_Y$, and (2) for any other pair of morphisms $q_X\colon Q \to X$ and $q_Y\colon Q \to Y$ with $f \circ q_X = g \circ q_Y$, there is a morphism $q\colon Q \to P$ such that $p_X \circ q = q_X$ and $p_Y \circ q = q_Y$. This pullback is "weak" because we are not requiring q to be unique.

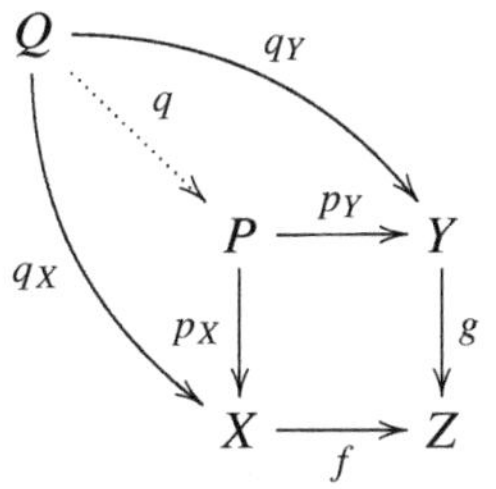

Saying that $\mathrm{T}\colon \mathbf{Set} \to \mathbf{Set}$ preserves weak pullbacks means that if $p_X\colon P \to X$ and $p_Y\colon P \to Y$ form a weak pullback of $f\colon X \to Z$ and $g\colon Y \to Z$, then $\mathrm{T}\,p_X\colon \mathrm{T}\,P \to \mathrm{T}\,X$ and $\mathrm{T}\,p_Y\colon \mathrm{T}\,P \to \mathrm{T}\,Y$ form a weak pullback of $\mathrm{T}\,f\colon \mathrm{T}\,X \to \mathrm{T}\,Z$ and $\mathrm{T}\,g\colon \mathrm{T}\,Y \to \mathrm{T}\,Z$. Examples of weak pullback-preserving endofunctors on the category of sets include the identity functor, constant functors, the covariant powerset functor, the multiset functor, the distribution functor, and finite products and sums of such functors.

We will now define the notion of relation lifting.

Definition 8. Let $\mathrm{T}\colon \mathbf{Set} \to \mathbf{Set}$ be a weak pullback-preserving functor, and let $R \subseteq X \times Y$ be a binary relation between sets X and Y. We denote the left and right projections of R as $\pi_X\colon R \to X$ and $\pi_Y\colon R \to Y$, respectively. Let $\alpha \in \mathrm{T}\,X$ and $\beta \in \mathrm{T}\,Y$; we now define

$$\alpha\,\overline{\mathrm{T}}\,R\,\beta :\Leftrightarrow \exists\delta \in \mathrm{T}\,R,\ \mathrm{T}\,\pi_X(\delta) = \alpha \text{ and } \mathrm{T}\,\pi_Y(\delta) = \beta.$$

We call $\overline{\mathrm{T}}\,R$ the T-*lifting of* R.

Observe that $\overline{\mathrm{T}}\,R$ is simply the binary relation between $\mathrm{T}\,X$ and $\mathrm{T}\,Y$ induced by the span

$$\mathrm{T}\,X \xleftarrow{\mathrm{T}\,\pi_X} \mathrm{T}\,R \xrightarrow{\mathrm{T}\,\pi_Y} \mathrm{T}\,Y.$$

Example 1. Recall that the covariant powerset functor is an example of a weak pullback-preserving functor. Now for any binary relation $R \subseteq X \times Y$, the P-lifting of R is precisely the Egli-Milner lifting $R_{EM} \subseteq \mathrm{P}\,X \times \mathrm{P}\,Y$. In other words, if $\alpha \in \mathrm{P}\,X$ and $\beta \in \mathrm{P}\,Y$, then $\alpha\,\overline{\mathrm{P}}\,R\,\beta$ iff

$$\forall x \in \alpha,\ \exists y \in \beta \text{ s.t. } x\ R\ y \text{ and } \forall y \in \beta,\ \exists x \in X \text{ s.t. } x\ R\ y.$$

Recall that **Set** can be embedded in the category **Rel** of sets and binary relations, using the functor $\mathrm{Graph}\colon \mathbf{Set} \to \mathbf{Rel}$, defined as $\mathrm{Graph}\colon X \mapsto X$ and

$$\mathrm{Graph}\colon X \xrightarrow{f} Y \mapsto \{(x, f(x)) \mid x \in X\},$$

where we view the right-hand-side above as a binary relation between X and Y. The desired property that turns $\overline{\mathrm{T}}$ into a *lifting* is that it makes the following diagram commute:

$$\begin{array}{ccc} \mathbf{Rel} & \xrightarrow{\overline{\mathrm{T}}} & \mathbf{Rel} \\ {\scriptstyle \mathrm{Graph}}\uparrow & & \uparrow{\scriptstyle \mathrm{Graph}} \\ \mathbf{Set} & \xrightarrow[\mathrm{T}]{} & \mathbf{Set} \end{array}$$

The condition that the functor T preserves weak pullbacks is needed to ensure that $\overline{\mathrm{T}}$ is indeed a *functor*.

Fact 16. *Let* T: **Set** $\to$ **Set** *be a functor. Then* $\overline{\mathrm{T}}$ *is a lifting in the sense described above, that is:*

$$\mathrm{Graph} \circ \mathrm{T} = \overline{\mathrm{T}} \circ \mathrm{Graph}\,.$$

Moreover, $\overline{\mathrm{T}}$ *is a functor on* **Rel**, *the category of sets and binary relations, iff* T *preserves weak pullbacks.*

6.4.2.3 Syntax and Semantics of Moss' Coalgebraic Logic

We will now present the syntax and semantics of Moss' coalgebraic logic for an arbitrary weak pullback-preserving functor T: **Set** $\to$ **Set**. We will make additional assumptions about T. Firstly, we assume that T is *standard*; in the case that T preserves weak pullbacks we can take this to mean that T preserves *inclusions* (that is, if $\iota : X \hookrightarrow Y$ is an inclusion map, then $\mathrm{T}\iota : \mathrm{T}X \hookrightarrow \mathrm{T}Y$ is the inclusion map witnessing that $\mathrm{T}X$ is a subset of $\mathrm{T}Y$). This assumption is innocuous from the viewpoint of **Set**-coalgebras, because for any T: **Set** $\to$ **Set** there is a standard T': **Set** $\to$ **Set** such that the category of T-coalgebras is equivalent to the category of T'-coalgebras.

If we would leave it at this, only assuming T: **Set** $\to$ **Set** is standard and weak pullback-preserving, we could already define Moss' language, and indeed this is what he does in [30]. A downside of this approach is, however, that one might obtain formulas with *infinitely* many subformulas. This can be avoided by requiring that T satisfies the following condition for all sets X:

$$\mathrm{T}X = \bigcup\{\mathrm{T}X' \mid X' \subseteq X, X'\text{finite}\}. \tag{6.15}$$

We say T is *finitary* if it satisfies (6.15).

If the coalgebra functor T: **Set** $\to$ **Set** one happens to be interested in is not finitary, this can be remedied. For each set X, we can define

$$\mathrm{T}_\omega X := \bigcup\{\mathrm{T}X' \mid X' \subseteq X, X'\text{finite}\}.$$

Using the assumption that T is standard, this gives us a functor $\mathrm{T}_\omega\colon \mathbf{Set} \to \mathbf{Set}$. For the covariant powerset functor $\mathrm{P}\colon \mathbf{Set} \to \mathbf{Set}$, the above definition of T_ω yields precisely the finite powerset functor $\mathrm{P}_\omega\colon \mathbf{Set} \to \mathbf{Set}$.

From the general viewpoint of coalgebraic logic, one would want to consider both T and T_ω when understanding Moss' logic. Our current viewpoint, however, is focused on the Carioca derivation system, and there we only really need T_ω. To simplify our presentation and notation, we will therefore assume from here on that $\mathrm{T} = \mathrm{T}_\omega$, i.e., that T is finitary.

We will now define the finitary, Boolean version of Moss' coalgebraic language. Note that again for simplicity, we are working with the closed fragment.

Definition 9. Let $\mathrm{T}\colon \mathbf{Set} \to \mathbf{Set}$ be a finitary, standard, weak pullback-preserving functor. We define $\mathscr{L}_\mathrm{T}$, the *closed (0-variable) Moss language for* T, to be the smallest set such that (1) $\top, \bot \in \mathscr{L}_\mathrm{T}$, (2) if $\varphi \in \mathscr{L}_\mathrm{T}$ then also $\neg\varphi \in \mathscr{L}_\mathrm{T}$, (3) if $\varphi, \psi \in \mathscr{L}_\mathrm{T}$ then also $\varphi \wedge \psi \in \mathscr{L}_\mathrm{T}$ and $\varphi \vee \psi \in \mathscr{L}_\mathrm{T}$, and (4) if $\alpha \in \mathrm{T}\mathscr{L}_\mathrm{T}$, then $\nabla\alpha \in \mathscr{L}_\mathrm{T}$.

The *coalgebraic semantics of* $\mathscr{L}_\mathrm{T}$ is defined as follows. Suppose we have a T-coalgebra $\sigma : S \to \mathrm{T}\,S$; we will define a satisfaction relation $\Vdash_\sigma$ between S (the set of states of our coalgebra) and $\mathscr{L}_\mathrm{T}$. Let $x \in S$; then we inductively define

1. $x \Vdash_\sigma \top$ and $x \nVdash_\sigma \bot$;
2. For all $\varphi, \psi \in \mathscr{L}_\mathrm{T}$, $x \Vdash_\sigma \varphi \wedge \psi$ iff $x \Vdash_\sigma \varphi$ and $x \Vdash_\sigma \psi$ (and similarly for $\varphi \vee \psi$ and $\neg\varphi$);
3. For all $\alpha \in \mathrm{T}\left(\mathscr{L}_\mathrm{T}\right)$, $x \Vdash_\sigma \nabla\alpha$ iff $\sigma(x)\, \overline{\mathrm{T}}\,(\Vdash_\sigma)\, \alpha$.

Note that if we choose $\mathrm{T} = \mathrm{P}_\omega$, the finite powerset functor, then the semantics in Definition 9 gives us precisely the syntax and semantics for the nabla we saw above in (6.14), since $\mathrm{P}_\omega\left(\mathscr{L}_{\mathrm{P}_\omega}\right)$ is the collection of finite sets of $\mathscr{L}_{\mathrm{P}_\omega}$-formulas, and the P_ω-lifting of $\Vdash_\sigma$ is precisely the Egli-Milner lifting of $\Vdash_\sigma$.

Notes

A classic reference for the theory of coalgebras is Rutten [34]. For a recent overview to the area of coalgebraic logic, with pointers to introductory literature, we suggest [12] or [29]. The idea to use nabla as a primitive modality plays an important role in the work of both Barwise and Moss [7] and Janin and Walukiewicz [21]. The idea to use nabla as a coalgebraic modality is due to Moss [30]. For a more detailed discussion of the material in this subsection, including detailed proofs of the technical results, we refer to [28]. The observation in Fact 16, that $\overline{\mathrm{T}}$ is a functor on **Rel** iff T preserves weak pullbacks, goes back to Trnková [36].

6.4.3 The Carioca Derivation System

We will now introduce the Carioca derivation system. The aim of this derivation system is to enable us to derive exactly those *inequalities* of formulas in Moss' language that are valid on all T-coalgebras. In order to state the axioms and rules of the

Carioca system, we will first have to introduce two new concepts: lifted conjunctions and disjunctions, and slim redistributions.

The inequalities we are considering are those of the form $\varphi \preccurlyeq \psi$, for $\varphi, \psi \in \mathcal{L}_{\mathrm{T}}$. We say $\varphi \preccurlyeq \psi$ is valid on a coalgebra $\sigma : S \to \mathrm{T}\,S$ if for all $x \in S$ such that $x \Vdash_\sigma \varphi$, it is also the case that $x \Vdash_\sigma \psi$. If $\varphi \preccurlyeq \psi$ is valid on *all* T-coalgebras, we write $\varphi \Vdash_{\mathrm{T}} \psi$.

When writing the Carioca axioms, we think of the formation of disjunctions and conjunctions as *functions* from $\mathrm{P}_\omega \mathcal{L}_{\mathrm{T}}$ to $\mathcal{L}_{\mathrm{T}}$, i.e., one can consider the maps $\bigvee : \mathrm{P}_\omega \mathcal{L}_{\mathrm{T}} \to \mathcal{L}_{\mathrm{T}}$ and $\bigwedge : \mathrm{P}_\omega \mathcal{L}_{\mathrm{T}} \to \mathcal{L}_{\mathrm{T}}$ as maps in **Set**. Consequently, we can also apply T to $\bigvee$ and $\bigwedge$, which gives us maps

$$\mathrm{T}\bigvee : \mathrm{T}\,\mathrm{P}_\omega \mathcal{L}_{\mathrm{T}} \to \mathrm{T}\,\mathcal{L}_{\mathrm{T}} \text{ and } \mathrm{T}\bigwedge : \mathrm{T}\,\mathrm{P}_\omega \mathcal{L}_{\mathrm{T}} \to \mathrm{T}\,\mathcal{L}_{\mathrm{T}}.$$

If $\Phi \in \mathrm{T}\,\mathrm{P}_\omega \mathcal{L}_{\mathrm{T}}$, we call $\mathrm{T}\bigvee(\Phi)$ and $\mathrm{T}\bigwedge(\Phi)$ a T-*lifted* disjunction and conjunction, respectively.

Example 2. If $\mathrm{T} = \mathrm{P}_\omega$ and we apply the P_ω-lifted disjunction operation $\mathrm{P}_\omega \bigvee$ to an element $\Phi = \{S_1, \ldots, S_n\} \in \mathrm{P}_\omega\,\mathrm{P}_\omega(\mathcal{L}_{\mathrm{P}_\omega})$, we obtain a forward image:

$$\mathrm{P}_\omega \bigvee\big(\{S_1, \ldots, S_n\}\big) = \{\bigvee S_1, \ldots, \bigvee S_n\}.$$

The final concept we will now introduce is that of *slim redistributions*.

Definition 10. Let T: **Set** $\to$ **Set** be a finitary, standard, weak pullback-preserving functor and let X be a set. If $\alpha \in \mathrm{T}\,X$, then we define the *base* of α to be the following intersection:

$$\mathrm{Base}(\alpha) := \bigcap\big\{X' \subseteq X \mid \alpha \in \mathrm{T}\,X'\big\}.$$

Now if $C \in \mathrm{P}_\omega\,\mathrm{T}\,X$ is a finite collection of elements of $\mathrm{T}\,X$, then we define a *slim redistribution of* C to be an element Ψ such that

$$\Psi \in \mathrm{T}\,\mathrm{P}_\omega\left(\textstyle\bigcup_{\alpha\in C} \mathrm{Base}(\alpha)\right) \text{ and for all } \alpha \in C,\ \alpha\,\overline{\mathrm{T}}{\in}\,\Psi.$$

We denote the set of all slim redistributions of C by $\mathrm{SRD}_{\mathrm{T}}(C)$.

Intuitively, the idea is that $\mathrm{Base}(\alpha)$ is the smallest set $X' \subseteq X$ such that $\alpha \in \mathrm{T}\,X'$.

Fact 17. *Let* T: **Set** $\to$ **Set** *be a finitary, standard, weak pullback-preserving functor and let X be a set. Then for all $\alpha \in \mathrm{T}\,X$ and all sets Y it holds that $\alpha \in \mathrm{T}\,Y$ iff* $\mathrm{Base}(\alpha) \subseteq Y$. *In fact,* Base *is a natural transformation from* T *to* P_ω.

Example 3. In the case that $\mathrm{T} = \mathrm{P}_\omega$, Definition 10 can be simplified as follows. Firstly, if $\mathrm{T} = \mathrm{P}_\omega$ and $\alpha \in \mathrm{P}_\omega\,X$ is simply a finite subset of X, then the smallest subset $X' \subseteq X$ such that $\alpha \in \mathrm{T}\,X'$ is α itself; in other words, Base is the identity if $\mathrm{T} = \mathrm{P}_\omega$.

Secondly, if $C \in \mathrm{P}_\omega\,\mathrm{P}_\omega\,X$ is a finite collection of finite subsets of X, then

$$\mathrm{SRD}_{\mathrm{P}_\omega}(C) = \left\{\Psi \in \mathrm{P}_\omega \mathrm{P}_\omega(\bigcup C) \mid \forall \alpha \in C,\ \alpha\, \overline{\mathrm{P}}{\in}\, \Psi\right\},$$

where $\overline{\mathrm{P}}{\in}$ is the Egli-Milner lifting of the element relation, viewed as a binary relation $\in \subseteq X \times \mathrm{P}_\omega X$. It is now not hard to see that if $\Psi = \{S_1, \ldots, S_m\}$ then

$$\alpha\, \overline{\mathrm{P}}{\in}\, \{S_1, \ldots, S_m\} \text{ iff } \alpha \subseteq \bigcup_{i=1}^m S_i \text{ and } \forall i \leq m, \alpha \cap S_i \neq \emptyset.$$

Thus, for $\mathrm{T} = \mathrm{P}_\omega$, we see that Ψ is a slim redistribution of $C \in \mathrm{P}_\omega \mathrm{P}_\omega X$ iff

$$\Psi \in \mathrm{P}_\omega \mathrm{P}_\omega X \text{ such that } \bigcup C = \bigcup \Psi \text{ and } \forall \alpha \in C,\ \forall S \in \Psi,\ \alpha \cap S \neq \emptyset. \qquad (6.16)$$

We are now ready to define the Carioca derivation system. For the sake of simplicity, we will present a simplified version in which all disjunctions and conjunctions are finite. The simplification we use to achieve this, is to assume that T maps finite sets to finite sets.

Definition 11. Let $\mathrm{T}\colon \mathbf{Set} \to \mathbf{Set}$ be a standard, finitary, weak pullback-preserving functor. Additionally, we assume that $\mathrm{T}X$ is finite whenever X is finite. The *Carioca derivation system* $\vdash_{\mathrm{T}}$ consists of a complete set of axioms and rules for all Boolean inequalities, combined with the following rule and axioms:

$$\begin{array}{lll}
(\nabla 1) & \alpha\, \overline{\mathrm{T}}{\preccurlyeq}\, \beta \Rightarrow \vdash_{\mathrm{T}} \nabla\alpha \preccurlyeq \nabla\beta & (\alpha, \beta \in \mathrm{T}\mathscr{L}_{\mathrm{T}}); \\
(\nabla 2) & \vdash_{\mathrm{T}} \bigwedge_{\alpha \in C} \nabla\alpha \preccurlyeq \bigvee \left\{\nabla \mathrm{T} \bigwedge(\Psi) \mid \Psi \in \mathrm{SRD}_{\mathrm{T}}(C)\right\} & (C \in \mathrm{P}_\omega \mathrm{T}\mathscr{L}_{\mathrm{T}}); \\
(\nabla 3.f) & \vdash_{\mathrm{T}} \nabla \mathrm{T} \bigvee(\Phi) \preccurlyeq \bigvee \left\{\nabla\beta \mid \beta\, \overline{\mathrm{T}}{\in}\, \Phi\right\} & (\Phi \in \mathrm{T}\mathrm{P}_\omega \mathscr{L}_{\mathrm{T}}).
\end{array}$$

(Note that it is provable in $\vdash_{\mathrm{T}}$ that $(\nabla 2)$ and $(\nabla 3)$ are in fact equations rather than inequalities.)

Example 4. We will make the rule and axioms above more concrete for the case $\mathrm{T} = \mathrm{P}_\omega$.

Starting with $(\nabla 1)$, suppose that $\alpha, \beta \in \mathrm{P}_\omega \mathscr{L}_{\mathrm{P}_\omega}$ are finite sets of formulas. Rule $(\nabla 1)$ says that in case that $\alpha\, \overline{\mathrm{P}}{\preccurlyeq}\, \beta$, i.e., in case that $\alpha \preccurlyeq_{EM} \beta$, then $\vdash_{\mathrm{P}_\omega} \nabla\alpha \preccurlyeq \nabla\beta$. In other words, if

$$\forall \varphi \in \alpha,\ \exists \psi \in \beta \text{ s.t. } \vdash_{\mathrm{P}_\omega} \varphi \preccurlyeq \psi \text{ and } \forall \psi \in \beta,\ \exists \varphi \in \alpha \text{ s.t. } \vdash_{\mathrm{P}_\omega} \varphi \preccurlyeq \psi, \qquad (6.17)$$

then $\vdash_{\mathrm{P}_\omega} \nabla\alpha \preccurlyeq \nabla\beta$. Intuitively, this means that we can derive that $\nabla\beta$ is $\preccurlyeq$-related to (follows from) $\nabla\alpha$, provided that the elements of β are $\preccurlyeq$-related to those of α in an "Egli-Milner" way.

Moving on to $(\nabla 2)$, we see that if $C \in \mathrm{P}_\omega \mathrm{P}_\omega \mathscr{L}_{\mathrm{P}_\omega}$ is a finite collection of finite sets of formulas, then

$$\vdash_{\mathrm{P}_\omega} \bigwedge_{\alpha \in C} \nabla\alpha \preccurlyeq \bigvee \left\{\nabla\{\bigwedge S_1, \ldots, \bigwedge S_n\} \mid \{S_1, \ldots, S_n\} \in \mathrm{SRD}_{\mathrm{P}_\omega}(C)\right\}, \qquad (6.18)$$

where we refer the reader to (6.16) for a description of $\text{SRD}_{\text{P}_\omega}(C)$. Intuitively, this means that any conjunction of ∇-formulas is equivalent to a disjunction of nablas of lifted conjunctions.

Finally, $(\nabla 3.f)$ can be simplified as follows. Suppose we have $\Phi = \{S_1, \ldots, S_n\} \in \text{P}_\omega \text{P}_\omega(\mathscr{L}_{\text{P}_\omega})$; then $(\nabla 3.f)$ boils down to the axiom

$$\vdash_{\text{P}_\omega} \nabla\{\bigvee S_1, \ldots, \bigvee S_n\} \preccurlyeq \bigvee \left\{\nabla\beta \mid \beta \subseteq \textstyle\bigcup_{i=1}^n S_i \text{ and } \forall i \leq n,\ \beta \cap S_i \neq \emptyset\right\}. \tag{6.19}$$

As a further simplification, one could also write the following:

$$\vdash_{\text{P}_\omega} \nabla(\alpha \cup \{\bigvee S\}) \preccurlyeq \bigvee \left\{\nabla(\alpha \cup \beta) \mid \beta \subseteq S \text{ and } \beta \neq \emptyset\right\}. \tag{6.20}$$

One can inductively derive (6.19) from (6.20). Regardless of how we look at $(\nabla 3.f)$, the intuitive content of this axiom is that finite disjunctions "under" nablas distribute to disjunctions of nablas.

Fact 18. *Let* T: **Set** → **Set** *be a standard, finitary, weak pullback-preserving functor, with the added property that* T X *is finite whenever* X *is finite.*

The Carioca derivation system for T *is sound and complete with respect to* T*-validity: for all* $\varphi, \psi \in \mathscr{L}_\text{T}$,

$$\vdash_\text{T} \varphi \preccurlyeq \psi \textit{ iff } \varphi \Vdash_\text{T} \psi.$$

Notes

A first axiomatization of the nabla modality (in the power set case) was given by Palmigiano and Venema [32]; this calculus was streamlined by Bílková et al. [11] into a formulation admitting a generalization to the arbitrary case in the Carioca system. (The name 'Carioca' refers to the fact that this version of the axiomatization was formulated in Rio de Janeiro.) Fact 18, the completeness of the Carioca system, was proved by Kupke et al. [27, 28]; the latter work also contains a discussion (with proof) of Fact 17.

6.4.4 The T-Powerlocale

Having acquainted ourselves with Moss' coalgebraic logic and the Carioca derivation system, we now introduce the T-powerlocale construction. This is a generalization of the Vietoris powerlocale construction, using techniques from coalgebraic logic. Because the Carioca axioms are parametric in their coalgebra type functor T, so is the T-powerlocale construction. We will see that certain properties of the Vietoris functor can be proved at the more general level of the T-powerlocale, and as a corollary, we show how the Vietoris powerlocale can be presented using nablas as generators, rather than boxes and diamonds. Recall that locales have finite meets and arbitrary joins; in a locale $\mathbb{A}$ we represent these maps as $\bigwedge : \mathrm{P}_\omega A \to A$ and $\bigvee : \mathrm{P} A \to A$, respectively.

Definition 12. Let T: **Set** $\to$ **Set** be a standard, finitary, weak pullback-preserving functor and let $\mathbb{A}$ be a locale with an underlying set of opens A. We define $\mathrm{V}_\mathrm{T}\,\mathbb{A}$, the T-*powerlocale of* $\mathbb{A}$, to be the locale generated by the set $\{\nabla\alpha \mid \alpha \in \mathrm{T}\,A\}$, subject to the following relations:

$$(\nabla 1)\quad \nabla\alpha \leq \nabla\beta \text{ if } \alpha\,\overline{\mathrm{T}}{\leq}\,\beta \qquad (\alpha, \beta \in \mathrm{T}\,A);$$

$$(\nabla 2)\quad \textstyle\bigwedge_{\alpha\in C}\nabla\alpha \leq \bigvee\left\{\nabla\mathrm{T}\bigwedge(\Psi) \mid \Psi \in \mathrm{SRD}_\mathrm{T}(C)\right\} \qquad (C \in \mathrm{P}_\omega\,\mathrm{T}\,A);$$

$$(\nabla 3)\quad \nabla\mathrm{T}\bigvee(\Phi) \leq \bigvee\left\{\nabla\beta \mid \beta\,\overline{\mathrm{T}}{\in}\,\Phi\right\} \qquad (\Phi \in \mathrm{T}\,\mathrm{P}\,A).$$

Note that the only real difference between the Carioca axioms in Definition 11 and the relations in Definition 12 above is the difference between $(\nabla 3.f)$ and $(\nabla 3)$. We will later see how this corresponds to the difference between finite disjunctions (as found in Boolean algebras and distributive lattices) and infinite disjunctions (as found in locales).

Fact 19. *The construction described in Definition 12 defines a functor* V_T: **Loc** $\to$ **Loc**.

The functor V_T we have just introduced has several additional properties which can be proved at an abstract level. As an example, note the following fact.

Fact 20. *Let* T: **Set** $\to$ **Set** *be a standard, finitary, weak pullback-preserving functor.*

1. *The functor* V_T: **Loc** $\to$ **Loc** *preserves regularity.*
2. *If we further assume that* $\mathrm{T}\,X$ *is finite for every finite set* X*, then* V_T *preserves the combination of compactness and zero-dimensionality.*

In the introduction to Sect. 6.4, we motivated our discussion of nabla expressions and Moss' coalgebraic logic with the question of whether we could describe the Vietoris powerlocale using nabla. The following fact asserts that the Carioca axioms indeed allow us to do this.

Fact 21. *The* P_ω*-powerlocale is the Vietoris powerlocale.*

Proof Sketch Suppose that $\mathbb{A}$ is a locale; we must show that $\mathrm{V}\,\mathbb{A} \simeq \mathrm{V}_{\mathrm{P}_\omega}\,\mathbb{A}$. This is achieved by defining frame morphisms in both directions, and showing that these morphisms are mutually inverse. From $\mathrm{V}_{\mathrm{P}_\omega}\,\mathbb{A}$ to $\mathrm{V}\,\mathbb{A}$, we send

$$\nabla\alpha \mapsto \Box\left(\bigvee\alpha\right) \wedge \textstyle\bigwedge_{a\in\alpha}\Diamond a.$$

From $\mathrm{V}\,\mathbb{A}$ to $\mathrm{V}_{\mathrm{P}_\omega}\,\mathbb{A}$, we use the following assignments:

$$\Diamond a \mapsto \nabla\{a, 1\} \text{ and } \Box a \mapsto \nabla\{a\} \vee \nabla\emptyset.$$

For details about the rest of the proof, we refer the reader to [32] or [37]. □

Below in Fact 23 we will look in more detail at nabla presentations of the Vietoris powerlocale. Before we do so, however, we will introduce an alternative presentation of V_T, in which we exploit the fact that in the language of locales we can use *infinite* disjunctions.

Fact 22. *The relation* (∇2) *in Definition 12 can equivalently be replaced by the following pair of relations:*

$$(\nabla 2.0) \quad 1 \leq \bigvee \{\nabla\alpha \mid \alpha \in \mathrm{T}A\};$$
$$(\nabla 2.2) \quad \nabla\alpha \wedge \nabla\beta \leq \bigvee \{\nabla\gamma \mid \gamma \;\overline{\mathrm{T}}{\leq}\; \alpha \text{ and } \gamma \;\overline{\mathrm{T}}{\leq}\; \beta\}.$$

Note that the suffixes ".0" and ".2" indicate nullary and binary conjunctions, respectively. Returning to the case $\mathrm{T} = \mathrm{P}_\omega$, we will now give a concrete nabla-presentation of V.

Fact 23. *Let* $\mathbb{A}$ *be a locale. We can present* $\mathrm{V}\,\mathbb{A}$*, the Vietoris powerlocale of* $\mathbb{A}$*, as the locale generated by the set* $\{\nabla\alpha \mid \alpha \in \mathrm{P}_\omega A\}$*, subject to the following relations:*

$$(\nabla 1) \quad \nabla\alpha \leq \nabla\beta \quad (\text{if } \alpha \leq_{EM} \beta);$$
$$(\nabla 2) \quad \bigwedge_{\alpha \in C} \nabla\alpha \leq \bigvee \{\nabla\{\bigwedge S_1, \ldots, \bigwedge S_n\} \mid \{S_1, \ldots, S_n\} \in \mathrm{SRD}_{\mathrm{P}_\omega}(C)\},$$

where C ranges over the finite subsets of $\mathrm{P}_\omega A$*, also see* (6.16)*; and*

$$(\nabla 3) \quad \nabla\{\bigvee S_1, \ldots, \bigvee S_n\} \leq \bigvee \{\nabla\beta \mid \beta \subseteq \bigcup_{i \leq n} S_i \text{ and } \forall i \leq n,\ \beta \cap S_i \neq \emptyset\},$$

where the S_i *range over (possibly infinite) subsets of A. Moreover, the* (∇2) *relation can be replaced by the following pair of relations:*

$$(\nabla 2.0) \qquad 1 \leq \bigvee\{\nabla\alpha \mid \alpha \in \mathrm{P}_\omega A\};$$
$$(\nabla 2.2) \qquad \nabla\alpha \wedge \nabla\beta \leq \bigvee\{\nabla\gamma \mid \gamma \leq_{EM} \alpha \text{ and } \gamma \leq_{EM} \beta\},$$

and the (∇3) *relation can be replaced by the following inductive version:*

$$(\nabla 3.ind) \quad \nabla(\alpha \cup \{\bigvee S\}) \leq \bigvee \{\nabla(\alpha \cup \beta) \mid \beta \in \mathrm{P}_\omega S \text{ and } \beta \neq \emptyset\} \quad (S \in \mathrm{P}A).$$

We have now seen how to present the Vietoris powerlocale, and more generally the T-powerlocale, using nablas. We can improve on this still, by showing that "every element of $\mathrm{V}_\mathrm{T}\,\mathbb{A}$ is a disjunction of nablas" in a rather strong sense. Recall that a *suplattice* is a complete join-semilattice, and that any locale has an underlying suplattice.

Definition 13. Let $\mathrm{T}\colon \mathbf{Set} \to \mathbf{Set}$ be a standard, finitary, weak pullback-preserving functor and let $\mathbb{L}$ be a suplattice. We define $\mathrm{W_T}\,\mathbb{L}$, the T-*powerlattice of* $\mathbb{L}$, to be the suplattice generated by the $\{\nabla\alpha \mid \alpha \in \mathrm{T}\,L\}$, subject to the following relations:

$$(\nabla 1)\quad \nabla\alpha \leq \nabla\beta \text{ if } \alpha\,\overline{\mathrm{T}}{\leq}\,\beta \qquad\qquad (\alpha, \beta \in \mathrm{T}\,L);$$

$$(\nabla 3)\quad \nabla\,\mathrm{T}\bigvee(\Phi) \leq \bigvee\{\nabla\beta \mid \beta\,\overline{\mathrm{T}}{\in}\,\Phi\} \qquad\qquad (\Phi \in \mathrm{T\,P}\,L).$$

If we now let U denote the (contravariant) forgetful functor from **Loc** to **SupLat**, the category of suplattices and suplattice morphisms, we can draw the following picture:

$$\begin{array}{ccc} \mathbf{Loc} & \xrightarrow{\mathrm{V_T}} & \mathbf{Loc} \\ {\scriptstyle \mathrm{U}}\downarrow & & \downarrow{\scriptstyle \mathrm{U}} \\ \mathbf{SupLat} & \xrightarrow[\mathrm{W_T}]{} & \mathbf{SupLat} \end{array}$$

We would like to emphasize that the following result, like Facts 19, 20 and 21, holds not only for the Vietoris powerlocale but for the T-powerlocale in general.

Fact 24. *Let* $\mathrm{T}\colon \mathbf{Set} \to \mathbf{Set}$ *be a standard, finitary, weak pullback-preserving functor. Then there exists a natural transformation such that for all locales* $\mathbb{A}$, $\mathrm{U}(\mathrm{V_T}\,\mathbb{A}) \simeq \mathrm{W_T}(\mathrm{U}\,\mathbb{A})$.

The proof of Fact 24 uses flat sites, a technique from formal topology [13], which is meant to capture the notion of a *basis* of a topological space. From a logical viewpoint, Fact 24 tells us that (1) any $(\wedge, \bigvee)$-formula in Moss' coalgebraic language for T is $((\nabla 1), (\nabla 2), (\nabla 3))$-equivalent to a $\bigvee$-formula, and that (2) for any inequality between $\bigvee$-formulas derived using $((\nabla 1), (\nabla 2), (\nabla 3))$, there is a $((\nabla 1), (\nabla 3))$-derivation which proves that inequality.

Notes

The T-powerlocale was introduced by Venema et al. [37]; this is also where one can find the above results. (An early version can be found in [42, Ch. 5].) For more information on the method of using sup-lattices to obtain results like our Fact 24 the reader is referred to [24].

Acknowledgments We would like to thank Liang-Ting Chen and Steve Vickers for providing helpful pointers to the literature. Additionally, we would like to thank the anonymous referee, whose thorough criticism contributed significantly to the quality of this chapter, and in particular to the presentation and choice of contents in Sect. 6.3.2.3. Finally, we are grateful to Guram Bezhanishvili, who provided many suggestions for improving the presentation of the chapter.

References

1. Abramsky S (1991) A domain equation for bisimulation. Inform Comput 92(2):161–218
2. Abramsky S (1991) Domain theory in logical form. Ann Pure Appl Logic, 51(1–2):1–77. Second annual IEEE symposium on logic in computer science (Ithaca, NY, 1987)
3. Abramsky S (2005) A Cook's tour of the finitary non-well-founded sets. In: Artemov S, Barringer H, d'Avila Garcez A, Lamb L, Woods J (eds) We will show them: essays in honour of Dov Gabbay, vol 1. College Publications, pp 1–18
4. Abramsky S, Jung A (1994) Domain theory. In: Handbook of logic in computer science, vol 3. Clarendon Press, Oxford, pp 1–168
5. Aczel P (1988) Non-well-founded sets. CSLI lecture notes, vol 14. Stanford University Center for the Study of Language and Information, Stanford, CA
6. Banaschewksi B, Mulvey C (1980) Stone-Čech compactification of locales. I. Houston J Math 6:301–312
7. Barwise J, Moss L (1996) Vicious circles. CSLI lecture notes, vol 60. CSLI Publications, Stanford, CA
8. Bezhanishvili G, Bezhanishvili N, Gabelaia D, Kurz A (2010) Bitopological duality for distributive lattices and Heyting algebras. Math Struct Comput Sci 20(3):359–393
9. Bezhanishvili G, Bezhanishvili N, Harding J (2012) Modal compact Hausdorff spaces. J Logic Comput. doi:10.1093/logcom/exs030
10. Bezhanishvili N, Kurz A (2007) Free modal algebras: a coalgebraic perspective. In: Mossakowski T, Montanari U, Haveraaen M (eds) Algebra and coalgebra in computer science. Lecture notes in computer science, vol 4624. Springer, Berlin, pp 143–157
11. Bílková M, Palmigiano A, Venema Y (2008) Proof systems for the coalgebraic cover modality. In: Areces C, Goldblatt R (eds) Advances in modal logic 7, Nancy 2008. College Publications, pp 1–21
12. Cîrstea C, Kurz A, Pattinson D, Schröder L, Venema Y (2011) Modal logics are coalgebraic. Comput J 54:524–538
13. Coquand T, Sambin G, Smith J, Valentini S (2003) Inductively generated formal topologies. Ann Pure Appl Logic 124(1–3):71–106
14. Cornish W (1975) On H. Priestley's dual of the category of bounded distributive lattices. Mat Vesnik 12(27):329–332
15. Davey B, Galati J (2003) A coalgebraic view of Heyting duality. Studia Logica 75(3):259–270
16. Esakia L (1974) Topological Kripke models. Dokl Akad Nauk SSSR 214:298–301
17. Fine K (1975) Normal forms in modal logic. Notre Dame J Formal Logic 16(2):229–237
18. Gierz G, Hofmann K, Keimel K, Lawson J, Mislove M, Scott D (2003) Continuous lattices and domains. Cambridge University Press, Cambridge
19. Hochster M (1969) Prime ideal structure in commutative rings. Trans Am Math Soc 142:43–60
20. Isbell J (1972) Atomless parts of spaces. Math Scand 31:5–32
21. Janin D, Walukiewicz I (1995) Automata for the modal μ-calculus and related results. In: Mathematical foundations of computer science 1995 (Prague). Lecture notes in computer science, vol 969. Springer, Berlin, pp 552–562
22. Johnstone P (1982) Stone spaces. Cambridge studies in advanced mathematics, vol 3. Cambridge University Press, Cambridge
23. Johnstone P (1985) Vietoris locales and localic semilattices. In: Continuous lattices and their applications (Bremen, 1982). Lecture notes in pure and applied mathematics, vol 101. Dekker, New York, pp 155–180
24. Jung A, Moshier M, Vickers S (2008) Presenting dcpos and dcpo algebras. In: Proceedings of the 24th conference on the mathematical foundations of programming semantics (MFPS XXIV). Electronic notes in theoretical computer science, vol 218, pp 209–229
25. Kegelmann M (2002) Continuous domains in logical form. Electron Notes Theor Comput Sci 49:1–166
26. Kupke C, Kurz A, Venema Y (2004) Stone coalgebras. Theor Comput Sci 327(1–2):109–134

27. Kupke C, Kurz A, Venema Y (2008) Completeness of the finitary Moss logic. In: Areces C, Goldblatt R (eds) Advances in modal logic 7, Nancy 2008. College Publications, pp 193–217
28. Kupke C, Kurz K, Venema Y (2012) Completeness for the coalgebraic cover modality. Logical Methods Comput Sci 8(3)
29. Kupke C, Pattinson D (2011) Coalgebraic semantics of modal logics: an overview. Theor Comput Sci 412:5070–5094
30. Moss L (1999) Coalgebraic logic. Ann Pure Appl Logic 96:277–317 (Erratum published Ann Pure Appl Logic 99:241–259, 1999)
31. Palmigiano A (2004) A coalgebraic view on positive modal logic. Theor Comput Sci 327(1–2):175–195
32. Palmigiano A, Venema Y (2007) Nabla algebras and Chu spaces. In: Mossakowski T, Montanari U, Haveraaen M (eds) Algebra and coalgebra in computer science. Second international conference, CALCO 2007, Bergen, Norway, 20–24 Aug 2007. LNCS, vol 4624. Springer, pp 394–408
33. Robinson E (1986) Power-domains, modalities and the Vietoris monad. Technical Report UCAM-CL-TR-98, University of Cambridge, Computer Laboratory, Oct 1986
34. Rutten J (2000) Universal coalgebra: a theory of systems. Theor Comput Sci 249(1):3–80. Modern algebra and its applications (Nashville, TN, 1996)
35. Simmons H (2004) The Vietoris modifications of a frame. Unpublished manuscript, 79 pp, available online at http://www.cs.man.ac.uk/~hsimmons/
36. Trnková V (1980) General theory of relational automata. Fundamenta Informaticae 3(2):189–234
37. Venema Y, Vickers S, Vosmaer J (2013) Generalized powerlocales via relation lifting. Math Struct Comput Sci 23:142–199
38. Vickers S (1989) Topology via logic. Cambridge tracts in theoretical computer science, vol 5. Cambridge University Press, Cambridge
39. Vickers S (1997) Constructive points of powerlocales. Math Proc Cambridge Philos Soc 122:207–222
40. Vickers S (1993) Information systems for continuous posets. Theor Comput Sci 114(2):201–229
41. Vietoris L (1922) Bereiche zweiter Ordnung. Monatsh Math Phys 32(1):258–280
42. Vosmaer J (2010) Logic, algebra and topology. Investigations into canonical extensions, duality theory and point-free topology. PhD thesis, Institute for Logic, Language and Computation, Dec 2010 (Promotores: M. Gehrke and Y. Venema)

Chapter 7
Logic KM: A Biography

Alexei Muravitsky

To the memory of my esteemed friend, Leo Esakia

Abstract This chapter is an attempt to collect under one roof all currently available facts related to logic **KM**. Discovered as an equational class of the corresponding algebras, it has been developed as a natural intuitionistic counterpart of provability logic **GL**. We also outline the background, the work of thought, which had preceded and eventually had led to the birth of **KM**. Where the results are new, the proofs are provided. Sometimes we derive conclusions, if they can be easily obtained from key results.

Keywords Intuitionistic logic · Provability logic · Logic KM · KM-algebra · Lattice of normal extensions of a propositional calculus

7.1 Introduction

Logic **KM** belongs to the numerous group of propositional modal logics on intuitionistic base.[1] Born in the late 1970s, at first **KM** was hardly noticeable among the members of the group. She was not among the first logics introduced into consideration, nor was she among the most attractive ones in the group. This should not be surprising, for the pioneers of the field, on the one hand, were taking into account axiomatic similarities with classical counterparts and, on the other, were investigating plausible correlations between two relations in a Kripke-style semantics—one that governs intuitionistic implication and the other that governs modality. If for the former researchers **KM** might not have been attractive, or even striking, because of

[1] See the following comprehensive surveys [47, 71, 77, 80], as well as [82].

A. Muravitsky (✉)
Louisiana Scholars' College, Northwestern State University, Natchitoches, LA 71457, USA
e-mail: alexeim@nsula.edu

G. Bezhanishvili (ed.), *Leo Esakia on Duality in Modal and Intuitionistic Logics*, Outstanding Contributions to Logic 4, DOI: 10.1007/978-94-017-8860-1_7,

the axiom $A \to \Box A$, for the latter group the Kripke semantics of **KM** (see Sect. 7.4.2) might have seemed too simple, or, to put it better, not expressive enough.[2]

The above axiom by itself does not cause a problem, for $\Box A$ could be renamed as $\Diamond A$. However, this move hardly leads to a rescue, since another axiom of **KM** is $(\Box A \to A) \to A$. The reader will find a full axiomatization of **KM** in Sect. 7.3, as well as in Sect. 7.4.1.

Thus it is clear that these formulae in one axiomatic system must be well motivated. For better or for worse, **KM** has not had a good philosophical motivation as a system of propositional monomodal logic, where modality meets requirements of the intuitionistic standpoint. (A brief discussion of this issue is deferred to Sect. 7.5, when enough information about **KM** will be available to the reader.) However, the very birth of **KM** was not due to good or bad "philosophical life" (expression of Harvey Friedman) in the field, but due to a close relationship between intuitionistic propositional logic and provability logic, which was discovered in the late 1970s.

My late teacher Alexander Kuznetsov used to say that a logic is more fully understood if it is considered along with its extensions. Continuing this thought, we can say that the relationship between two logics is more fully understood if the relationship between the families of their extensions is understood. It turned out that the lattices of the extensions of **KM** and of those of provability logic are isomorphic. This isomorphism and other mappings involving other lattices are discussed in Sect. 7.4.8.

In addition to the interconnection between **KM** and provability logic, there has been found an interconnection between **KM** and intuitionistic logic known as *Kuznetsov's Theorem*. The latter is just an extension of the conservativity of **KM** over intuitionistic logic. This conservativity can be obtained in a relatively simple way. Kuznetsov's Theorem prolongs it to any modal-free extension of **KM** and the corresponding intermediate logic. On the other hand, Kuznetsov's Theorem has an interesting interpretation in terms of enrichable Heyting algebras. This theorem also establishes an interesting mapping of the lattice of extensions of **KM** onto the lattice of intermediate logics, thereby it had been shown that each intermediate logic can be regarded as the superintuitionistic fragment of some extension of provability logic. All this will be explained in Sect. 7.4.6.

Thus, even if not motivated well philosophically, **KM** has a strong mathematical support. In this light, its relatively simple Kripke semantics is a plus. Other properties of **KM** such as the separation property (Sect. 7.4.1), the finite model property (Sect. 7.4.2), the disjunction property (Sect. 7.4.3), the fixed point property (Sect. 7.4.4), in its stronger form (existence and strong uniqueness), and the Craig interpolation property (Sect. 7.4.7) will be also presented. Section 7.4.5 is devoted to topological semantics of **KM**. Here we have to note the main contribution of Leo Esakia in discovering that scattered topological spaces are an adequate semantics for provability logic. Then, this semantics was easily transmitted to **KM** due to a close

---

[2] Although unrelated to the subject matter, which will be discussed below, the axiom $A \to \Box A$ had been used by Gödel in his ontological proof; see [32], pp. 431 and 435.

relationship between Magari algebras, an algebraic semantics for provability logic, and **KM**-algebras, which is shown in Sect. 7.3.

In conclusion, we want to briefly discuss **KM**'s nearest neighbors. If we ignore for now the only common consistent extension of **KM** and provability logic, which will appear in Sect. 7.3, there remains only one candidate that should not be missed—the system **mHC** defined by Leo Esakia in [26]. Since **mHC** can be obtained from the axioms of **KM** simply by replacing the second formula mentioned above with $\Box(A \to B) \to (\Box A \to \Box B)$, which is a theorem of **KM**, the latter is clearly a proper extension of **mHC**. (It should be noted that Extensionality Conditional, $(A \to B) \to (\Box A \to \Box B)$, is derivable in **KM** but not in **mHC**.) Esakia has observed ([26], Proposition 3) that, among the extensions of **mHC**, only the extensions of **KM** have the fixed point property (existence). Also, he established ([26], Corollary III.14) that **mHC** is embedded fully and faithfully into modal logic **K4.Grz**. This property is somewhat weaker in comparison to an analogous embedding of **KM** and its extensions, which will be discussed briefly in Sect. 7.4.8. More information about the interconnection of **mHC** with other logics can be found in Sect. 7.4.6 and especially in another chapter of this book, [47, Sect. 4].

Besides the topics listed above, we will outline the history of the thought work that has led to the birth of **KM** (Sect. 7.3).

7.2 Preliminaries

In what follows, we present an exposition of relations radiating from modal logic **KM**. Although an interesting star in our galaxy, **KM** is not located in its focus, two of which are occupied by intuitionistic propositional logic **Int** and provability logic **GL**.[3]

We set off with defining a propositional *modal language* $\mathscr{L}_\Box$ based on a denumerable set of propositional variables, **Var**, denoting them by $p, q, \ldots$ (with or without indices) and connectives $\wedge$ (conjunction), $\vee$ (disjunction), $\neg$ (negation), $\to$ (conditional) and $\Box$ (necessity). Well-formed formulae, or simply formulae, in $\mathscr{L}_\Box$ will be denoted by letters $A, B, \ldots$ (with or without indices). Omitting modality $\Box$ from $\mathscr{L}_\Box$, we obtain *assertoric language* $\mathscr{L}_a$. Formulae of $\mathscr{L}_a$ will be denoted by letters $a, b, \ldots$. Also, we will need another modal language, $\mathscr{L}_\bigcirc$, which differs from $\mathscr{L}_\Box$ in only one respect—modality $\bigcirc$ of $\mathscr{L}_\bigcirc$ replaces modality $\Box$ of $\mathscr{L}_\Box$. Finally, we will use language $\mathscr{L}_{\Box\bigcirc}$ which is the extension of $\mathscr{L}_\Box$ with modality $\bigcirc$. Arbitrary formulae of $\mathscr{L}_\bigcirc$ and $\mathscr{L}_{\Box\bigcirc}$ will be denoted by $\alpha, \beta, \ldots$. Occasionally, we will use in definitions the last notation for formulae of unspecified (propositional) language. Context will allow to avoid confusion. Finally, in all the languages under consideration, we denote

$$\alpha \leftrightarrow \beta = (\alpha \to \beta) \wedge (\beta \to \alpha),$$

for any formulae α and β, and

---

[3] See a comprehensive account of provability logic in [4].

$$\top = p \to p.$$

Given two languages $\mathscr{L}$ and $\mathscr{L}'$, we call the latter a *restriction* of the former, symbolically $\mathscr{L}' \subseteq \mathscr{L}$, if all $\mathscr{L}'$-connectives are included in the set of the $\mathscr{L}$-connectives with the same **Var** for both languages. For example, $\mathscr{L}_a$ is a restriction of both $\mathscr{L}_\Box$ and $\mathscr{L}_\bigcirc$, i.e. $\mathscr{L}_a \subseteq \mathscr{L}_\Box$ and $\mathscr{L}_a \subseteq \mathscr{L}_\bigcirc$.

Before getting started, we advise the reader that all calculi that will be under consideration have postulated rule of (uniform) substitution. Also, we remind the reader that a *normal extension* of a logic L is such a set of formulae which contains all the formulae of L and is closed under all rules of inference postulated in L. In other words, for a normal extension of L, it matters by which calculus L is defined. We write $L \vdash \alpha$ if formula α is derivable in L. Here we have to make the following remark. Given a formal proof, say $L \vdash \alpha$, where substitution is allowed, there is a *reduced* formal proof (of the same formula) which we denote by $L \vdash_r \alpha$ and in which substitution applies only to axioms of L, and then only the other postulated rules of inference may apply.[4] This property was used in Kuznetsov's proof of Proposition 16.

Our basis system is **K4**. Actually, we need two **K4** systems—one formulated in language $\mathscr{L}_\Box$ and the other in $\mathscr{L}_\bigcirc$. The first will be denoted by $\mathbf{K4}_\Box$ and the second by $\mathbf{K4}_\bigcirc$. Both systems have substitution and modus ponens as their postulated rules of inference. In addition, they have necessitation rule—the former has $\frac{\vdash A}{\vdash \Box A}$ and the latter has $\frac{\vdash \alpha}{\vdash \bigcirc \alpha}$. Then, we define

$$\mathbf{S4} = \mathbf{K4}_\bigcirc \oplus \bigcirc p \to p.$$

Intuitionistic logic will also be formulated in two languages—**Int** in language $\mathscr{L}_a$ and $\mathbf{Int}_\Box$ in language $\mathscr{L}_\Box$. Changing language we do not need to change the axioms and rules of inference of **Int**. For the axioms we take the formulae corresponding to the schemata for **Int** in [37]; for the postulated rules we take substitution and modus ponens. Other systems will be appearing as the story goes on.

7.3 Background

The birth of **KM** was accidental. It is worth telling the story of who her parents are and how they met.

The modal logic community knows that the year of 1976 was marked, among other wonderful events, with the birth of provability logic **GL**. However, as it often happens in the development of a scientific discipline, **GL**'s story has not been going straightforward. And this story, as any other, will be understood better if we put right accents in right places in its exposition. Since provability logic **GL** is defined as

---

[4] This was observed by Lindenbaum in 1934; see [75] for the proof of this property in a general setting. Church in his discussion [16], §29, on the history of substitution rule from Frege (1879) to von Neumann (1927), who proposed axiom schemata, does not mention this fact. However, see [16], §27, propositions *270 and *271.

$$\mathbf{GL} = \mathbf{K4}_{\Box} \oplus \Box(\Box p \to p) \to \Box p, \qquad (\textit{Gödel-Löb Logic})$$

the formula

$$\mathbf{gl} = \Box(\Box p \to p) \to \Box p \qquad (\textit{Löb Formula})$$

is the key one in the story of **GL**.

George Boolos and Giovanni Sambin note in [14] (see also [12]) that **gl** had already appeared in print in Smiley [72]. It seems that, when Krister Segerberg discussed his system **K4W** in the volume 2 of his doctoral thesis [66], using name **W** for **gl**, he did not know about Smiley's work, as well as did he not know about the interest in **gl** in the philosophical circle of Cambridge, Massachusetts, in the mid 1960s with Boolos and Kripke among the main participants. To be more precise, according to [14], Boolos and Kripke talked about interconnection between Gödel's second incompleteness theorem and Löb's theorem,[5] the modal counterpart of which is the following rule:

$$\frac{\vdash \Box A \to A}{\vdash A}. \qquad (\textit{Löb's Rule})$$

The equivalence of **gl** and Löb's Rule was proved later in the 1970s (see below).[6]

In the first half of the next decade, two groups in Europe, one in Italy with a strong algebraic background under the direction of Roberto Magari ("the Siena group") and the other group of mathematical logicians in Amsterdam, including Löb, de Jongh and Smoryňski who visited Amsterdam at the time, got involved in issues of self-reference exemplified by Gödel's incompleteness theorems. The idea was to work with arithmetic provability predicate as modality. This approach was clearly suggested by Derivability Conditions.[7] As Smoryński [73] puts it, "Gödel's Theorems and Löb's Theorem were propositional in character, that is they used propositional logic with an additional operator and some familiar laws—i.e. modal logic."

In the spring of 1973 Giovanni Sambin (of Siena at the time) came up with **gl**, while working on an example of what would be called later Magari (or diagonalizable) algebra. For his part, Dick de Jongh witnesses:

> The crucial point [...] was to see that Löb's theorem holds not only as a rule, but also as the fully formalized statement.

(cf. [14, pp. 14 and 17], respectively).

To the best of our knowledge, the self-reference issues were grouped around two main goals. The first was the fixpoint theorem for **GL** and the second was arithmetical completeness for it. Sambin and de Jongh succeeded in the former, Robert Solovay published in [76] a proof of the latter, thereby having certified the birth of **GL**.

---

[5] Kripke's result about this interconnection was reproduced in [13].

[6] In the sequel, we will be omitting the sign $\vdash$ while formulating a rule of inference.

[7] See [74], Section 0.1, about the transformation of Hilbert-Bernays Derivability Conditions to Löb Derivability Conditions.

At this point it is worth reminding that Gödel was perhaps the first who proposed in 1933 to interpret a unary modality as provability. More than that, he meant to interpret intuitionistic logic in terms of provability, known today as the Gödel-McKinsey-Tarski translation of modal-free, i.e. *assertoric*, propositional language into a modal one. By means of such a translation, Gödel expected to obtain an embedding, known nowadays as the Gödel-McKinsey-Tarski embedding, of intuitionistic logic into a modal system, which was assumed to be an axiomatic definition of provability. It was desirable to have for such a definition also a sound interpretation of the modality through the provability predicate of a sufficiently powerful first order theory, for example, Peano Arithmetic. However, for the modal system Gödel introduced in his 1933 note, **S4**, and Peano Arithmetic, this turned out to be impossible, as he himself mentioned at the end of his note, because of Gödel's second incompleteness theorem (cf. [31, pp. 296–303]). As we can comment on Gödel's observation today, the reason is that Löb's rule is not admissible in $\mathbf{S4}_\Box = \mathbf{K4}_\Box \oplus \Box p \to p$, or equivalently **gl** is not derivable in it.

It seems that no one of the groups mentioned above thought of connecting intuitionistic logic and provability logic, yet unborn. Here is where Chisinau's story of **GL** starts.

In 1955 Petr S. Novikov taught at Moscow State University (MGU) a course on *constructive mathematical logic*, under which he meant intuitionistic propositional logic. Most of the course was devoted to proving the Gödel-McKinsey-Tarski embedding theorem [54]. The idea goes back to Gödel's 1933 historic note [29], where he proposed to interpret **Int** in **S4**. Also, he proposed to read modality B of the latter as provability. However, as Gödel himself had noted, a direct interpretation of B via arithmetical provability predicate failed. To emphasize the distinction between modality B and provability predicate, we will be using modality $\Box$ with the intended interpretation as this predicate and formulate modal logic **S4** and its extensions in language $\mathscr{L}_\bigcirc$. In the sequel we will be dealing with several embeddings. So we proceed with the following definition.

Definition 1 (Gödel-McKinsey-Tarski translation) Translation t: $\mathscr{L}_a \to \mathscr{L}_\bigcirc$ is defined as follows.

(1) $t(p) = \bigcirc p$, for any $p \in$ **Var**
(2) $t(a \vartriangle b) = t(a) \vartriangle t(b)$, for any connective $\vartriangle \in \{\wedge, \vee\}$
(3) $t(\neg a) = \bigcirc \neg t(a)$
(4) $t(a \to b) = \bigcirc (t(a) \to t(b))$

Then what Gödel envisaged and McKinsey and Tarski proved is:

$$\mathbf{Int} \vdash a \Leftrightarrow \mathbf{S4} \vdash t(a), \tag{7.1}$$

for any assertoric formula a.

In 1955, Moscow's story towards relationship between intuitionistic and modal deduction machineries, which is an indispensable part of Chisinau's story, might have started (but did not), if the question were raised, whether the following equivalence is also true:

$$\mathbf{Int} + a \vdash b \Leftrightarrow \mathbf{S4} + t(a) \vdash t(b), \tag{7.2}$$

for any assertoric formulae a and b. However, it is not the last equivalence above, which was proved to be true (see e.g. [15, 19]), but its spirit that plays an essential role in Chisinau's story.

Martin Löb announced his solution of Henkin's problem at the International Congress of Mathematicians in Amsterdam in 1954 with a subsequent publication of a two-page abstract in the proceedings of the congress, where Löb's Rule, though in arithmetical context, appeared in print for the first time. In 1955, a short article [48] with a full proof was published. It seems very likely that Novikov was not familiar with Löb's result, while teaching the 1955 course. The latter was published posthumously as a book [63]. However, the book was not based on Novikov's manuscript but on the notes of some of those who attended the course, among which Kuznetsov's notes were perhaps the most complete scores.[8] Apparently, Novikov was not concerned with the problem of interpreting $\bigcirc$ as provability in arithmetic, but rather drew audience's attention to the question of finding a constructive interpretation of **Int** through constructively interpreted $\bigcirc$ and (7.1). Taking as an example the process of finding approximations of values of amounts such as weights, Novikov proposed to interpret "provability" of a correlation p of amount values, e.g. weights, symbolically $\bigcirc p$, if the truth value of p is stable in an experimental course of measurement.[9]

Having in mind interpretation of **Int** through the notion of provability, Kuznetsov and Muravitsky had formulated in the spring of 1975 the following calculi: **GL**, as well as

$$\begin{aligned}
\mathbf{GL}^- &= \mathbf{K4}_\Box \oplus \frac{\Box A \to A}{A}, \\
\mathbf{GL}^+ &= \mathbf{K4}_\Box \oplus \frac{\Box A \to A}{A} \oplus \frac{\Box A}{A} \ (\textit{Reflexivity Rule}), \\
\mathbf{GL}^* &= \mathbf{GL} \oplus \frac{\Box A}{A}.
\end{aligned}$$

(cf. [42] and the footnote on p. 211 of [44]).

Definition 2 (equivalence and equipollence of two calculi)[10] Given two languages $\mathscr{L}_1 \subseteq \mathscr{L}_2$ and two calculi C_1 in $\mathscr{L}_1$ and C_2 in $\mathscr{L}_2$, we say that C_1 and C_2 are $\mathscr{L}_1$-equivalent if for any $\mathscr{L}_1$-formula α,

$$C_1 \vdash \alpha \Leftrightarrow C_2 \vdash \alpha.$$

[8] Kuznetsov was not formally a student at MGU and attended classes there as a freelance; see [61].

[9] Unfortunately, Novikov's approach along this line had been merely outlined in Kuztnetsov's notes and was not included in the book at all.

[10] All calculi in this chapter are assumed to define structural monotonic consequence operator.

The calculi C_1 and C_2 are $\mathscr{L}_1$-equipollent if for any two $\mathscr{L}_1$-formulae α and β,

$$C_1 + \alpha \vdash \beta \Leftrightarrow C_2 + \alpha \vdash \beta.$$

If $\mathscr{L}_1 = \mathscr{L}_2$, we simply say that the calculi are equivalent or equipollent, respectively.

We note that if calculi C_1 and C_2 are equipollent and there is a formula α such that both $C_1 \vdash \alpha$ and $C_2 \vdash \alpha$ are true then C_1 and C_2 are also equivalent.

In view of the observation that $\mathbf{GL}^- \vdash \mathbf{gl}$ (it follows from the fact that $\mathbf{K4} \vdash \Box(\Box(\Box p \to p) \to \Box p) \to (\Box(\Box p \to p) \to \Box p)$, cf. [44]) and the remark above, we obtain the following.

Proposition 1 ([44]) *The following statements hold.*

(*a*) **GL** *and* $\mathbf{GL}^-$ *are equipollent and equivalent.*
(*b*) $\mathbf{GL}^*$ *and* $\mathbf{GL}^+$ *are equipollent and equivalent.*
(*c*) **GL** *and* $\mathbf{GL}^*$ *are equivalent but not equipollent.*

On the ground of (*a*) and (*b*) of Proposition 1, calculi $\mathbf{GL}^-$ and $\mathbf{GL}^+$ can be abandoned. Being equivalent to one another, all these **GL**-calculi are sound and complete with respect to arithmetic interpretation. Also, they are complete with respect to relational, alias Kripke, semantics, which is the class of irreflexive transitive frames with the ascending chain condition, ***GL-frames***. (Cf. [10, 66, 74, 76].) As to extensions of **GL**, there are those which are not Kripke complete; see Sect. 7.4.8 below. Therefore, it makes sense to employ algebraic semantics into consideration. As we will see below, the class of algebraic models of $\mathbf{GL}^*$ as a calculus is more complicated than that of **GL**.

To comprise all the systems being considered here, we take for algebraic semantics *bounded relatively pseudo-complemented lattices*, [34] enriching them, when necessary, with additional unary operations. In particular, in case of **GL** and $\mathbf{GL}^*$ we need $\neg$ and $\Box$, where the former is complementation. (This makes the algebras for all **GL**-systems above Boolean.) Since each such algebra $\mathfrak{A}$ has a top element, **1**, we can define *validity* of a formula in the algebra as usual, that is a formula α is *valid* in $\mathfrak{A}$, symbolically $\mathfrak{A} \models \alpha$, if for each valuation v on $\mathfrak{A}$, $v(\alpha) = \mathbf{1}$. Also, given class of algebras $\mathfrak{M}$, we write $\mathfrak{M} \models \alpha$ if $\mathfrak{A} \models \alpha$, for any $\mathfrak{A} \in \mathfrak{M}$. The *logic of an algebra* $\mathfrak{A}$, in symbols $L(\mathfrak{A})$, is the set of formulae valid in $\mathfrak{A}$.

Now following [44], with a calculus C (which is an extension of **Int**) we associate two classes of algebras, MC and ΣC, where the former is the class of all algebras on which all formulae derivable in C are valid, and the latter is all the algebras whose logics are normal extensions of C. Given C, it is obvious that $\Sigma C \subseteq \mathrm{M}C$.

Definition 3 **(on correspondence; cf.** [44]) We say that a class of algebras $\mathfrak{M}$ corresponds to a calculus C if for any formula α,

$$C \vdash \alpha \Leftrightarrow \mathfrak{M} \models \alpha,$$

and $\mathfrak{M}$ fully corresponds to C if for any two formulae α and β,

$$C + \alpha \vdash \beta$$

if and only if for any $\mathfrak{A} \in \mathfrak{M}$,

$$\mathfrak{A} \models \alpha \Rightarrow \mathfrak{A} \models \beta.$$

These two notions of correspondence are supposed to capture from an algebraic point of view the distinction between two understandings of $\vdash$, on the one hand, as a unary predicate of deducibility and, on the other, as a consequence (binary) relation. As we will see below, applying these notions to **GL** and **GL***, this algebraic view is especially interesting, when one and the same logic can be defined via two distinct calculi.

It is obvious that if a class $\mathfrak{M}$ fully corresponds to a calculus C and there is a formula α such that both $C \vdash \alpha$ and $\mathfrak{M} \models \alpha$ are true, then $\mathfrak{M}$ also corresponds to C. However, the converse may not be the case. For instance, we show below that the class of Magari algebras corresponds, though not fully, to the calculus **GL***.

Definition 4 (Lindenbaum-Tarski algebra; cf. [64]) We say that a calculus C in a language with $\rightarrow$ **admits** the Lindenbaum-Tarski algebra if the binary relation "$C \vdash \alpha \rightarrow \beta$ and $C \vdash \beta \rightarrow \alpha$" is a congruence on the algebra of all formulae and all the formulae derivable in C form a congruence class. Then the quotient algebra of the algebra of formulae is called the Lindenbaum-Tarski algebra of C.

It is easy to notice the following.

Proposition 2 ([44, **p. 215**]) *Let a calculus C be an extension of* **Int**. *Then the following hold.*

(*a*) *If C admits the Lindenbaum-Tarski algebra, then MC corresponds to C and is a variety.*
(*b*) *ΣC fully corresponds to C and if $\mathfrak{M}$ fully corresponds to C then $\mathfrak{M} \subseteq \Sigma C$.*
(*c*) *If a variety $\mathfrak{M}$ fully corresponds to C then $\mathfrak{M} = MC = \Sigma C$.*

Taking into account that both **GL** and **GL*** admit the Lindenbaum-Tarski algebra and applying Proposition 2, when necessary, we obtain the following.

Proposition 3 *The following statements hold.*

(*a*) *$MGL = \Sigma GL = MGL^*$. This class is a variety and fully corresponds to* **GL** (*Cf*. [44]).
(*b*) *$\Sigma GL^* \subset \Sigma GL$* (*Cf*. [44]).
(*c*) *The class ΣGL^* is not universally axiomatizable* (*Cf*. [56]).

Algebras of class **MGL** are called Magari algebras. More exactly we define them as follows.

Definition 5 (modal algebra, [15, 46] **Magari algebra** [15, 49]) An algebra ($\mathscr{A}$, $\wedge, \vee, \rightarrow, \neg, \mathbf{1}, \Box$) is called modal if with respect to assertoric operations it is a Boolean algebra, and a unary operation $\Box$ is subject to the following conditions:

$$(a)\ \Box(x \wedge y) = \Box x \wedge \Box y$$
$$(b)\ \Box \mathbf{1} = \mathbf{1}.$$

A modal algebra is called a Magari (*ordiagonalizable*) algebra if, in addition, the identity

$$(c)\ \Box(\Box x \rightarrow x) = \Box x$$

holds on it.[11]

Thus the class of Magari algebras corresponds to $\mathbf{GL}^*$ and fully corresponds to **GL**. This correlation will be preserved for all normal extensions of **GL**, on the one hand, and for the corresponding subvarieties of **MGL** on the other. The picture for the normal extensions of $\mathbf{GL}^*$ is more complicated. As an obvious observation, one can note that the quasivariety of Magari algebras satisfying the quasi-identity

$$\Box x = \mathbf{1} \supset x = \mathbf{1}$$

fully corresponds to $\mathbf{GL}^*$.

On this note Chisinau's story of $\mathbf{GL}^*$ almost comes to an end. In the sequel we will mention $\mathbf{GL}^*$ merely two more times. Here we only observe [42, 44] that the modal logic $\mathbf{D}^\star$ of the frame $(\omega, >)$ is the largest consistent extension of $\mathbf{GL}^*$.[12]

Let us return to Gödel's original idea of interpreting **Int** through provability in some sense of this notion. At this point we can see the two alternatives. Either we continue searching for another provability interpretation of $\Box$ axiomatized by **S4** and then use (7.1) to interpret **Int** in terms of provability or we explicitly distinguish two provabilities, one provability that is axiomatized by **GL** and another one which we would like to understand in a more general sense of the word, namely in the sense of informal deducibility. Gödel himself realized [30] how the first task could be conducted. This approach has been materialized independently by Artemov [3, 4] in his logic of proofs.

Here we want to examine an alternative path. Since modality $\Box$ axiomatized by **GL** cannot serve for the purpose of interpretation of **Int**, one can assume that another modality will do. However, the following question can be raised: How does the new modality correlate with $\Box$ in **GL**? Addressing this question, as a first step of

---

[11] Sambin proved that the identity $\Box x \leq \Box\Box x$ is derivable from (7.1)–(7.3) (cf. [50]). Compare this with the elimination of $\Box p \rightarrow \Box\Box p$ as an axiom of **GL** in [12], Theorem 1.18.

[12] Compare $\mathbf{D}^\star$ with D^* of [51].

investigation, we can suggest that not a new modality but a formula $\bigcirc p$ of a single variable p of language $\mathscr{L}_\Box$ can play this role.[13]

Proposition 4 ([44, 45]) *Let $\odot p$ denote a formula of a single variable p of language $\mathscr{L}_\Box$ satisfying the following conditions:*

$$(a)\ \mathbf{GL} \vdash \odot p \to p$$
$$(b)\ \mathbf{GL} \vdash \odot p \to \Box p$$
$$(c)\ \mathbf{GL} \vdash \odot \top.$$

Then $\mathbf{GL} \vdash \odot p \leftrightarrow (p \wedge \Box p)$. *The converse is also true.*

The last theorem inspires one to propose the following definitions.

Definition 6 (splitting) Splitting s: $\mathscr{L}_{\Box\bigcirc} \to \mathscr{L}_\Box$ is defined as follows.

$$(a)\ s(p) = p, \text{ for any } p \in \mathbf{Var}$$
$$(b)\ s(\alpha \vartriangle \beta) = s(a) \vartriangle s(b), \text{ for any connective } \vartriangle \in \{\wedge, \vee, \to\}$$
$$(c)\ s(\neg\alpha) = \neg s(\alpha)$$
$$(d)\ s(\Box\alpha) = \Box s(\alpha)$$
$$(e)\ s(\bigcirc\alpha) = s(\alpha) \wedge \Box s(\alpha)$$

This definition suggests, among other things, to regard **GL** as a fragment of a bimodal logic.[14]

The next definition perhaps is not important by itself but is important as an intermediate step.

Definition 7 ($\bigcirc$-Magari algebra An algebra $(\mathscr{A}, \wedge, \vee, \to, \neg, \mathbf{1}, \bigcirc)$ is called $\bigcirc$-Magari algebra if it is obtained from the Magari algebra $(\mathscr{A}, \wedge, \vee, \to, \neg, \mathbf{1}, \Box)$ by replacement of the signature operation $\Box x$ in the latter with

$$\bigcirc x = x \wedge \Box x. \tag{7.3}$$

To explain the part that $\bigcirc$-Magari algebras play in Chisinau's story we have to remind successively the following two-in-one definitions.

Definition 8 (S4-algebra, Grz-algebra) A modal algebra (with modal operation $\bigcirc$) is called an **S4**-algebra (alias interior algebra) if it satisfies the identities:

$$(a)\ \bigcirc x \wedge x = \bigcirc x$$
$$(b)\ \bigcirc\bigcirc x = \bigcirc x.$$

An **S4**-algebra is called a **Grz**-algebra if, in addition, the identity

[13] Compare with *Basic Working Hypothesis* in [44, 45].

[14] Indeed, one can prove that for any $\mathscr{L}_{\Box\bigcirc}$-formula α, $\mathbf{GL} \vdash s(\alpha)$ if and only if $\mathbf{GL}^{\bigcirc} \vdash \alpha$, where $\mathbf{GL}^{\bigcirc} = \mathbf{GL}_{\Box\bigcirc} \oplus \bigcirc p \leftrightarrow (p \wedge \Box p)$.

$$(c)\ \bigcirc(\bigcirc(x \to \bigcirc x) \to x) = \bigcirc x$$

holds.

Proposition 5 ([43, 44]) *Any* $\bigcirc$*-Magari algebra is a* **Grz***-algebra.*[15]

Let us denote by **Grz** the modal calculus (known as *Grzegorczyk logic*) in the language $\mathscr{L}_\bigcirc$, obtained from the system **S4**, understood also in $\mathscr{L}_\bigcirc$, by endowing it with the new axiom:

$$\mathbf{grz} = \bigcirc(\bigcirc(p \to \bigcirc p) \to p) \to p. \qquad \textit{(Grzegorczyk Formula)}$$

Corollary 1 ([11, 12, 33, 43, 44]) *For any formula* α *of* $\mathscr{L}_\bigcirc$,

$$\mathbf{Grz} \vdash \alpha \Leftrightarrow \mathbf{GL} \vdash s(\alpha).$$

For any formula a,

$$\mathbf{Int} \vdash a \Leftrightarrow \mathbf{GL} \vdash s(t(a)).$$

The last equivalence can be refined as follows. Let us define in the language $\mathscr{L}_\Box$

$$\mathbf{GL^i} = \mathbf{Int}_\Box \oplus \Box(p \to q) \to (\Box p \to \Box q) \oplus \mathbf{gl} \oplus \frac{A}{\Box A}.$$

Taking into account that $\mathbf{GL^i} \vdash \Box p \to \Box\Box p$ (see [65], Lemma 0.2), $\mathbf{GL^i}$ is simply **GL** on intuitionistic basis.[16] Then

$$\mathbf{Int} \vdash a \Leftrightarrow \mathbf{GL^i} \vdash s(t(a)), \tag{7.4}$$

for any formula $a \in \mathscr{L}_a$ (cf. [44, p. 222]).

Corollary 1 immediately raises the question: Are the following two equivalencies true?

$$\mathbf{Grz} + \alpha \vdash \beta \Leftrightarrow \mathbf{GL} + s(\alpha) \vdash s(\beta) \tag{7.5}$$

$$\mathbf{Int} + a \vdash b \Leftrightarrow \mathbf{GL} + s(t(a)) \vdash s(t(b)) \tag{7.6}$$

We address this question in Sect. 7.4.8. Now let us turn again to $\bigcirc$-Magari algebras.

---

[15] This result was obtained in 1976 and first was announced in the abstract [43] with a subsequent publication in full detail in [44]. The preparation of the article [44] took eight months from October 1976 to June 1977, mainly due to Kuznetsov's health instability and in part due to obtaining new results in the course of writing. Thus [44] had been submitted for publication in the summer of 1977; however, it took three more years for the editor of the collection to get it printed.

[16] This system first had been considered by Sambin [65] in relation to effective fixed points in Magari algebras.

Given a Magari algebra $\mathfrak{A} = (\mathfrak{B}, \Box)$, where $\mathfrak{B}$ is the Boolean reduct of $\mathfrak{A}$, we denote by $\mathfrak{A}^\circ$ the Heyting algebra of the open elements of $(\mathfrak{B}, \bigcirc)$. Now the next observation was a key one, though accidental. Given an open element $\bigcirc x \in \mathfrak{A}$, we notice that

$$\Box \bigcirc x = \Box(x \wedge \Box x) = \Box x \wedge \Box\Box x = \bigcirc\Box x,$$

that is $\mathfrak{A}^\circ$ is closed under operation $\Box$. Thus, endowing the Heyting algebra $\mathfrak{A}^\circ$ with operation $\Box$ that was first defined on Magari algebra $\mathfrak{A}$, we define the algebra $\mathfrak{A}^\Box = (\mathfrak{A}^\circ, \Box)$. This definition can be extended over any class Σ of Magari algebras:

$$\Box\Sigma = \{\mathfrak{A}^\Box \mid \mathfrak{A} \in \Sigma\}. \tag{7.7}$$

By using Birkhoff's criterion, it was pointed out in [44], Footnote on p. 224, that all algebras $\mathfrak{A}^\Box$ form a variety axiomatized by the identities:

$$\begin{aligned} &(a)\ x \leq \Box x \\ &(b)\ \Box x \to x = x \\ &(c)\ \Box x \leq y \vee (y \to x). \end{aligned}$$

The identities (7.1) and (7.2) were induced by the transitivity axiom, $\Box p \to \Box\Box p$, and Löb's Formula of **GL**, respectively. The identity (7.3) is a result of transformation of the formula $\Box p \to (q \vee \neg q)$ which was borrowed by Kuznetsov and me from [6].

This inspired Kuznetsov [39] to formulate in language $\mathscr{L}_\Box$ the "proof-intuitionistic calculus" I^Δ,[17] nowadays known as the logic **KM**:

$$\mathbf{KM} = \mathbf{Int}_\Box \oplus p \to \Box p \oplus (\Box p \to p) \to p \oplus \Box p \to (q \vee (q \to p)).$$

Thus, if **GL** is the mother of **KM**, her father is **Int**.[18]

One can easily prove that

$$\mathbf{GL^i} = \mathbf{GL} \cap \mathbf{KM}.$$

On the other hand, defining

$$\mathbf{KM}^c = \mathbf{Cl}_\Box \oplus p \to \Box p \oplus (\Box p \to p) \to p \oplus \frac{A}{\Box A},$$

one can observe that

$$\mathbf{KM}^c = \mathbf{GL} \oplus \mathbf{KM}$$

and $\mathbf{KM}^c$ is the greatest consistent extension of both **GL** and **KM** (cf. [60]).

[17] This notation was adapted by Kuznetsov from [6].

[18] After having learned the story behind the footnote on p. 224 of [44], Leo Esakia began using the new name, **KM**, instead of I^Δ, the first time in [26]. Thus Esakia can be regarded as the godfather of this logic.

Definition 9 (KM-algebra) An algebra $(\mathscr{A}, \wedge, \vee, \rightarrow, \neg, \mathbf{1}, \Box)$ is called a **KM**-algebra if it is a Heyting algebra according to its assertoric operations and also satisfies the identities (a)–(c) above.

The following observation will be useful in Sect. 7.4.5

Proposition 6 ([41]) *In the above definition, identity* (*b*) *can be replaced by*

$$(d)\ \Box(x \wedge y) = \Box x \wedge \Box y \text{ and}$$
$$(e)\ \Box x = x \ \Rightarrow \ x = \mathbf{1}.$$

We conclude this section with the following.

Proposition 7 *The variety of* **KM**-*algebras fully corresponds to* **KM**. *Hence, this class coincides with M***KM** *and* Σ**KM**.

Proof is obvious, since for any formula A, the calculus $\mathbf{KM} \oplus A$ admits the Lindenbaum-Tarski algebra which belongs to **MKM**. Then we apply Proposition 2. □

7.4 Other Axiomatizations and Some Properties

In this section we consider the properties that are usually asked about a logic or calculus.

7.4.1 The Separation Property

As a calculus the logic **KM** was formulated originally as follows:

$$I^{\Delta} = \mathbf{Int}_{\Box} \oplus p \rightarrow \Box p \oplus (\Box p \rightarrow p) \rightarrow p \oplus ((p \rightarrow q) \rightarrow p) \rightarrow (\Box q \rightarrow p)$$

(cf. [39, 41, 45]). The reason for such an axiomatization was the possibility to raise the question about the ***separation property*** for I^{Δ}, that is whether for any formula a, if $I^{\Delta} \vdash a$ then there is a derivation, which uses only those axioms that contain $\rightarrow$ and the connectives actually occurring in a.[19] The positive answer to the question was given in [58, 68, 70].[20]

There is another axiomatization proposed by Kuznetsov [41], though implicitly:

$$\mathbf{KM}^{+} = \mathbf{Int}_{\Box} \oplus \Box(p \rightarrow q) \rightarrow (\Box p \rightarrow \Box q) \oplus p \rightarrow \Box p \oplus \Box p \rightarrow (q \vee (q \rightarrow p)) \oplus \frac{\Box A \rightarrow A}{A}.$$

[19] A general setting for the separation property can be found in [17].

[20] All these proofs were obtained independently and about the same time. The proof in [58] is algebraic, while the proof in [68, 70] uses syntactic means. Two last papers differ only in style.

This axiomatization will be illustrated in Sect. 7.4.5. We merely note here that the last axiomatization is in accordance with Proposition 6.

7.4.2 Kripke Semantics and the Finite Model Property

As the reader has seen, **KM** came out of the flash of **GL**, more precisely from the algebraic interpretation for the latter. Thus Kripke semantics has not been a matter of justification for **KM**. However, the question of Kripke semantics for **KM** can be and has been considered.

Definition 10 **(KM-frame, KM-model)** Let W be a nonempty set, $\leq$ be a partial ordering on W with the ascending chain condition [34] and $<$ be the reflexive reduction [79] of $\leq$. Then $\mathfrak{F} = (W, \leq, <)$ is called a **KM**-frame. If, in addition, $(W, \leq)$ is a tree, the corresponding frame is called a **KM**-tree-frame. $(\mathfrak{F}, v)$, where $\mathfrak{F}$ is a **KM**-frame, is a **KM**-model if $(W \leq, v)$ is an intuitionistic model with a valuation v (see [15]) and forcing relation $\models$ is stipulated for $\Box$ in the following way:

$$(v, x) \models \Box A \Leftrightarrow \forall y. \, x < y \Rightarrow (v, y) \models A.$$

It is easily seen that **KM** is sound with respect to **KM**-frames. On the other hand, we have the following.

Proposition 8 ([55]) *Logic* **KM** *is determined by finite* **KM**-*frames, that is, enjoys the finite model property and hence is decidable.*

In 1984 Guram Darjania gave a sequential formalization of **KM** and proved the completeness of **KM** with respect to the finite **KM**-tree-frames [18].

7.4.3 The Disjunction Property

In view of Proposition 8, it is clear that **KM** is determined by the finite rooted **KM**-frames. This implies that **KM** enjoys the disjunction property:

$$\mathbf{KM} \vdash A \vee B \Leftrightarrow \mathbf{KM} \vdash A \text{ or } \mathbf{KM} \vdash B.$$

Proof is based on the fact that each **KM**-frame is irreflexive.

Darjania also proved the disjunction property for **KM**. His proof is proof-theoretic; see [18], Proposition 3.

7.4.4 The Fixed Point Property

We start with the following definition.

Definition 11 (relative fixed point property)[21] Let $\Delta(q)$ be a set of propositional formulae containing variable q. A propositional logic L enjoys the fixed point property relative to $\Delta(q)$ if for any formula $\alpha(p_1, \ldots, q) \in \Delta(q)$, there is a formula β (a fixed point of α with respect to q), which contains only variables that occur in α, does not contain q, and such that

$$(a)\; L \vdash \beta \leftrightarrow \alpha(p_1, \ldots, \beta) \quad \text{(existence)}$$
$$(b)\; L \vdash \gamma \leftrightarrow \alpha(p_1, \ldots, \gamma) \quad \text{for some } \gamma, \;\Rightarrow\; L \vdash \gamma \leftrightarrow \beta. \quad \text{(uniqueness)}$$

The fixed point property is called strong if (b) above can be replaced with

$$(c) L \vdash (\gamma \leftrightarrow \alpha(p_1, \ldots, \gamma)) \to (\gamma \leftrightarrow \beta) \quad \text{for any} \gamma. \quad \text{(strong uniqueness)}$$

Also, we call the fixed point property effective if a fixed point β can be obtained for any $\alpha(p_1, \ldots, q) \in \Delta(q)$ by an algorithm.

In this section we will be considering logics $\mathbf{GL^i}$, $\mathbf{GL}$ and $\mathbf{KM}$, that is logics of language $\mathscr{L}_\Box$. Let us define the following classes of formulae.

Definition 12 (sentences modalized in a variable; [12]**)** Given a propositional variable q, a formula A is called modalized in q if every occurrence of q in A is in the scope of an occurrence of $\Box$. The class of all formulae modalized in q is denoted by $\Delta(\Box q)$.

Proposition 9 ([65], **Theorem 3.5 (existence) and Corollary 2.5 (uniqueness))** *Given a variable q, logic $\mathbf{GL^i}$ has the effective fixed point property relative to $\Delta(\Box q)$.*

We note that the rule $\frac{A}{\Box A}$ is obviously admissible in $\mathbf{KM}$. Let us define

$$\mathbf{KM}^{++} = \mathbf{KM} \oplus \frac{A}{\Box A}.$$

Next we observe that $\mathbf{GL^i}$ is a normal extension of $\mathbf{KM}^{++}$. This is because $\mathbf{KM} \vdash (p \to q) \to (\Box p \to \Box q)$. (See [45].) Therefore, $\mathbf{KM} \vdash \mathbf{gl}$. On the other hand, $\Box(p \to q) \to (\Box p \to \Box q)$ cannot be refuted on a $\mathbf{KM}$-frame and hence, in view of Proposition 8, is derivable in $\mathbf{KM}$.[22]

Next we consider language $\mathscr{L}_\Box$ as a language of first order logic with connectives of $\mathscr{L}_\Box$ as functional constants. This allows to read the statement of uniqueness in [65], Theorem 2.4, as a quasi-identity:

---

[21] Compare the fixed point property relative to all formulae with the Beth definability property for propositional logic (see e.g. [28]).

[22] This is also a straightforward consequence of Lemma 2 in [60]: For any $\mathbf{KM}$-algebra $\mathfrak{A}$, there is a Magari algebra $\mathfrak{B}$ such that $\mathfrak{A} = \mathfrak{B}^\Box$.

$$A = \top \supset B = \top.$$

Now let $\mathfrak{M}$ and $\mathfrak{N}$ be the varieties of **KM**- and $\mathbf{GL^i}$-algebras, respectively. It is clear that $\mathfrak{M} \subseteq \mathfrak{N}$. Sambin proves that $\mathfrak{N} \models A = \top \supset B = \top$. Therefore, $\mathfrak{M} \models A = \top \supset B = \top$. However,

$$\mathfrak{M} \models A = \top \supset B = \top \Leftrightarrow \mathfrak{M} \models A \to B = \top.$$

The latter holds because of the correspondence between Heyting filters and congruences on each **KM**-algebra, for each Heyting filter on a **KM**-algbera is closed under $\Box$. Therefore, when $\mathfrak{A} \not\models A \to B = \top$, for some $\mathfrak{A} \in \mathfrak{M}$, there is a homomorphic image $\mathfrak{B}$ of $\mathfrak{A}$ such that $\mathfrak{B} \not\models A = \top \supset B = \top$.[23] This leads to the following conclusion.

Proposition 10 *Given variable q, logic* **KM** *has the effective strong fixed point property relative to $\Delta(\Box q)$.*

The fixed point property for **GL** has been discussed in detail in [74, Chap. 1, Sect. 3], and [12, Chap. 8]. In the latter, the fixed point property for **GL** is presented in a very elegant form: For any formula $A \in \Delta(\Box q)$, there is a formula A^* containing variables contained in A, not containing variable q, and such that $\mathbf{GL} \vdash ((q \leftrightarrow A) \wedge \Box(q \leftrightarrow A)) \leftrightarrow ((q \leftrightarrow A^*) \wedge \Box(q \leftrightarrow A^*))$. Smoryński [74, Chap. 1, Theorem 3.5], gives another variant of the fixed point property for **GL**. In both cases it is not the strong form of uniqueness as indicated above.

7.4.5 Topological Semantics

First we consider topological interpretations of **GL** and **KM**, which are interconnected. Topological notions used in this section can be found in any textbook on point-set topology. Consult e.g. [20].

Topological semantics of **KM**, as well as **GL**, uses scattered topological spaces. It is customary to define a *topological space* (or simply *space*) as a pair $(X, \mathcal{O})$, where X is a nonempty set and $\mathcal{O}$ is a family of its subsets, called *open*, that is subject to some well-known conditions. Given $x \in X$, by U_x we denote an *open neighborhood* of x, that is an open subset of X, which contains x. Given a set $Y \subseteq X$, a point $x \in X$ is called a *limit point* (or *cluster point*) of Y if for any U_x, $(U_x \cap Y) \setminus \{x\} \neq \emptyset$. By $\mathrm{d}(Y)$ we denote the set of all limit points of Y.[24] We have $Y \in \mathcal{O}$ if and only if $d(X \setminus Y) \subseteq X \setminus Y$.

---

[23] This property for the varieties of Heyting algebras has been known since the 1970s; see it implicitly (and with the use of Zorn's lemma) in [83], Lemma 1, or explicitly (and without Zorn's lemma) in [52], Lemma 1.

[24] $\mathrm{d}(Y)$ is called the *derived set* of Y.

A topological space $(X, \mathcal{O})$ is called *scattered* if X is a *scattered set* in $(X, \mathcal{O})$, that is X does not contain any nonempty subset that is *dense-in-itself.* If γ is an ordinal then the space $T(\gamma)$ of the ordinals not exceeding γ with its interval topology is scattered. A finite space is scattered if and only if it is a T_0-space (cf. [78] and also [5]).[25]

Given a topological space $(X, \mathcal{O})$, let us define:

$$\Box Y = X \setminus d(X \setminus Y),$$

for any $Y \subseteq X$. It is clear that

$$x \in \Box Y \Leftrightarrow \exists U_x.\ U_x \setminus \{x\} \subseteq Y.$$

Kuznetsov [40] had noted that a space $(X, \mathcal{O})$ is scattered if and only if for any $Y \in \mathcal{O}$,

$$\Box Y = Y \Rightarrow Y = X. \tag{7.8}$$

With any space $(X, \mathcal{O})$ we associate the algebra $(\mathcal{P}(X), \Box)$, where $\mathcal{P}(X)$ is the Boolean algebra of all subsets of X and $\Box$ is the operation defined above.

A comprehensive account of topological interpretation of **GL** the reader can find in [5]. Here we mention merely some of them which will be used for topological interpretation of **KM**.

Proposition 11 ([22–24, 67]) *Given a space $(X, \mathcal{O})$, $(\mathcal{P}(X), \Box)$ is a Magari algebra if and only if the space is scattered.*

Corollary 2 ([22, 23]) *Logic* **GL** *is determined by the algebras $(\mathcal{P}(X), \Box)$ of scattered spaces.*

Proposition 12 ([1, 2, 7]) *The algebra $(\mathcal{P}(T(\omega^\omega)), \Box)$ is an adequate model for* **GL**.

Corollary 3 ([1, 2]) *If a formula A is unprovable in* **GL**, *then it is invalid in $(\mathcal{P}(T(\omega^n))$ for some $n \in \omega$.*

Given a space $(X, \mathcal{O})$, we associate the Heyting algebra of the open sets $\mathcal{O}$ along with additional operation $\Box$ defined as above, denoting this expansion by $(\mathcal{H}(X), \Box)$

Proposition 13 ([40]) *Given a space $(X, \mathcal{O})$, $(\mathcal{H}(X), \Box)$ is a* **KM**-*algebra if and only if the space is scattered.*

The last proposition shows that the axiomatization $\mathbf{KM}^+$ is in accordance with (7.8).

Corollary 4 ([40]) *Logic* **KM** *is determined by algebras $(\mathcal{H}(X), \Box)$ of scattered spaces.*

---

[25] This observation can be used in the proof of Corollaries 4.4.1 and 4.6.1.

More about **KM** and scattered spaces can be found in [47], Sect. 4.2.

Corollary 5 *The algebra* $(\mathscr{H}(T(\omega^\omega)), \Box)$ *is an adequate model for* **KM**.

Proof By virtue of Proposition 4.6, if **KM** $\vdash A$ then A is valid in $(\mathscr{H}(T(\omega^\omega)), \Box)$. On the other hand, if A is valid in $(\mathscr{H}(T(\omega^\omega)), \Box)$ then $s(t(A))$ is valid in $(\mathscr{P}(T(\omega^\omega)), \Box)$. Then, according to Corollary 4.5, **GL** $\vdash s(t(A))$. By virtue of Proposition 20 below, **KM** $\vdash A$. □

A topos-theoretic interpretation of **KM** is considered in [27]; also, see [47], Sect. 5.2, for detail.

7.4.6 The $\mathscr{L}_a$-equipollence of KM and Int

In this section we will discuss Kuznetsov's Theorem on the $\mathscr{L}_a$-equipollence of **KM** and **Int** and related issues. We start with the following observation.

Proposition 14 *Let* $\mathfrak{A}$ *be a Heyting algebra and* $a \in \mathfrak{A}$. *Suppose for some element* $b \in \mathfrak{A}$, *the following inequalities hold:*

$$(a)\ a \le b,$$
$$(b)\ b \to a \le a,$$
$$(c)\ b \le x \vee (x \to a), \textit{for any } x \in \mathfrak{A}.$$

Then the element b satisfying the properties (a)–(c) is unique.

Proof Indeed, suppose an element b' also satisfies (a)–(c). Then we obtain:

$$b \le b' \vee (b' \to a) = b' \vee a = b'.$$

Similarly, we get $b' \le b$. □

The last proposition suggests the following definition.

Definition 13 (enrichable Heyting algebras) An element a of a Heyting algebra $\mathfrak{A}$ is called enriched by an element a^*, or a^* enriches a, if the following conditions are fulfilled:

$$(a)\ a \le a^*,$$
$$(b)\ a^* \to a \le a,$$
$$(c)\ a^* \le x \vee (x \to a), \text{ for any } x \in \mathfrak{A}.$$

A Heyting algebra is called enrichable if each element is enrichable.

Thus "forgetting" operation □ in a **KM**-algebra, we obtain an enrichable Heyting algebra. On the other hand, each finite Heyting algebra is enrichable. This is a consequence of the following.

Given a Heyting algebra $\mathfrak{A}$ and an element $a \in \mathfrak{A}$, we define:

$$F_a = \{x \vee (x \rightarrow a) \mid x \in \mathfrak{A}\}.$$

Lemma 1 *Given a Heyting algebra $\mathfrak{A}$ and an element $a \in \mathfrak{A}$, the following conditions are equivalent:*

(a) $y \in F_a$;
(b) $y \rightarrow a = a$ *and* $a \leq y$;
(c) $y \rightarrow a \leq y$.

Proof We prove that $(a) \Rightarrow (b) \Rightarrow (c) \Rightarrow (a)$.

$(a) \Rightarrow (b)$. Let $y \in F_a$. Then for some $x \in \mathfrak{A}$, $y = x \vee (x \rightarrow a)$. It is clear that $a \leq y$. Also,

$$\begin{aligned} y \rightarrow a &= (x \rightarrow a) \wedge ((x \rightarrow a) \rightarrow a) \\ &= (x \rightarrow a) \wedge a \\ &= a. \end{aligned}$$

$(b) \Rightarrow (c)$. Obvious, by transitivity of $\leq$.

$(c) \Rightarrow (a)$. Obvious again, for $y \rightarrow a \leq y$ implies $y = y \vee (y \rightarrow a)$. □

Proposition 15 *Let $\mathfrak{A}$ be a Heyting algebra and $[a)$ be the filter in $\mathfrak{A}$ generated by $a \in \mathfrak{A}$. If $[a)$ is an atomic lattice with a finite set of atoms, say $a_1, \ldots, a_n$, then a is enriched by $a^* = a_1 \vee \ldots \vee a_n$. If $a = \mathbf{1}$ then a is enriched by* **1**.

Proof The last statement of the proposition is readily seen.

Next assume that $a \neq \mathbf{1}$. It is obvious that $a \leq a^* \rightarrow a$. For contradiction, assume that $a < a^* \rightarrow a$. Then for some i, $a_i \leq a^* \rightarrow a$. The latter implies $a_i \wedge a^* \leq a$ and hence $a_i \leq a$. A contradiction. As $a = a^* \rightarrow a$, we have $a^* = a^* \vee (a^* \rightarrow a)$, so $a^* \in F_a$. Now let x be any element of F_a. We have to prove that $a^* \leq x$. According to Lemma 1, $a \leq x$ and $x \rightarrow a = a$. For contradiction, assume that for some i, $a_i \not\leq x$. Then $a_i \wedge x \leq a$. The latter implies $a_i \leq x \rightarrow a = a$. A contradiction. Therefore, $a^* \leq x$. □

Corollary 6 *Any finite Heyting algebra is enrichable.*

Corollary 7 *Logic* **KM** *is a conservative extension of* **Int**. *In other words,* **KM** *and* **Int** *are $\mathscr{L}_a$-equivalent.*

Proof Suppose for some formula A, **Int** $\nvdash A$. Then A can be refuted on a finite Heyting algebra, say $\mathfrak{A}$. Since $\mathfrak{A}$ is enrichable (Corollary 6), we define $\Box$ on $\mathfrak{A}$ to make it a **KM**-algebra.

The remaining part of the proof is obvious. □

The last corollary allows the following generalization.

Proposition 16 **(Kuznetsov's Theorem,** [41]) *Logics* **KM** *and* **Int** *are $\mathscr{L}_a$-equipollent.*

In [26] Esakia introduced

$$\mathbf{mHC} = \mathbf{Int}_{\Box} \oplus \Box(p \to q) \to (\Box p \to \Box q) \oplus p \to \Box p \oplus \Box p \to (q \vee (q \to (q \to p))$$

and proved that **Int** and **mHC** are $\mathscr{L}_a$-equivalent. Since

$$\mathbf{KM} \vdash (p \to q) \to (\Box p \to \Box q),$$

(see [45]) we conclude that

$$\mathbf{KM} = \mathbf{mHC} \oplus (\Box p \to p) \to p.$$

Thus the following question seems natural to ask.

Problem 1 *Are* **Int** *and* **mHC** *also* $\mathscr{L}_a$*-equipollent?*

When Kuznetsov was working on a "thin" paper, where Proposition 16 appeared in print for the first time, the author pointed out to him that this proposition is equivalent to the following.[26]

Proposition 17 ([41]) *Any Heyting algebra is embedded into some enrichable Heyting algebra so that both generate the same variety.*

The proof of the equivalence of Propositions 16 and 17 can be found in [62, p. 53].

7.4.7 The Craig Interpolation Property

Jónsson [36] perhaps was the first who drew attention to connection between the interpolation property in logic (in the form of the interpolation principle for equalities of terms [28]) and the amalgamation property of the corresponding variety of algebras. Maksimova [52] (see also [28]) established the equivalence between the two properties for all intermediate (alias superintuitionistic) logics, on the one hand, and the corresponding varieties of Heyting algebras, on the other.

By the ***Craig interpolation property***, or simply the ***interpolation property***, for a logic L, is meant the following: If $\alpha \to \beta$ is valid in L, that is $L \vdash \alpha \to \beta$, and the set of common variables of α and β is not empty, then there is a formula γ such that both $\alpha \to \gamma$ and $\gamma \to \beta$ are valid in L and γ contains only those variables which occur in both α and β.

A class Σ of similar algebras has the ***amalgamation property*** if given $\mathfrak{A}, \mathfrak{B}, \mathfrak{C} \in \Sigma$ and embeddings $f : \mathfrak{A} \to \mathfrak{B},\ g : \mathfrak{A} \to \mathfrak{C}$, there exists $\mathfrak{D} \in \Sigma$ and embeddings $f' : \mathfrak{B} \to \mathfrak{D},\ g' : \mathfrak{C} \to \mathfrak{D}$ such that $f'f = g'g$. Often, it is convenient to regard

[26] The next proposition appeared as Corollary 2 in [41].

$\mathfrak{A}$ as a common subalgebra of $\mathfrak{B}$ and $\mathfrak{C}$, symbolically $\mathfrak{A} \subseteq \mathfrak{B} \cap \mathfrak{C}$, in which case the triple $(\mathfrak{A}, \mathfrak{B}, \mathfrak{C})$ is called an ***amalgam*** (in Σ). The amalgam $(\mathfrak{A}, \mathfrak{B}, \mathfrak{C})$ can be amalgamated if there is an algebra $\mathfrak{D} \in \Sigma$ and two embeddings $f : \mathfrak{B} \to \mathfrak{D}$ and $g : \mathfrak{C} \to \mathfrak{D}$ such that $f \upharpoonright \mathfrak{A} = g \upharpoonright \mathfrak{A}$. It is obvious that a class Σ satisfies the amalgamation property if and only if each amalgam in Σ can be amalgamated.

Proposition 18 *Any normal extension L of* **KM** *enjoys the interpolation property if and only if the corresponding variety Σ_L has the amalgamation property.*

Proof literally repeats the proof of Theorem 1 in [52], since the theory of filters and ideals for **KM**-algebras is the same as for Heyting algebras. □

Proposition 19 *The variety of* **KM**-*algebras has the amalgamation property. Therefore,* **KM** *has the interpolation property.*

Proof Let $(\mathfrak{A}, \mathfrak{B}_1, \mathfrak{B}_2)$ be an amalgam in the variety of **KM**-algebras. We denote by $S_{\mathfrak{B}_i}$, where $i \in \{1, 2\}$, the set of all prime filters of the algebra $\mathfrak{B}_i$. Then we define:

$\mathscr{F}_i = \{\Phi \in S_{\mathfrak{B}_i} \mid \forall \Phi' \in S_{\mathfrak{B}_i}.\ \Phi \subset \Phi' \Rightarrow \Phi' \cap \mathfrak{A} \neq \Phi \cap \mathfrak{A}\}$, where $i \in \{1, 2\}$,
$S = \{(\Phi_1, \Phi_2) \in \mathscr{F}_1 \times \mathscr{F}_2 \mid \Phi_1 \cap \mathfrak{A} = \Phi_2 \cap \mathfrak{A}\}$,
$(\Phi_1, \Phi_2) \leq (\Phi_1', \Phi_2') \Leftrightarrow \Phi_1 \subseteq \Phi_1'$ and $\Phi_2 \subseteq \Phi_2'$, where $(\Phi_1, \Phi_2), (\Phi_1', \Phi_2') \in S$,
$\mathscr{H}(S)$ is the Heyting algebra of subsets of S, upward closed w.r.t. $\leq$. □

We observe:

$$\begin{aligned}&\textit{Given } i, j \in \{1,2\},\ \textit{where } i \neq j,\\ &(a)\ \forall \Phi_i \in \mathscr{F}_i \exists \Phi_j \in \mathscr{F}_j.\ (\Phi_1, \Phi_2) \in S.\\ &(b)\ (\Phi_1, \Phi_2) \in S \textit{ and } \Phi_i \subseteq \Phi_i' \Rightarrow \exists \Phi_j' \in \mathscr{F}_j.\ (\Phi_1', \Phi_2') \in S \textit{ and } \Phi_j \subseteq \Phi_j'.\end{aligned} \tag{7.9}$$

We prove (7.9), following routinely the proof of Lemma 7 in [52].
Now, given $i \in \{1, 2\}$, we define $\varphi_i : \mathfrak{B}_i \to \mathscr{H}(S)$ as follows:

$$\varphi_i(x) = \{(\Phi_1, \Phi_2) \in S \mid x \in \Phi_i\}$$

and then,

$$\Box\varphi_i(x) = \{(\Phi_1, \Phi_2) \in S \mid \forall \Phi_i' \in S_{\mathfrak{B}_i}.\ \Phi_i \subset \Phi_i' \Rightarrow x \in \Phi_i'\}.$$

It is clear that, given $x \in \mathfrak{B}_i$, $\Box\varphi_i(x) \in \mathscr{H}(S)$.

Next, given $x \in \mathfrak{B}_1$ (or $x \in \mathfrak{B}_2$), we further define:

$$\begin{aligned}&\mathscr{A}_x = \{\Phi \in S_{\mathfrak{B}_1} \mid x \in \Phi\},\\ &\mathscr{A}_{\overline{x}} = \{\Phi \in S_{\mathfrak{B}_1} \mid x \notin \Phi\},\\ &\max \mathscr{A}_{\overline{x}} = \{\Phi \in \mathscr{A}_{\overline{x}} \mid \forall \Phi' \in S_{\mathfrak{B}_1}.\ \Phi \subset \Phi' \Rightarrow \Phi' \in \mathscr{A}_x\}.\end{aligned}$$

We will need the following property.

If an element x of a Heyting algebra $\mathfrak{B}$ is enriched by an element y, then $\mathscr{A}_y = \mathscr{A}_x \cup \max \mathscr{A}_{\overline{x}}$. ([58], Lemma 4) (7.10)

Now let us fix $i \in \{1, 2\}$, for instance $i = 1$, and prove that for any $x \in \mathfrak{B}_1$,

$$\varphi_1(\Box x) = \Box\varphi_1(x). \tag{7.11}$$

Indeed, assume first that $(\Phi_1, \Phi_2) \in \varphi_1(\Box x)$. Then $\Box x \in \Phi_1$. Let $\Phi_1' \in S_{\mathfrak{B}_1}$ with $\Phi_1 \subset \Phi_1'$. Since $\Box x$ enriches x in $\mathfrak{B}_1$, by virtute of (7.10), $\Phi_1 \in \mathscr{A}_x \cup \max \mathscr{A}_{\overline{x}}$. Therefore, $x \in \Phi_1'$. Thus we conclude that $(\Phi_1, \Phi_2) \in \Box\varphi_1(x)$.

Now suppose $(\Phi_1, \Phi_2) \in \Box\varphi_1(x)$. For contradiction, assume that $\Box x \notin \Phi_1$. Then, according to (7.10), $\Phi_1 \notin \mathscr{A}_x \cup \max \mathscr{A}_{\overline{x}}$. Then there is $\Phi_1' \in \max \mathscr{A}_{\overline{x}}$ such that $\Phi_1 \subseteq \Phi_1'$. More precisely, $\Phi_1 \subset \Phi_1'$, since $\Phi_1 \notin \max \mathscr{A}_{\overline{x}}$. But then $(\Phi_1, \Phi_2) \notin \Box\varphi_1(x)$. A contradiction.

Next we show that for $Y \in \mathscr{H}(S)$ and $x \in \mathfrak{B}_1$,

$$\Box\varphi_1(x) \subseteq Y \cup (Y \to \varphi_1(x)). \tag{7.12}$$

Indeed, let $(\Phi_1, \Phi_2) \in \Box\varphi_1(x)$ and $(\Phi_1, \Phi_2) \notin Y$. Also, assume for some $(\Phi_1', \Phi_2') \in S$, $(\Phi_1, \Phi_2) \leq (\Phi_{1'}, \Phi_2')$ and $(\Phi_1', \Phi_2') \in Y$. We have to consider two cases.

Case 1: $\Phi_1 \subset \Phi_1'$. It implies that $x \in \Phi_1'$, i.e. $(\Phi_1', \Phi_2') \in \varphi_1(x)$.

Case 2: $\Phi_1 = \Phi_1'$ and $\Phi_2 \subset \Phi_2'$. Then, if $\Phi_2 \cap \mathfrak{A} = \Phi_2' \cap \mathfrak{A}$ then $\Phi_2 \notin \mathscr{F}_2$. If $\Phi_2 \cap \mathfrak{A} \subset \Phi_2' \cap \mathfrak{A}$ then $\Phi_1' \cap \mathfrak{A} \neq \Phi_2' \cap \mathfrak{A}$, i.e. $(\Phi_1', \Phi_2') \notin S$.

We conclude that, according to [52], Lemma 6, (7.9) ensures that the maps φ_i are Heyting embeddings of each $\mathfrak{B}_1$ and $\mathfrak{B}_2$ into $\mathscr{H}(S)$ so that $\varphi_1(x) = \varphi_2(x)$, for any $x \in \mathfrak{A}$.

On the other hand, (7.11) and (7.12) ensure that the images of the elements of $\mathfrak{B}_1$ and $\mathfrak{B}_2$ are enrichable in $\mathscr{H}(S)$. Further, by virtue of [58], Proposition 1, $\mathscr{H}(S)$ can be embedded into an enrichable Heyting algebra, in which, according to [58], Corollary 2, if y enriches x in $\mathscr{H}(S)$, this relation will be preserved for the corresponding images.

Simonova [69] had constructed a continuum of normal extensions of **KM** having the interpolation property, as well as an example of an extension without it.

7.4.8 The Lattice of Normal Extensions

Comparing Kuznetsov's Theorem with (7.6), we raise the question: Is the following equivalence true?

$$\mathbf{KM} + a \vdash b \Leftrightarrow \mathbf{GL} + s(t(a)) \vdash s(t(b))$$

Before answering this question, first we extend the Gödel-McKinsey-Tarski translation as follows.

Definition 14 (translation T: $\mathscr{L}_{\Box} \to \mathscr{L}_{\Box\bigcirc}$)

$$
\begin{array}{ll}
(a) & T(p) = \bigcirc p, \text{ for any } p \in \textbf{Var} \\
(b) & T(A \vartriangle B) = T(A) \vartriangle T(B), \text{ for any connective } \vartriangle \in \{\wedge, \vee\} \\
(c) & T(\neg A) = \bigcirc \neg T(A) \\
(d) & T(A \to B) = \bigcirc (T(A) \to T(B)) \\
(e) & T(\Box A) = \bigcirc \Box T(A).
\end{array}
$$

This definition, among other things, leads to an embedding of **KM** into the bimodal logic $\mathbf{GL}^{\bigcirc}$ of Footnote 13.[27]

The next proposition answers the question above.

Proposition 20 ([45, 57, 60]) *For any set Δ of $\mathscr{L}_{\Box}$-formulae and a formula B,*

$$\mathbf{KM} + \Delta \vdash B \Leftrightarrow \mathbf{GL} + \Delta^{*} \vdash s(T(B)),$$

where $\Delta^{*} = \{s(T(A)) \mid A \in \Delta\}$.

Corollary 8 *The equivalence* (7.6) *holds.*

Eventually, Proposition 20 leads to the following conclusion. We define:

$$\tau \colon \mathbf{KM} \oplus \Delta \mapsto \mathbf{GL} \oplus \Delta^{*}.$$

Proposition 21 ([45, 57, 60]) *The map τ establishes an isomorphism between the lattices of normal extensions of* **KM** *and* **GL**.

Let **H**, **S** and **P** be as usual the class-operators of forming homomorphic images, subalgebras and direct products. Then the map τ can be interpreted in algebraic terms as follows.

Proposition 22 ([57, 60]) *For any class Σ of Magari algebras, the following equalities hold*:

$$
\begin{array}{l}
\Box\mathbf{H}(\Sigma) = \mathbf{H}(\Box\Sigma), \\
\Box\mathbf{S}(\Sigma) = \mathbf{S}(\Box\Sigma), \\
\Box\mathbf{P}(\Sigma) = \mathbf{P}(\Box\Sigma).
\end{array}
$$

Therefore, $\Box\mathbf{HSP}(\Sigma) = \mathbf{HSP}(\Box\Sigma)$, *where operation $\Box$ is defined in (7.7).*

Definition 15 (tabular, pretabular logic) A logic is called tabular if it is the logic of a finite algebra. A logic is pretabular if it is not tabular but all its proper normal extensions are tabular.

---

27 Indeed, one can prove that $\mathbf{KM} \vdash A$ if and only if $\mathbf{GL}^{\bigcirc} \vdash T(A)$.

By using Propositions 4.14 and 4.15 we derive the following properties which we include in one proposition.

Proposition 23 ([57, 60]) *The following properties hold: For any $L = \mathbf{KM} \oplus \Delta$,*

(*a*) *L is finitely axiomatizable* $\Leftrightarrow$ *$\tau(L)$ is finitely axiomatizable*;
(*b*) *L is tabular* $\Leftrightarrow$ *$\tau(L)$ is tabular*;
(*c*) *L is pretabular* $\Leftrightarrow$ *$\tau(L)$ is pretabular*;
(*d*) *L is complete w.r.t.* $\mathbf{KM}$ *– frames* $\Leftrightarrow$ *$\tau(L)$ is complete w.r.t.* $\mathbf{GL}$ *– frames*;
(*e*) *Reflexivity Rule* (*Sect.* 3) *is admissible in L* $\Leftrightarrow$ *this rule is admissible in $\tau(L)$*.

From the fact that there are exactly countably many pretabular normal extension of **GL**, (cf. [9]) we derive the following.

Corollary 9 ([57, 60]) *There are exactly countably many pretabular normal extensions of* **KM**.

Since there exists a finitely axiomatizable extension of **KM**, which is incomplete with respect to **KM**-frames, (cf. [55]) we obtain the following.

Corollary 10 ([57, 60]) *There is a finitely axiomatizable normal extension of* **GL** *which is incomplete with respect to* **GL***-frames.*

(Chagrov and Zakharyaschev [15], Sect. 6.5, give a somewhat systematic way for obtaining examples of incomplete extensions of **GL**.)

Since there is a continuum of normal extensions of **GL**, in which the Reflexivity Rule is admissible, (cf. [56]) we have the following.

Corollary 11 ([57, 60]) *There is a continuum of normal extensions of* **KM**, *where the Reflexivity Rule is admissible.*

Let us introduce the following maps.

$$\sigma : \mathbf{Int} \oplus \Gamma \mapsto \mathbf{Grz} \oplus \Gamma^{\circ},$$

where Γ is any set of $\mathscr{L}_a$-formulae and $\Gamma^{\circ} = \{t(a) \mid a \in \Gamma\}$, is known to be an isomorphism of NExt**Int** onto NExt**Grz** (cf. [8, 15, 21, 82]).

Also, let NExt**GL** and NExt**KM** be the lattices of normal extensions of **GL** and **KM**, respectively. Next we define:

$$\lambda: \mathbf{KM} \oplus \Delta \mapsto \{A \in \mathscr{L}_{\Box} \mid \mathbf{KM} \oplus \Delta \vdash A\},$$
$$\mu: \mathbf{GL} \oplus \Delta \mapsto \{\alpha \in \mathscr{L}_{\bigcirc} \mid \mathbf{GL} \oplus \Delta \vdash s(\alpha)\},$$

where Δ is any set of $\mathscr{L}_{\Box}$-formulae.

Proposition 24 ([45]) *The diagram below, in which* λ *and* μ *are join-epimorphisms, is commutative.*

$$\begin{array}{ccc} \text{NExt}\mathbf{KM} & \xrightarrow{\ \tau\ } & \text{NExt}\mathbf{GL} \\ \lambda\big\downarrow & & \mu\big\downarrow \\ \text{NExt}\mathbf{Int} & \xrightarrow{\ \sigma\ } & \text{NExt}\mathbf{Grz} \end{array}$$

Taking into account Proposition 24, the next proposition does not seem surprising; also, it is a generalization of equivalence (7.5).

Proposition 25 ([59]) *For any set* Θ *of* $\mathscr{L}_{\bigcirc}$*-formulae,*

$$\mathbf{Grz} \oplus \Theta \vdash \beta \Leftrightarrow \mathbf{GL} \oplus \{s(\alpha) \mid \alpha \in \Theta\} \vdash s(\beta).$$

From Proposition 4.17 it follows that for any $L \in \text{NExt}\mathbf{Int}$, there is $M \in \text{NExt}\mathbf{GL}$ such that $L = \lambda \circ \tau^{-1}(M)$; cf. [45], Corollary 3. Also, since it is well known [35] that NExt**Int** forms a continuum and that there is a continuum of normal extensions of $\mathbf{GL}^*$ [56], the following question can be raised: If we restrict $\tau^{-1} \circ \lambda$ to NExt$\mathbf{GL}^*$, which is a subsemilattice (with respect to join) of NExt**GL**, will the range of this mapping be all NExt**Int**? The answer is negative [44], since

$$\lambda \circ \tau^{-1}(\mathrm{d}) = \mathbf{LC},$$

where $\mathbf{D}^\star$ is the logic of the frame $(\omega, >)$, which was briefly discussed in Sect. 7.3, and $\mathbf{LC} = \mathbf{Int} \oplus (p \to q) \vee (q \to p)$.

In [26] Esakia introduced the system

$$\mathbf{K4.Grz} = \mathbf{K4}_{\Box} \oplus \Box(\Box(p \to \Box p) \to p) \to \Box p.$$

It has been announced in [26], Sect. 3, and proved in [47], Corollary 6, that the lattices NExt**mHC** and NExt**K4.Grz** are isomorphic. Thus the diagram in Proposition 4.17 can be extended as follows:

$$\begin{array}{ccc} \text{NExt}\mathbf{mHC} & \xrightarrow{\ \tau'\ } & \text{NExt}\mathbf{K4.Grz} \\ \lambda'\big\downarrow & & \mu'\big\downarrow \\ \text{NExt}\mathbf{Int} & \xrightarrow{\ \sigma\ } & \text{NExt}\mathbf{Grz} \end{array}$$

Here τ' is the isomorphism mentioned above and λ' and μ' are defined as follows:

$$\begin{aligned} \lambda' &: \mathbf{mHC} \oplus \Delta \mapsto \{A \in \mathscr{L}_{\Box} \mid \mathbf{mHC} \oplus \Delta \vdash A\}, \\ \mu' &: \mathbf{K4.Grz} \oplus \Delta \mapsto \{\alpha \in \mathscr{L}_{\bigcirc} \mid \mathbf{K4.Grz} \oplus \Delta \vdash s(\alpha)\}, \end{aligned}$$

where, as above, Δ is any set of $\mathscr{L}_{\Box}$-formulae.

Now the question in Problem 1 acquires its importance. If the answer to it is positive, then λ' and μ' are surjective join-homomorphisms and the last diagram is commutative.

Problem 2 *Are the maps λ' and μ' surjective join-homomorphisms?*

7.5 Conclusion

The last touch in the presented biography of **KM** is the following. In [26] Esakia made an attempt to embed **mHC** into intuitionistic propositional quantification logic **IntQ**. As he pointed out, interpreting $\Box$ as

$$\Box A = \forall p(p \vee (p \to A)), \tag{7.13}$$

where p does not occur freely in A, all axioms of **mHC** are derivable in **IntQ** and modus ponens preserves this derivability. Thus the embedding is sound with respect to the above translation. The question about faithfulness remains open.

For **KM** however this embedding is not even sound. Indeed, let us consider the instance of the **KM**-axiom

$$(\Box A \to A) \to A$$

when $A = \bot$, that is

$$(\Box\bot \to \bot) \to \bot.$$

The translation of the last formula is equivalent in **IntQ** to the formula

$$\neg\neg\forall p(p \vee \neg p).$$

In his 1969 abstract [38], Kuznetsov noted that the last formula, although is valid in all finite Heyting algebras with two additional operations for quantifiers, is invalid in any segment of rational numbers and hence, by virtue Theorem 1 of [38], is not derivable in **IntQ**. In the light of the Kuznetsov observation, we conclude with the following.

Problem 3 *Does the translation* (7.13) *generate an embedding of* **KM** *into the fragment of those formulas of* **IntQ** *which are valid in all finite Heyting algebras endowed with operations for quantifiers?*

Acknowledgments Over all period of the preparation of this text I have been in close contact with Alex Citkin. Discussions with him helped me understand better some key points and he provided me with useful references. Also, I am indebted to Grigori Mints for informing me that the last formula in Sect. 7.5 is not derivable in **IntQ**. It happened before I noticed Kuznetsov's remark in [38]. Also, I am grateful to Srećko Kovač for drawing my attention to Gödel's ontological proof (see Footnote 2).

References

1. Abashidze M (1985) On one adequate interpretation of provability logic. In: Intensional logics and the logical structure of theories, abstracts of the fourth Soviet-Finish colloquium, Tbilisi, pp 3–4 (Russian)
2. Abashidze M (1988) Ordinal completeness of the Gödel-Löb modal system. In: Intensional logics and the logical structure of theories (Telavi, 1985). Metsniereba, Tbilisi, pp 49–73 (Russian)
3. Artemov S (2001) Explicit provability and constructive semantics. Bull Symbol Logic 7(1):1–36
4. Artemov S, Beklemishev L (2005) Provability logic. In: Gabbay D, Guenthner F (eds) Handbook of philosophical logic, vol 13, 2nd edn. Springer, Dordrecht, pp 189–360
5. Beklemishev L, Gabelaia D (2014) Topological interpretations of provability logic (this volume)
6. Bessonov A (1977) New operations in the intuitionistic calculus. Mat Zametki 22(1):23–28 (Russian). English edition: Bessonov A (1977) Math Notes 22(1):503–506
7. Blass A (1990) Infinitary combinatorics and modal logic. J Symbol Logic 55(2):761–778
8. Blok W (1976) Varieties of interior algebras. PhD thesis, University of Amsterdam
9. Blok W (1980) Pretabular varieties of modal algebras. Studia Logica 39(2/3):101–124
10. Boolos G (1979) The unprovability of consistency (an essay in modal logic). Cambridge University Press, Cambridge, Mass
11. Boolos G (1980) Provability in arithmetic and a schema of Grzegorczyk. Fund Math 106(1): 41–45
12. Boolos G (1993) The logic of provability. Cambridge University Press, Cambridge, Mass
13. Boolos G, Jeffrey R (1989) Computability and logic, 3rd edn. Cambridge University Press, Cambridge, Mass
14. Boolos G, Sambin G (1991) Provability: the emergence of a mathematical modality. Studia Logica 50(1):1–23
15. Chagrov A, Zakharyaschev M (1997) Modal logic. Oxford logic guides, vol 35. The Clarendon Press, Oxford University Press, New York
16. Church A (1996) Introduction to mathematical logic. Princeton University Press, Princeton, NJ
17. Curry H (1963) The foundations of mathematical logic. McGraw-Hill Book Co., New York
18. Darjania G (1984) A sequential variant of modal system I^{Δ}. In: Metalogical Studies. Metsniereba, Tbilisi (Russian)
19. Dummett M, Lemmon E (1959) Modal logics between s4 and s5. Zeitschrift für mathematische logik und grundlagen der mathematik 5:250–264
20. Engelking R (1989) General topology, sigma series in pure mathematics, vol 6, 2nd edn. Heldermann Verlag, Berlin
21. Esakia L (1976) On modal "counterparts" of superintuitionistic logics. In: Abstracts of the 7th all-union symposium on logic and methodology of science, Kiev, pp 135–136 (Russian)
22. Esakia L (1978) Diagonal constructions, Löb's formula, and Cantor's scattered spaces. In: Abstracts of the coordinating meeting, modal and intensional logics, Moscow, Moscow, pp 177–179 (Russian)
23. Esakia L (1981) Diagonal constructions, Löb's formula and Cantor's scattered spaces. In: Studies in logic and semantics. Metsniereba, Tbilisi, pp 128–143 (Russian)
24. Esakia L (1985) Heyting algebras. Duality Theory, I. Metsniereba, Tbilisi (Russian)
25. Esakia L (2004) Intuitionistic logic and modality via topology. Ann Pure Appl Logic 127(1–3):155–170
26. Esakia L (2006) The modalized heyting calculus: a conservative modal extension of the intuitionistic logic. J Appl Non-Classical Logics 16(3–4):349–366
27. Esakia L, Jibladze M, Pataraia D (2000) Scattered toposes. Ann Pure Appl Logic 103(1–3):97–107

28. Gabbay D, Maksimova L (2005) Interpolation and definability, Oxford logic guides, vol 46. The Clarendon Press, Oxford University Press, Oxford
29. Gödel K (1933) Eine Interpretation des intuitionistischen Aussagenkalküls. In: Ergebnisse eines mathematischen Kolloquiums 4, s. 39–40 (in the German) (English translation is available in [31])
30. Gödel K (1938) Vortrag bei Zilsel. In: Gödel's Nachlass (in the German) (English translation is available in [32])
31. Gödel K (1986) In: Feferman S (ed) Collected works, vol 1, publications 1929–1936. With a preface by Solomon Feferman. The Clarendon Press, Oxford University Press, New York
32. Gödel K (1995) In: Feferman S, Dawson J Jr, Goldfarb W, Parsons C, Solovay R (eds) Collected works, vol 3. Unpublished essays and lectures. With a preface by Solomon Feferman. The Clarendon Press, Oxford University Press, New York
33. Goldblatt R (1978) Arithmetical necessity, provability and intuitionistic logic. Theoria 44(1):38–46
34. Grätzer G (1978) General lattice theory. Birkhäuser Verlag, Basel
35. Jankov V (1968) The construction of a sequence of strongly independent superintuitionistic propositional calculi. Doklady Akad. Nauk SSSR 181:33–34 (Russian). English edition: Jankov V (1968) Soviet Math Doklady 9:806–807
36. Jónsson B (1965) Extensions of relational structures. Theory of models (proceedings of the 1963 international symposium at Berkeley). North-Holland, Amsterdam, pp 146–157
37. Kleene S (1952) Introduction to metamathematics. D. Van Nostrand Co., Inc., New York
38. Kuznetsov A (1969) On propositional logics with quantifiers over propositional variables. In: The all-union symposium on mathematical logic, abstracts of communications, Alma-Ata, 2–7 June 1969, pp 23–26. Alma-Ata (Russian)
39. Kuznetsov A (1978) Proof-intuitionistic logic. In: Modal and intensional logics, abstracts of the coordinating meeting, Moscow, pp 75–79 (Russian)
40. Kuznetsov A (1979) On algebras of open sets. In: Abstracts of the fourth Tiraspol symposium on general topology and its applications, Chisinau, pp 72–75 (Russian)
41. Kuznetsov A (1985) The proof-intuitionistic propositional calculus. Dokl Akad Nauk SSSR 283(1):27–30 (Russian). English edition: Kuznetsov AV (1985) Soviet Math Doklady 32(1): 27–30
42. Kuznetsov A, Muravitsky A (1976) The logic of provability. In: Abstracts of the 4th all-union conference on mathematical logic. Shtiintsa, Chisinau, p 73 (Russian)
43. Kuznetsov A, Muravitsky A (1977) Magari algebras. In: Abstracts of the 14th algebraic conference. Institute of Mathematics, Novosibirsk, pp 105–106 (Russian)
44. Kuznetsov A, Muravitsky A (1980) Provability as modality. In: Current problems in logic and methodology of sciences. Naukova Dumka, Kiev, pp 193–230 (Russian)
45. Kuznetsov A, Muravitsky A (1986) On superintuitionistic logics as fragments of proof logic extensions. Studia Logica 45(1):77–99
46. Lemmon E (1966) Algebraic semantics for modal logics i. J Symb Logic 31:46–65
47. Litak T (2014) Constructive modalities with provability smack (this volume)
48. Löb M (1955) Solution of a problem of leon henkin. J Symb Logic 20:115–118
49. Magari R (1975) The diagonalizable algebras. Boll Unione Mat Ital (4) 12(3):117–125
50. Magari R (1975) Representation and duality for diagonalizable algebras. Studia Logica 34:305–313
51. Makinson D (1966) There are infinitely many diodorean modal functions. J Symb Logic 31:406–408
52. Maksimova L (1977) Craig's theorem in superintuitionistic logics and amalgamable varieties. Algebra i Logika 16(6):643–681 (Russian). English edition: Maksimova LL (1977) Algebra Logic 16(6):427–455
53. Maksimova L, Rybakov V (1974) On the lattice of normal modal logics. Algebra i Logika 13(2):188–216 (Russian). English edition: Maksimova L, Rybakov V (1974) Algebra Logic 13(2):105–122

54. McKinsey J, Tarski A (1948) Some theorems about the sentential calculi of lewis and heyting. J Symb Logic 13:1–15
55. Muravitsky A (1981) Finite approximability of the calculus I^{Δ} and the nonmodelability of some of its extensions. Mat Zametki 29(6):907–916 (Russian). English edition: Muravitsky A (1981) Math Notes 29(5–6):463–468
56. Muravitsky A (1983) Extensions of the logic of provability. Mat Zametki 33(6):915–927 (Russian). English edition: Muravitsky A (1983) Math Notes 33(5–6):469–475
57. Muravitsky A (1985) Correspondence of proof-intuitionistic logic extensions to proof-logic extensions. Dokl Akad Nauk SSSR 281(4):789–793 (Russian). English edition: Muravitsky A (1985) Soviet Math Doklady 31(2):345–348
58. Muravitsky A (1988) Algebraic proof of the separation property for the proof-intuitionistic calculus. Mat Sb (N.S.) 131 (173)(3):403–412 (Russian). English edition: Muravitsky A (1988) Math USSR-Sbornik 59(2):397–406
59. Muravitsky A (1988) Embedding of extensions of Grzegorczyk's logic into extensions of provability logic. In: Intensional logics and the logical structure of theories (Telavi, 1985). Metsniereba, Tbilisi, pp 74–80 (Russian)
60. Muravitsky A (1989) Correspondence of proof-intuitionistic logic extensions to proof-logic extensions. Mat Logika i Algoritm Probl, vol 12. Trudy Instituta Matematiki, Novosibirsk, pp 104–120 (Russian)
61. Muravitsky A (2008) Alexander vladimirovich kuztnetsov. Logic Log Philos 17(1–2):9–21
62. Muravitsky A (2008) The contribution of A. V. Kuznetsov to the theory of modal systems and structures. Logic Log Philos 17(1–2):41–58
63. Novikov P (1977) Konstruktivnaya matematicheskaya logika s tochki zreniya klassicheskoi. Nauka, Moscow (Russian) [Constructive mathematical logic from the point of view of classical logic]
64. Rasiowa H, Sikorski R (1970) The mathematics of metamathematics, Monografie Matematyczne, 3rd ed., Tom 41. PWN–Polish Scientific Publishers, Warsaw
65. Sambin G (1976) An effective fixed-point theorem in intuitionistic diagonalizable algebras (the algebraization of the theories which express theor, ix). Studia Logica 35(4):345–361
66. Segerberg K (1971) An essay in classical modal logic, vol 1–3. Filosofiska Föreningen och Filosofiska Institutionen vid Uppsala Universitet, Filosofiska Studier, Uppsala
67. Simmons H (1975) Topological aspects of suitable theories. Proc Edinburgh Math Soc 19(2):383–391
68. Simonova I (1987) The separation property for the provability-intuitionistic calculus. Mat Issled Neklass Logiki 98:121–133 (Russian)
69. Simonova I (1990) On the interpolation property for extensions of proof-intuitionistic logic. Mat Zametki 47(5):88–99 (Russian). English edition: Simonova I (1990) Math Notes 47(5–6):483–490
70. Simonova I (2008) The separation property for the provability-intuitionistic calculus. Vīsn Odes Nats Unīv, Matematika i Mekhanika 13(17):73–85 (Russian)
71. Simpson A (1994) The proof theory and semantics of intuitionistic modal logic. PhD thesis. The University of Edinburgh, Department of Computer Science
72. Smiley T (1963) The logical basis of ethics. Acta Philos Fenn Fasc 16:237–246
73. Smoryński C (1984) Modal logic and self-reference. Handbook of philosophical logic, vol II, Synthese Library, vol 165. Reidel, Dordrecht, pp 441–495
74. Smoryński C (1985) Self-reference and modal logic. Universitext, Springer, New York
75. Sobociński B (1974) A theorem concerning a restricted rule of substitution in the field of propositional calculi, I, II. Notre Dame J Formal Logic 15:465–476; ibid. 15:589–597 (1974)
76. Solovay R (1976) Provability interpretations of modal logic. Israel J Math 25(3–4):287–304
77. Sotirov V (1980) Modal theories with intuitionistic logic. Mathematical Logic (Sofia, 1980). Publishing House of the Bulgarian Academy of Science, Sofia, pp 139–171
78. Steen L, Seebach J Jr (1970) Counterexamples in topology. Holt, Rinehart and Winston Inc., New York
79. Weisstein E (2003) CRC concise encyclopaedia of mathematics. CRC Press, Boca Raton

80. Wolter F, Zakharyaschev M, (1999) Intuitionistic modal logic. In: Logic and foundations of mathematics (Florence (1995) Synthese library, vol 280. Kluwer Academic Publishers, Dordrecht, pp 227–238
81. Wolter F, Zakharyaschev M (1999) Intuitionistic modal logics as fragments of classical bimodal logics. In: Orlowska E (ed) Logic at Work. Springer, New York, pp 168–186
82. Wolter F, Zakharyaschev M (2014) On the Blok-Esakia theorem (this volume)
83. Wroński A (1973) Intermediate logics and the disjunction property. Rep Math Logic 1:39–51

Chapter 8
Constructive Modalities with Provability Smack

Tadeusz Litak

To the memory of Leo Esakia and Dito Pataraia

Abstract I overview the work of the Tbilisi school on intuitionistic modal logics of well-founded/scattered structures and its connections with contemporary Theoretical Computer Science. Fixed-point theorems and their consequences are of particular interest.

Keywords Constructive fixpoints · Intuitionistic modal logic · Point-free derivative · Topos of trees · Scattered toposes

8.1 Introduction

Readers of this volume are probably aware that much of Leo Esakia's research concentrated on semantics for the intuitionistic logic **IPC**, the modal logic $\mathbf{GL}^{\mathsf{cl}}$ of Löb, its weakening $\mathbf{wGrz}^{\mathsf{cl}}$ and intuitionistic-modal systems like the logic **KM** or its weakening **mHC**; see Table 8.2 for all definitions. $\mathbf{GL}^{\mathsf{cl}}$ is also known as the *Gödel-Löb* logic, but this name may suggest more personal involvement with the system than Gödel ever had; **KM** or **mHC** will be discussed in Sect. 8.4. A central feature of semantics for such systems is well-foundedness or *scatteredness*. While in the case of **IPC** well-foundedness is a sufficient, but not necessary condition—intuitionistic logic is complete wrt well-founded or even finite partial orders, but sound wrt much bigger class of structures—$\mathbf{GL}^{\mathsf{cl}}$ and **KM** require it even for soundness. This is due to the fact that the latter two systems include a form of an explicit induction axiom: in the case of $\mathbf{GL}^{\mathsf{cl}}$ the well-known Löb axiom (which here will be called the *weak*

T. Litak (✉)
Friedrich-Alexander-Universität, Martensstraße 3, 91058, Erlangen, Germany
e-mail: tadeusz.litak@gmail.com

G. Bezhanishvili (ed.), *Leo Esakia on Duality in Modal and Intuitionistic Logics*, Outstanding Contributions to Logic 4, DOI: 10.1007/978-94-017-8860-1_8,

Löb axiom) and in the case of **KM** the *strong Löb axiom*—less well-known to modal logicians, but as we are going to see, better known to type theorists. Scatteredness is the topological generalization of well-foundedness; Simmons [69] provided tools necessary to define its point-free counterpart and the Tbilisi school noted that this notion also makes sense in the topos setting. In fact, the most generic way of defining scateredness is via modal syntax: as validity of the Löb principle for a suitable "later" modality.

The interplay of relational, topological, point-free and algebraic aspects in the above paragraph should not feel unnatural to anybody familiar with Leo's attitude to research. Let us look at an important example how results can travel from one setting to another. In the mid-1970s, it was established that Löb-like logics enjoy the so-called Fixpoint Theorem. At first, the intention was to grasp the algebraic content of Gödel's Diagonalization Lemma. Yet in its own right it turned out to be one of the most fascinating results ever proved about such systems. Section 8.3 gives an overview of some of its applications and consequences. For now, let us just mention that Leo Esakia used it, e.g., to characterize algebras for **KM**, see Theorem 4 and Corollary 1 here. Furthermore, it seems to have inspired the work on scattered toposes: [30, Sect. 3] claims to present its *topos-theoretic counterpart*. However, as the result central for the topos version (Theorem 8 here) does not even include modalities in its formulation, the word *counterpart* has to be understood rather loosely.

As we will see, in hindsight [30] turns out to be closely connected to very recent developments in Theoretical Computer Science, in particular the work of Birkedal et al. on the *topos of trees* [14], itself an example of a scattered topos. Thus, it seems particularly regrettable that the spadework of the Tbilisi school has not been carried further and is not more widely known.

The chapter is structured as follows. Section 8.2 recalls syntactic and semantic basics of *intuitionistic* normal modal logics. Section 8.3 focuses on fixpoint results for Löb-like systems. Section 8.4 introduces the work of the Georgian school on extensions of **mHC**. Finally, Sect. 8.5 discusses scattered toposes, beginning with an overview of the topos logic.

While the chapter is intended as an overview and claims to novelty are minimal, they are perhaps not entirely non-existent. Theorem 4 is the most general form of [31, Proposition 3] I could think of and Sect. 8.4.3 reproves results on extensions of **mHC** using the framework of [81, 82]; in fact, it seems that Corollary 6 is the first published proof of the corresponding extension of the Blok-Esakia Theorem announced in [31].

Remark 1 As a part of a larger project, I formalized most of syntactic derivations in the paper—in particular those relevant for Sect. 8.5.2— in the Coq proof assistant. Readers interested in this ongoing project are welcome to contact me.

Table 8.1 Axioms for $\mathbf{K}^{\mathsf{i}}$

Axioms of the intuitionistic propositional calculus, see, e.g., [23, Sect. 1.3, (A1)–(A9)]	
Axiom for $\Box$	
(nrm)	$\Box(A \to B) \to (\Box A \to \Box B)$
Inference rule for $\mathcal{L}_{\mathsf{int}}$-fragment	Inference rule for modality
MP $\frac{A \to B,\ A}{B}$	**NEC** $\frac{A}{\Box A}$

8.2 A Primer on Intuitionistic Modalities

Modal formulas over a supply of propositional variables Σ are defined by

$$A, B ::= \bot \mid p \mid A \to B \mid A \wedge B \mid A \vee B \mid \Box A$$

where $p \in \Sigma$. The set is denoted by $\mathcal{L}_{\Box\mathsf{int}}\Sigma$, but unless explicitly stated otherwise, I will keep Σ fixed throughout and drop it from the notation. The purely intuitionistic language (i.e., without 2) will be denoted by $\mathcal{L}_{\mathsf{int}}$. Note that the syntax extended with a $\Diamond$ operator, intuitionistically *not* definable from $\Box$, is of no interest for us here.

$\Gamma \subseteq \mathcal{L}_{\Box\mathsf{int}}$ is *a normal $\mathcal{L}_{\Box\mathsf{int}}$-logic* or an *intuitionistic normal modal logic* if it is closed under rules and axioms from Table 8.1 plus the rule of substitution. For any $\Gamma, \Delta \subseteq \mathcal{L}_{\Box\mathsf{int}}$, $\Gamma \oplus \Delta$ will denote the closure of $\Gamma \cup \Delta$ under substitution and the rules **MP** and **NEC**. In the case of $\Delta = \{\alpha\}$, I will also write $\Gamma \oplus \alpha$. Occasionally, I will write $\Gamma + \Delta$ for the closure under substitution and **MP**, but without **NEC**. This notation is analogous to the one used in [23].

$\mathbf{K}^{\mathsf{i}}$ is the smallest intuitionistic normal modal logic, i.e., **IPC** $\oplus$ **(nrm)**. **IPC**—the intuitionistic propositional calculus—can be thus defined as the intersection of $\mathbf{K}^{\mathsf{i}}$ and $\mathcal{L}_{\mathsf{int}}$. Table 8.2 provides a list of additional axioms and logics which will be of interest to us. $\mathbf{GL}^{\mathsf{i}}$, $\mathbf{SL}^{\mathsf{i}}$, **mHC** and **KM** are of particular importance. As we see in Table 8.2, there are several ways in which these and related systems can be axiomatized. In particular, we have

Theorem 1 (Ursini [77], **following Smorynski for the classical case)** *The following formalisms have the same set of theorems:*

1. $\mathbf{GL}^{\mathsf{i}}$ *as defined in Table 8.2*
2. $\mathbf{K4}^{\mathsf{i}} \oplus \frac{\Box A \to A}{A}$
3. $\mathbf{K4}^{\mathsf{i}} \oplus \frac{\Box A \to A}{\Box A}$
4. $\mathbf{K4}^{\mathsf{i}} \oplus$ **(ufp)** $= \boxdot(B \leftrightarrow A[B/p]) \to (\boxdot(C \leftrightarrow A[C/p]) \to (B \leftrightarrow C))$
5. $\mathbf{K4}^{\mathsf{i}} \oplus$ **(henk)** $= \boxdot(A \leftrightarrow \Box A) \to A$

A variable $p \in \Sigma$ is $\Box$-*guarded* in $A \in \mathcal{L}_{\Box\mathsf{int}}$ if all its occurrences are within the scope of $\Box$. This notion will be used repeatedly in connection with $\mathbf{GL}^{\mathsf{i}}$ and its extensions.

Table 8.2 $\mathcal{L}_{\Box\mathsf{int}}$ axioms and logics. See, e.g., [75, 81, 82] for more (also in the syntax extended with a $\Diamond$ operator). $\boxdot A$ below abbreviates $A \wedge \Box A$

(cl)	$((B \to A) \to B) \to B$	**(em)**	$A \vee \neg A$
	$\mathbf{Cl} = \mathbf{IPC} + \textbf{(cl)} = \mathbf{IPC} + \textbf{(em)}$		
(nrm)	$\Box(A \to B) \to (\Box A \to \Box B)$	**(opr)**	$\Box(A \wedge B) \leftrightarrow (\Box A \wedge \Box B)$
	$\mathbf{K}^{\mathsf{i}} = \mathbf{IPC} \oplus \textbf{(nrm)} = \mathbf{IPC} \oplus \textbf{(opr)}$		$\mathbf{K}^{\mathsf{cl}} = \mathbf{K}^{\mathsf{i}} \oplus \mathbf{Cl}$
(trns)	$\Box A \to \Box\Box A$		
	$\mathbf{K4}^{\mathsf{i}} = \mathbf{K}^{\mathsf{i}} \oplus \textbf{(trns)}$		$\mathbf{K4}^{\mathsf{cl}} = \mathbf{K4}^{\mathsf{i}} \oplus \mathbf{Cl}$
(bind)	$\Box\Box A \to \Box A$		$\mathbf{C4}^{\mathsf{i}} = \mathbf{K}^{\mathsf{i}} \oplus \textbf{(bind)}$
(r)	$A \to \Box A$	**(fmap)**	$(A \to B) \to (\Box A \to \Box B)$
	$\mathbf{R}^{\mathsf{i}} = \mathbf{K}^{\mathsf{i}} \oplus \textbf{(r)} = \mathbf{K}^{\mathsf{i}} \oplus \textbf{(fmap)}$	Note that above $\mathbf{R}^{\mathsf{i}}$, $\oplus$ is the same as $+$	
	In using the symbol **R**, I follow [32]		
(refl)	$\Box A \to A$		$\mathbf{S4}^{\mathsf{i}} = \mathbf{T}^{\mathsf{i}} \oplus \mathbf{K4}^{\mathsf{i}}$
	$\mathbf{T}^{\mathsf{i}} = \mathbf{K}^{\mathsf{i}} \oplus \textbf{(refl)}$		$\mathbf{Triv}^{\mathsf{i}} = \mathbf{T}^{\mathsf{i}} + \mathbf{R}^{\mathsf{i}}$
(pll)	$(A \vee \Box\Box A) \to \Box A$		
	$\mathbf{PLL}^{\mathsf{i}} = \mathbf{K}^{\mathsf{i}} \oplus \textbf{(pll)} = \mathbf{C4}^{\mathsf{i}} \oplus \mathbf{R}^{\mathsf{i}}$		
(wlöb)	$\Box(\Box A \to A) \to \Box A$	**(henk)**	$\boxdot(A \leftrightarrow \Box A) \to A$
(ufp)	$\boxdot(B \leftrightarrow A[B/p]) \to (\boxdot(C \leftrightarrow A[C/p]) \to (B \leftrightarrow C))$		
	$\mathbf{GL}^{\mathsf{i}} = \mathbf{K}^{\mathsf{i}} \oplus \textbf{(wlöb)} = \mathbf{K4}^{\mathsf{i}} \oplus \textbf{(henk)} = \mathbf{K4}^{\mathsf{i}} \oplus \textbf{(ufp)}$		
	(see Theorem 1 below)		$\mathbf{GL}^{\mathsf{cl}} = \mathbf{GL}^{\mathsf{i}} \oplus \mathbf{Cl}$
(slöb)	$(\Box A \to A) \to A$	**(glb)**	$(\Box A \to A) \to \Box A$
	$\mathbf{SL}^{\mathsf{i}} = \mathbf{K}^{\mathsf{i}} \oplus \textbf{(slöb)} = \mathbf{K}^{\mathsf{i}} \oplus \textbf{(glb)} = \mathbf{GL}^{\mathsf{i}} + \mathbf{R}^{\mathsf{i}}$		
	The form **(glb)** comes from Goldblatt [42]		
(grz)	$\Box(\Box(A \to \Box A) \to A) \to \Box A$	**(sgrz)**	$\Box(\Box(A \to \Box A) \to A) \to A$
	$\mathbf{wGrz}^{\mathsf{cl}} = \mathbf{K4}^{\mathsf{cl}} \oplus \textbf{(grz)}$		$\mathbf{sGrz}^{\mathsf{cl}} = \mathbf{K}^{\mathsf{cl}} \oplus \textbf{(sgrz)}$
	Note we only consider here classical variants of **(grz)**		
(next)	$\Box A \to (((B \to A) \to B) \to B)$	**(derv)**	$\Box A \to ((B \to A) \vee B)$
	$\mathbf{CB}^{\mathsf{i}} = \mathbf{K}^{\mathsf{i}} \oplus \textbf{(next)} = \mathbf{K}^{\mathsf{i}} \oplus \textbf{(derv)}$		
	$\mathbf{CB}^{\mathsf{i}}$ stands for *Cantor-Bendixson*, see Sect. 8.4		$\mathbf{mHC} = \mathbf{R}^{\mathsf{i}} + \mathbf{CB}^{\mathsf{i}}$
	$\mathbf{CBL}^{\mathsf{i}} = \mathbf{CB}^{\mathsf{i}} \oplus \mathbf{GL}^{\mathsf{i}}$		$\mathbf{KM} = \mathbf{CB}^{\mathsf{i}} \oplus \mathbf{SL}^{\mathsf{i}}$
	$\mathbf{KM} = \mathbf{SL}^{\mathsf{i}} + \mathbf{CB}^{\mathsf{i}} = \mathbf{SL}^{\mathsf{i}} + \mathbf{mHC}$		
(ver)	$\Box A$	**(boxbot)**	$\Box\bot$
	$\mathbf{Ver}^{\mathsf{i}} = \mathbf{K}^{\mathsf{i}} \oplus \textbf{(ver)} = \mathbf{K}^{\mathsf{i}} \oplus \textbf{(boxbot)}$		
(nnv)	$\neg\neg\Box\bot$	**(nv)**	$\neg\Box\bot$
	$\mathbf{NNV}^{\mathsf{i}} = \mathbf{K}^{\mathsf{i}} \oplus \textbf{(nnv)}$		$\mathbf{NV}^{\mathsf{i}} = \mathbf{K}^{\mathsf{i}} \oplus \textbf{(nv)}$

Remark 2 Equalities in Table 8.2 should be in fact treated as a large lemma on interderivability for a number of intuitionistic normal modal axioms. Of particular interest for this chapter are: the derivability of **(r)** from **(slöb)**, which mirrors derivability of **(trns)** from **(wlöb)**; interderivability between **(glb)** and **(slöb)**; equivalence between either of these and the conjunction of **(wlöb)** and **(r)**; two different ways of axiomatizing $\mathbf{CB}^{\mathsf{i}}$ by using **(next)** and **(derv)**. All these statements are made assuming $\mathbf{K}^{\mathsf{i}}$.

All normal $\Box\mathsf{int}$-logics as defined above—more precisely, their associated global consequence relations—are strongly finitely algebraizable; see standard references like [17, 34–36, 67] for basic notions of algebraic logic and a more detailed discussion. Strong finite algebraizability also applies to normal logics in any fragment of $\Box\mathsf{int}$ containing $\to$. This is due to the fact that all these systems are *implicative logics* in the sense of Rasiowa [67]. Given a normal logic Γ, I will call the corresponding class of algebras obtained via the algebraization process *Γ-algebras*, e.g., $\mathbf{K}^{\mathsf{i}}$-algebras, $\mathbf{GL}^{\mathsf{i}}$-algebras, $\mathbf{wGrz}^{\mathsf{cl}}$-algebras etc. $\mathbf{K}^{\mathsf{i}}$-algebras are obviously special cases of HAOs—Heyting algebras with operators—namely HAOs with a single unary operator. Recall that an *operator* on a Heyting algebra is an operation preserving $\top$ and finite meets. An operator on a Heyting algebra which turns it into a Γ-algebra will be called a *Γ-operator.*

Finally, recall that for any algebra $\mathfrak{H}$, a *$\mathfrak{H}$-polynomial* is a term in the similarity type of $\mathfrak{H}$ *enriched with a separate constant for each element of* $\mathfrak{H}$ [21, Definition 13.3]. In my notation for polynomials, I will not distinguish between an element of $\mathfrak{H}$ and its corresponding constant. Moreover, I will use elements of Σ (i.e., propositional variables) as variables of polynomials, consistent with the general policy of blurring the distinction between logical formulas and algebraic terms. The notion of $\Box$-guardedness for polynomials will be used in the same way as for formulas.

A Kripke frame or *a relational structure* is of the form $(W, \trianglelefteq, \prec)$, where

- $\trianglelefteq$ is a poset order used to interpret intuitionistic connectives
- $\prec$ is the modal accessibility relation used to interpret $\Box$ and
- $\trianglelefteq\, ;\, \prec \subseteq \prec$, where; is relation composition.

A *valuation* is a mapping $V{:}\Sigma \to \mathtt{Up}_{\trianglelefteq}(W)$, where $\mathtt{Up}_{\trianglelefteq}(W)$ is the Heyting algebra of upward closed sets of W and the inductive extension to $\mathcal{L}_{\Box\mathsf{int}}\Sigma$ is standard. See [20, 22, 42, 75, 81, 82] and in particular [71, Sect. 3.3] for more on the subject of necessary and sufficient conditions to be imposed on the interplay of $\prec$ and $\trianglelefteq$. In brief: even a weaker interplay condition found first in [20] would suffice to ensure that denotations of all modal formulas are upward closed. However, one is lead to the condition assumed above and standard in most references by, for example, the canonical model construction, see [20, 75]. Furthermore, this condition has the advantage of simplifying correspondence theory and yet is harmless from the point of view of validity, see [20, 42]. The situation would be different if our language included $\Diamond$.

Remark 3 When at least one of lattice connectives is removed, the situation at first sight appears more complicated. While the papers proving the separation property of **IPC** [44, 58, 66] showed that its reducts remain complete wrt relational semantics, a Stone-type representation theorem for arbitrary algebras would seem more problematic; the one for Heyting algebras relies crucially on the fact that they have distributive lattice reducts. However, a series of papers beginning with [48] and finishing with [12] established that Brouwerian semilattices enjoy in fact Stone-, Priestley- and Esakia-type dualities.

Table 8.3 lists semantic conditions corresponding to modal axioms. For $\mathbf{GL}^{\mathsf{i}}$ in particular, we have:

Table 8.3 Semantic counterparts of intuitionistic modal axioms. See, e.g., [29, 75, 81, 82] for more. $\Delta := \{(w, w) \mid w \in W\}$ and $\lhd := \unlhd - \Delta$, i.e., it is the strict version of the intuitionistic poset order. For $R \in \{\prec, \unlhd\}$, $V \subseteq W$, $V^R\uparrow := \{w \in W \mid \exists v \in V.vRw\}$ and $V^R\downarrow := \{w \in W \mid \exists v \in V.wRv\}$

Axiom	Semantic condition	Axiom	Semantic condition	Axiom	Semantic condition
(cl)	$\unlhd = \Delta$	**(trns)**	$\prec ; \prec \subseteq \prec$	**(bind)**	$\prec \subseteq \prec ; \prec$
(r)	$\prec \subseteq \unlhd$	**(refl)**	$\unlhd \subseteq \prec$	**(wlöb)**	see Theorem 5
(next)	$\lhd \subseteq \prec$	**(ver)**	$\prec = \emptyset$	**(nv)**	$W^{\prec}\downarrow = W$
(nnv)	$(W - W^{\prec}\downarrow)^{\unlhd}\downarrow = W$				

Theorem 2 ([77]) *A structure* $(W, \unlhd, \prec)$ *validates* $\mathbf{GL}^{\mathsf{i}}$ *iff*

- $\prec$ *is transitive, i.e.,* $\prec ; \prec \subseteq \prec$ *and*
- $\prec$ *is* $\mathtt{Up}_{\unlhd}(W)$-Noetherian*: for any* $A \in \mathtt{Up}_{\unlhd}(W)$*, if* $A \neq W$*, then there is* $w \in \Box A - A$*, i.e., a* $\Box$*-maximal non-*A *point.*

Proof (Sketch) We show only the "if" direction. Assuming **(wlöb)** fails under a valuation V, take B to be the extension of **(wlöb)** under V and show that $\Box B \subseteq B$. This means that $W - B$ witnesses the failure of Noetherianity. □

Reference [77] provides an interesting motivation for this semantics of $\mathbf{GL}^{\mathsf{i}}$ in terms of *projects* and *streamlines* in a research. It also provides other important results, such as the finite model property and decidability.

8.3 The Fixpoint Theorem

Theorem 1 above gave us several equivalent axiomatizations of $\mathbf{GL}^{\mathsf{i}}$. In particular, **(ufp)** forces *uniqueness of fixed points.*

Definition 1 *Let* B *be a formula of* $\mathcal{L}_{\Box\mathsf{int}}$ *(its denotation in a given algebra and under a given valuation).* B *is a* fixed point *of (the term function associated with)* $A \in \mathcal{L}_{\Box\mathsf{int}}$ *relative to* p *in a given normal logic* Γ *(here always an extension of* $\mathbf{GL}^{\mathsf{i}}$*) if* $B \leftrightarrow A[B/p] \in \Gamma$ *and* p *does not occur in* B.

According to [73], the fact that **(ufp)** holds in $\mathbf{GL}^{\mathsf{cl}}$ was discovered independently by Bernardi, Sambin and de Jongh. We know thus that, surprisingly, in $\mathbf{GL}^{\mathsf{i}}$ a syntactic fixed point of an expression is unique up to equivalence whenever it exists; same applies to all of its extensions, such as $\mathbf{SL}^{\mathsf{i}}$ or **KM**. But do they exist at all? An even more amazing fact is that they not only do exist—under the assumption of $\Box$-guardedness on p—but are effectively computable. This is guaranteed by the following algebraic (or propositional, if one prefers) variant of Gödel's Diagonalization Lemma. Sambin [68] proved it for $\mathbf{GL}^{\mathsf{i}}$ itself and de Jongh proved it for $\mathbf{GL}^{\mathsf{cl}}$ building on an earlier result by Smorynski, another proof being found soon afterwards by Boolos; the reader is referred to [18, 19, 61, 73] for more on its history and the connection with Gödel's result:

Theorem 3 (Diagonalization) *For any A and p, there exists a constructively obtained formula* $\mathsf{diag}_p A$ *s.t.*

1. $\mathsf{diag}_p A \leftrightarrow B \in \mathbf{GL}^{\mathsf{i}}$, *where B is obtained from A by replacing all occurrences of p under $\Box$ by* $\mathsf{diag}_p A$
2. *A and $\mathsf{diag}_p A$ have provably the same fixed points with respect to p, that is, for any C not containing p we have*

$$\boxdot(C \leftrightarrow A[C/p]) \leftrightarrow \boxdot(C \leftrightarrow \mathsf{diag}_p A[C/p]) \in \mathbf{GL}^{\mathsf{i}}$$

Clearly, if p is $\Box$-guarded in A, then B in the first clause is precisely $A[\mathsf{diag}_p A/p]$ and $\mathsf{diag}_p A$ does not contain p, hence being trivially its own fixed point wrt p. Thus, in such a situation $\mathsf{diag}_p A$ itself is also the unique fixed-point of A with respect to p!

Proof (of Theorem 3, sketch) We only give a sketch of how $\mathsf{diag}_p A$ is constructed. Any formula $A(p, \overline{q}) \in \mathcal{L}_{\Box\mathsf{int}}$ in variables $p, \overline{q} \in \Sigma$ can be written as $B(\Box C_1(p, \overline{q}), \ldots, \Box C_k(p, \overline{q}), p, \overline{q})$, where $B \in \mathcal{L}_{\mathsf{int}}$ (i.e., is a formula without $\Box$) and $\overline{C} \in \mathcal{L}_{\Box\mathsf{int}}$. Clearly, if $k = 0$, then A itself belongs to $\mathcal{L}_{\mathsf{int}}$ and, in particular, there are no occurrences of p under $\Box$. Hence we can take $\mathsf{diag}_p A$ to be A itself. Otherwise, the proof can be conducted by induction on k, as we already have the base step. For any $i \leq k$, set

$$A_i := B(\Box C_1(p, \overline{q}), \ldots, \Box C_{i-1}(p, \overline{q}), \top, \Box C_{i+1}(p, \overline{q}), \ldots, \Box C_k(p, \overline{q}), p, \overline{q}).$$

By definition, the inductive hypothesis applies to A_i. Now we set

$$\mathsf{diag}_p A := B(\Box C_1(\mathsf{diag}_p A_1/p, \overline{q}), \ldots, \Box C_k(\mathsf{diag}_p A_k/p, \overline{q}), p, \overline{q}). \qquad \Box$$

Remark 4 In fact, extensions of $\mathbf{SL}^{\mathsf{i}}$—in particular **KM**—allow a much simpler proof of Theorem 3 and a much simpler algorithm for computing these fixpoints: it is enough to substitute $\top$ for p. This follows already from observations made by Smorynski in [73, Lemma 2.3] and has been discussed explicitly in [26, Propositions 4.2–4.6]. De Jongh and Visser describe $\mathbf{SL}^{\mathsf{i}}$ as *a kind of Kindergarten Theory in which all the well-known syntactical results of Provability Logic have extremely simple versions.*

Remark 5 It is known that at least in the case of $\mathbf{GL}^{\mathsf{cl}}$ a non-constructive and non-explict form of Theorem 3 can be obtained already from uniqueness of fixed-points combined with the Beth definability theorem for $\mathbf{GL}^{\mathsf{cl}}$, see, e.g., [18, 73] for more information. However, as should be clear from the discussion below, the very fact that fixed points are obtained explicitly and constructively seems an advantage not to be given up lightly.

Theorem 3 has a nice algebraic corollary. I present it here as a more general version of [31, Proposition 3].

Theorem 4 *A* $\mathbf{K4}^{\mathsf{i}}$*-algebra $\mathfrak{H}$ is a* $\mathbf{GL}^{\mathsf{i}}$*-algebra iff every $\mathfrak{H}$-polynomial $t(p)$ in one $\Box$-guarded variable $p \in \Sigma$ has a fixed point.*

Proof The "only if" direction. This is a direct corollary of Theorem 3.

The "if" direction. Given any $\mathsf{h} \in \mathfrak{H}$, consider the polynomial $t(p) = \Box p \to \mathsf{h}$. As p is $\Box$-guarded in it, it has a fixed point $\mathsf{i_h} \in \mathfrak{H}$; that is, $\mathsf{i_h} = \Box\mathsf{i_h} \to \mathsf{h}$. By the fact that $\to$ is conjugate (or residual) to $\wedge$, one half of this equality is equivalent to $\mathsf{i_h} \wedge \Box\mathsf{i_h} \leq \mathsf{h}$. On the other hand, $\mathsf{h} \leq \Box\mathsf{i_h} \to \mathsf{h} = \mathsf{i_h}$ by general implication laws. Taken together, these two inequalities imply $\mathsf{i_h} \wedge \Box\mathsf{i_h} = \mathsf{h} \wedge \Box\mathsf{h}$: the $\leq$ direction from the first inequality, normality and (**trns**), the $\geq$ direction from the second inequality and monotonicity of $\Box$. Using normality again, we get $\Box\mathsf{i_h} \wedge \Box\Box\mathsf{i_h} = \Box\mathsf{h} \wedge \Box\Box\mathsf{h}$ and using (**trns**) again, we arrive at $\Box\mathsf{i_h} = \Box\mathsf{h}$. Then we get $\Box(\Box\mathsf{h} \to \mathsf{h}) = \Box(\Box\mathsf{i_h} \to \mathsf{h}) = \Box\mathsf{i_h} = \Box\mathsf{h}$. As $\mathsf{h} \in \mathfrak{H}$ was chosen arbitrarily, we have that $\mathfrak{H}$ is a $\mathbf{GL}^{\mathsf{i}}$-algebra. □

There is an analogy between the above result and alternative axiomatizations for $\mathbf{GL}^{\mathsf{i}}$ presented in [77].

Corollary 1

- *A* $\mathbf{R}^{\mathsf{i}}$*-algebra* $\mathfrak{H}$ *is a* $\mathbf{SL}^{\mathsf{i}}$*-algebra iff every* $\mathfrak{H}$*-polynomial* $t(p)$ *in one* $\Box$*-guarded variable* $p \in \Sigma$ *has a fixed point.*
- *[31, Proposition 3] A* **mHC***-algebra* $\mathfrak{H}$ *is a* **KM***-algebra iff every* $\mathfrak{H}$*-polynomial* $t(p)$ *in one* $\Box$*-guarded variable* $p \in \Sigma$ *has a fixed point.*

As stated above, Theorem 3 occurred first as an algebraization of Gödel's Diagonalization Lemma. While the connection between $\mathbf{GL}^{\mathsf{i}}$ and Heyting Arithmetic HA is not as tight as the one between $\mathbf{GL}^{\mathsf{cl}}$ and Peano Arithmetic PA established by the completeness result of Solovay [74] (see also [18, Chap. 9]), Sambin [68] notes that Theorem 3 yields a counterpart of the Diagonalization Lemma for *any intuitionistic first-order theory with a canonical derivability predicate*, including obviously HA. At any rate, the relevance of fixpoint results for Löb-like logics is not limited to arithmetic.

Remark 6 It is worth mentioning here that—unlike the case of PA—the search for a complete axiomatization of the provability logic of HA is not over yet; [45] gives a fascinating account. Regarding the arithmetical interpretation of $\mathbf{SL}^{\mathsf{i}}$, see the discussion of HA^* in [26, Sect. 4–5].

To begin with, in the classical case one can use Theorem 7 to show that explicit smallest or greatest fixed-point operators are eliminable over $\mathbf{GL}^{\mathsf{cl}}$. In other words, adding μ or ν does not increase the expressive power of the classical modal logic of transitive and conversely well-founded structures; see [4, 79, 80]. Note that this includes all correctly formed expressions with μ, without assuming that all occurrences of p are $\Box$-guarded: as usual, they only have to be positive. ten Cate [76, Sect. 3] discusses an application in the context of expressivity of navigational fragments of XML query languages.

While I am not aware of an exact analogue of the results in [4, 79, 80] in the intuitionistic context, Löb-like modalities—more specifically, variants of systems $\mathbf{GL}^{\mathsf{i}}$ and $\mathbf{SL}^{\mathsf{i}}$—have recently become rather popular in type theory. Examples include:

- *modality for recursion* [63, 64]
- *approximation modality* [7]
- *tomorrow idiom* [8]
- *next-step modality/next clock tick* [49, 50]
- *later operator* [11, 14, 15, 46].

One of reasons is precisely that such modalities guarantee existence and uniqueness of fixed-points of suitably guarded type expressions. However, the modal spadework of 1970's seems rarely acknowledged. In [63], which may be credited with introducing intuitionistic Löb-like modalities to the attention of this community, we find the following statement:

> Similar results concerning the existence of fixed points of proper type expressions ... *could historically go back to the fixed point theorem of the logic of provability ... The difference is that our logic is intuitionistic*, and fixed points are treated as sets of realizers [the emphasis is mine–T.L.].

This formulation suggests that Nakano was not aware that the *intuitionistic* fixed-point theorem had been already proved in [68], not to mention improvements possible above $\mathbf{SL}^{\mathsf{i}}$ (cf. Remark 4). The only related references quoted in [63] focus on classical $\mathbf{GL}^{\mathsf{cl}}$—e.g, [18]—and in later papers even these are omitted. A valuable part of the logical tradition seems lost this way. Let us see what insights can be found in the work of the Tbilisi school.

8.4 Leo Esakia and Extensions of mHC

8.4.1 mHC *and Topological Derivative*

Leo Esakia and collaborators devoted special attention to the system **mHC** and its extensions. Reference [31] is an excellent overview. The abbreviation **mHC** stood for *modalized Heyting calculus*. The reader may find the name surprising; after all, many natural intuitionistic modal systems are not subsystems of **mHC**. Esakia [31] was perfectly aware of that:

> The postulate (**r**) is not typical, while the postulate (**derv**) stresses even more "nonstandardness" of the chosen basic system **mHC** and of its extension **KM**, which enables one to draw a conventional "demarcation line" between **mHC** and the standard intuitionistic modal logics.

Remark 7 Both axioms seem "nonstandard" mostly if one focuses on these intuitionistic modal logics which are obtained from popular classical systems. It is enough to look at Table 8.3 to realize why it must be so: (**derv**) is trivially derivable in $\mathbf{K}^{\mathsf{cl}}$ (its consequent being a classical tautology), while the combination of (**cl**) and (**r**) yields that $\prec \subseteq \Delta$. That is, the only classical frames for $\mathbf{R}^{\mathsf{cl}}$ are disjoint unions of reflexive and irreflexive points. However, (**r**) is nowhere as pathological in a properly intuitionistic setting. There are many references on systems different from **mHC** and **KM** where nevertheless (**r**) is still derivable or even explicitly included as an axiom. A short and inexhaustive list includes [1, 2, 10, 14, 25, 32, 37, 38, 42, 43, 50, 56, 63, 64]. They can be split roughly into two main groups. The first one—e.g., [14, 50, 63, 64]—concerns $\mathbf{SL}^{\mathsf{i}}$ and has already been mentioned in Sect. 8.3. The second one—e.g., [2, 10, 25, 32, 37, 38, 42, 43]—concerns the system which is denoted here as $\mathbf{PLL}^{\mathsf{i}}$ (a.k.a. **CL**, see [10]). Reference [40, Sect. 7.6] and [43] are good if incomplete overviews of most relevant papers on this system—the most important omissions being perhaps [65, Sect. 7] and [1, 38]. See also [56] for a discussion of type systems with $\mathbf{R}^{\mathsf{i}}$ modalities from programmer's point of view.

Reference [31] gives the following reasons for the importance of **mHC**:

- Its connection to **KM**: **mHC** is **KM** minus the Löb axiom (**slöb**). Note that (**wlöb**) and (**slöb**) are equivalent in the presence of (**r**)
- The connection with *intuitionistic temporal logic "Always & Before" possessing rich expressive possibilities*
- The fact that **mHC** can be obtained as a fragment of **QINT** (or, in Esakia's notation, **QHC**)—quantified intuitionistic propositional calculus. This is similar to the encoding of **mHC** in the internal language of a topos, see the last point
- The topological connection with *Cantor's scattered spaces, notions of the limit and isolated point*. This will be discussed at length in this section
- Finally, as mentioned above, **mHC** turns out to be a natural fragment of the logic of toposes. This last point builds on all the preceding ones and will be discussed in Sect. 8.5.

As we can see in Table 8.3, the conditions on the accessibility relation $\prec$ imposed by the axioms of **mHC**—the combination of (**r**) and (**derv**)—is $\lhd \subseteq \prec \subseteq \unlhd$. A natural question to ask is whether it is possible to enforce syntactically that $\prec$ is even more closely determined by $\unlhd$ as one of the two borderline cases, i.e., either $\prec = \unlhd$ or $\prec = \lhd$.

For $\prec = \unlhd$, the answer is obviously positive. This is achieved precisely by the axioms of the logic $\mathbf{Triv}^{\mathsf{i}}$, strengthening (**derv**) to (**refl**). In fact, this is a semantic counterpart of an observation in [31] that enriching any Heyting algebra with a trivial operator $\Box^{\mathbf{Triv}^{\mathsf{i}}} x := x$ yields an **mHC**-algebra. Note here that whenever $\Box$ is an **mHC**- or even $\mathbf{R}^{\mathsf{i}}$-operator, its associated $\boxdot$ is a $\mathbf{Triv}^{\mathsf{i}}$-operator.

For $\prec = \lhd$, the answer is obviously negative. Irreflexivity is a typical example of a condition which cannot be defined by any purely modal axiom, see [16]. Here is perhaps the most natural proof.

Example 1 Consider a frame for **mHC** defined as $(\omega, \unlhd, \lhd)$ where $\unlhd$ is the natural order $\leq$ on ω. The dual algebra contains as a subalgebra the two-element Boolean algebra with $\Box\bot = \bot$, which is the dual of a single $\prec$-reflexive point. Hence, no modal axiom can define the class of $\prec$-irreflexive frames over the class of frames for **mHC**.

However, an "irreflexive" **mHC**-operator is clearly definable on any Heyting algebra obtained as the dual of an intuitionistic Kripke frame $(W, \trianglelefteq)$: for any $A \in \mathtt{Up}_{\trianglelefteq}(W)$, we take $\ast A := \{w \in W \mid \{w\}^{\trianglelefteq} \uparrow -\{w\} \subseteq A\}$; again, see Table 8.3 for notation. It is straightforward to note that for any $w \in W$, $w \in \ast A$ iff for any $B \in \mathtt{Up}_{\trianglelefteq}(W)$, $w \in (B \to A) \cup B$. This observation actually explains the shape of axiom (**derv**). Hence, given any complete Heyting algebra $\mathfrak{H}$, define its *point-free coderivative* [30, 69, 70] as $\ast \mathsf{h} := \bigwedge_{\mathsf{i} \in \mathfrak{H}} (\mathsf{i} \vee (\mathsf{i} \to \mathsf{h}))$.

Proposition 1 *For any complete Heyting algebra, its point-free coderivative is an* **mHC***-operator.*

Proof A rather easy exercise for the reader; can be also extracted from the proof of [31, Proposition 5]. □

There is a slightly different description of $\ast$. Take a Heyting algebra $\mathfrak{H}$ and $\mathsf{h}, \mathsf{i} \in \mathfrak{H}$ s.t. $\mathsf{i} \leq \mathsf{h}$. h is i-*dense* or *dense in* $[\mathsf{i}, \top]$ if for any $\mathsf{j} \in \mathfrak{H}$, we get that $\mathsf{h} \wedge \mathsf{j} = \mathsf{i}$ implies $\mathsf{j} = \mathsf{i}$. Note that the standard topological notion of density can be considered a special case: an open set is topologically dense iff it is $\bot$-dense in the Heyting algebra of open sets of the space. The following was observed, e.g., in [30]:

Fact 1 *For any Heyting algebra $\mathfrak{H}$ and any $\mathsf{h} \geq \mathsf{i} \in \mathfrak{H}$, h is i-dense iff there exists $\mathsf{j} \in \mathfrak{H}$ s.t. $\mathsf{h} = \mathsf{j} \vee (\mathsf{j} \to \mathsf{i})$.*

Corollary 2 *For any complete Heyting algebra $\mathfrak{H}$ and any $\mathsf{i} \in \mathfrak{H}$,*

$$\ast \mathsf{i} = \bigwedge \{\mathsf{h} \in \mathfrak{H} \mid \mathsf{h} \geq \mathsf{i} \, and \, \mathsf{h} \, is \, \mathsf{i} - dense\}.$$

Why *coderivative*? The reader is referred to a detailed account by Simmons [70] in this volume. Briefly, recall that in topology the *Cantor-Bendixson derivative* of a set A is the set of those x whose every neighbourhood contains a point of A *other than* x; the dual operator (hence *co-derivative*) consists of those x which have an open neighbourhood *entirely contained in* $A \cup \{x\}$ [31]. As it turns out, this indeed coincides with $\ast$ for practically all sensible topological spaces:

Theorem 5 **(Simmons)** *For any T_0-space, its co-derivative operator coincides with the point-free coderivative $\ast$ on the Heyting algebra of open sets.*

Proof For $\ast$ defined as in Corollary 17, this was proved in [69]. □

Obviously, as any intuitionistic Kripke frame with the Alexandroff topology given by $\mathtt{Up}_{\trianglelefteq}(W)$ is T_0, we get that $\ast$ coincides with the dual of topological derivative of this topology. It is, in fact, easier to prove directly than by Simmons' result.

Remark 8 One observation from [69] is worth quoting here:

> for non-T_0-spaces the usual definition of isolated point does not quite capture the intended notion

and hence for arbitrary spaces, the point-free definition of derivative given by $\circledast$ seems in fact *more adequate* than the standard one. The reader can verify this by extending the definition of intuitionistic Kripke frames to qosets rather than just posets and checking how both notions would fare in such a setting.

8.4.2 KM and Scatteredness

A complete Heyting algebra will be called *scattered* if its coderivative $\circledast$ is not only an **mHC**-operator, but a **KM**-operator. Recall Corollary 1 as an algebraic characterization of such a situation.

Proposition 2

- *For any topological space, its point-free coderivative is a* **KM**-*operator* ***if*** *the space is scattered in the usual sense: that is, if each non-empty subset has an isolated point.*
- *For any* T_0 *topological space, its point-free coderivative is a* **KM**-*operator* ***only if*** *the space is scattered.*

Proof A non-T_0-space can never be scattered, and for T_0-spaces, the point-free coderivative coincides with ordinary one as stated in Theorem 5. The remaining calculations are an exercise in point-set topology; in fact, the point-set part of this result has been shown first by Kuznetsov [52, 61]. One can use an alternative characterization of scatteredness here: for any open set A distinct from the whole space, $\circledast A - A$ is non-empty. □

Let us summarize some of the results above:

Corollary 3 *A topological space* $\mathfrak{T}$ *is scattered iff in the complete Heyting algebra of open sets of* $\mathfrak{T}$*, every polynomial in one* $\circledast$*-guarded variable has a fixed point.*

Corollary 4 *The following are equivalent for any* $(W, \trianglelefteq, \prec)$:

- $\prec = \triangleleft$ *and the Alexandroff topology* $(W, \mathrm{Up}_{\trianglelefteq}(W))$ *is scattered*
- $(W, \trianglelefteq, \prec)$ *is a frame for* **KM**
- $\prec = \triangleleft$ *and* $\trianglelefteq$ *contains no infinite ascending chains.*

Proof We only need to prove the equivalence of the last two conditions. $\triangleleft \subseteq \prec$ is, as observed, enforced by **mHC**. Theorem 2 gives the corresponding semantic condition for (**wlöb**). Quite obviously, $\mathrm{Up}_{\trianglelefteq}(W)$-Noetherianity forces irreflexivity of $\prec$. Thus, whenever $(W, \trianglelefteq, \prec)$ is a frame for **KM** (that is, the join of **mHC** and $\mathbf{GL}^{\mathrm{i}}$), we have $\prec = \triangleleft$. Moreover, rewriting the condition of $\mathrm{Up}_{\trianglelefteq}(W)$-Noetherianity for **mHC**-frames, we obtain that for any $A \in \mathrm{Up}_{\trianglelefteq}(W) - \{W\}$, there is $w \in A^{\triangleleft}\uparrow -A$. Rewriting further, we obtain that any $B \neq \emptyset$ s.t. $B = B^{\trianglelefteq}\downarrow$ has a maximal element wrt $\trianglelefteq$. But this means that *any* nonempty subset of W has a maximal $\trianglelefteq$-element. □

8.4.3 Completeness, Lattice Isomorphisms and Bimodal Translations

Two important kinds of results have been missing from this overview so far. First, while I discussed Kripke *correspondence* for modal logics (Table 8.3, Theorem 2 and Corollary 4), I have not discussed *completeness*. Second, I have not said much about lattices of extensions of Löb-like logics and their relatives—in particular, about generalizations of the Blok-Esakia Theorem.

This section fixes both oversights. Rather than using original proofs of Kuznetsov, Muravitsky (for **KM**) and Esakia (for **mHC**), we are going to use corollaries of Wolter and Zakharyaschev's results on bimodal translations [81, 82], briefly discussed also in this very volume [83, Sect. 4]. Their techniques allow to interpret intuitionistic modal logics as fragments of classical polymodal ones (cf. the discussion of *implict vs. explicit epistemics* in [78]). In the case of the Blok-Esakia theorem for **mHC**, we will be able to see why axioms of **mHC** and $\mathbf{wGrz}^{\mathsf{cl}}$ have to look the way they look in order to allow the classical counterpart to be unimodal rather than polymodal, as it happens in the more general framework of [81, 82].

Take the bimodal language $\mathcal{L}_{\mathbf{i,m}}$ with operators $[\,\mathbf{i}\,]$ and $[\mathbf{m}]$. For any formula $A \in \mathcal{L}_{\Box\mathsf{int}}$, I will write $A_{[\,\mathbf{i}\,]}$ (respectively $A_{[\mathbf{m}]}$) for A with all occurrences of $\Box$ replaced with $[\,\mathbf{i}\,]$ (respectively $[\mathbf{m}]$). $[\mathbf{m}]$ is the default counterpart of the original modality $\Box$ and $[\,\mathbf{i}\,]$ encodes the intuitionistic poset order, hence the notation.[1] The logic $\mathbf{S4}^{\mathsf{cl}}_{\mathbf{i,m}}$ is the normal logic axiomatized by the following axioms: (**cl**), $(\mathbf{nrm})_{[\,\mathbf{i}\,]}$, $(\mathbf{refl})_{[\,\mathbf{i}\,]}$, $(\mathbf{trns})_{[\,\mathbf{i}\,]}$ and $(\mathbf{nrm})_{[\mathbf{m}]}$; in other words, it is what modal logicians would describe as the *fusion* of $\mathbf{S4}^{\mathsf{cl}}_{\mathbf{i}}$ and $\mathbf{K}^{\mathsf{cl}}_{\mathbf{m}}$. The logic $\mathbf{S4M}^{\mathsf{cl}}_{\mathbf{i,m}}$ is $\mathbf{S4}^{\mathsf{cl}}_{\mathbf{i,m}}$ extended with

$$(\mathbf{mix}): \qquad [\mathbf{m}]A \leftrightarrow [\,\mathbf{i}\,][\mathbf{m}][\,\mathbf{i}\,]A.$$

The logic $\mathbf{sGrzM}^{\mathsf{cl}}_{\mathbf{i,m}}$ is $\mathbf{S4}^{\mathsf{cl}}_{\mathbf{i,m}}$ extended with (**mix**) and $(\mathbf{sgrz})_{[\,\mathbf{i}\,]}$. The translation $\flat : \mathcal{L}_{\Box\mathsf{int}} \to \mathcal{L}_{\mathbf{i,m}}$ prefixes every subformula in $(\cdot)_{[\mathbf{m}]}$ with $[\,\mathbf{i}\,]$. Of course, many occurrences of $[\,\mathbf{i}\,]$ in the translation $\flat A$ can be removed relative to logics defined above:

Fact 2 *The following equivalences belong to* $\mathbf{S4}^{\mathsf{cl}}_{\mathbf{i,m}}$*:* $\flat A \leftrightarrow [\,\mathbf{i}\,]\flat A$, $\flat(A \wedge B) \leftrightarrow (\flat A \wedge \flat B)$, $\flat(A \vee B) \leftrightarrow (\flat A \vee \flat B)$*; in* $\mathbf{S4M}^{\mathsf{cl}}_{\mathbf{i,m}}$*, we moreover have* $\flat(\Box A) \leftrightarrow [\mathbf{m}]\flat A$.

For any intuitionistic normal modal logic $\Gamma \supseteq \mathbf{S4}^{\mathsf{cl}}_{\mathbf{i,m}}$ and any bimodal normal logic $\Delta \subseteq \mathcal{L}_{\mathbf{i,m}}$, let

- $\tau^{\mathbf{mix}}\Gamma := \mathbf{S4M}^{\mathsf{cl}}_{\mathbf{i,m}} \oplus \{\flat A \mid A \in \Gamma\}$
- $\sigma^{\mathbf{mix}}\Gamma := \tau^{\mathbf{mix}}\Gamma \oplus (\mathbf{sgrz})_{[\,\mathbf{i}\,]}$
- $\rho_{\mathbf{in}}\Delta := \{A \in \mathcal{L}_{\Box\mathsf{int}} \mid \flat A \in \Delta\}$.

[1] The reader has to be warned that the notation in this section differs somewhat from that in references like [81–83].

Δ is *a* $\mathcal{L}_{\mathbf{i},\mathbf{m}}$-*companion of* Γ if for any $A \in \mathcal{L}_{\Box\mathsf{int}}$, $A \in \Gamma$ iff $\flat A \in \Delta$, i.e., iff $\rho_{\mathbf{in}}\Delta = \Gamma$.

Theorem 6 ([81–83]) *Let* $\Delta \supseteq \mathbf{S4}^{\mathsf{cl}}_{\mathbf{i},\mathbf{m}}$ *be a normal bimodal logic and* $\Gamma \subseteq \mathcal{L}_{\Box\mathsf{int}}$ *be an intuitionistic normal logic. Then*

(A) $\rho_{\mathbf{in}}\Delta$ *is an intuitionistic normal modal logic*
(B) $\tau^{\mathbf{mix}}\Gamma$ *and* $\sigma^{\mathbf{mix}}\Gamma$ *are, respectively, the smallest and the greatest* $\mathcal{L}_{\mathbf{i},\mathbf{m}}$-*companions of* Γ *containing* (**mix**)
(C) $\rho_{\mathbf{in}}$ *preserves decidability, Kripke completeness and the finite model property. If* (**mix**) $\in \Delta$, $\rho_{\mathbf{in}}$ *also preserves canonicity*
(D) $\tau^{\mathbf{mix}}$ *preserves canonicity*
(E) $\sigma^{\mathbf{mix}}$ *preserves the finite model property*
(F) $\sigma^{\mathbf{mix}}$ *is an isomorphism from the lattice of normal extensions of* $\mathbf{K}^{\mathsf{i}}$ *onto the lattice of normal extensions of* $\mathbf{sGrzM}^{\mathsf{cl}}_{\mathbf{i},\mathbf{m}}$
(G) $\Gamma \supseteq \mathbf{K4}^{\mathsf{i}}$ *has the finite model property whenever its* $\mathcal{L}_{\mathbf{i},\mathbf{m}}$-*companions over* $\mathbf{S4M}^{\mathsf{cl}}_{\mathbf{i},\mathbf{m}} \oplus (\mathbf{trns})_{[\mathbf{m}]}$ *include a canonical subframe logic.*

Proof (A) can be easily proved from Fact 2; note that we need the assumption we are above $\mathbf{S4}^{\mathsf{cl}}_{\mathbf{i},\mathbf{m}}$. (B) is a consequence of Theorem 27 in [81]. (C), (D) and (E) are consequences of Proposition 29 and Theorem 30 in [81] and Theorems 11 and 12 in [82].(F) is a consequence of Corollary 28 in [81]. (G) is a consequence of Corollary 18 in [82]. □

References [81, 82] illustrate on many examples how powerful these results are. As it turns out, they also have corollaries of immediate interest for us.

Corollary 5 **mHC** *is canonical and has the finite model property.*

Proof First, note that

$$\mathbf{S4M}^{\mathsf{cl}}_{\mathbf{i},\mathbf{m}} \oplus \flat(\mathbf{r}) = \mathbf{S4M}^{\mathsf{cl}}_{\mathbf{i},\mathbf{m}} \oplus ([\,\mathbf{i}\,]A \to [\mathbf{m}]A) \supseteq \mathbf{S4M}^{\mathsf{cl}}_{\mathbf{i},\mathbf{m}} \oplus (\mathbf{trns})_{[\mathbf{m}]}.$$

Clearly, $[\,\mathbf{i}\,]A \to [\mathbf{m}]A$ is a Sahlqvist formula with an universal FO counterpart. Furthermore, $\mathbf{S4M}^{\mathsf{cl}}_{\mathbf{i},\mathbf{m}} \oplus \flat(\mathbf{derv})$ is the same logic as the extension of $\mathbf{S4M}^{\mathsf{cl}}_{\mathbf{i},\mathbf{m}}$ with

$$[\mathbf{m}]B \wedge \langle\mathbf{i}\rangle C \to [\,\mathbf{i}\,](\langle\mathbf{i}\rangle C \vee [\,\mathbf{i}\,]B). \tag{8.1}$$

The latter is a simple Sahlqvist implication (cf. e.g., [16, Definition 3.47]). Applying the algorithm in the proof of Theorem 3.49 in [16] and doing some FO-preprocessing, we get an universal formula

$$\forall y, z, w.(x \trianglelefteq y \wedge x \trianglelefteq z \to (z \trianglelefteq y \vee (z \trianglelefteq w \to x \prec w))) \tag{8.2}$$

(where $\trianglelefteq$ is the accessibility relation corresponding to $[\,\mathbf{i}\,]$ and $\prec$ is the accessibility relation corresponding to $[\mathbf{m}]$). Thus, $\tau^{\mathbf{mix}}\mathbf{mHC}$ is a canonical subframe logic over

$\mathbf{S4M}^{\mathsf{cl}}_{\mathbf{i,m}} \oplus (\mathbf{trns})_{[\mathbf{m}]}$. Now, canonicity of **mHC** follows from (C) and the fmp from (G) of Theorem 6. □

Remark 9 It is worth noting that the semantic counterpart of (**next**) from Table 8.3, i.e., $\lhd \subseteq \prec$ is equivalent to 8.2 above. For one direction, substitute $x = y$ in 8.2 and use poset properties. For the other direction, note that whenever $x \unlhd y$ and $x \unlhd z$ but $\neg(z \unlhd y)$, then $x \lhd z$, ergo $x \prec z$. Now whenever $z \unlhd w$, we can use the interaction between $\prec$ and $\unlhd$ as expressed by (**mix**) (in fact, even a weaker axiom would do).

Canonicity of **mHC** has been noted , e.g., in [31, 42]. I was unable to locate references where the finite model property has been explicitly claimed. The following corollary can appear more surprising, as bimodal logics over $\mathcal{L}_{\mathbf{i,m}}$ do not even occur in its statement.

Corollary 6 *The lattice of normal extensions of* **mHC** *is isomorphic to the lattice of normal extensions of* $\mathbf{wGrz}^{\mathsf{cl}}$. *The sublattice of normal extensions of* **KM** *is isomorphic to the lattice of normal extensions of* $\mathbf{GL}^{\mathsf{cl}}$.

Proof The heart of the proof is to notice that $[\mathbf{i}]A \leftrightarrow [\mathbf{m}]A \wedge A$ and $(\mathbf{grz})_{[\mathbf{m}]}$ are derivable in $\sigma^{\mathbf{mix}}\mathbf{mHC}$; in fact, these two formulas axiomatize this logic over $\mathcal{L}_{\mathbf{i,m}} \oplus (\mathbf{trns})_{[\mathbf{m}]}$. Let us derive the first of them. For convenience, we will do it in the algebraic setting:

$$
\begin{aligned}
[\mathbf{m}]A \wedge A \wedge \langle \mathbf{i} \rangle \neg A &\leq && \text{by } (\mathbf{refl})_{[\mathbf{i}]}\\
[\mathbf{m}]A \wedge \langle \mathbf{i} \rangle (A \wedge \langle \mathbf{i} \rangle \neg A) &\leq && \text{by } (1)\\
[\mathbf{i}](\langle \mathbf{i} \rangle (A \wedge \langle \mathbf{i} \rangle \neg A) \vee [\mathbf{i}]A) &\leq && \text{by } (\mathbf{refl})_{[\mathbf{i}]}\\
[\mathbf{i}](\langle \mathbf{i} \rangle (A \wedge \langle \mathbf{i} \rangle \neg A) \vee A) &= &&\\
[\mathbf{i}]([\mathbf{i}](A \to [\mathbf{i}]A) \to A) &\leq [\mathbf{i}]A && \text{by } (\mathbf{grz})_{[\mathbf{i}]}.
\end{aligned}
$$

We get that $\sigma^{\mathbf{mix}}\mathbf{mHC}$ is just a notational variant of $\sigma^{\mathbf{mix}}\mathbf{wGrz}^{\mathsf{cl}}$, with $[\mathbf{m}]$ being $\Box$ and $[\mathbf{i}]$ being $\boxdot$. This yields the first statement by clause (F) of Theorem 24. For the second, it is enough to add the observation that over $\sigma^{\mathbf{mix}}\mathbf{mHC}$, adding $\flat(\mathbf{sl\ddot{o}b})$ is equivalent to adding $(\mathbf{wl\ddot{o}b})_{[\mathbf{m}]}$. □

The second statement of Corollary 6 above was first proved by Kuznetsov and Muravitsky in mid-1980's, see [51, 61]. The first statement was announced in [31] as follows:

> Finally let us note that ...the lattice Lat(mHC) of all extensions of mHC is isomorphic to the lattice Lat(K4.Grz) of all normal extensions of the modal system K4.Grz. However, a proof of this result requires additional considerations as the above algebraic machinery does not suffice for it.

It seems that the proof has not been published so far.

Corollary 7 *([42, 51, 60])* **KM** *has the finite model property.*

Proof The proof of Corollary 6 has established that $\sigma^{\mathbf{mix}}\mathbf{KM}$ is just a notational variant of $\mathbf{GL}^{\mathsf{cl}}$, with $[\mathbf{m}]$ being $\Box$ and $[\,\mathbf{i}\,]$ being $\boxdot$. Now use clause (C) of Theorem 6 and the finite model property for $\mathbf{GL}^{\mathsf{cl}}$ (see, e.g., [16, 18, 23, 33, 59] for references). □

Remark 10 Note that we could also prove Corollary 5 in an analogous way to Corollary 7, using the fmp of $\mathbf{wGrz}^{\mathsf{cl}}$ established explicitly by Amerbauer [6]. The latter is actually a direct consequence of $\mathbf{wGrz}^{\mathsf{cl}}$ being a transitive subframe logic [33, 54]. However, I believe that the proof of Corollary 5 provided above has some additional value: we obtained a convenient form of $\flat(\mathbf{derv})$—which we actually used in the proof of Corollary 6—together with its FO translation, which also provides some additional insight, as discussed in Remark 9.

Remark 11 It could be an interesting exercise—and very much in the spirit of the Tbilisi school—to show that the above-discussed results of [81–83] survive when the base bimodal logic is weakened from $\mathbf{S4}^{\mathsf{cl}}_{\mathbf{i},\mathbf{m}}$ to the fusion of $\mathbf{K}^{\mathsf{cl}}_{\mathbf{i}} \oplus (A \wedge [\,\mathbf{i}\,]A \rightarrow [\,\mathbf{i}\,][\,\mathbf{i}\,]A)$ with $\mathbf{K}^{\mathsf{cl}}_{\mathbf{m}}$ and the translation $\flat$ is modified to $\flat^*$ replacing every subformula A with $\flat^* A \wedge [\,\mathbf{i}\,]\flat^* A$. On the other hand, it is not obvious how much generality would be really gained in this way. Note that using Wolter and Zakharyaschev's original results we were able to investigate lattices of logics which are **not** extensions of $\mathbf{S4}^{\mathsf{cl}}$, such as $\mathbf{wGrz}^{\mathsf{cl}}$ in Corollary 6 above.

8.5 Scattered Toposes

We are ready to discuss the *topos of trees* of [14], *scattered toposes* of [30] and the relationship between fixpoint results in both papers.

8.5.1 Preliminaries on the Logic of a Topos

Just like Sect. 8.4.3 assumed certain familiarity with technicalities of modal logic, this section in turn assumes some familiarity with basics of category theory—mostly with the notions of a ccc (cartesian closed category), a functor and a natural transformation. Those readers who know more than that, in particular understand well the internal logic of a topos, can probably skip this subsection. Due to obvious space constraints, the presentation has to be rather abstract and example-free; see [41, 47, 55] for more examples and motivation. Furthermore, like any presentation of topos theory by logicians and for logicians, it can be accused of neglecting spatial intuitions. See, e.g., [57] for a passionate polemic with the view that toposes were invented to generalize set-theoretical foundations of mathematics.[2] Nevertheless, applications

[2] Speaking of [57], footnote 4 provides an argument that the plural form intended by Grothendieck was *toposes* rather than *topoi*. I stick to the same convention, also because—as a quick Google search shows—the form *toposes* is used mostly by mathematicians, whereas *topoi* seems prevalent for unrelated notions in the humanities. Besides, this was the form used by Leo, Mamuka and Dito.

of toposes in fields like algebraic geometry or foundations of physics or their actual historical origins are not directly relevant here. My aim is a minimalist presentation focusing on the contrast between the logic of a topos and that of a ccc, but also making clear how the Beth-Kripke-Joyal semantics is related to more familiar ones for the intuitionistic predicate logic. Of all accounts in the literature, the one in [53] is probably closest to this goal.

Let $\mathscr{E}$ be a ccc with the terminal object $\mathbf{1}$ and for any $Y \in \mathscr{E}$, let $\mathtt{fin!}_Y$ be the unique element of $\mathscr{E}[Y, \mathbf{1}]$. I use the obvious notation for (finite) products, coproducts (whenever they exist, but in a topos they always do), their associated morphisms and I denote the ccc evaluation mapping $B^A \times A \to B$ as $\mathsf{eval}_{A,B}$. Recall that $\mathscr{E}$ is an *elementary topos* if there exists an object $\Omega \in \mathscr{E}$ s.t. $(\Omega, \mathbf{1} \overset{\mathtt{true}}{\to} \Omega)$ is a *subobject classifier*, i.e., for any *monic* (left-cancellable morphism) $Y \overset{f}{\rightarrowtail} X$ there exists exactly one mapping $X \overset{\chi_f}{\to} \Omega$ s.t. we have a pullback diagram:

$$\begin{array}{ccc} Y & \overset{f}{\rightarrowtail} & X \\ {\scriptstyle \mathtt{fin!}_Y}\big\downarrow & \lrcorner & \big\downarrow{\scriptstyle \chi_f} \\ \mathbf{1} & \xrightarrow{\mathtt{true}} & \Omega \end{array}$$

As observed by C. Juul Mikkelsen, this definition already implies that $\mathscr{E}$ is *bicartesian closed*, where the latter notion is defined as in, e.g., [53]; see [41, Sect. 4.3] for references.

Before we proceed with formal definitions, some general discussion can be helpful. In every category, topos or not, (equivalence classes of) monics into X are abstract counterparts of subsets of X; in fact, they are called *subobjects of* X, just like morphisms $\mathbf{1} \to X$ are *global elements* of X. Global elements (or equivalence classes thereof) can be considered as special cases of subobjects: think of the usual identification of an element $x \in X$ with the subset $\{x\}$[3]. We can go further and define a *generalized element* of X as *any* morphism $A \to X$, which is then called A*-based* or *defined over* A. See [55, Sect. V] for a lucid and brief discussion of those notions.

In particular, the global elements of Ω can be identified with logical constants, X-based generalized elements of Ω with predicates over X (i.e., formulas with a single free variable from X) and n-ary propositional connectives with morphisms $\Omega^n \to \Omega$. Set $\mathtt{false} := \chi_{\mathtt{fin!}_0}$, $\neg := \chi_{\mathtt{false}}$, $\wedge := \chi_{\langle \mathtt{true}, \mathtt{true} \rangle}$ and $\vee := \chi_{[\langle \mathtt{true}_\Omega, \mathsf{id}_\Omega \rangle, \langle \mathsf{id}_\Omega, \mathtt{true}_\Omega \rangle]}$. Recall that for any $X \in \mathscr{E}$, $\mathtt{true}_X$ stands for $\mathtt{true} \circ \mathtt{fin!}_X$ and $\mathtt{eq}_X$ stands for $\chi_{\langle \mathsf{id}_X, \mathsf{id}_X \rangle}$. The latter allows to define *internal equality for generalized elements of type* X as $\sigma \approx \tau := \mathtt{eq}_X \circ \langle \sigma, \tau \rangle$. That is, if $A \overset{\sigma}{\to} X$ and $B \overset{\tau}{\to} X$ are generalized elements of X, then $\sigma \approx \tau$ is a generalized element of Ω defined over $A \times B$. For Ω, we can define not only $\mathtt{eq}_\Omega$, but also $\mathtt{leq}_\Omega$ as the equalizer of $\Omega \times \Omega \underset{\Pi_1}{\overset{\wedge}{\rightrightarrows}} \Omega$. Implication, the only remaining intuitionistic connective, can be now defined as $\to := \chi_{\mathtt{leq}_\Omega}$.

[3] Note that toposes very rarely happen to mimic sets in having enough global elements to determine all subobjects; such special toposes are called *well-pointed*.

Thus, in toposes one can reduce reasoning about the poset of subobjects of any given object $X \in \mathscr{E}$ (in fact, whenever $\mathscr{E}$ is a topos, this poset is always a lattice and even a Heyting algebra—see [55, Theorem IV.8.1]) to reasoning about $\mathscr{E}[X, \Omega]$ and further still to reasoning about *an internal Heyting algebra* provided by a suitable exponential object. What this means is: in any category, monics into X have a natural preorder defined as $f \subseteq g$ if f factors through g, i.e., there is a morphism h s.t. $f = g \circ h$. Dividing by equivalence classes with respect to $\subseteq$, we get a category-theoretic generalization of the poset of subsets of X ordered by inclusion. In general, without understanding the global structure of $\mathscr{E}$, we are not likely to learn much about these posets of subobjects. But in a topos, the poset of subobjects of X is isomorphic to something more tangible: namely, to $\mathscr{E}[X, \Omega]$, i.e., X-based generalized elements of Ω. Think of the usual identification of subsets of X with elements of 2^X. Here is also where first- and higher-order aspects of the internal logic come into play.

If $\mathscr{E}$ is a category with products, a *power object* of $X \in \mathscr{E}$ is a pair $(\mathcal{P}X, \ni_X \xrightarrow{\ni_X} \mathcal{P}X \times X)$ s.t. for any $(Y, R \xrightarrow{r} Y \times X)$ there exists exactly one $Y \xrightarrow{f_r} \mathcal{P}X$ for which there is a pullback

$$\begin{array}{ccc} R & \xrightarrow{r} & Y \times X \\ \downarrow & \lrcorner & \downarrow f_r \times \mathsf{id}_X \\ \ni_X & \xrightarrow{\ni_X} & \mathcal{P}X \times X \end{array}$$

As shown, e.g., in [41, Theorem 4.7.1], in any topos we can take $\mathcal{P}X$ to be Ω^X and the subobject $\ni_X \xrightarrow{\ni_X} \Omega^X \times X$ can be obtained by pulling back $\mathtt{true}$ along $\Omega^X \times X \xrightarrow{\mathsf{eval}_{X,\Omega}} \Omega$. Thus, we see that in a topos, the notions of power object, subobject classifier and exponential object are indeed well-matched and we can define the membership predicate $\sigma \in \tau$ for a pair of generalized elements $(A \xrightarrow{\sigma} X, B \xrightarrow{\tau} \mathcal{P}X)$ as $\mathsf{eval}_{X,\Omega} \circ \langle \tau, \sigma \rangle$. We are now ready for a single definition formalizing the whole discussion above and more (see [55, Sect. VI.5–7] and also [24, 53]):

Definition 2 (The Mitchell-Bènabou languague) *Consider a topos $\mathscr{E}$. The* collection of ground types *and the* signature *of the Mitchell-Bènabou language of $\mathscr{E}$ are defined, respectively, as*

$$\mathsf{Ground}(\mathscr{E}) := \{\underline{E} \mid E \in \mathscr{E}\}$$
$$\mathsf{Sg}(\mathscr{E}) := \{\underline{f} : \underline{F_1}, \ldots, \underline{F_n} \to \underline{E} \mid f \in \mathscr{E}[F_1 \times \cdots \times F_n, E]\}$$

(instead of $\underline{k} : \underline{\mathbf{1}} \to \underline{E}$ I will write $\underline{k} : \underline{E}$) and the full collection of types $\mathsf{Types}(\mathscr{E})$ *is $A, B ::= \underline{E} \mid \underline{\mathbf{1}} \mid \underline{\Omega} \mid A \underline{\times} B \mid B^A$ where $\underline{E} \in \mathsf{Ground}(\mathscr{E})$. $\underline{\Omega}^A$ can be also written as $\underline{\mathcal{P}}A$. Fix, moreover, a supply of term variables $x, y, z \cdots \in \mathsf{tVar}$.* The collection of terms $\mathsf{Terms}(\mathscr{E})$ *over* $\mathsf{Sg}(\mathscr{E})$ *is defined as*

$$M, N ::= x \mid \underline{f}\overline{M} \mid \underline{\mathtt{fin!}} \mid M \approx N \mid \underline{\langle} M, N \underline{\rangle} \mid \underline{\pi_1} M \mid \underline{\pi_2} M \mid \lambda x : A.M \mid M \underline{\wr} N$$

where $x \in \mathsf{tVar}$ *and* $\underline{f} \in \mathsf{Sg}(\mathscr{E})$ *is of suitable arity. The* typing rules *and some standard abbreviations (including all logical connectives) of the language are defined in Table 8.4. Interpretation of types, contexts and terms-in-context in* $\mathscr{E}$ *is given in Table 8.5.*

Definition 3 (Forcing for an elementary topos) *Assume* $\Gamma = x_1 : \underline{F_1}, \ldots, x_n : \underline{F_n}$ *and* $\Gamma \vdash \phi : \underline{\Omega}$*. By* $\{\langle x_1 \ldots x_n\rangle \in F_1 \times \cdots \times F_n \mid \phi\}$*, I will denote the pullback of the following diagram:*

$$\begin{array}{ccc} & & [\Gamma] \\ & & \Big\downarrow {\scriptstyle [\Gamma \vdash \phi : \underline{\Omega}]} \\ \mathbf{1} & \xrightarrow{\texttt{true}} & \Omega \end{array}$$

Now for $F \xrightarrow{f_1} F_1, \ldots, F \xrightarrow{f_n} F_n$ *write* $F, f_1, \ldots, f_n \Vdash \phi$ *if* $F \xrightarrow{\langle f_1, \ldots, f_n\rangle} [\Gamma]$ *factors through* $\{\langle x_1 \ldots x_n\rangle \in F_1 \times \cdots \times F_n \mid \phi\} \to [\Gamma]$*. In what follows,* $f_1 \circ g, \ldots, f_n \circ g$ *will be denoted by* $\overline{f} \circ g$*. Moreover, let* $[\Gamma \vdash \phi : \underline{\Omega}] = \texttt{true}_{[\Gamma]}$ *be written as* $\Gamma \vDash_{\mathscr{E}} \phi$*.*

Fact 3

- $F, \overline{f} \Vdash \phi$ *iff* $[\Gamma \vdash \phi : \underline{\Omega}] \circ \overline{f} = \texttt{true}_F$
- $\Gamma \vDash_{\mathscr{E}} \phi$ *iff for any* $F \xrightarrow{\overline{f}} [\Gamma]$*, it holds that* $F, \overline{f} \Vdash \phi$

The following result, which can be found as Theorem VI.6.1 in [55] or Theorem II.8.4 in [53], connects the definition of forcing given above with more standard intuitionistic semantics:

Theorem 7 (Beth-Kripke-Joyal semantics in an elementary topos) *Assume* $F \xrightarrow{f_1} E_1, \ldots, F \xrightarrow{f_n} E_n$ *and* $\Gamma = x_1 : \underline{E}_1 \ldots x_n : \underline{E}_n$*.*

- $F, \overline{f} \Vdash \phi \wedge \psi$ *iff* $F, \overline{f} \Vdash \phi$ *and* $F, \overline{f} \Vdash \psi$
- $F, \overline{f} \Vdash \phi \vee \psi$ *iff there are arrows* $G \xrightarrow{g} F$ *and* $H \xrightarrow{h} F$ *s.t.* $G + H \overset{[g,h]}{\twoheadrightarrow} F$ *is epi,* $G, \overline{f} \circ g \Vdash \phi$ *and* $H, \overline{f} \circ h \Vdash \psi$
- $F, \overline{f} \Vdash \phi \rightsquigarrow \psi$ *iff for any* $G \xrightarrow{g} F$ *it holds that* $G, \overline{f} \circ g \Vdash \psi$ *whenever* $G, \overline{f} \circ g \Vdash \phi$
- $F, \overline{f} \Vdash \neg\phi$ *iff for any* $G \xrightarrow{g} F$*, it holds that* $G \cong \mathbf{0}$ *whenever* $G, \overline{f} \circ g \Vdash \phi$
  For the case of quantified formulas, note that $\Gamma \vdash \forall x_{n+1} : \underline{E}_{n+1}.\phi$ *iff* $\Gamma, x_{n+1} : \underline{E}_{n+1} \vdash \phi$*. Same holds for* $\Gamma \vdash \exists x_{n+1} : \underline{E}_{n+1}.\phi$*. Then we have:*
- $F, \overline{f} \Vdash \forall x_{n+1} : \underline{E}_{n+1}.\phi$ *iff for every* $G \xrightarrow{g} F$ *and every* $G \xrightarrow{g'} E_{n+1}$ *it holds that* $G, f_1 \circ g, \ldots, f_n \circ g, g' \Vdash \phi$
- $F, \overline{f} \Vdash \exists x_{n+1} : \underline{E}_{n+1}.\phi$ *iff there exist* $G \xrightarrow{g'} E_{n+1}$ *and an epi* $G \overset{g}{\twoheadrightarrow} F$ *s.t.* $G, f_1 \circ g, \ldots, f_n \circ g, g' \Vdash \phi$
- $F, \overline{f} \Vdash \sigma \approx \tau$ *iff* $[\Gamma \vdash \sigma : \underline{E}] \circ \overline{f} = [\Gamma \vdash \tau : \underline{E}] \circ \overline{f}$

Table 8.4 Typing rules and defined abbreviations of the Mitchell-Bènabou language of $\mathscr{E}$

$\dfrac{}{\Gamma, x : A \vdash x : A}$	$\dfrac{\Gamma \vdash M : A \quad \Gamma \vdash N : A}{\Gamma \vdash M \approx N : \underline{\Omega}}$	$\dfrac{}{\Gamma \vdash \underline{\mathtt{fin!}} : \underline{\mathbf{1}}}$	$\dfrac{\Gamma \vdash M : A \quad \Gamma \vdash N : B}{\Gamma \vdash \underline{\langle} M, N \underline{\rangle} : A \underline{\times} B}$
$\dfrac{\Gamma, x : A \vdash M : B}{\Gamma \vdash \lambda x : A.M : B^A}$	$\dfrac{\Gamma \vdash M : B^A \quad \Gamma \vdash N : A}{\Gamma \vdash M \underline{\wr} N : B}$	$\dfrac{\Gamma \vdash M : A \underline{\times} B}{\Gamma \vdash \underline{\pi_1} M : A}$	$\dfrac{\Gamma \vdash M : A \underline{\times} B}{\Gamma \vdash \underline{\pi_2} M : B}$
	$\dfrac{\underline{f} : \underline{F_1}, \ldots, \underline{F_n} \to \underline{E} \in \mathsf{Sg}(\mathscr{E}) \quad \Gamma \vdash M_1 : \underline{F_1} \ldots \Gamma \vdash M_n : \underline{F_n}}{\Gamma \vdash \underline{f} M_1 \ldots M_n : \underline{E}}$		
$\underline{\mathtt{true}} := \underline{\mathtt{fin!}} \approx \underline{\mathtt{fin!}}$	$\phi \wedge \psi := \underline{\langle} \phi, \psi \underline{\rangle} \approx \underline{\langle} \underline{\mathtt{true}}, \underline{\mathtt{true}} \underline{\rangle}$		
$\forall x : A.\phi := \lambda x : A.\phi \approx \lambda x : A.\underline{\mathtt{true}}$	$\phi \rightsquigarrow \psi := \phi \wedge \psi \approx \phi$		
$\exists x : A.\phi := \forall t : \underline{\Omega}.((\forall x : A.\phi \rightsquigarrow t) \rightsquigarrow t)$	$\underline{\mathtt{false}} := \forall t : \underline{\Omega}.t$		
$\phi \vee \psi := \forall t : \underline{\Omega}.((\phi \rightsquigarrow t) \wedge (\psi \rightsquigarrow t) \rightsquigarrow t)$	$\neg \phi := \phi \rightsquigarrow \underline{\mathtt{false}}$		

Table 8.5 Interpretation of types, contcxts and terms-in-context

$$[\underline{E}] := E \qquad [\underline{\mathbf{1}}] := \mathbf{1} \qquad [\underline{\Omega}] := \Omega \qquad [A \underline{\times} B] := [A] \times [B] \qquad [B^A] := [B]^{[A]}$$

$$[x_1 : A_1, \ldots, x_n : A_n] := [A_1] \times \cdots \times [A_n]$$

$$\frac{}{[\Gamma, x : A \vdash x : A] = \pi : [\Gamma] \times [A] \to [A]} \qquad \frac{}{[\Gamma \vdash \underline{\mathtt{fin!}} : \underline{\mathbf{1}}] = \mathtt{fin!}_{[\Gamma]}}$$

$$\frac{f \in \mathscr{E}[A_1 \times \cdots \times A_n, B] \quad [\Gamma \vdash M_1 : A_1] = \sigma_1 : [\Gamma] \to [A_1] \ldots [\Gamma \vdash M_n : A_n] = \sigma_n : [\Gamma] \to [A_n]}{[\Gamma \vdash \underline{f} M_1 \ldots M_n : B] := f \circ \langle \sigma_1, \ldots, \sigma_n \rangle : [\Gamma] \to B}$$

$$\frac{[\Gamma \vdash M : A] = \sigma : [\Gamma] \to [A] \quad [\Gamma \vdash N : A] = \tau : [\Gamma] \to [A]}{[\Gamma \vdash M \approx N : \underline{\Omega}] := \vdash m \langle \mathsf{id}_{[\Gamma]}, \mathsf{id}_{[\Gamma]} \rangle \circ \langle \sigma, \tau \rangle}$$

$$\frac{[\Gamma \vdash M : A] = \sigma : [\Gamma] \to [A] \quad [\Gamma \vdash N : B] = \tau : [\Gamma] \to [B]}{[\Gamma \vdash \underline{\langle} M, N \underline{\rangle} : A \underline{\times} B] := \langle \sigma, \tau \rangle : [\Gamma] \to [A] \times [B]}$$

$$\frac{[\Gamma \vdash M : A \times B] = \sigma : [\Gamma] \to [A] \times [B]}{[\Gamma \vdash \underline{\Pi_1} M : A] := \Pi_1 \circ \sigma : [\Gamma] \to [A]} \qquad \frac{[\Gamma \vdash M : A \times B] = \sigma : [\Gamma] \to [A] \times [B]}{[\Gamma \vdash \underline{\Pi_2} M : B] := \Pi_2 \circ \sigma : [\Gamma] \to [B]}$$

$$\frac{[\Gamma, x : A \vdash M : B] = \sigma : [\Gamma] \times [A] \to [B]}{[\Gamma \vdash \lambda x : A . M : B^A] := \mathsf{curry}^{[\Gamma] \times [A], [B]}(\sigma) : [\Gamma] \to [B]^{[A]}}$$

$$\frac{[\Gamma \vdash M : B^A] = \sigma : [\Gamma] \to [B]^{[A]} \quad [\Gamma \vdash N : A] = \tau : [\Gamma] \to [A]}{[\Gamma \vdash M \underline{{}^{\wr}} N : B] := \mathsf{eval}_{[A],[B]} \circ \langle \tau, \sigma \rangle}$$

The clauses for $\exists$ and $\vee$ above resemble those of intuitionistic Beth semantics. This is why "Beth-Kripke-Joyal" seems a more appropriate name in the general case of an arbitrary elementary topos; see, e.g., [41, Sect. 14.6]. However, when the topos happens to be the topos of *presheaves*, i.e., covariant functors into **Set**, on a given small category $\mathscr{R}$—in particular, a poset taken as a category—the definition of forcing can be significantly simplified.[4]

Perhaps the most straightforward account of this simplification can be found in [53]. First, the clause for disjunction can be "kripkefied" for *indecomposable* objects and the clause for existential quantifiers—for *projective* ones [53, Proposition 8.7]. Second, the second clause of Fact 3 suggests that to check validity of a given judgement-in-context $\Gamma \vdash \phi : \underline{\Omega}$ in a topos, it is enough to restrict attention to those F which belong to *a generating set* for a given topos. Third, by the Yoneda Lemma, in a topos of presheaves $\mathbf{Set}^{\mathscr{R}}$ for an arbitrary small category $\mathscr{R}$, objects of the form $\mathsf{hom}^C_{\mathscr{R}} := \mathscr{R}[C, -]$ for any given $C \in \mathscr{R}$ satisfy all these conditions: they are projective, indecomposable and do form a generating set. Moreover, also by the Yoneda Lemma, elements of $\mathbf{Set}^{\mathscr{R}}[\mathsf{hom}^C_{\mathscr{R}}, F]$ are in $1-1$ correspondence with elements of $F(C)$:

$$\mathbf{Set}^{\mathscr{R}}[\mathsf{hom}^C_{\mathscr{R}}, F] \ni f \to \check{f} := f_C(\mathsf{id}_C)$$
$$F(C) \times \mathsf{hom}^C_{\mathscr{R}} \ni (c, h) \to \hat{c}(h) := Fh(c)$$

Note also that in clauses like the one for $\rightsquigarrow$, we can restrict attention to those $G \overset{g}{\to} F$ whose source G lies in the generating set. In the case of $\mathbf{Set}^{\mathscr{R}}$, this means replacing $G \overset{g}{\to} \mathsf{hom}^C_{\mathscr{R}}$ with elements of $\mathbf{Set}^{\mathscr{R}}[\mathsf{hom}^D_{\mathscr{R}}, \mathsf{hom}^C_{\mathscr{R}}]$. But, by the Yoneda Lemma again, these can be replaced with arrows in $\mathscr{R}[C, D]$ (note the change of direction!).

[4] Reader should be warned that in most of categorical literature, presheaves are assumed to be *contra*variant, but see, e.g., [39] for an example of the covariant convention.

Taking all this into account, we can obtain the following modified version of the semantics—this time properly "Kripkean" (see [53, Proposition 9.3]).

Corollary 8 (Kripke-Joyal semantics in a topos of presheaves) *Let $\mathscr{R}$ be a small category, $F_1, \ldots, F_n \in \mathbf{Set}^{\mathscr{R}}$, $C \in \mathscr{R}$, $c_1 \in F_1(C)$, $\ldots c_n \in F_n(C)$, $\Gamma = c_1 : \underline{F}_1, \ldots, c_n : \underline{F}_n$ and $\Gamma \vdash \phi : \underline{\Omega}$. Write $C, \overline{c} \Vdash \phi$ for $\mathsf{hom}^C_{\mathscr{R}}, \hat{c}_1, \ldots, \hat{c}_n \Vdash \phi$. Given any $f \in \mathscr{R}[C, D]$, write $f(\overline{c})$ for $\hat{c}_1(f), \ldots, \hat{c}_n(f)$—that is, $Ff(c_1), \ldots, Ff(c_n)$. Then we have:*

- $C, \overline{c} \Vdash \phi \wedge \psi$ *iff* $C, \overline{c} \Vdash \phi$ *and* $C, \overline{c} \Vdash \psi$
- $C, \overline{c} \Vdash \phi \vee \psi$ *iff* $C, \overline{c} \Vdash \phi$ *or* $C, \overline{c} \Vdash \psi$
- $C, \overline{c} \Vdash \phi \rightsquigarrow \psi$ *iff for any* $f \in \mathscr{R}[C, D]$, $D, f(\overline{c}) \Vdash \psi$ *whenever* $D, f(\overline{c}) \Vdash \phi$
- $C, \overline{c} \Vdash \neg\phi$ *iff for any* $f \in \mathscr{R}[C, D]$, *it does not hold that* $D, f(\overline{c}) \Vdash \phi$
- $C, \overline{c} \Vdash \forall x_{n+1} : \underline{F}_{n+1}.\phi$ *iff for every* $f \in \mathscr{R}[C, D]$ *and* $d \in F_{n+1}(D)$, *it holds that* $D, f(\overline{c}), d \Vdash \phi$
- $C, \overline{c} \Vdash \exists x_{n+1} : \underline{F}_{n+1}.\phi$ *iff there exist* $d \in F_{n+1}(C)$ *s.t.* $C, \overline{c}, d \Vdash \phi$.

Reference [39] uses toposes of presheaves as a generalization of Kripke semantics for the intuitionistic first-order logic to prove incompleteness results. Of numerous follow-ups of that work, let me just mention [62, 72]. Let us also note that the derivation of Corollary 8 from Theorem 7 takes a somewhat more roundabout route in [55]: toposes of presheaves are handled there as a subclass of toposes of *sheaves on a site*.

8.5.2 Non-expansive Morphisms, Fixed Points and Scattered Toposes

Let $\mathscr{E}$ be an elementary topos. Call an endomorphism $f \in \mathscr{E}[X, X]$ *unchanging* [30] or *non-expansive* if

$$\models_{\mathscr{E}} \forall x, y : \underline{X}.(\underline{f}x \approx \underline{f}y \rightsquigarrow x \approx y) \rightsquigarrow x \approx y.$$

As noted in [30], in a boolean setting *non-expansive* means just *constant*: negate the sentence and play with boolean laws. Obviously then a *classical* proof that a non-expansive endomorphism on a non-empty set has a unique fixed point does not carry much computational content. In a constructive setting, however, the situation is different.

Assume $\vdash \phi, \psi : \underline{\Omega}^{\underline{X}}$ and $f \in \mathscr{E}[X, X]$ and define:

$$\begin{aligned}
\mathsf{SubTe}(\phi) &:= \forall x, y : \underline{X}(\phi \wr x \wedge \phi \wr y \rightsquigarrow x \approx y)\\
\phi \subseteq \psi &:= \forall x : \underline{X}.(\phi \wr x \rightsquigarrow \psi \wr x)\\
\mathsf{MaxST}(\phi) &:= \mathsf{SubTe}(\phi) \wedge \forall \alpha : \underline{\Omega}^{\underline{X}}.(\mathsf{SubTe}(\alpha) \wedge \phi \subseteq \alpha \rightsquigarrow \alpha \subseteq \phi)\\
\mathsf{Non_exp}(\underline{f}) &:= \forall x, y : \underline{X}.(\underline{f}x \approx \underline{f}y \rightsquigarrow x \approx y) \rightsquigarrow x \approx y\\
\mathsf{fix_so}_{\underline{f}} &:= \lambda x : \underline{X}.(x \approx \underline{f}x)
\end{aligned}$$

With this apparatus, we can state the main Theorem of Section 3 of [30]:

Theorem 8 *Assume $\mathscr{E}$ is an elementary topos and $f \in \mathscr{E}[X, X]$. Then*

$$\vDash_{\mathscr{E}} \mathsf{Non\_exp}(\underline{f}) \rightsquigarrow \mathsf{MaxST}(\mathsf{fix\_so}_{\underline{f}}).$$

Remark 12 A proof formalized in the Coq proof assistant is available from the author, see Remark 1. Those who would like to try a manual yet rigorous proof in the Mitchell-Bènabou language should do first Exercise 5 in [53, p. 139] and then formalize the proof in [30, p. 105] using all the abbreviations given above.

In words, this result says: *the fixpoints of a non-expansive endomorphism form a maximal subterminal subobject.*[5] The syntactic shape of $\mathsf{SubTe}\, f$ easily suggests that subterminality is the internal counterpart of "being of cardinality at most one", i.e., uniqueness of fixpoints. However, the situation with existence is more complicated. First of all, toposes of presheaves can differ significantly from the topos of sets in having non-trivial objects with *no global elements whatsoever*. More importanly, even being inhabited is not enough to ensure maximal subterminal objects are global elements.

Example 2 ([30]) Consider the topos of presheaves on $(\omega + 1, \trianglerighteq)$, where $\trianglerighteq$ is the converse of the standard ordinal order. Presheaf X defined as $X(n) = n + 1$ and $X(\omega) = \omega$ with $X(\beta \trianglerighteq \alpha)(n) = min(n, \alpha)$ is clearly inhabited. Furthermore, $f : X \to X$ defined as $f_n(i) = min(i + 1, n)$ and $f_\omega(i) = i + 1$ is a non-expansive endomorphism. Yet it fails to have a fixpoint—i.e., a global element $\mathbf{1} \overset{c}{\to} X$ s.t. $f \circ c = c$.

Of course, we can do better in special cases.

Corollary 9 *Whenever $X \in \mathscr{E}$ is s.t. any maximal subterminal subobject of X is a global element (for example, X is an injective object), there exists* `fin!` $\overset{c}{\to} X$ *s.t. $f \circ c = c$ for any non-expansive $f \in \mathscr{E}[X, X]$.*

We could try to express unique existence in the internal logic using the standard abbreviation $\exists!$ for "exists exactly one". However, as kindly pointed out by Thomas Streicher, this abbreviation works as intended in toposes of presheaves, but not necessarily in arbitrary ones.

But where is the place for a modality in all this? Say that $\circledcirc : \Omega \to \Omega$ is a *strong Löb operator* if $\vDash_{\mathscr{E}} \forall p : \underline{\Omega}.(\underline{\circledcirc} p \rightsquigarrow p) \rightsquigarrow p$. Also, call a morphism $f \in \mathscr{E}[X, Z]$ *$\circledcirc$-contractive* if $\vDash_{\mathscr{E}} \forall x, y : \underline{X}.\underline{\circledcirc}(x \approx y) \rightsquigarrow (\underline{f} x \approx \underline{f} y)$:

Corollary 10 *Let $\circledcirc : \Omega \to \Omega$ be a strong Löb operator, $f \in \mathscr{E}[X, X,]$ and assume that f is $\circledcirc$-contractive. Then f is non-expansive and hence its subobject of fixed points is a maximal subterminal one.*

---

[5] The corresponding theorem in [30] contained also an additional statement about density of the support of the fixed-point subobject, but this does not seem essential for us here.

Proof We have that $\vDash_{\mathscr{E}} \forall p, q : \underline{\Omega}.(\circledcirc p \rightsquigarrow (p \vee q)) \rightsquigarrow ((q \rightsquigarrow p) \rightsquigarrow p)$. In fact, this is an equivalent form of (**slöb**)—cf. the proof of Theorem 2(iv) in [30]. Now substitute $x \approx y$ for p and $\underline{f} x \approx \underline{f} y$ for q to get the result. □

Reference [30] states the result only for a special case of contractiveness and a special subclass of toposes (introduced below) but this generalization is straightforward. As before, we can derive the conclusion about the existence of unique fixed points *whenever every maximal subterminal object of X happens to be a global element*—e.g., whenever X is injective.

Define $\underline{\odot}\phi := \forall t : \underline{\Omega}.(t \vee (t \rightsquigarrow \phi))$, i.e., an internalized coderivative. We have the following counterpart of Proposition 1:

Proposition 3 *In any elementary topos* $\mathscr{E}$, *we have* $\vDash_{\mathscr{E}} \forall p : \underline{\Omega}.p \rightsquigarrow \underline{\odot} p$ *and* $\vDash_{\mathscr{E}} \forall p, q : \underline{\Omega}.\underline{\odot} p \rightsquigarrow (q \vee (q \rightsquigarrow p))$.

A *scattered topos* is defined analogously to scattered locales or Heyting algebras in Sect. 8.4.2 by the validity of the only remaining **KM** law, i.e., the axiom $\forall p : \underline{\Omega}.(\underline{\odot} p \rightsquigarrow p) \rightsquigarrow p$. Thus, scattered toposes are those where $\underline{\odot}$ is a strong Löb operator. This notion turns out to have several equivalent characterizations, see [30]. Let us discuss in detail here another one for the special case of $\mathbf{Set}^{\mathscr{R}}$:

Theorem 9 *Let* $\mathscr{R} = (W, \trianglelefteq)$ *be a poset. Then* $\mathbf{Set}^{\mathscr{R}}$ *is scattered iff* $(W, \trianglelefteq, \triangleleft)$ *satisfies any of the equivalent conditions in Corollary 4.*

Proof For a direct proof, it is useful to compute the semantic meaning of $\underline{\odot}$. Define $\odot : \Omega \to \Omega$ as $[p : \underline{\Omega} \vdash \underline{\odot} p : \underline{\Omega}]$. Recall also that in a topos of the form $\mathbf{Set}^{\mathscr{R}}$ for $\mathscr{R} = (W, \trianglelefteq)$, $\Omega(w)$ is equal to $\{A \cap \{w\}^{\trianglelefteq}\uparrow \mid A \in \mathrm{Up}_{\trianglelefteq}(W)\}$. A morphism $f : \Omega \to \Omega$ is a natural transformation: a family of mappings $\{f_w : \Omega(w) \to \Omega(w) \mid w \in W\}$ satisfying

$$f_z(A \cap \{z\}^{\trianglelefteq}\uparrow) = f_w(A) \cap \{z\}^{\trianglelefteq}\uparrow \text{ for any } A \in \Omega(w), z \trianglerighteq w.$$

Now let us note the following

Fact 4 *For any topos of the form* $\mathbf{Set}^{\mathscr{R}}$ *where* $\mathscr{R} = (W, \trianglelefteq)$, *for any* $w \in W$ *and for any* $A \in \Omega(w)$ *(i.e., A an upward closed subset of* $\{w\}^{\trianglelefteq}\uparrow$*),*

$$\odot_w(A) = \{z \trianglerighteq w \mid \{z\}^{\triangleleft}\uparrow \subseteq A\}.$$

The reader may want to consult Sect. 8.4 and Table 8.3 for the notation used above; in particular, recall that $\{z\}^{\triangleleft}\uparrow = \{z\}^{\trianglelefteq}\uparrow - \{z\}$. Note also that we can add an atomic clause to Corollary 8 in the preceding section:

Fact 5 $w, A \Vdash t$ *(where* $\Gamma = t : \underline{\Omega}$ *and* $A \in \Omega(w)$*) iff* $A = \{w\}^{\trianglelefteq}\uparrow$.

This fact, while rather basic, is worth an explicit proof, as it helps to put together several definitions and propositions above:

Proof (of Fact 5) $w, A \Vdash t$ is an abbreviation for $\mathsf{hom}^{w}_{\mathscr{R}}, \hat{A} \Vdash t$, while this in turn can be reformulated as $\hat{A} = \texttt{true} \circ \texttt{fin}!_{\mathsf{hom}^{w}_{\mathscr{R}}}$ (Fact 3). In particular, $\hat{A}(w \trianglelefteq w) = \{w\}^{\trianglelefteq}\uparrow$. But $\hat{A}(w \trianglelefteq w) = A$. □

Fact 5 can be generalized with variable t on the right hand side of the turnstile replaced with arbitrary $\phi(t)$. Somewhat informally speaking, $w, A \Vdash \phi(t)$ iff the value of $\phi(A)$ contains $\{w\}^{\trianglelefteq}\uparrow$ (think of ϕ here as a polynomial on the Heyting algebra of upward closed subsets of $\{w\}^{\trianglelefteq}\uparrow$).

Putting all this together, we get that $w, A \Vdash (\underline{\odot} p \rightsquigarrow p) \rightsquigarrow p$ iff for any $z \trianglerighteq w$, it holds that $z, A \cap \{z\}^{\trianglelefteq}\uparrow \Vdash (\underline{\odot} p \rightsquigarrow p)$ implies $\{z\}^{\trianglelefteq}\uparrow \subseteq A$. That is, $\{z' \trianglerighteq z \mid \{z'\}^{\triangleleft}\uparrow \subseteq A\} \subseteq A$ only if $\{z\}^{\trianglelefteq}\uparrow \subseteq A$ and in order for $\mathbf{Set}^{(W,\trianglelefteq)}$ to be scattered this has to hold for any $w \in W$, any $z \trianglerighteq w$ and any $A \in \Omega(w)$. But then the reasoning can be completed just like in the case of Corollary 4. This finishes the proof of Theorem 9. □

Theorem 9 shows that the *topos of trees* (or *forests*) $\mathcal{S}$ in [14]—i.e., the topos of presheaves on $(\omega, \trianglerighteq)$, where $\trianglerighteq$ is the converse of usual order on ω—is scattered. Ω-endomorphism "$\triangleright$" (this notation here would risk clashing with the one for a strict partial order and its converse) defined in [14] is easily seen to coincide with $\odot$ interpreting $\underline{\odot}$ in $\mathbf{Set}^{(\omega,\trianglerighteq)}$ as specified by Fact 4 above. The Internal Banach Fixpoint Theorem 2.9 of [14] shows that $\odot$-contractive mappings on arbitrary inhabited objects in $\mathcal{S}$ do have (unique) fixpoints.

Now, Example 2 above shows that such a strong statement is not valid in arbitrary scattered toposes of presheaves, even quite similar to $\mathcal{S}$. The crucial Lemma 2.10 in [14] is not amenable to far-reaching generalizations.

However, [14, Sect. 8] discusses a whole class of toposes together with a notion of a contractiveness guaranteeing fixpoint's existence. The class in question are *sheaves* on complete Heyting algebras with a well-founded basis [28] rather than just presheaves on Noetherian partial orders—crucially, $\mathcal{S}$ can be also seen as such a sheaf topos—and the required notion of contractiveness is stronger than the one expressible in the internal logic.

Let us elaborate on the last point. As we saw, toposes allow an internal interpretation of modalities as morphisms $\Omega \to \Omega$. Actually, from the "propositions as predicates" perspective, any operation on subobjects of a given object is a "local" candidate for a modality. However, constructive or categorical logic is mostly about "propositions as types"; see, e.g., [5, 9, 13, 27, 65] for modal aspects. This perspective works even with mild assumptions about the underlying category. In particular, *algebraic type theories* require only finite products, whereas ccc's correspond to *functional type theories* [24]: those whose type system is in fact that of Brouwerian semilattices of Remark 3 above. To see the details of this correspondence, just remove the rules for $\underline{\Omega}$, $\approx$ and all abbreviations using these from Tables 8.4 and 8.5, then interpret conjunctions as products and implication as exponentation.

From this perspective, modalities correspond to endofunctors. In particular, ND systems for $\mathbf{PLL}^{\mathsf{i}}$ and $\mathbf{S4}^{\mathsf{i}}$ are interpreted by, respectively, *monads* and *comonads*—see, e.g., references in Remark 7—and $\mathbf{SL}^{\mathsf{i}}$ yields a special subclass of *pointed* or *applicative* functors [56]. More precisely, one obtains a variant of [14, Definition 6.1]. Possible differences are: the modal assumption of normality forces only being monoidal wrt cartesian structure (cf. [9, 27]) rather than preservation of all finite limits as in the second clause of that definition; furthermore, the assumption of uniqueness in the first clause would rely on exact reduction and conversion rules of the proof system. Possible names for such endofunctors include *contraction*, *delay*, *(strong) Löb*, $\mathbf{SL}^{\mathsf{i}}$ and *MGRT*, the last being an abbreviation of the original name in [14].

One can relate these two views on modalities. Whenever $F : \mathscr{E} \to \mathscr{E}$ is monic-preserving and $\mathscr{E}$ has pullbacks, associate with a F-coalgebra $C \xrightarrow{\gamma} FC$ a modality $[F.\gamma]$ on subobjects $M \overset{m}{\rightarrowtail} C$:

$$\begin{array}{ccc} [F.\gamma]M & \overset{[F.\gamma]m}{\rightarrowtail} & C \\ \downarrow & \lrcorner & \downarrow \gamma \\ FM & \overset{Fm}{\rightarrowtail} & FC \end{array}$$

(see [3] for the history of this diagram in papers on well-founded coalgebras). Furthermore, whenever F is pointed (applicative), i.e., a $\mathbf{R}^{\mathsf{i}}$-endofunctor with $\boldsymbol{(F)} : 1 \to F$ being the *point* or *unit* of F, $\boldsymbol{(F)}_M$ is a subcoalgebra of $\boldsymbol{(F)}_C$ for any $M \overset{m}{\rightarrowtail} C$ and hence $m \leq [F.\boldsymbol{(F)}_C]m$, i.e., the "local" translation of $(\mathbf{r})$ is universally valid. Birkedal et al. [14, Theorem 6.8] allows to extract additional assumptions needed to ensure that $[F.\boldsymbol{(F)}_C]$ is a $\mathbf{SL}^{\mathsf{i}}$-modality for F a $\mathbf{SL}^{\mathsf{i}}$-endofunctor: F has to preserve pullbacks and, importantly, $\mathscr{E}$ has to be a topos. Otherwise, there are natural counterexamples.

The operation on subobjects of $C \in \mathcal{S}$ induced by $\odot : \Omega \to \Omega$ from Fact 4 is defined in an alternative way in [14]: as $[\bullet.\boldsymbol{(\bullet)}_C]$ for a delay endofunctor $\bullet : \mathcal{S} \to \mathcal{S}$, whose action on objects is $(\bullet C)(0) := \mathbf{1}$ and $(\bullet C)(n+1) := (\bullet C)(n)$. In a sense, $\bullet$ can be called the *Cantor-Bendixson endofunctor*. Factoring through it is the desired "external" notion of contractivity ensuring fixpoint's existence. Both notions nicely complement each other:

> …the external notion provides for a simple algebraic theory of fixed points for not only morphisms but also functors (see Sect. 2.6), whereas the internal notion is useful when working in the internal logic [14].

Could [30] have had more impact if the authors had (a) employed the external perspective on modalities in addition to the internal one and (b) had the hindsight of [28]? This is rather too counterfactual a question to consider. Note also that what matters from the point of view of [14]—and Theoretical CS in general—is the use made of these external and internal Löb modalities. Birkedal et al. [14, Sect. 3] constructs a model of a programming language with higher-order store and recursive types entirely inside the internal logic of $\mathcal{S}$. Birkedal et al. [14, Sect. 4] provides semantic foundation for dependent type theories extended with a $\mathbf{SL}^{\mathsf{i}}$ modality and guarded recursive types; this can be regarded as an extension of fixpoint results along the lines of Sect. 8.3 above to predicate and higher-order constructive logics. Birkedal et al. [14, Sect. 5] shows that a class of (ultra-)metric spaces commonly used in modelling corecursion on streams is equivalent to a subcategory of $\mathcal{S}$.

Clearly, this is a large area rather overlooked by researchers on (intuitionistic) modal logic side. There is no space here to discuss my own work in progress, e.g., on the Curry-Howard interpretation of **mHC**, but let us conclude with two other questions. The first was posed to me by Lars Birkedal: what are additional logical principles which would allow a scattered topos to model not only guarded (co-)recursion, but also, e.g., countable nondeterminism? The second comes from participants of ToLo III: is there a natural subclass of internal modalities in toposes (endomorphisms $\Omega \rightarrow \Omega$) inducing external modalities (endofunctors) in a generic way?

Acknowledgments Special thanks are due to: the anonymous referee for careful reading and many helpful comments; Guram Bezhanishvili, for an invitation to write this chapter, but also his tolerance, understanding and patience during the whole process; Alexander Kurz for his support, interest and numerous discussions; Mamuka Jibladze and the late Dito Pataraia, particularly for all the discussions we had in Kutaisi. Apart from this, the list of people who provided feedback and/or suggestions include (in no particular order) Lars Birkedal, Rasmus Møgelberg, Stefan Milius, Thomas Streicher, Sam Staton, Alex Simpson, Gordon Plotkin, Johan van Benthem, Ramon Jansana, Achim Jung, Andrew Pitts, Conor McBride, Nick Benton, Guilhem Jaber, Dirk Pattinson, Drew Moshier, Nick Bezhanishvili, Alexei Muravitsky, Fredrik Nordvall Forsberg, Ben Kavanagh, Oliver Fasching, participants of Domains X, TYPES 2011, MGS 2012 and seminars at the University of Leicester. During most of the write-up, I was supported by the EPSRC grant EP/G041296/1.

References

1. Abadi M (2007) Access control in a core calculus of dependency. Electr. Notes Theor. Comput. Sci. 172:5–31
2. Aczel P (2001) The Russell-Prawitz modality. Math Struct Comput Sci 11(4):541–554
3. Adámek J, Milius S, Moss L, Sousa L (2012) Well-pointed coalgebras (extended abstract). In: Proceedings of 15th international conference on foundations of software science and computation structures (FoSSaCS'12). Lecture notes in computer science (ARCoSS), vol 7213, pp 89–103
4. Alberucci L, Facchini A (2009) On modal μ-calculus and Gödel-Löb logic. Studia Logica 91(2):145–169
5. Alechina N, Mendler M, de Paiva V, Ritter E (2001) Categorical and Kripke semantics for constructive S4 modal logic. In: Fribourg L (ed) CSL. Lecture notes in computer science, vol 2142. Springer, Berlin, pp 292–307
6. Amerbauer M (1996) Cut-free tableau calculi for some propositional normal modal logics. Studia Logica 57(2/3): 359–372
7. Appel A, Melliès P, Richards C, Vouillon J (2007) A very modal model of a modern, major, general type system. In: Hofmann M, Felleisen M (eds) POPL. ACM, pp 109–122
8. Atkey R, McBride C (2013) Productive coprogramming with guarded recursion. In: ICFP 2013 Proceedings, pp 197–208
9. Bellin G, de Paiva V, Ritter E (2001) Extended Curry-Howard correspondence for a basic constructive modal logic. In: Proceedings of methods for modalities
10. Benton P, Bierman G, de Paiva V (1998) Computational types from a logical perspective. J Funct Program 8(2):177–193
11. Benton N, Tabareau N (2009) Compiling functional types to relational specifications for low level imperative code. In: Kennedy A, Ahmed A (eds) TLDI. ACM, pp 3–14
12. Bezhanishvili G, Jansana R (2013) Esakia style duality for implicative semilattices. Appl Categorical Struct 21:181–208

13. Bierman G, de Paiva V (2000) On an intuitionistic modal logic. Studia Logica 65(3):383–416
14. Birkedal L, Møgelberg R, Schwinghammer J, Støvring K (2012) First steps in synthetic guarded domain theory: step-indexing in the topos of trees. Logical Meth Comput Sci 8:1–45
15. Birkedal L, Møgelberg R (2013) Intensional type theory with guarded recursive types qua fixed points on universes. In: Proceedings of LICS (2013)
16. Blackburn P, de Rijke M, Venema Y (2001) Modal logic. Cambridge University Press, Cambridge
17. Blok W, Pigozzi D (1989) Algebraizable logics. Memoirs AMS 77:396
18. Boolos G (1993) The logic of provability. Cambridge University Press, Cambridge
19. Boolos G, Sambin G (1991) Provability: the emergence of a mathematical modality. Studia Logica 50:1–23
20. Božić M, Došen K (1984) Models for normal intuitionistic modal logics. Studia Logica 43(3):217–245
21. Burris S, Sankappanavar H (1981) A course in universal algebra. Graduate texts in mathematics. Springer, New York
22. Celani S, Jansana R (1999) Priestley duality, a Sahlqvist theorem and a Goldblatt-Thomason theorem for positive modal logic. Logic J IGPL 7(6):683–715
23. Chagrov A, Zakharyaschev M (1997) Modal logic. No. 35 in Oxford logic guides. Clarendon Press, Oxford
24. Crole R (1993) Categories for types. Cambridge mathematical textbooks. Cambridge University Press, Cambridge
25. Curry H (1952) The elimination theorem when modality is present. J Symbolic Logic 17(4):249–265
26. de Jongh D, Visser A (1996) Embeddings of Heyting algebras. In: Logic: from foundations to applications (Staffordshire, 1993), 187–213, Oxford Sci Publ, Oxford University Press, New York
27. de Paiva V, Ritter E (2011) Basic constructive modality. In: Beziau J, Coniglio M (eds) Logic without frontiers—Festschrift for Walter Alexandre Carnielli on the occasion of his 60th birthday. College Publications, pp 411–428
28. Di Gianantonio P, Miculan M (2004) Unifying recursive and co-recursive definitions in sheaf categories. In: Walukiewicz I (ed) Foundations of software science and computation structures. Lecture notes in computer science, vol 2987. Springer, Berlin/Heidelberg, pp 136–150
29. Došen K (1985) Models for stronger normal intuitionistic modal logics. Studia Logica 44(1):39–70
30. Esakia L, Jibladze M, Pataraia D (2000) Scattered toposes. Ann Pure Appl Logic 103(1–3):97–107
31. Esakia L (2006) The modalized Heyting calculus: a conservative modal extension of the intuitionistic logic. J Appl Non-Class Logics 16(3–4):349–366
32. Fairtlough M, Mendler M (1997) Propositional lax logic. Inf Comput 137(1):1–33
33. Fine K (1985) Logics containing K4. part II. J Symbolic Logic 50(3):619–651
34. Font J (2006) Beyond Rasiowa's algebraic approach to non-classical logics. Studia Logica 82(2):179–209
35. Font J, Jansana R, Pigozzi D (2003) A survey of abstract algebraic logic. Studia Logica 74(1–2):13–97
36. Font J, Jansana R, Pigozzi D (2009) Update to "A survey of abstract algebraic logic". Studia Logica 91(1):125–130
37. Garg D, Abadi M (2008) A modal deconstruction of access control logics. In: Amadio R (ed) FoSSaCS. Lecture notes in computer science, vol 4962. Springer, Berlin, pp 216–230
38. Garg D, Pfenning F (2006) Non-interference in constructive authorization logic. In: IEEE computer society CSFW, pp 283–296
39. Ghilardi S (1989) Presheaf semantics and independence results for some non-classical first-order logics. Arch Math Logic 29:125–136
40. Goldblatt R (2003) Mathematical modal logic: a view of its evolution. J Appl Logic 1(5–6):309–392

41. Goldblatt R (2006) Topoi: the categorial analysis of logic. In: Dover books on mathematics. Dover Publications, New York
42. Goldblatt R (1981) Grothendieck topology as geometric modality. Math Logic Q 27(31–35):495–529
43. Goldblatt R (2010) Cover semantics for quantified lax logic. J Logic Comput 1035–1063
44. Horn A (1962) The separation theorem of intuitionist propositional calculus. J Symbolic Logic 27(4):391–399
45. Iemhoff R (2001) Provability logic and admissible rules. PhD thesis, University of Amsterdam
46. Jaber G, Tabareau N, Sozeau M (2012) Extending type theory with forcing. In: Proceedings of LICS'12, Dubrovnik, Croatia, pp. 395–404
47. Johnstone P (2002) Sketches of an elephant: a topos theory compendium. In: Oxford logic guides. Clarendon Press, Oxford
48. Köhler P (1981) Brouwerian semilattices. Trans Am Math Soc 268(1):103–126
49. Krishnaswami N, Benton N (2011) A semantic model for graphical user interfaces. In: Chakravarty M, Hu Z, Danvy O (eds) ICFP. ACM, pp 45–57
50. Krishnaswami N, Benton N (2011) Ultrametric semantics of reactive programs. In: LICS. IEEE Computer Society, pp 257–266
51. Kuznetsov A, Muravitsky A (1986) On superintuitionistic logics as fragments of proof logic extensions. Studia Logica 45(1):77–99
52. Kuznetsov A (1979) On algebras of open sets. In: The 4th Tiraspol symposium on general topology and its applications, Kishinev. Abstracts, Kishinev (Russian), pp 72–75
53. Lambek J, Scott P (1986) Introduction to higher order categorical logic. No. 7 in Cambridge studies in advanced mathematics. Cambridge University Press, Cambridge
54. Litak T (2007) The non-reflexive counterpart of Grz. Bull Sect Logic 36:195–208
55. MacLane S, Moerdijk I (1992) Sheaves in geometry and logic: a first introduction to topos theory, corrected edn. Springer, Berlin
56. McBride C, Paterson R (2008) Applicative programming with effects. J Funct Program 18(1):1–13
57. McLarty C (1990) The uses and abuses of the history of topos theory. Br J Philos Sci 41(3):351–375
58. Milne P (2010) Subformula and separation properties in natural deduction via small Kripke models. Rev Symbolic Logic 3(02):175–227
59. Moss L (2007) Finite models constructed from canonical formulas. J Philos Logic 36:605–640
60. Muravitsky A (1981) Finite approximability of the calculus I^{Δ} and the nonmodelability of some of its extensions. Mat Zametki 29(6):907–916
61. Muravitsky A (2014) Logic KM: a biography (this volume)
62. Nagaoka K, Isoda E (1997) Incompleteness results in Kripke bundle semantics. Math Logic Q 43(4):485–498
63. Nakano H (2000) A modality for recursion. In: LICS. IEEE Computer Society, pp 255–266
64. Nakano H (2001) Fixed-point logic with the approximation modality and its Kripke completeness. In: Kobayashi N, Pierce B (eds) TACS. Lecture notes in computer science, vol 2215. Springer, Berlin, pp 165–182
65. Pfenning F, Davies R (2001) A judgmental reconstruction of modal logic. Math Struct Comput Sci 11(4):511–540
66. Prawitz D (1965) Natural deduction: a proof-theoretical study. PhD thesis, Almqvist & Wiksell
67. Rasiowa H (1974) An algebraic approach to non-classical logics. North Holland
68. Sambin G (1976) An effective fixed-point theorem in intuitionistic diagonalizable algebras (The algebraization of the theories which express Theor, IX). Studia Logica 35(4):345–361
69. Simmons H (1982) An algebraic version of Cantor-Bendixson analysis. In: Banaschewski B (ed) Categorical aspects of topology and analysis. Lecture notes in mathematics, vol 915. Springer, Berlin / Heidelberg, pp 310–323
70. Simmons H (2014) Cantor-Bendixson properties of the assembly of a frame (this volume)
71. Simpson A (1994) The proof theory and semantics of intuitionistic modal logic. PhD thesis, University of Edinburgh

72. Skvortsov D (2012) Kripke sheaf completeness of some superintuitionistic predicate logics with a weakened constant domains principle. Studia Logica 100:361–383
73. Smorynski C (1979) Calculating self-referential statements, I: explicit calculations. Studia Logica 38(1):17–36
74. Solovay R (1976) Provability interpretations of modal logic. Israel J Math 25:287–304
75. Sotirov V (1984) Modal theories with intuitionistic logic. In: Mathematical logic, proc. conf. math. logic dedicated to the memory of A. A. Markov (1903–1979), Sofia, 22–23 Sept 1980, pp 139–171
76. ten Cate B, Fontaine G, Litak T (2010) Some modal aspects of XPath. J Appl Non-Class Logics 20(3):139–171
77. Ursini A (1979) A modal calculus analogous to K4W, based on intuitionistic propositional logic, I-. Studia Logica: Int J Symbolic Logic 38(3):297–311
78. van Benthem J (1991) Reflections on epistemic logic. Logique et Anal 34:5–14
79. van Benthem J (2006) Modal frame correspondences and fixed-points. Studia Logica 83(1–3):133–155
80. Visser A (2005) Löb's logic meets the μ-calculus. In: Middeldorp A, van Oostrom V, van Raamsdonk F, de Vrijer R (eds) Processes, terms and cycles. Lecture notes in computer science, vol 3838. Springer, Berlin, pp 14–25
81. Wolter F, Zakharyaschev M (1997) On the relation between intuitionistic and classical modal logics. Algebra and Logic 36:121–125
82. Wolter F, Zakharyaschev M (1998) Intuitionistic modal logics as fragments of classical modal logics. In: Orlowska E (ed) Logic at work, essays in honour of Helena Rasiowa. Springer, Berlin, pp 168–186
83. Wolter F, Zakharyaschev M (2014) On the Blok-Esakia theorem (this volume)

Chapter 9
Cantor-Bendixson Properties of the Assembly of a Frame

Harold Simmons

In memory of Leo Esakia

Abstract In the sense used here a frame A is the algebraic generalization of a topology, the family of open sets of a topological space. It is a complete Heyting algebra, although from that perspective frame morphisms are not quite what you expect. The category of complete Boolean algebras sits inside the category of frames. However, a frame need not have a Boolean reflection. It seems that it does have a Boolean reflection precisely when it is 'nearly pathological' in some sense. For instance, the topology of a T_0 space has Boolean reflection precisely when the space is scattered. Each frame A has an assembly NA which collects together all the quotients of A, and this assembly is itself a frame. Since NA is a frame it has its own assembly N^2A, which has its assembly N^3A, and so on. This generates the assembly tower of A. It is known that A has Boolean reflection precisely when some member of this tower is Boolean, and then that is the Boolean reflection. It seems that the nature of this tower is somehow connected with a generalization of the Cantor-Bendixson process on a topological space. In this chapter I investigate this idea.

Keywords Frame · Nucleus · Assembly · Ranking

9.1 Introduction

In various parts of mathematics we come across a 1-placed operation d carried by a lattice (which is often a Boolean algebra) for which one side of

H. Simmons (✉)
School of Mathematics, The University of Manchester, Manchester, England
e-mail: Harold.Simmons@manchester.ac.uk

G. Bezhanishvili (ed.), *Leo Esakia on Duality in Modal and Intuitionistic Logics*, Outstanding Contributions to Logic 4, DOI: 10.1007/978-94-017-8860-1_9,

$d(a \vee b) = d(a) \vee d(b)$	$d(a \wedge b) = d(a) \wedge d(b)$
nearly $d^2(a) \leq d(a)$	nearly $d^2(a) \geq d(a)$

holds for all elements a, b. Here the 'nearly' means that the comparison doesn't quite hold, but some minor modification does make it true. For instance, with a topological space S the

closure operation   interior operation

on the power set of S satisfy the indicated side, exactly. The Cantor-Bendixson derivative almost satisfies the left hand condition. In the study of modules there are several dimension devices that satisfy the right hand condition. These include the gadgets for measuring socle length, the Krull and the Gabriel dimension, as well as other less well known measuring devices. The localization of module categories and the sheafifying of presheaf categories also make use of the right hand gadgets. These properties also occur in various modal logics, especially those concerning the modal logic related to Gödel's incompleteness of arithmetic.

When first seen such a gadget may look like a mere curiosity and hardly worth thinking about. However, the examples mentioned suggest that perhaps there is something more general going on, and perhaps such gadgets should be investigated in more depth.

Leo Esakia instigated such an investigation, and built up a school with this and other objectives in mind. Together with several collaborators he uncovered many unexpected results. The papers [4–9] record some of these investigations. A full bibliography is given in [10].

For many years I have been looking at one small part of the wide investigation. The way the Cantor-Bendixson gadget appears in point-free topology. This chapter is a survey of much of what I know. I haven't done much on the modal and proof theoretic aspects. The two papers [15, 16] are all I can offer.

9.2 Preamble

After some pre-history, the study of frames began in the Ehresmann seminar in Paris during the 1950s. I should say that as used here a frame is nothing to do with a Kripke frame as used in modal logic. The idea for this kind of frame was to study the algebraic properties of the topology of a topological space. The topic was taken up by Dowker and Papert (Strauss) who in [3] proved the first fundamental result, namely that the set of quotients of a frame is itself a frame. Because they used congruences they got the frame of quotients upside-down but that was corrected later. The gadgets controlling the quotients of a frame A are the nuclei on A. These form the assembly NA of A, and this is itself a frame. Each frame A is canonically embedded into its assembly NA. Since this assembly is also a frame it has its own assembly, and this construction can be iterated to generate the assembly tower of A.

$$A \longrightarrow NA \longrightarrow N^2A \longrightarrow N^3A \longrightarrow \cdots \tag{9.1}$$

This tower can go on for ever, even 'beyond the infinite'.

Most of the early work was done in an entirely algebraic fashion. In [11] Isbell started to 'organize' the categorical aspects. He showed that the assembly tower need never stabilize. The tower stabilizes if, and only if, some level is a complete Boolean algebra. For some frames the tower does stabilize, for others it doesn't. In [18] I showed that for a T_0 topological space S the assembly of the topology is Boolean (and hence the tower stabilizes at that level) precisely when S is scattered. In [19] I observed that there is a Cantor-Bendixson ranking technique that can be applied to any frame and the various levels of the assembly. This was developed in a bit more detail in [20, 21]. Further analysis can be found in [26] and, with some effort, in [13].

In this chapter I try to uncover the Cantor-Bendixson properties of the first few levels of the tower. There is some kind of natural extension as we move up the tower, and this provides new ranking techniques for the parent frame. It becomes clear that there is something going on, which should be investigated in greater depth (or perhaps I mean height).

Section 9.3 contains the various algebraic information that we need. In Sect. 9.4 I show how the point-sensitive, that is standard topological, Cantor-Bendixson process can be extended to a point-free version, that is to the frame-theoretic setting. Section 9.5 shows how this gadget can be lifted all the way up the assembly tower, although what is happening from N^3A is a complete mystery. Section 9.6 investigates the nature of this lifted Cantor-Bendixson gadget on NA, and then Sect. 9.7 begins to investigate what is happening on N^2A. One of the main results of the section is that N^2A is Boolean precisely when NA is Artinian relative to a certain comparison. The proof of that result uses a rather intricate construction which is developed and investigated in Sect. 9.8. Finally, in Sect. 9.9 I make a few remarks on what all this may mean and perhaps what could happen in the future.

9.3 Background Material

A *frame*

$$(A, \wedge, \top, \bigvee, \bot)$$

is a complete lattice with the distinguished attributes, as indicated, and which satisfies the Frame Distributive Law (FDL)

$$a \wedge \bigvee X = \bigvee \{a \wedge x \mid x \in X\}$$

for each $a \in A$ and $X \subseteq A$. A *frame morphism* $A \xrightarrow{\ f\ } B$ between frames is a function that respects the distinguished attributes. In this section we gather together the relevant background material. The book [12] is about frames and their various ramifications. It is now a little old. A more recent, but shorter, account is given in [14]. Sections 2 and 3 of [22] contain material relevant to the analysis given here.

As well as its distinguished attributes each frame A also carries an implication, a 2-placed operations $(\cdot \succ \cdot)$ such that

$$x \le (b \succ a) \Longleftrightarrow x \wedge b \le a$$

for all $a, b, x \in A$. The symbol $\rightarrow$ is often used for the implication on a frame, but that can be confused with other notions, so here I will use $\succ$ instead.

As a particular case we set

$$\neg b = (b \succ \bot)$$

to obtain the 1-placed negation operation on A. These operations need not be preserved by morphisms, but play a major part in many calculations.

For each topological space S the topology $\mathcal{O}S$ (of open sets) is a frame. These are the initial examples of frames. Each complete Boolean algebra is a frame, and need not be a topology. There are also other examples which are nothing like topologies.

The FDL immediately gives the following.

Lemma 1. *For each frame A we have*

$$(\bigvee X) \succ a = \bigwedge \{x \succ a \mid x \in X\}$$

for each subset $X \subseteq A$ and element $a \in A$.

Each quotient of a frame (surjective frame morphism) can be controlled by a kind of congruence on the frame. There is also a neater controlling gadget.

Definition 1. Let A be a frame.

An *inflator* on A is a inflationary and monotone function $f : A \longrightarrow A$ that is

$$x \le f(x) \le f(y)$$

for all comparable elements $x \le y$ of A.

A *pre-nucleus* on A is an inflator f for which

$$f(x \wedge y) = f(x) \wedge f(y)$$

for all $x, y \in A$.

A *nucleus* on A is a pre-nucleus j which is idempotent, that is $j^2 = j$.

In fact, there are two kinds of pre-nuclei that are used, but rarely in the same place. I tend to call the one above a binary pre-nucleus. There is also a unary pre-nucleus which has the weaker property

$$f(x \wedge y) = x \wedge f(y)$$

for all $x, y \in A$. Fortunately, only the binary version is needed here, so I will call it a pre-nucleus, as above.

There are three standard examples of nuclei.

Definition 2. For each frame A and element $a \in A$ we set

$$\mathsf{u}_a(x) = (a \vee x) \qquad \mathsf{v}_a(x) = (a \succ x) \qquad \mathsf{w}_a(x) = ((x \succ a) \succ a)$$

for each $x \in A$.

The nuclei and inflators on a frame A are partially ordered pointwise.

$$f \leq g \iff (\forall x \in A)[f(x) \leq g(x)]$$

Both nuclei and inflators form complete posets (with infima easy to compute). We let

$$NA$$

be the complete poset of all nuclei on A. This is the *assembly* of A.

The fact that the assembly NA is a complete lattice with pointwise infima is almost immediate. However, NA has a much better property. A proof of the following well-known fact can be found in the notes on frames on my web-page [25]. I hope to organize those notes into a coherent account in the near future.

Theorem 1. *For each frame A the assembly NA is also a frame and the following assignment is a frame embedding.*

$$\begin{array}{ccc} A & \longrightarrow & NA \\ a & \longmapsto & \mathsf{u}_a \end{array}$$

This embedding is an isomorphism precisely when A is Boolean.

Since the assembly NA is a frame, it has its own assembly N^2A, and this has its assembly N^3A, and so on. This gives the *assembly tower* (9.1) of A. The tower stabilizes precisely when the parent frame has a Boolean reflection. For set theoretic reasons, some frames don't have such a reflection. At the other extreme, for some frames the tower stabilizes quickly. We attempt to gain some understanding of the first few levels of the tower.

The next two lemmas are well known. Their proofs can be found in the notes on my web-page.

Lemma 2. *For each frame A, nucleus $j \in NA$, element $a \in A$, with $b = j(\bot)$, we have*

$$\begin{array}{llll} \mathsf{u}_a \leq j & \iff a \leq j(\bot) & \mathsf{v}_a \leq j & \iff j(a) = \top \\ j \leq \mathsf{w}_a & \iff j(a) = a & \mathsf{w}_a \leq j & \implies j = \mathsf{w}_b \end{array}$$

and u_a and v_a are complementary elements of NA.

The nuclei $\mathsf{u}_{(\cdot)}, \mathsf{v}_{(\cdot)}, \mathsf{w}_{(\cdot)}$ produce all nuclei, in one of two ways.

Lemma 3. *For each frame A we have*

$$\bigvee \left\{\mathsf{u}_{j(a)} \wedge \mathsf{v}_a \mid a \in A\right\} = j = \bigwedge \{\mathsf{w}_a \mid a \in A_j\}$$

for each nucleus $j \in NA$.

We need a refinement of the left hand construction, and this does provide a proof of that equality.

Consider any inflator f on the frame A. We may iterate this through the ordinals to produce an ascending chain of inflators

$$\boldsymbol{id} = f^0 \leq f = f^1 \leq f^2 \leq \cdots \leq f^\alpha \leq \cdots$$

under the pointwise comparison. Thus we set

$$f^0 = \boldsymbol{id} \quad f^{\alpha+1} = f \circ f^\alpha \quad f^\lambda(a) = \bigvee \{f^\alpha(a) \mid \alpha < \lambda\}$$

for each ordinal α, each limit ordinal λ and each element $a \in A$ at the limit stage. Here, and below, $\boldsymbol{id}$ is the identity function on A. This generates an ascending chain of inflators. On cardinality ground this must stabilize at some level. In other words there is some ordinal θ such that

$$f^\alpha = f^\theta$$

for each ordinal $\alpha \geq \theta$. It doesn't take too long to see that f^θ is the smallest closure operation (idempotent inflator) above f. For what we do here we don't need the value of θ, thus we write f^∞ for this closure. In other words we think of ∞ as a sufficiently large ordinal. (In a more delicate analysis we would need to determine θ. That takes us into the realm of ranking or dimension techniques which is why ordinals were first invented.)

Construction 1. Let j be a given nucleus, and suppose $j = f^\infty$ for some inflator f. For a given $a \in A$ let

$$a(\alpha) = f^\alpha(a)$$

for each ordinal α. Thus

$$a(0) = a \qquad a(\alpha + 1) = f(a(\alpha)) \qquad a(\lambda) = \bigvee \{a(\alpha) \mid \alpha < \lambda\}$$

for each ordinal α and limit ordinal λ. Similarly set

$$j_{a,0} = \boldsymbol{id} \qquad j_{a,\alpha+1} = \left(\mathsf{u}_{a(\alpha+1)} \wedge \mathsf{v}_{a(\alpha)}\right) \vee j_{a,\alpha} \qquad j_{a,\lambda} = \bigvee \{j_{a,\alpha} \mid \alpha < \lambda\}$$

for each ordinal α and limit ordinal λ.

This construction produces an ascending chain of elements

$$a = a(0) \leq \cdots \leq a(\alpha) \leq \cdots \qquad \boldsymbol{id} = j_{a,0} \leq \cdots \leq j_{a,\alpha} \leq \cdots \leq j$$

and an ascending chain of nuclei attached to j and a. Each will stabilize at a sufficiently large ordinal. We know that

$$a(\infty) = f^{\infty}(a) = j(a)$$

but we need some information about

$$j_a = \bigvee \{j_{a,\alpha} \mid \alpha \in \mathbb{O}\mathrm{rd}\} = \bigvee \left\{\mathsf{u}_{a(\alpha+1)} \wedge \mathsf{v}_{a(\alpha)} \mid \alpha \in \mathbb{O}\mathrm{rd}\right\}$$

the stable limit of the ascending chain of nuclei. This second equality holds since for each sufficiently large ordinal α we have $a(\alpha+1) = a(\alpha) = j(a)$ so that

$$\mathsf{u}_{a(\alpha+1)} \wedge \mathsf{v}_{a(\alpha)} = \boldsymbol{id}$$

and hence $j_{a,\alpha+1} = j_{a,\alpha}$. The whole family of nuclei j_a combine as follows.

Lemma 4. *We have*

$$\bigvee \{j_a \mid a \in A\} = \bigvee \{\mathsf{u}_{f(a)} \wedge \mathsf{v}_a \mid a \in A\}$$

for the chain of nuclei given by Construction 1.

Proof. From the description of j_a given just above we have

$$\begin{aligned} \bigvee \{j_a \mid a \in A\} &= \bigvee \left\{\mathsf{u}_{a(\alpha+1)} \wedge \mathsf{v}_{a(\alpha)} \mid \alpha \in \mathbb{O}\mathrm{rd}, a \in A\right\} \\ &= \bigvee \left\{\mathsf{u}_{f(a(\alpha))} \wedge \mathsf{v}_{a(\alpha)} \mid \alpha \in \mathbb{O}\mathrm{rd}, a \in A\right\} \end{aligned}$$

where the second equality follows by the definition of $a(\alpha+1)$. Each compound

$$\mathsf{u}_{f(a(\alpha))} \wedge \mathsf{v}_{a(\alpha)} \quad \text{has the form} \quad \mathsf{u}_{f(a)} \wedge \mathsf{v}_a$$

for some $a \in A$, so we may forget the indexing by ordinals, to obtain the required result. □

Since $f \leq j$ this result shows that

$$\bigvee \{j_a \mid a \in A\} = \bigvee \{\mathsf{u}_{f(a)} \wedge \mathsf{v}_a \mid a \in A\} \leq \bigvee \{\mathsf{u}_{j(a)} \wedge \mathsf{v}_a \mid a \in A\} = j$$

and our aim is to improve this comparison to an equality. To do that we look at the components $j_{a,\alpha}$ that build up j_a.

Lemma 5. *For the chain of nuclei given by Construction 1 we have*

$$j_{a,\alpha} = \mathsf{u}_{a(\alpha)} \wedge \mathsf{v}_a$$

for each ordinal α.

Proof. We proceed by induction on α.

For the base case, $\alpha = 0$, we have $a(0) = a$ and $j_{a,0} = \boldsymbol{id}$.

For the induction step, $\alpha \mapsto \alpha + 1$, since $a \leq a(\alpha) \leq a(\alpha + 1)$, we have

$$\begin{aligned} j_{a,\alpha+1} &= \big(\mathsf{u}_{a(\alpha+1)} \wedge \mathsf{v}_{a(\alpha)}\big) \vee j_{a,\alpha} \\ &= \big(\mathsf{u}_{a(\alpha+1)} \wedge \mathsf{v}_{a(\alpha)}\big) \vee \big(\mathsf{u}_{a(\alpha)} \wedge \mathsf{v}_a\big) \\ &= \big(\mathsf{u}_{a(\alpha+1)} \vee \mathsf{u}_{a(\alpha)}\big) \wedge \big(\mathsf{u}_{a(\alpha+1)} \vee \mathsf{v}_a\big) \wedge \big(\mathsf{v}_{a(\alpha)} \vee \mathsf{u}_{a(\alpha)}\big) \wedge \big(\mathsf{v}_{a(\alpha)} \vee \mathsf{v}_a\big) \\ &= \mathsf{u}_{a(\alpha+1)} \wedge \big(\mathsf{u}_{a(\alpha+1)} \vee \mathsf{v}_a\big) \wedge \top_N \wedge \mathsf{v}_a \\ &= \big(\mathsf{u}_{a(\alpha+1)} \vee \mathsf{v}_a\big) \end{aligned}$$

as required. The second equality uses the induction hypothesis, and the others various (finitary) distributive laws.

For the induction leap to a limit ordinal λ we see that $j_{a,\lambda}$ is

$$\bigvee \{j_\alpha \mid \alpha < \lambda\} = \bigvee \{\mathsf{u}_{a(\alpha)} \wedge \mathsf{v}_a \mid \alpha < \lambda\} = \bigvee \{\mathsf{u}_{a(\alpha)} \mid \alpha < \lambda\} \wedge \mathsf{v}_a = \mathsf{u}_{a(\lambda)} \wedge \mathsf{v}_a$$

as required. This follows by the induction hypothesis, the frame distributive law (on NA), and the morphism properties of $\mathsf{u}_{(\cdot)}$. □

This result gives

$$j_a = \bigvee \{j_{a,\alpha} \mid \alpha \in \mathbb{O}\mathrm{rd}\} = \bigvee \{\mathsf{u}_{a(\alpha)} \wedge \mathsf{v}_a \mid \alpha \in \mathbb{O}\mathrm{rd}\}$$

which improve the earlier description of j_a. For all sufficiently large ordinals α we have $a(\alpha) = j(a)$ so that

$$j_a = \mathsf{u}_{j(a)} \wedge \mathsf{v}_a \qquad \text{so that} \qquad j = \bigvee \{j_a \mid a \in A\}$$

by the standard decomposition of j given by Lemma 3.

We combine these various parts to obtain the main result of this section.

Theorem 2. *For a nucleus* j *and inflator* f *on* A *with* $j = f^\infty$ *we have*

$$j = \bigvee \{\mathsf{u}_{f(a)} \wedge \mathsf{v}_a \mid a \in A\}$$

Proof. We have just seen that

$$j = \bigvee \{j_a \mid a \in A\}$$

so that Lemma 4 give the required result. □

Table 9.1 The point-sensitive and point-free Cantor-Bendixson gadgets

(Point-sensitive) Let S be a topological space. For closed sets Y, X of S we write $Y \sqsubset X$ and say Y is an *inessential part* of X if $Y \subseteq X$ and $X = (X - Y)^-$ holds. For each $X \in \mathscr{C}S$ we set $\boldsymbol{lim}_S(X) = \left(\bigcup\{Y \in \mathscr{C}S \mid Y \sqsubset X\}\right)^-$ to produce an operation $\boldsymbol{lim}_S$ on $\mathscr{C}S$. We call $\boldsymbol{lim}_S$ the *CB-process* on S	(Point-free) Let A be a frame. For elements $a, x \in A$ we write $a \lessdot x$ and say x is *essentially above* a if $a \leq x$ and $(x \succ a) = a$ holds. For each $a \in A$ we set $\boldsymbol{der}^A(a) = \bigwedge\{x \in A \mid a \lessdot x\}$ to produce an operation $\boldsymbol{der}^A$ on A. We call $\boldsymbol{der}^A$ the *CB-derivative* on A

9.4 The CB-Derivative

We set up the CB-derivatives on a frame and its assemblies. Although our approach here is slightly different, this section is much the same as Sect. 3 of [22]. We cite that document for details of various proofs. We set up two related gadgets, a relation $\lessdot$ and a pre-nucleus ***der*** on an arbitrary frame A. These have point-sensitive ancestors, and it is instructive to look at the point-sensitive and the point-free version in parallel.

Let S be an arbitrary topological space. We write $\mathscr{O}S$ and $\mathscr{C}S$ for the families of open subsets U and closed subsets X of S. We also write $(\cdot)^-$, $(\cdot)^\circ$, $(\cdot)'$ for the closure operation, interior operation, and complementation operation on subsets of S.

Definition 3. The *CB-process* $\boldsymbol{lim}_S$ for a space S and the *CB-derivative* $\boldsymbol{der}_A$ on a frame A are defined in Table 9.1.

Remembering that $(\cdot)^{-\prime} = (\cdot)^{\prime\circ}$ we obtain the following.

Lemma 6. *For a space S the two relations $\sqsubset$ and $\lessdot$ and the two operators $\boldsymbol{lim}_S$ and $\boldsymbol{der}^{\mathscr{O}S}$, are connected by*

$$Y \sqsubset X \iff X' \lessdot Y' \qquad U \lessdot V \iff V' \sqsubset U'$$

$$\boldsymbol{lim}_S(X)' = \boldsymbol{der}^{\mathscr{O}S}(X') \qquad \boldsymbol{der}^{\mathscr{O}S}(U)' = \boldsymbol{lim}_S(U')$$

for each $X \in \mathscr{C}S$ and $U \in \mathscr{O}S$.

The operation ***lim*** is standard but is not usually introduced in this way. A proof of the following can be found in [22].

Lemma 7. (Point-sensitive) *Let S be a T_0 space. For each closed set $X \in \mathscr{C}S$ the subset $\boldsymbol{lim}_S(X) \subseteq X$ is the set of limit points of X.*

(Point-free) *Let A be a frame. For each element a of A the interval $[a, \boldsymbol{der}^A(a)]$ is the largest Boolean interval above a.*

The operation $\boldsymbol{der}^A$ has been investigated by Esakia and his collaborators, but from a slightly different perspective. For instance it appears as τ in [6–8]. It also appears in [16] which I still find a little curious.

The following simple result will motivate a construction in Sect. 9.9.

Lemma 8. *For each frame A both*

$$b \le a \lessdot x \le y \Longrightarrow b \lessdot y \qquad \left.\begin{matrix} a \lessdot x \\ b \lessdot y \end{matrix}\right\} \Longrightarrow a \wedge b \lessdot x \wedge y$$

hold for all elements a, b, x, y.

Proof. For the left hand implication we are given

$$b \le a \le x \le y \Longrightarrow b \lessdot y \quad \text{with} \quad x \succ a \le a$$

and we require $y \succ b \le b$. To this end let

$$z = (y \succ b) \quad \text{so that} \quad z \wedge y \le b$$

and we required $z \le b$. We have

$$z \wedge x \le z \wedge y \le b \le a \quad \text{and hence} \quad z \le a \le y$$

since $a \lessdot x$. But now

$$z = z \wedge y \le b$$

as required.

For the second implication we are given

$$a \lessdot x \quad b \lessdot y \quad \text{that is} \quad x \succ a \le a \quad y \succ b \le b$$

with $a \le x$ and $b \le y$. Let

$$z = \big((x \wedge y) \succ (a \wedge b)\big)$$

so that $z \le a \wedge b$ is required. We have

$$z \wedge x \wedge y \le a \wedge b \le b \quad \text{so that} \quad z \wedge x \le b \le y$$

since $b \lessdot y$. But now

$$z \wedge x \le z \wedge x \wedge y \le a \quad \text{so that} \quad z \le a \le x$$

since $a \lessdot x$. But now

$$z \le a \quad \text{and} \quad z = z \wedge x \le b$$

to give the required result. □

To analyse the CB-derivative $\boldsymbol{der}^A$ we produce a generalization of its construction. For a nucleus $j \in NA$ consider the induced quotient

$$A \xrightarrow{\ j^*\ } A_j$$

to the fixed set of j. We have two CB-derivatives

$$\boldsymbol{der}^A \qquad \boldsymbol{der}^{A_j}$$

on the source frame and target frame respectively. Remember that infima computed in A_j agree with those computed in A. Consider the composite function

$$A \xrightarrow{\ j^*\ } A_j \xrightarrow{\ \boldsymbol{der}^{A_j}\ } A_j \hookrightarrow A$$

using the CB-derivative on A_j as the central component, and the insertion of A_j into A as the right hand component. This gives us a new operation on A.

Definition 4. For each frame A and nucleus j on A the operator

$$\boldsymbol{der}_j^A : A \longrightarrow A$$

is given by

$$\boldsymbol{der}_j^A(a) = \boldsymbol{der}^{A_j}(j(a)) = \bigwedge \{x \in A_j \mid j(a) \lessdot x\}$$

for each $a \in A$.

This gives two distinct operations.

$$\boldsymbol{der}_j^A \qquad \boldsymbol{der}^{A_j}$$

The left hand one is on A and the right hand one is on A_j. We need not distinguish between the relation $\lessdot_A$ on A and the relation $\lessdot_{A_j}$ on A_j since they agree on A_j.

A proof of the following crucial result is given by Lemma 3.3 of [22].

Theorem 3. *For each nucleus $j \in NA$, the inflator $\boldsymbol{der}_j^A$ is a pre-nucleus on A.*

The assembly NA of a frame is also a frame, and so carries an implication operation. This is used in the following important property of $\boldsymbol{der}$, as given by Lemma 3.5 of [22].

Lemma 9. *For each frame A we have*

$$\boldsymbol{der}_j^A(a) = \big(\mathsf{w}_{j(a)} \succ j\big)(a)$$

for each nucleus $j \in NA$ and element $a \in A$.

Each frame A carries its CB-derivative $\boldsymbol{der}^A$. The assembly NA carries its own CB-derivative $\boldsymbol{der}^{NA}$. Once we have fixed the frame A we write

$$\boldsymbol{der} \text{ for } \boldsymbol{der}^A \qquad \boldsymbol{Der} \text{ for } \boldsymbol{der}^{NA}$$

to avoid too many affixes. Each nucleus $j \in NA$ gives us a refined version $\boldsymbol{der}_j = \boldsymbol{der}_j^A$ of $\boldsymbol{der}$, where again we may drop the affix A. This is also a pre-nucleus, and so its closure $\boldsymbol{der}_j^\infty$ is a nucleus on A. The following fundamental result is proved as Theorem 3.6 of [22].

Theorem 4. *For each frame A we have*

$$\boldsymbol{der}_j^\infty = \boldsymbol{Der}(j)$$

for each $j \in NA$.

The gadget $\boldsymbol{der}_j$, the relative point-free Cantor-Bendixson derivative, its lifting $\boldsymbol{Der}$ to the assembly, and the connection given by Theorem 4 was first investigated in [20, 21].

9.5 The Battery of Derivatives

For a frame A consider the assembly tower and the carried CB-derivatives.

$$\begin{array}{cccccc} (T) & A & NA & N^2A & N^3A & \cdots \\ (D) & \boldsymbol{der}^A & \boldsymbol{der}^{NA} & \boldsymbol{der}^{N^2A} & \boldsymbol{der}^{N^3A} & \cdots \end{array}$$

In this chapter we do not get beyond level 3, so we let

$$\boldsymbol{der}^A = \boldsymbol{der}^A \qquad \boldsymbol{Der}^A = \boldsymbol{der}^{NA} \qquad \boldsymbol{DER}^A = \boldsymbol{Der}^{NA} = \boldsymbol{der}^{N^2A}$$

and even drop the affix when the parent frame is known. We let

$$\delta^A = (\boldsymbol{der}^A)^\infty \qquad \Delta^A = (\boldsymbol{Der}^A)^\infty \qquad \boldsymbol{\Delta}^A = (\boldsymbol{DER}^A)^\infty$$

to obtain derivatives and nuclei on levels 0, 1, and 2.

$$\boldsymbol{der},\ \delta \text{ on } A \qquad \boldsymbol{Der},\ \Delta \text{ on } NA \qquad \boldsymbol{DER},\ \boldsymbol{\Delta} \text{ on } N^2A \qquad \ldots$$

With the decorations we have the following.

$$\delta^A = \delta^A \qquad \Delta^A = \delta^{NA} \qquad \boldsymbol{\Delta}^A = \Delta^{NA} = \delta^{N^2A}$$

Theorem 4 shows that for an arbitrary frame A we have

$$\delta^A = \boldsymbol{Der}^A(\boldsymbol{id}_A)$$

so we may apply this all the way up the assembly tower. We write

$$\begin{array}{ccc} A & NA & N^2A \\ \boldsymbol{id}_A & \boldsymbol{Id}_A = \boldsymbol{id}_{NA} & \boldsymbol{ID}_A = \boldsymbol{Id}_{NA} = \boldsymbol{id}_{N^2A} \end{array}$$

for the identity function on the indicated level. Each of these is the bottom of the assembly at the next level up. Theorem 4 gives

$$\delta = \boldsymbol{der}^\infty = \boldsymbol{Der}(\boldsymbol{id}) \qquad \Delta = \boldsymbol{Der}^\infty = \boldsymbol{DER}(\boldsymbol{Id}) \qquad \boldsymbol{\Delta} = \boldsymbol{DER}^\infty = \cdots$$

and so on. Since $\boldsymbol{Der}$ is a pre-nucleus on NA it can be iterated to obtain

$$\boldsymbol{id} \leq \delta = \boldsymbol{Der}(\boldsymbol{id}) \leq \boldsymbol{Der}^2(\boldsymbol{id}) \leq \boldsymbol{Der}^3(\boldsymbol{id}) \leq \cdots \leq \boldsymbol{Der}^\alpha(\boldsymbol{id}) \leq \cdots$$

all of which are nuclei on A. Eventually this process closes off at some nucleus

$$\theta = \boldsymbol{Der}^\infty(\boldsymbol{id}) = \Delta(\boldsymbol{id})$$

above δ. Even in the spatial case the closure ordinal of $\boldsymbol{Der}$ can be arbitrarily large. Examples are given in [22]. Continuing with this trick we set

$$\begin{array}{llll} NA & N^2A & N^3A & \cdots \\ \delta = \boldsymbol{Der}(\boldsymbol{id}) & \Delta = \boldsymbol{DER}(\boldsymbol{Id}) & \boldsymbol{\Delta} = \cdots & \cdots \\ \theta = \Delta(\boldsymbol{id}) & \Theta = \boldsymbol{\Delta}(\boldsymbol{Id}) & \boldsymbol{\Theta} = \cdots & \cdots \\ \xi = \Theta(\boldsymbol{id}) & \Xi = \boldsymbol{\Theta}(\boldsymbol{Id}) & \boldsymbol{\Xi} = \cdots & \cdots \\ \vdots & \vdots & \vdots & \cdots \end{array}$$

to obtain nuclei on A together with higher level analogues.

$$\delta \leq \theta \leq \xi \leq \Xi(\boldsymbol{id}) \leq \boldsymbol{\Xi}(\boldsymbol{Id})(\boldsymbol{id}) \leq \cdots$$

These nuclei are a better version of the obstructions used in [13]. Each of the generated nuclei on A can be evaluated at $\bot$, the bottom of the parent frame, to produce an ascending chain of elements of the frame.

$$d_0 = \boldsymbol{der}(\bot) \leq d_1 = \delta(\bot) \leq d_2 = \theta(\bot) \leq d_3 = \xi(\bot) \leq d_4 = \Xi(\boldsymbol{id})(\bot) \leq d_5$$

In [19–21] I called this the backbone of the frame.

Theorem 5. *For each frame A we have a chain of equivalences.*

$$\begin{array}{lllll}
(0) & A \textit{ is Boolean} & & \Longleftrightarrow & \boldsymbol{der}(\bot) = \top \\
(1) & NA \textit{ is Boolean} & \Longleftrightarrow \delta = \top_N & \Longleftrightarrow & \delta(\bot) = \top \\
(2) & N^2A \textit{ is Boolean} & \Longleftrightarrow \theta = \top_N & \Longleftrightarrow & \theta(\bot) = \top \\
(3) & N^3A \textit{ is Boolean} & \Longleftrightarrow \xi = \top_N & \Longleftrightarrow & \xi(\bot) = \top \\
 & & \vdots & &
\end{array}$$

Proof. (0) Lemma 7 ensures that $[\bot, \boldsymbol{der}(\bot)]$ is the largest Boolean lower section of A. Thus A is Boolean precisely when this interval is all of A.

$(1, 2, 3, \ldots)$ Applying the previous part to NA gives

$$\begin{array}{lll}
NA \text{ is Boolean} & \Longleftrightarrow \delta = \boldsymbol{Der}(\boldsymbol{id}) = \top_N & \Longleftrightarrow \delta(\bot) = \top \\
N^2A \text{ is Boolean} & \Longleftrightarrow \theta = \Delta(\boldsymbol{id}) = \top_N & \Longleftrightarrow \theta(\bot) = \top \\
N^3A \text{ is Boolean} & \Longleftrightarrow &
\end{array}$$

and so on. □

This result shows that the generated gadgets have some intrinsic interest. We gather together some information about the first few of these.

Theorem 6. *For each frame A and $j \in NA$ the following are equivalent.*

$$(i)\ \ \boldsymbol{der}_j = j \qquad (ii)\ \ \boldsymbol{Der}(j) = j \qquad (iii)\ \ \Delta(j) = j$$

Proof. Since $\boldsymbol{Der}(j) = \boldsymbol{der}_j^{\infty}$, the two inflators $\boldsymbol{der}_j$ and $\boldsymbol{Der}(j)$ fix the same elements. For each $a \in A$ we have the following.

$$\boldsymbol{der}_j(a) = a \Longleftrightarrow \boldsymbol{Der}(j)(a) = a$$

(i)⇒(ii). This is an immediate consequence of the above equivalence.

(ii)⇒(iii). Applying the above equivalence to NA with j replaced by $\boldsymbol{id}$ and a replaced by j we have the following.

$$\boldsymbol{Der}(j) = j \Longrightarrow \Delta(j) = \boldsymbol{DER}(\boldsymbol{id})(j) = j$$

(iii)⇒(i). Since $\boldsymbol{der}_j \leq \boldsymbol{Der}(j) \leq \Delta(j)$, this is immediate. □

Setting $j = \boldsymbol{id}$ gives the following.

Corollary 1. *For each frame A the following conditions are equivalent.*

$$(i)\ \ \boldsymbol{der} = \boldsymbol{id} \qquad (ii)\ \ \delta = \boldsymbol{id} \qquad (iii)\ \ \theta = \boldsymbol{id}$$

In the spatial case we are familiar with $\boldsymbol{der}$ and δ. For some spaces $\boldsymbol{Der}$ seems a natural gadget. In [23] there are examples of spaces with

$$\boldsymbol{DER}(\boldsymbol{Id})(\boldsymbol{id}) = \theta = \boldsymbol{id} \qquad \boldsymbol{DER}^2(\boldsymbol{Id})(\boldsymbol{id}) = \top_N$$

to show that Corollary 1 does not extend in the obvious way.

9.6 First Level

The first characterization of when NA is Boolean was given by Beazer and Macnab as Theorem 2 in [1]. The topological content was extracted in [18]. The topology of a T_0 space has a Boolean assembly precisely when the space is scattered. This led to the notion of the CB-derivative ***der*** on an arbitrary frame, first presented in [17, 19] and developed later in [20–22].

Since ***Der*** is a pre-nucleus on NA, the set of nuclei j with $\boldsymbol{Der}(j) = \top_N$ is a filter on NA. By Lemma 7, a nucleus j belongs to this filter precisely when the interval $[j, \top_N]$ of NA is Boolean. In general this filter is quite big.

Lemma 10. *For each frame A and element $a \in A$ we have $\boldsymbol{Der}(\mathsf{w}_a) = \top_N$.*

Proof. For $a \in A$, let $j = \mathsf{w}_a$, so that $j(\bot) = a$ and Lemma 9 gives

$$\boldsymbol{Der}(j)(\bot) \geq \boldsymbol{der}_j(\bot) = \big(\mathsf{w}_a \succ j\big)(\bot) = \top$$

for the required result. □

Anything true about every frame is true about every assembly. By the lifting of Corollary 1, if $\boldsymbol{Der} = \boldsymbol{Id}$ then for each element $a \in A$ we have

$$\mathsf{w}_a = \boldsymbol{Id}(\mathsf{w}_a) = \boldsymbol{Der}(\mathsf{w}_a) = \top_N$$

and hence $a = \top$ by evaluation at a (or at any element below a). This proved the non-obvious part of the following.

Corollary 2. *For a frame A we have $\boldsymbol{Der} = \boldsymbol{Id}$ if and only if A is trivial.*

By Lemma 3 we have

$$\boldsymbol{Der}(j) = \bigwedge \{\mathsf{w}_a \mid a \in A_{\boldsymbol{Der}(j)}\}$$

where the indexing ranges over $a \in A$ with $\boldsymbol{Der}(j) \leq \mathsf{w}_a$. Each frame A carries its essentially above relation $\lessdot$, so each assembly NA carries its own essentially above relation.

Lemma 11. *For a frame A, nucleus $j \in NA$, and element $a \in A$, the four conditions are equivalent.*

$$(i)\ \ \boldsymbol{der}_j(a) = a \qquad (ii)\ \ \boldsymbol{der}_j \leq \mathsf{w}_a \qquad (iii)\ \ \boldsymbol{Der}(j) \leq \mathsf{w}_a \qquad (iv)\ \ j \lessdot \mathsf{w}_a$$

Proof. $(i) \Rightarrow (ii)$. This is immediate.

$(ii) \Rightarrow (iii)$. Assuming (ii) a use of Theorem 4 gives the required result.

$$\boldsymbol{Der}(j) = \boldsymbol{der}_j^{\infty} \leq \mathsf{w}_a$$

$(iii) \Rightarrow (iv)$. Assuming (iii) we have $j \leq \boldsymbol{Der}(j) \leq \mathsf{w}_a$ so that $j(a) = a$. Let

$$k = (\mathsf{w}_a \succ j)$$

so that $j \leq k$ and we required $k \leq j$ to give $k = j$. By Lemma 9 we have

$$k(a) = (\mathsf{w}_a \succ j)(a) = \boldsymbol{der}_j(a) \leq \boldsymbol{Der}(j)(a) \leq a$$

so that $k \leq \mathsf{w}_a$, and hence we obtain the required result.

$$k \leq \mathsf{w}_a \wedge (\mathsf{w}_a \succ j) \leq j$$

$(iv) \Rightarrow (i)$. Assuming (iv) we have $j(a) = a$ (since $j \leq \mathsf{w}_a$) and

$$(\mathsf{w}_a \succ j) = j$$

so that Lemma 9 gives

$$\boldsymbol{der}_j(a) = (\mathsf{w}_a \succ j)(a) = j(a) = a$$

as required. □

Of these conditions it is the equivalence of (iii, iv) which is most useful, but other parts do have some applications. We have

$$\boldsymbol{der}(\bot) = \bot \iff \delta \leq \neg\neg$$

by setting $j = \boldsymbol{id}$ and $a = \bot$ in $(i) \Leftrightarrow (iii)$. Lifting this adds to Corollary 1.

Corollary 3. *For each frame A the four conditions are equivalent.*

$$(i)\ \ \boldsymbol{der} = \boldsymbol{id} \qquad (ii)\ \ \delta = \boldsymbol{id} \qquad (iii)\ \ \theta = \boldsymbol{id} \qquad (iv)\ \ \Delta \leq \neg\neg_{(1)}$$

Proof. The equivalence of (i, ii, iii) is just Corollary 1. From the lift of the observation above we have

$$\delta = \boldsymbol{Der}(\boldsymbol{id}) = \boldsymbol{id} \iff \Delta \leq \neg\neg_{(1)}$$

which gives $(ii) \Leftrightarrow (iv)$. □

Lemma 11 also suggests we might want to characterize a comparison

$$J(j) \leq \mathsf{w}_a$$

where J is a given derivative on NA. We look at the particular case $J = \Delta$ in Sect. 9.9.

The implication operation on A is used to define the relation $\lessdot$ on A, and this is used to obtain the family of derivatives $\boldsymbol{der}_j$ (for $j \in NA$). By Lemma 9 these gadgets can be evaluated using the implication operation on the assembly. Conversely, knowing how to evaluate these derivatives helps us to calculate certain implications on NA.

Lemma 12. *Let j, k be nuclei on a frame A. With $l = (k \succ j)$ we have*

$$\mathsf{v}_a \wedge \mathsf{u}_d \leq l \leq \mathsf{w}_a \vee \mathsf{u}_d$$

where $a = k(\perp)$ and $d = l(a)$.

Proof. For each $x \in A$ we have

$$\mathsf{u}_d(x) = d \vee x \leq l(a \vee x) = \big(l \vee \mathsf{u}_a\big)(x)$$

so that $\mathsf{u}_d \leq l \vee \mathsf{u}_a$ for the left hand comparison. For the other we have

$$\big(l \wedge \mathsf{v}_d\big)(a) = d \wedge (d \succ a) \leq a$$

so that $l \wedge \mathsf{v}_d \leq \mathsf{w}_a$ and hence $l \leq \mathsf{w}_a \vee \mathsf{u}_d$. □

When $k = \mathsf{w}_a$ we can take these calculations quite a bit further.

Lemma 13. *For a frame A, nucleus $j \in NA$, and element $a \in A_j$ with*

$$d = \boldsymbol{der}_j(a) = \big(\mathsf{w}_a \succ j\big)(a) \qquad b = \mathsf{w}_a(d)$$

we have the following.

$$\begin{array}{llll} (a) & \big(\mathsf{w}_a \succ j\big) = \big(\mathsf{v}_a \wedge \mathsf{u}_d\big) \vee j = \big(\mathsf{v}_a \vee j\big) \wedge \big(j \vee \mathsf{u}_d\big) \\ (b) & \Big(\big(\mathsf{w}_a \succ j\big) \succ j\Big) = \mathsf{v}_d \vee j \vee \mathsf{u}_a = \mathsf{w}_{d \succ a} \\ (c) & \mathsf{w}_a \vee \big(\mathsf{w}_a \succ j\big) = \mathsf{w}_a \vee \mathsf{u}_d = \mathsf{w}_b \end{array}$$

Proof. (a) We use Lemma 12 with $k = \mathsf{w}_a$. We have $a = k(\perp)$ and $l = \big(\mathsf{w}_a \succ j\big)$, so that $d = l(a) = \boldsymbol{der}_j(a)$ (by Lemma 9). Lemma 12 gives

$$\mathsf{v}_a \wedge \mathsf{u}_d \leq l \leq \mathsf{w}_a \vee \mathsf{u}_d$$

and $j \leq l$, so that the left hand comparison gives

$$\big(\mathsf{v}_a \vee j\big) \wedge \big(j \vee \mathsf{u}_d\big) = \big(\mathsf{v}_a \wedge \mathsf{u}_d\big) \vee j \leq l$$

so the converse comparison will suffice. The right hand comparison gives

$$l \le l \wedge (\mathsf{w}_a \vee \mathsf{u}_d) = (l \wedge \mathsf{w}_a) \vee (l \wedge \mathsf{u}_d) \le j \vee \mathsf{u}_d$$

which is halfway there. But $\mathsf{u}_a \le \mathsf{w}_a$ so that $l \wedge \mathsf{u}_a \le j$ to give $l \le \mathsf{v}_a \vee j$.

(b) By part (a) we have

$$\big((\mathsf{w}_a \succ j) \succ j\big) = \big((\mathsf{u}_d \wedge \mathsf{v}_a) \vee j\big) \succ j = (\mathsf{u}_d \wedge \mathsf{v}_a) \succ j = \mathsf{v}_d \vee \mathsf{u}_a \vee j$$

for the first equality. But $\mathsf{w}_a \le \big((\mathsf{w}_a \succ j) \succ j\big)$ so that Lemma 2 gives

$$\big((\mathsf{w}_a \succ j) \succ j\big) = \mathsf{w}_b$$

with b as follows.

$$b = \big((\mathsf{w}_a \succ j) \succ j\big)(\bot) = (\mathsf{v}_d \vee \mathsf{u}_a \vee j)(\bot) = (d \succ j(a)) = (d \succ a)$$

(c) From part (a) we have

$$\mathsf{w}_a \vee (\mathsf{w}_a \succ j) = (\mathsf{u}_d \wedge \mathsf{v}_a) \vee \mathsf{w}_a = (\mathsf{u}_d \vee \mathsf{w}_a) \wedge (\mathsf{v}_a \vee \mathsf{w}_a)$$

since $j \le \mathsf{w}_a$. But

$$(\mathsf{v}_a \vee \mathsf{w}_a) \ge (\mathsf{v}_a \vee \mathsf{u}_a) = \top_N$$

to give the first equality. The second follows by Lemma 2. □

The $\bigwedge$-closure of the filter of nuclei j where $[j, \top_N]$ is Boolean is the whole of the assembly. What about those nuclei j where $[j, \top_N]$ is not Boolean?

Definition 5. Let A be a frame and let j be some nucleus on A. We say a nucleus $k \in NA$ is, respectively,

$$j\text{-complemented} \qquad j\text{-regular}$$

if $j \le k$ and

$$k \vee (k \succ j) = \top_N \qquad k = ((k \succ j) \succ j)$$

as appropriate.

In other words, we first locate the complemented or regular members of NA_j, and then view these as members of the interval $[j, \top_N]$ of NA.

Lemma 14. *For each frame A, each nucleus $j \in NA$, and each element $a \in A$ with $j \le \mathsf{w}_a$ and $d = \mathbf{der}_j(a)$, the following conditions are equivalent.*

(i) $a \lessdot d$
(ii) *The nucleus* w_a *is* j*-complemented.*
(iii) *The nucleus* w_a *is* j*-regular.*

Furthermore, when these hold we have $\mathsf{w}_a = \mathsf{v}_d \vee j \vee \mathsf{u}_a = \mathsf{w}_{d \succ a}$.

Proof. (i)⇒(ii). Assuming (i) a use of Lemma 13(c) leads to (ii).

$$\big(\mathsf{w}_a \vee (\mathsf{w}_a \succ j)\big)(\bot) = \mathsf{w}_a(d) = \top$$

(ii)⇒(iii). This is trivial.
(iii)⇒(i). Assuming (iii) a use of Lemma 13(b) gives

$$\mathsf{w}_a = \big((\mathsf{w}_a \succ j) \succ j\big) = \mathsf{w}_{d \succ a}$$

so evaluation at $\bot$ gives $a = (d \succ a)$.
The final description of w_a follows by Lemma 13(b). □

Setting $j = \boldsymbol{id}$ in Lemma 14 gives Lemma 10 of [1]. To go further we introduce some notation.

Definition 6. For a frame A and nucleus $j \in NA$ we use

$$a \in B_j \Longleftrightarrow j(a) = a \qquad a \in C_j \Longleftrightarrow j(a) = a \lessdot \boldsymbol{der}_j(a)$$

$$\mathscr{B}_j = \{\mathsf{w}_a \mid a \in B_j\} \qquad \mathscr{C}_j = \{\mathsf{w}_a \mid a \in C_j\}$$

to extract subsets of A with $C_j \subseteq B_j$ and subsets of NA with $\mathscr{C}_j \subseteq \mathscr{B}_j$.

Of course, $B_j = A_j$ and $\mathscr{B}_j$ is the corresponding final section of NA. These sets are named to give a comparison with the sets C_j and $\mathscr{C}_j$. We know that

$$a \in B_j \Longleftrightarrow j \leq \mathsf{w}_a \qquad a \in C_j \Longleftrightarrow \mathsf{w}_a \text{ is } j\text{-complemented}$$

where the first is routine and the second is Lemma 14.

Theorem 7. *For each nucleus* $j \in NA$ *the following conditions are equivalent.*

(i) $\boldsymbol{Der}(j) = \top_N$.
(ii) $B_j = C_j$.
(iii) *For each* $a \in A_j$ *we have* $a \lessdot \boldsymbol{der}_j(a)$.

Proof. (i)⇒(ii). Assuming (i), the interval $[j, \top_N]$ of NA is Boolean. Since $C_j \subseteq B_j$, it suffices to show the converse. Consider any $a \in B_j$. Then $j(a) = a$, so that $j \leq \mathsf{w}_a$, and hence w_a is j-complemented, to give $a \in C_j$.
(ii)⇒(iii). This follows by Lemma 14.
(iii)⇒(i). Assuming (iii) let $a = \boldsymbol{Der}(j)(\bot)$, so that $j(a) = a$. By (iii)

$$a \lessdot \boldsymbol{der}_j(a) \leq \boldsymbol{Der}(j)(a) = a$$

so that $a \lessdot a$, and hence $a = \top$, as required. □

By setting $j = \boldsymbol{id}$ we get the Beazer-Macnab characterization.

Theorem 8. *For each frame A the following conditions are equivalent.*

(i) NA is Boolean.
(ii) Each w-nucleus is complemented.
(iii) For each $a \in A$ we have $a \lessdot \boldsymbol{der}(a)$.

9.7 Second Level

The Beazer-Macnab methods analyse the first assembly NA. Similar methods give information about the second assembly N^2A. By Definition 6 and Lemma 3 we have $j = \bigwedge \mathscr{B}_j \leq \bigwedge \mathscr{C}_j$ and the infimum $\bigwedge \mathscr{C}_j$ is worth looking at. By Lemma 14 we have

$$\mathscr{C}_j = \{\mathsf{v}_d \vee j \vee \mathsf{u}_a \mid a \in A_j, d = \boldsymbol{der}_j(a)\}$$

which we use towards the end of the proof of the following.

Theorem 9. *For each frame A and nucleus $j \in NA$ we have the following.*

$$(\boldsymbol{Der}(j) \succ j) = \bigwedge \mathscr{C}_j$$

Proof. By Theorem 4 we have $\boldsymbol{Der}(j) = \boldsymbol{der}_j^{\infty}$ so Theorem 2 gives

$$\boldsymbol{Der}(j) = \bigvee \{\mathsf{u}_d \wedge \mathsf{v}_a \mid a \in A, d = \boldsymbol{der}_j(a)\}$$

and hence Lemma 1 gives

$$(\boldsymbol{Der}(j) \succ j) = \bigwedge \{(\mathsf{u}_d \wedge \mathsf{v}_a) \succ j \mid a \in A, d = \boldsymbol{der}_j(a)\}$$

so that

$$((\mathsf{u}_d \wedge \mathsf{v}_a) \succ j) = \mathsf{v}_d \vee j \vee \mathsf{u}_a$$

where, as yet, a is an arbitrary member of A. Since

$$d = \boldsymbol{der}_j(a) = \boldsymbol{der}_j(j(a)) \qquad j \vee \mathsf{u}_a = j \vee \mathsf{u}_{j(a)}$$

we may restrict to $a \in A_j$. By the observation above this infimum is $\bigwedge \mathscr{C}_j$. □

We have $j = \bigwedge \mathscr{B}_j = \bigwedge \mathscr{C}_j$ when $B_j = C_j$, in other words when $\boldsymbol{Der}(j) = \top_N$ by Theorem 7. By Theorem 9 this happens more generally.

Corollary 4. *For $j \in NA$ we have $j = \bigwedge \mathscr{C}_j$ precisely when $j \lessdot \boldsymbol{Der}(j)$.*

For a nucleus j the set C_j can be quite large, but for one particular nucleus it is always small. Recall that θ is the least nucleus with $\boldsymbol{Der}(\theta) = \theta$.

Lemma 15. *For each frame A we have $C_\theta = \{\top\}$.*

Proof. Consider any $a \in C_\theta$. By definition, we have $a \lessdot \boldsymbol{der}_\theta(a)$, so that

$$a \lessdot \boldsymbol{der}_\theta(a) \leq \boldsymbol{Der}(\theta)(a) = \theta(a) = a$$

to give $a \lessdot a$, and hence $a = \top$. □

With this we obtain a kind of lifted analogue of Theorem 8.

Theorem 10. *For each frame A the three conditions are equivalent.*

(i) $N^2 A$ is Boolean.
(ii) For each $j \in NA$ we have $j = \bigwedge \mathscr{C}_j$.
(iii) For each $j \in NA$ if $C_j = \{\top\}$ then $j = \top_N$.

Proof. (i)⇒(ii). Assuming (i), for $j \in NA$ Theorem 8 applied to NA gives $j \lessdot \boldsymbol{Der}(j)$. Thus Theorem 9 gives

$$j = (\boldsymbol{Der}(j) \succ j) = \bigwedge \mathscr{C}_j$$

as required.

(ii)⇒(iii). This is immediate.

(iii)⇒(i). This follows by Lemma 15. □

We now need a specialization of Lemma 14.

Lemma 16. *For each frame A we have*

$$\neg d = \bot \iff \mathsf{w}_\bot = \mathsf{v}_d$$

where $d = \boldsymbol{der}(\bot)$.

Proof. With this d we have

$$\neg d = \bot \iff \bot \lessdot d \iff \mathsf{w}_\bot \text{ is complemented} \iff \mathsf{w}_\bot = \mathsf{v}_d \vee \mathsf{u}_\bot = \mathsf{v}_d$$

where the second and third equivalences use Definition 6 (for $j = \boldsymbol{id}$). □

This result says that

$$\neg d = \bot \iff (\forall x)[(\neg\neg)(x) = (d \succ x)]$$

and lifting this to the assembly gives the following.

$$\neg_1\delta = \boldsymbol{id} \iff (\forall j)[(\neg\neg)_1(j) = (\delta \succ j)]$$

Using

$$\mathscr{D} = \mathscr{C}_{\boldsymbol{id}} = \{(\neg\neg)_1\mathsf{w}_a \mid a \in A\} = \{\mathsf{u}_a \vee \mathsf{v}_d \mid a \in A, d = \boldsymbol{der}(a)\}$$

we may set $j = \boldsymbol{id}$ in Theorem 9 to obtain the following.

Lemma 17. *For each frame A we have $\neg_1\delta = \bigwedge \mathscr{D}$.*

We lift Lemmas 16 and 17 to the second level assembly. To prepare for this we do a calculation using Lemma 17. We have

$$\neg_1\delta = \bigwedge \mathscr{D} = \bigwedge \{\mathsf{u}_a \vee \mathsf{v}_d \mid a \in A, d = \boldsymbol{der}(a)\} \leq \bigwedge \{\mathsf{u}_a \mid a \in A, \boldsymbol{der}(a) = \top\}$$

and hence

$$(\neg_1\delta)(x) \leq \bigwedge \{a \vee x \mid a \in A, \boldsymbol{der}(a) = \top\} \leq \bigwedge \{a \mid a \in A, x \leq a, \boldsymbol{der}(a) = \top\}$$

for each $x \in A$. We now lift this up a level.

Theorem 11. *For each frame A and $J \in N^2A$ we have the following.*

$$\neg_{(2)}\Delta = \boldsymbol{Id} \qquad (\neg\neg)_{(2)}J = (\Delta \succ J)$$

Proof. Consider any $j \in NA$ and any $a \in A$ with $j \leq \mathsf{w}_a$ (any $a \in A_j$). By Lemma 10 we have $\boldsymbol{Der}(\mathsf{w}_a) = \top_N$, and hence

$$(\neg_{(2)}\Delta)(j) \leq \bigwedge\{\mathsf{w}_a \mid a \in A_j\} = j$$

where the left hand comparison comes from the calculations above, and the right hand equality is standard. This gives $\neg_{(2)}\Delta = \boldsymbol{Id}$, the first part of the required result. The second part follows by Lemma 16 applied to N^2A. □

This result says that $(N^2A)_{\neg\neg}$ is canonically isomorphic to the interval $[\boldsymbol{Id}, \Delta]$ of N^2A, and so the second level assembly is a special kind of frame. A lot of information about $(N^2A)_{\neg\neg}$ is in [26].

Each nucleus j on A has a fixed set of A_j, and this is closed under arbitrary infima. A fixed set arising from $\Delta(j)$ has a stronger closure property.

Definition 7. A subset K of a frame A is *cohesive* if for each $a \in K$ there is some $X \subseteq K$ with $a = \bigwedge X$ and $a \lessdot x$ for each $x \in X$.

The fixed set $\{\top\}$ is cohesive and can be the only cohesive set.

Lemma 18. *Let A be a frame where $\lessdot$ has the Ascending Chain Condition (ACC). Then $\{\top\}$ is the only cohesive subset.*

Proof. Suppose the frame A has ACC and, by way of contradiction, suppose K is a non-trivial cohesive set, that is $K \neq \{\top\}$. There is some $a \in K$ with $a \neq \top$. But now, since K is cohesive, we have

$$a = \bigwedge X$$

for some $X \subseteq K$ with $a \lessdot x$ for each $x \in X$. We can not have $X = \{\top\}$ for otherwise

$$a = \bigwedge X = \bigwedge \{\top\} = \top$$

which is not the case. Thus there is some $a' \in K - \{\top\}$ with $a \lessdot a'$.

By iterating this construction we obtain an ascending chain

$$a = a_0 \lessdot a_1 \lessdot a_2 \lessdot \cdots$$

of members of $K - \{\top\}$. Since A has ACC this chain is eventually constant, so we obtain some $b \in K - \{\top\}$ with $b \lessdot b$. But now, by the definition of $\lessdot$, we have

$$b = b \succ b = \top$$

which is the contradiction. □

The union of any family of cohesive sets is itself cohesive. Thus each quotient A_j includes a largest cohesive subset. We locate this set.

Lemma 19. *Consider a frame A and nucleus $k \in NA$ with $\mathbf{Der}(k) = k$. Then the fixed set A_k is cohesive.*

Proof. Consider $a \in A_k$. Let X be the set of all $x \in A_k$ with $a \lessdot x$. Then

$$a \leq \bigwedge X = \mathbf{der}_k(a) \leq \mathbf{Der}(k)(a) = k(a) = a$$

to give the required result. □

This result can be rephrased as follows.

Corollary 5. *Consider any $j \in NA$ and let $k = \Delta(j)$. Then A_k is cohesive.*

Next we look at an arbitrary cohesive subset.

Lemma 20. *Suppose K is cohesive in A. Then we have*

$$K \subseteq A_j \Longrightarrow K \subseteq A_{\Delta(j)}$$

for each nucleus $j \in NA$.

Proof. We observe four implications of increasing strength

$$\begin{array}{ll} (1) & K \subseteq A_j \Longrightarrow (\forall a \in K)[\boldsymbol{der}_j(a) = a] \\ (2) & K \subseteq A_j \Longrightarrow (\forall a \in K)[\boldsymbol{der}_j^{\alpha}(a) = a] \\ (3) & K \subseteq A_j \Longrightarrow (\forall a \in K)[\boldsymbol{Der}(j)(a) = a] \\ (4) & K \subseteq A_j \Longrightarrow (\forall a \in K)[\boldsymbol{Der}^{\alpha}(j)(a) = a] \end{array}$$

and then setting $\alpha = \infty$ in (4) gives the required result. In (2) and (4) the α is an arbitrary ordinal, over which we proceed by induction. The proofs of (2, 3, 4) are routine. To prove (1) consider any $j \in NA$ with $K \subseteq A_j$. Consider any $a \in K$. We have

$$\boldsymbol{der}_j(a) = \bigwedge \{x \in A_j \mid a \lessdot x\} \leq \bigwedge \{x \in K \mid a \lessdot x\} = a$$

where the left hand equality is the definition of $\boldsymbol{der}_j$ and the right hand equality uses the cohesive property of K. □

With this we obtain the following.

Theorem 12. *For each frame A and $j \in NA$ we have the following.*

($\uparrow$) $\Delta(j) = \top_N \Longleftrightarrow \{\top\}$ *is the only cohesive subset of* A_j.

($\leftrightarrow$) $A_{\Delta(j)}$ *is the largest cohesive subset of* A_j.

($\downarrow$) $\Delta(j) = j \Longleftrightarrow A_j$ *is cohesive.*

Proof. Let $k = \Delta(j)$.

($\uparrow$) Suppose first that $k = \top_N$ and consider a cohesive $K \subseteq A_j$. Then by Lemma 20 we have $K \subseteq A_k = \{\top\}$. Conversely, by Lemma 19, the fixed set A_k is cohesive and so, if this must be $\{\top\}$, then $K = \top_N$.

($\leftrightarrow$) By Corollary 5 we see that $A_k \subseteq A_j$ is cohesive. Conversely, Lemma 20 ensures that each cohesive subset of A_j is included in A_k.

($\downarrow$) If $k = j$ then A_j is cohesive by Lemma 19. Conversely, if A_j is cohesive then $A_j \subseteq A_k$, by Lemma 20, to give $k \leq j$. □

Part ($\downarrow$) adds to Theorem 6, and gives a slight extension of Lemma 18.

Corollary 6. *For each frame A and nucleus $j \in NA$, if the relation $\lessdot$ on A_j has ACC then $\Delta(j) = \top_N$.*

For each frame A, nucleus $j \in NA$, and element $a \in A$, Lemma 11 gives

$$\boldsymbol{Der}(j) \leq \mathsf{w}_a \Longleftrightarrow j \lessdot \mathsf{w}_a$$

and we now replace $\boldsymbol{Der}$ by Δ. We strengthen the essentially above relation.

Definition 8. Let A be a frame. For elements $b, a \in A$ we write $b \lll a$ and say a is *substantially above* b if there is a descending ω-chain $\{x_r \mid r < \omega\}$ of elements with

$$b \leq x_{r+1} \lessdot x_r \leq a$$

for each $r < \omega$.

Trivially we have

$$b \lll a \Longrightarrow b \lessdot a \qquad a \lll a \Rightarrow a = \top$$

and $\lll$ has some simple properties. Here is the analogue of Lemma 8.

Lemma 21. *For each frame A and $a, b, x, y \in A$ we have the following.*

$$y \leq b \lll a \leq x \Longrightarrow y \lll x \qquad \left.\begin{matrix} x \lll a \\ y \lll b \end{matrix}\right\} \Longrightarrow x \wedge y \lll a \wedge b$$

Proof. The left hand implication follows from Lemma 8. For the right hand implication suppose $x \lll a$ and $y \lll b$ where the two chains

$$X = \{x_r \mid r < \omega\} \qquad Y = \{y_r \mid r < \omega\}$$

witness these separations. For each $r < \omega$ let

$$z_r = x_r \wedge y_r \quad \text{and set} \quad Z = \{z_r \mid r < \omega\}$$

to produce a third chain. For each $r < \omega$ we have

$$x \leq x_{r+1} \lessdot x_r \leq a \qquad y \leq y_{r+1} \lessdot y_r \leq b$$

so that Lemma 8 gives

$$x \wedge y \leq z_{r+1} \lessdot z_r \leq a \wedge b$$

and hence $x \wedge y \lll a \wedge b$. □

The descending chains give the relation $\lll$ its power.

Lemma 22. *For each frame A and $a, b \in A$ we have the following.*

$$b \lll a \Longrightarrow \delta(b) \lll a$$

Proof. Suppose $b \lll a$ and consider

$$x = \bigwedge \{x_r \mid r < \omega\}$$

using the witnessing chain between b and a. For each $r < \omega$ we have

$$x \leq x_{r+1} \lessdot x_r$$

so that $\mathbf{der}(x) \leq x_r$, and hence $\mathbf{der}(x) = x$. But now $\delta(b) \leq \delta(x) = x$ which, since $x \lll a$, gives the required result. □

We may apply this result to the assembly to get

$$j \lll \mathsf{w}_a \Longrightarrow \Delta(j) \leq \mathsf{w}_a$$

for each $j \in NA$ and $a \in A$. The next result is a strengthening of this.

Theorem 13. *For each frame A, element $a \in A$, and nucleus $j \in NA$, the following three conditions are equivalent.*

$$(i) \quad \Delta(j) \leq \mathsf{w}_a \qquad (ii) \quad \Delta(j) \lll \mathsf{w}_a \qquad (iii) \quad j \lll \mathsf{w}_a$$

The implication $(ii) \Rightarrow (iii)$ is trivial, and the implication $(iii) \Rightarrow (i)$ is the observation above. Thus the content of the result is the implication $(i) \Rightarrow (ii)$. The proof of that uses a splitting technique and takes some time. That is the topic of Sect. 9.8. For the remainder of this section we obtain some consequences of Theorem 13.

First we combine parts of Lemmas 2 and 11 with Theorem 13.

Theorem 14. *For each frame A we have*

$$\begin{aligned} j \leq \ & \mathsf{w}_a \Longleftrightarrow j(a) = a \\ j \lessdot \ & \mathsf{w}_a \Longleftrightarrow \mathbf{Der}(j) \leq \mathsf{w}_a \\ j \lll \ & \mathsf{w}_a \Longleftrightarrow \Delta(j) \leq \mathsf{w}_a \end{aligned}$$

for each $j \in NA$ and $a \in A$.

The result can be rephrased differently. By Lemma 3 we have

$$j = \bigwedge \{\mathsf{w}_a \mid a \in A \text{ with } j \leq \mathsf{w}_a\}$$

for each $j \in NA$. We apply this to $\mathbf{Der}(j)$ and $\Delta(j)$ to obtain the following.

Corollary 7. *For each frame A and $j \in NA$ we have the following.*

$$\begin{aligned} j &= \bigwedge \{\mathsf{w}_a \mid a \in A \textit{ with } j \leq \mathsf{w}_a\} \\ \mathbf{Der}(j) &= \bigwedge \{\mathsf{w}_a \mid a \in A \textit{ with } j \lessdot \mathsf{w}_a\} \\ \Delta(j) &= \bigwedge \{\mathsf{w}_a \mid a \in A \textit{ with } j \lll \mathsf{w}_a\} \end{aligned}$$

By definition $\Delta = \mathbf{Der}^{\infty}$. The use of $\lll$ leads to an explicit construction.

Theorem 15. *For each frame A and $j \in NA$ we have the following.*

$$\mathbf{Der}(j) = \bigwedge \{k \in NA \mid j \lessdot k\} \qquad \Delta(j) = \bigwedge \{k \in NA \mid j \lll k\}$$

Proof. The first part is just the definition of ***Der***. For the second part let

$$\Pi(j) = \bigwedge \{k \in NA \mid j \lll k\}$$

so that $\Pi(j) = \Delta(j)$ is required. But $\Pi(j)$ is

$$\bigwedge \{k \in NA \mid j \lll k\} \le \bigwedge \{\mathsf{w}_a \mid j \lll \mathsf{w}_a\} = \bigwedge \{\mathsf{w}_a \mid \Delta(j) \le \mathsf{w}_a\} = \Delta(j)$$

using Theorem 13 and the general representation of nuclei. For the converse comparison consider any nucleus k with $j \lll k$. By the lift of Lemma 22 we have $\Delta(j) \lll k$ and hence $\Delta(j) \le \Pi(j)$ since k is arbitrary. □

This gives the following.

Theorem 16. *For each nucleus j on a frame A we have $\Delta(j) = \top_N$ precisely when the relation $\lessdot$ on the interval $[j, \top_N]$ of NA has the Descending Chain Condition.*

The particular case $j = \boldsymbol{id}$ gives a characterization of $\theta = \Delta(\boldsymbol{id}) = \top_N$ in other words a characterization of when N^2A is Boolean.

Corollary 8. *For each frame A the second level assembly N^2A is Boolean precisely when the relation $\lessdot$ on NA has the Descending Chain Condition.*

9.8 A Splitting Technique

In this section we prove the implication $(i) \Rightarrow (ii)$ of Theorem 13 using a splitting technique. The proof is rather long so we begin with an overview.

We work in a fixed, but arbitrary frame A. We set up a 3-placed relation

$$l \propto j \preceq b$$

between a pair of nuclei $j, l \in NA$ and an element $b \in A$. By the construction of this relation we have

$$j = \bigwedge \{\mathsf{w}_{a(r)} \mid r \in R\}$$

over some index set R where $a(r) \in A_l$ for each index r. In particular, we have $l \le \mathsf{w}_{a(r)}$, so that $l \le j$. The set $\{a(r) \mid r \in R\}$ is wide in the sense that

$$\mathsf{w}_{a(r)} \vee \mathsf{w}_{a(s)} = \top_N$$

for distinct $r, s \in R$. It is part of a *spread* as in Definition 9. In due course we prove the following.

Lemma 23. *Working in a frame A suppose*

$$\Delta(l) = l \qquad b \in A_l$$

for some nucleus l and element b. For each $a \in A_l$ with $(b \succ a) = a$ we have

$$l \propto j \preceq b \qquad j \lessdot \mathsf{w}_a \qquad j(\bot) = a$$

for at least one nucleus j.

We merge many spreads by an intricate construction and then we prove the following.

Lemma 24. *Working in a frame A suppose*

$$\Delta(l) = l \qquad b \in A_l \qquad l \propto j \preceq b$$

for some nuclei l, j and element b. Then

$$l \propto k \preceq b \qquad k \lessdot j$$

for some nucleus k.

By an iterated use of Lemma 24 we obtain the following.

Corollary 9. *Working in a frame A suppose*

$$\Delta(l) = l \qquad b \in A_l$$

for some nucleus l and element b. Then

$$l \propto j \preceq b \Longrightarrow l \lll j$$

for each nucleus $j \in NA$.

Proof. We fix the nucleus l and the element b with the two given properties. Consider any nucleus j such that

$$l \propto j \preceq b$$

holds. We must generate a certain descending chain of nuclei between j and l. By a use of Lemma 24 we obtain

$$l \propto j' \propto b \qquad j' \lessdot j$$

for some nucleus j' (the nucleus k of the Lemma). The global conditions still hold, so a second application gives

$$l \propto j'' \preceq b \qquad j'' \lessdot j' \lessdot j$$

for some nucleus j''. By iteration, we generate a descending sequence

$$(i \in \mathbb{N}) \quad \cdots \lessdot j_{i+1} \lessdot j_i \lessdot \cdots \lessdot j_2 \lessdot j_1 \lessdot j_0 = j$$

where

$$l \propto j_i \preceq b$$

for each $i \in \mathbb{N}$. This gives the required result. □

In the 3-placed relation the element b is merely a convenience, and it is always possible to take $b = \top$, so we obtain the following.

Theorem 17. *For each frame A we have*

$$l \leq \mathsf{w}_a \Longrightarrow l \lll \mathsf{w}_a$$

for all nuclei l with $\Delta(l) = l$ and all $a \in A$.

Proof. Assuming $\Delta(l) = l$ and $l \leq \mathsf{w}_a$, we may take $b = \top$ to get

$$\Delta(l) = l \qquad a, b \in A_l \qquad (b \succ a) = a$$

so that Lemma 23 gives

$$l \propto j \preceq b \qquad j \lessdot \mathsf{w}_a$$

for some nucleus j. But now Corollary 9 gives

$$l \lll j \lessdot \mathsf{w}_a$$

and hence $l \lll \mathsf{w}_a$, as required. □

This leads to a proof of Theorem 13.

Corollary 10. *For each frame A we have*

$$\Delta(j) \leq \mathsf{w}_a \Longrightarrow \Delta(j) \lll \mathsf{w}_a$$

for all nuclei $j \in NA$ and elements $a \in A$.

Proof. Let $l = \Delta(j)$, so that $\Delta(l) = l$. Then

$$\Delta(j) \leq \mathsf{w}_a \Longrightarrow l \leq \mathsf{w}_a \Longrightarrow l \lll \mathsf{w}_a$$

by Theorem 17. □

That completes the overview, we now begin the harder work.

The following notion is a refinement of that of coheight used in [13].

Definition 9. For a frame A a *spread* is a pair

$$\mathsf{a} = (a(r) \mid r \in R) \qquad \mathsf{b} = (b(r) \mid r \in R)$$

of indexed families of elements of A, over the same index set R, such that

(sp 1) $(b(r) \succ a(r)) = a(r)$ equivalently $\mathsf{v}_{b(r)} \leq \mathsf{w}_{a(r)}$
(sp 2) $b(r) \wedge b(s) \leq a(r) \vee a(s)$

for all $r, s \in R$ with $r \neq s$ for (sp 2).

Condition (sp 1) can be expressed in different ways. Both are useful.

The important components of a spread are the elements of a. The elements of b play a secondary role. Their main job is to ensure wideness.

Lemma 25. *For a frame A let (a, b) be a spread indexed by R. Then*

$$\mathsf{w}_{a(r)} \vee \mathsf{w}_{(a(s)} = \mathsf{u}_{a(r)} \vee \mathsf{u}_{a(s)} \vee \mathsf{v}_{b(r)} \vee \mathsf{v}_{b(s)} = \top_N$$

for each pair $r \neq s$ of distinct indexes.

Proof. Let

$$a = a(r) \vee a(s) \qquad b = b(r) \wedge b(s)$$

using the given distinct indexes $r \neq s$. By (sp 2) we have $b \leq a$, so that

$$\mathsf{u}_{a(r)} \vee \mathsf{u}_{a(s)} \vee \mathsf{v}_{b(r)} \vee \mathsf{v}_{b(s)} = \mathsf{u}_a \vee \mathsf{v}_b \geq \mathsf{u}_b \vee \mathsf{v}_b = \top_N$$

by the properties of u- and v-nuclei. Now (sp 1) gives the equality. □

The job of a spread is to produce a nucleus.

Definition 10. For a frame A let (a, b) be a spread indexed by R. We set

$$j = \bigwedge \{\mathsf{w}_{a(r)} \mid r \in R\}$$

to obtain the nucleus *induced* by the spread.

This depends only on a. The b is there to ensure Lemma 25.

The official definition of a spread uses an arbitrary index set R. By well ordering this we may assume that R is an initial stretch of the ordinals $\mathbb{O}\mathrm{rd}$. Thus we may assume that the two witnessing families of elements are

$$(a(\alpha) \mid \alpha < \kappa) \qquad (b(\alpha) \mid \alpha < \kappa)$$

for some ordinal κ. For what we do here the size of κ doesn't matter, thus it is convenient to hide κ. We do this by extending the two families to ordinal indexed families

$$(a(\alpha) \mid \alpha \in \mathbb{O}\mathrm{rd}) \qquad (b(\alpha) \mid, \alpha \in \mathbb{O}\mathrm{rd})$$

both of which are eventually constant. Of course, we must ensure that the spread conditions (sp 1) and (sp 2) still hold (for *all* ordinals). To do this we first set $a(\alpha) = \top$ for all sufficiently large α (that is $\alpha \geq \kappa$). We also take $b(\alpha)$ sufficiently small for all large α.

Sometimes it is convenient to use an arbitrary index set R, sometimes it is convenient to use an ordinal indexing, and sometimes it is convenient to use a more exotic indexing. We see an example of this in the proof of Lemma 24.

There are trivial examples of spreads, and more interesting examples.

Consider any representation $a = \bigwedge X$ of an element a of the parent frame. By well-ordering X we may write

$$X = (x(\alpha) \mid \alpha < \kappa)$$

for some ordinal κ. We now set $x(\alpha) = \top$ for each ordinal $\alpha \geq \kappa$, and use the ordinal indexed family

$$X = (x(\alpha) \mid \alpha \in \mathbb{O}\mathrm{rd})$$

which still satisfies $a = \bigwedge X$.

Construction 2. Let A be a frame, let $a, b \in A$, and suppose

$$a = \bigwedge X \qquad X = (x(\alpha) \mid \alpha \in \mathbb{O}\mathrm{rd})$$

where the indexed family X is eventually $\top$. We set

$$\begin{aligned} b(0) &= b \\ b(\alpha+1) &= b(\alpha) \wedge x(\alpha) \\ b(\lambda) &= \bigwedge \{b(\alpha) \mid \alpha < \lambda\} \end{aligned} \qquad a(\alpha) = (b(\alpha) \succ x(\alpha))$$

for each ordinal α and limit ordinal λ.

This construction first generates a descending chain of elements of A

$$b = b(0) \geq b(1) \geq \cdots \geq b(\alpha) \geq \cdots \quad (\alpha \in \mathbb{O}\mathrm{rd})$$

and then produces the chain $a(\cdot)$. The chain $b(\cdot)$ will be eventually constant, and the chain $a(\cdot)$ will be eventually $\top$. Notice that

$$b(\alpha) = b \wedge \bigwedge \{x(\beta) \mid \beta < \alpha\}$$

for each ordinal α. This is sometimes useful, as in the later part of the proof of Lemma 26. In particular

$$b(\infty) = b \wedge a$$

where ∞ is any sufficiently large ordinal.

Observe that if l is a nucleus on A with

$$X \subseteq A_l \qquad b \in A_l$$

then $a \in A_l$ and the components of a and b are in A_l, since A_l is closed under implication and arbitrary infima. Later we use this to modify the following.

Lemma 26. *For a frame A, elements $a, b \in A$, and a representation $a = \bigwedge X$, the Construction 2 produces a spread* (a, b) *with*

$$b(0) = b \qquad b(\alpha + 1) = b(\alpha) \wedge a(\alpha) \qquad b(\lambda) = \bigwedge\{b(\alpha) \mid \alpha < \lambda\}$$

for each ordinal α and limit ordinal λ. Furthermore, we have

$$a \leq j(\perp) \leq (b \succ a)$$

where j is the nucleus induced by the spread.

Proof. We first verify (sp 1) and (sp 2) to show that we do have a spread.

(sp 1) For each ordinal α we have the following.

$$(b(\alpha) \succ a(\alpha)) = (b(\alpha) \succ (b(\alpha) \succ x(\alpha))) = (b(\alpha) \succ x(\alpha)) = a(\alpha)$$

(sp 2) Consider distinct ordinals α, β. By symmetry we may suppose $\alpha < \beta$. Observe that $b(\beta) \leq b(\alpha + 1) \leq x(\alpha)$, so that we have the following.

$$b(\alpha) \wedge b(\beta) = b(\beta) \leq x(\alpha) \leq a(\alpha) \leq a(\alpha) \vee a(\beta)$$

For the alternative description of b consider the chain $c(\cdot)$ generated by

$$c(0) = b \qquad c(\alpha + 1) = c(\alpha) \wedge a(\alpha) \qquad c(\lambda) = \bigwedge\{c(\alpha) \mid \alpha < \lambda\}$$

for each ordinal α and limit ordinal λ. We require $b(\alpha) = c(\alpha)$ and the obvious ordinal induction works. The base case, $\alpha = 0$, is trivial. For the induction step, $\alpha \mapsto \alpha + 1$, we see that $c(\alpha + 1)$ is

$$c(\alpha) \wedge a(\alpha) = b(\alpha) \wedge a(\alpha) = b(\alpha) \wedge (b(\alpha) \succ x(\alpha)) = b(\alpha) \wedge x(\alpha) = b(\alpha + 1)$$

as required. The induction leap to a limit ordinal λ is immediate.

The induced nucleus is

$$j = \bigwedge \{\mathsf{w}_{a(\alpha)} \mid \alpha \in \mathbb{O}\mathrm{rd}\} \quad \text{so that} \quad j(\bot) = \bigwedge \{a(\alpha) \mid \alpha \in \mathbb{O}\mathrm{rd}\}$$

for its least value. We have $x(\alpha) \leq a(\alpha)$ for each ordinal $\alpha \in \mathbb{O}\mathrm{rd}$, and hence

$$a = \bigwedge X \leq \bigwedge \{a(\alpha) \mid \alpha \in \mathbb{O}\mathrm{rd}\} = j(\bot)$$

for one of the required comparisons. For the other comparison observe that

$$b \wedge \bigwedge \{a(\beta) \mid \beta < \alpha\} = b(\alpha) = b \wedge \bigwedge \{x(\beta) \mid \beta < \alpha\}$$

for each ordinal α. The right hand equality is a consequence of the original construction of b, and the left hand equality follows in the same way from the modified construction given here. By taking α sufficiently large, we have

$$b \wedge j(\bot) = b(\infty) = b \wedge \bigwedge X = b \wedge a \leq a$$

which leads to the required result. □

Suppose we start from a pair of elements $a, b \in A$ and a representation.

$$(b \succ a) = a = \bigwedge X$$

Construction 2 gives us a spread (a, b), and the second part of Lemma 26 shows we have a second representation of a.

$$a = j(\bot) = \bigwedge \{a(\alpha) \mid \alpha \in \mathbb{O}\mathrm{rd}\}$$

We could now take this through Construction 2 to produce a second spread, but the first part of Lemma 26 shows that it is exactly the same spread.

We now begin to refine Lemma 26.

Lemma 27. *Consider a frame A and elements $a, b \in A$ where*

$$(b \succ a) = a \qquad \boldsymbol{der}(a) = a$$

hold. Then Construction 2 produces a spread (a, b) *with*

$$j \lessdot \mathsf{w}_a \qquad j(\bot) = a$$

where j is the induced nucleus.

Proof. Construction 2 requires a representation $a = \bigwedge X$ of the element a. Since $\boldsymbol{der}(a) = a$ there is such a representation where $a \lessdot x$ for each $x \in X$. We may assume that

$$X = (x(\alpha) \mid \alpha \in \mathbb{O}\mathrm{rd})$$

where the sequence $x(\cdot)$ is eventually $\top$. We now use Construction 2 to produce a spread and then the induced nucleus j. By Lemma 26 we have

$$a \le j(\bot) \le (b \succ a) = a$$

and hence $j(\bot) = a$. We have $j \le \mathsf{w}_a$, and it remains to show $j \lessdot \mathsf{w}_a$.

To this end let $k = (\mathsf{w}_a \succ j)$ so that $j \le k$, and we require $k \le j$. We have $k \wedge \mathsf{w}_a \le j$. We show that $k(a) = a$, so that $k \le \mathsf{w}_a$, and hence $k = k \wedge \mathsf{w}_a \le j$.

Consider any $x \in X$. We have $x = x(\alpha)$ for some ordinal α. Let $c = a(\alpha)$ be the corresponding element of the constructed sequence $a(\cdot)$. We have $a \lessdot x \le c$ so that $\mathsf{w}_a(x) = \top$ and

$$k(a) \le k(x) \wedge \mathsf{w}_a(x) \le j(x) \le \mathsf{w}_c(x) = c = a(\alpha)$$

for each ordinal α. This gives

$$k(a) \le \bigwedge \{a(\alpha) \mid \alpha \in \mathbb{O}\mathrm{rd}\} = j(\bot) = a$$

as required. □

We can now introduce the 3-placed relation of Lemmas 23 and 24.

Definition 11. For a frame A with nuclei $l, j \in NA$ and an element $b \in A$, we write

$$l \propto j \preceq b$$

to indicate that j is induced by some spread (a, b) where

$$a(r) \in A_l \qquad b(r) \in A_l \qquad b(r) \le b$$

for each index r.

Since $a(r) \in A_l$ we have

$$l \le \bigwedge \{\mathsf{w}_{a(r)} \mid r \in R\} = j$$

and we have a lower bound for j. We are going to show that in the appropriate circumstances the nucleus j is some way above l. In the end we take $b = \top$. The parameter b is in Definition 11 to help with various constructions.

Proof of Lemma 23. Since $\Delta(l) = l$, the fixed set A_l is cohesive. Since $a \in A_l$ we have

$$a = \bigwedge X$$

for some $X \subseteq A_l$ where $a \lessdot x$ for each $x \in X$. With $b \in A_l$, Construction 2 produces a spread (a, b) with a nucleus j. As in Lemma 27 we have

$$j \lessdot \mathsf{w}_a \qquad j(\bot) = a$$

for this induced nucleus. We have

$$X \subseteq A_l \qquad b \in A_l$$

and the fixed set is closed under implication and arbitrary infima. Thus, as observed just before Lemma 26, each component of a and b belongs to A_l. Finally each component of b is below b, so we have $l \propto j \preceq b$ as required. □

The proof of Lemma 24 is not so straight forward. We meld many sandwich spreads into one. The following holds for any frame, but we apply it to an assembly, which explains the notation.

Lemma 28. *Suppose we have an indexed family of elements of a frame*

$$p_r \le k_r \lessdot j_r \qquad (r \in R)$$

where each p_r is complemented. Suppose $p_r \vee p_s = \top$ for distinct $r, s \in R$. Then with

$$k = \bigwedge\{k_r \mid r \in R\} \qquad j = \bigwedge\{j_r \mid r \in R\}$$

we have $k \lessdot j$.

Proof. Let $l = (j \succ k)$, so that we have $l \wedge j \le k$, and we required $l \le k$. We show $l \le k_r$ for each r. Let

$$\overline{j_r} = \bigwedge\{j_s \mid r \ne s \in R\}$$

so that $j = \overline{j_r} \wedge j_r$. For distinct indexes r, s we have

$$p_r \vee j_s \ge p_r \vee p_s = \top$$

and hence $\neg p_r \le j_s$ (since p_r is complemented), to give $\neg p_r \le \overline{j_r}$. Thus

$$l \wedge \neg p_r \wedge j_r \le l \wedge \overline{j_r} \wedge j_r = l \wedge j \le k \le k_r$$

so that since $k_r \lessdot j_r$ we have

$$l \wedge \neg p_r \le (j_r \succ k_r) = k_r \quad \text{and hence} \quad l \le p_r \vee k_r = k_r$$

as required. □

We are now ready to obtain our main construction result, Lemma 24.

Proof of Lemma 24. We use Construction 2, in refined form, many times to produce a family of spreads. We then merge these into one spread, and use this to induce k.

Table 9.2 Various conditions used in the proof of Lemma 24

(1) $a(r) \in A_l$	(2) $b(r) \in A_l$	(3) $b(r) \leq b$
(4) $(b(r) \succ a(r)) = a(r)$ equivalently $\mathsf{v}_{b(r)} \leq \mathsf{w}_{a(r)}$		(5) $b(r) \wedge b(s) \leq a(r) \vee a(s)$
(6) $j = \bigwedge \{\mathsf{w}_{a(r)} \mid r \in R\}$	(7) $l \propto k_r \preceq b(r)$	
(8) $k_r \lessdot \mathsf{w}_{a(r)}$	(9) $k_r(\bot) = a(r)$	(10) $a(r) \leq a(r, \alpha)$
(11) $a(r, \alpha) \in A_l$	(12) $b(r, \alpha) \in A_l$	(13) $b(r, \alpha) \leq b(r)$
(14) $(b(r, \alpha) \succ a(r, \alpha)) = a(r, \alpha)$ equivalently $\mathsf{v}_{b(r,\alpha)} \leq \mathsf{w}_{a(r,\alpha)}$		
(15) $b(r, \alpha) \wedge b(r, \beta) \leq a(r, \alpha) \vee a(r, \beta)$		(16) $k_r = \bigwedge \{\mathsf{w}_{a(r,\alpha)} \mid \alpha \in \mathbb{O}\mathrm{rd}\}$

We need to take some care with the various conditions. These are listed in Table 9.2. Of course, there are some hidden quantifiers.

The given condition $l \propto j \preceq b$ is witnessed by a spread

$$\mathsf{a} = (a(r) \mid r \in R) \qquad \mathsf{b} = (b(r) \mid r \in R)$$

over some index set R. This is not an arbitrary spread for it satisfies (1, 2, 3) for all $r \in R$. As a spread it also satisfies (4, 5) for all $r, s \in R$ with $r \neq s$. Of course, the induced nucleus is given by (6).

For the time being fix $r \in R$.

For this $r \in R$ we are given $\Delta(l) = l$ and (1, 2, 4). Thus Lemma 23 provides a nucleus k_r where (7, 8, 9) hold.

Condition (7) is witnessed by an auxiliary spread

$$(a(r, \alpha) \mid \alpha \in \mathbb{O}\mathrm{rd}) \qquad (b(r, \alpha) \mid \alpha \in \mathbb{O}\mathrm{rd})$$

where (10, 11, 12, 13) hold for all $\alpha \in \mathbb{O}\mathrm{rd}$. Condition (10) holds since

$$a(r) = k_r(\bot) \leq a(r, \alpha)$$

for all $\alpha \in \mathbb{O}\mathrm{rd}$. As a spread it also satisfies (14, 15) for all $\alpha, \beta \in \mathbb{O}\mathrm{rd}$ with $\alpha \neq \beta$. The induced nucleus is given by (16).

We now release $r \in R$. By (11, 12) we have families of elements of A_l

$$a(r, \alpha) \qquad b(r, \alpha)$$

indexed by pairs $(r, \alpha) \in R \times \mathbb{O}\mathrm{rd}$.

By (13, 3) we have

(!0) $b(r, \alpha) \leq b(r) \leq b$

for all $(r, \alpha) \in R \times \mathbb{O}\mathrm{rd}$. We show that the two $(R \times \mathbb{O}\mathrm{rd})$-indexed families form a spread, that is

(!1) $(b(r, \alpha) \succ a(r, \alpha)) = a(r, \alpha)$

(!2) $b(r, \alpha) \wedge b(s, \beta) \leq a(r, \alpha) \vee a(s, \beta)$

for all $r, s \in R$ and $\alpha, \beta \in \mathbb{O}\mathrm{rd}$ with $(r, \alpha) \neq (s, \beta)$.

Condition (!1) is just (14).

To verify (!2) consider distinct pairs $(r, \alpha) \neq (s, \beta)$ from $R \times \mathbb{O}\mathrm{rd}$. If $r = s$ then $\alpha \neq \beta$, and then (15) gives (!2). If $r \neq s$ then we have

$$b(r, \alpha) \wedge b(s, \beta) \leq b(r) \wedge b(s) \leq a(r) \vee a(s) \leq a(r, \alpha) \vee a(s, \beta)$$

by (13, 5, 10). We now have a spread indexed by $R \times \mathbb{O}\mathrm{rd}$. Let

$$k = \bigwedge \{\mathsf{w}_{a(r,\alpha)} \mid (r, \alpha) \in R \times \mathbb{O}\mathrm{rd}\}$$

be the induced nucleus. By (11, 12, 13) we have $l \propto k \preceq b$ so it remains to show $k \lessdot j$. By uncoupling the indexing pair we see that k is

$$\bigwedge \{\mathsf{w}_{a(r,\alpha)} \mid (r, \alpha) \in R \times \mathbb{O}\mathrm{rd}\} = \bigwedge \{\bigwedge \{\mathsf{w}_{a(r,\alpha)} \mid \alpha \in \mathbb{O}\mathrm{rd}\} \mid r \in R\}$$

which is $\bigwedge \{k_r \mid r \in R\}$ by (16). We now use Lemma 28. For each $r \in R$ let

$$p_r = \mathsf{u}_{a(r)} \vee \mathsf{v}_{b(r)} \quad \text{so that} \quad p_r \leq \mathsf{w}_{a(r)}$$

by (4). For distinct $r, s \in R$ a use of (5) gives

$$p_r \vee p_s = \mathsf{u}_{a(r)} \vee \mathsf{v}_{b(r)} \vee \mathsf{u}_{a(s)} \vee \mathsf{v}_{b(s)} = \mathsf{u}_{a(r) \vee a(s)} \vee \mathsf{v}_{b(r) \wedge b(s)} = \top_N$$

(as in the proof of Lemma 25). From (9) we have $\mathsf{u}_{a(r)} \leq k_r$ and (13, 14) give

$$\mathsf{v}_{b(r)} \leq \mathsf{v}_{b(r,\alpha)} \leq \mathsf{w}_{a(r,\alpha)} \quad \text{so that} \quad \mathsf{v}_{b(r)} \leq k_r$$

and hence $p_r \leq k_r$ and we are almost there. We remember (8), and then apply Lemma 28.

This completes the proof of Theorem 13.

9.9 Final Remarks

I think we all understand that from a point-sensitive perspective the Cantor-Bendixson process is a way of measuring a certain kind of 'pathology' of a topological space, or a way of locating the 'useful' part of the space. In this chapter I describe a point-free version of the process, an analogous construction on an arbitrary frame. As I mentioned earlier the point-free operation ***der*** also occurs in other areas of mathematics such as the analysis of the modal systems connected with the 'formal provability' notion first used by Gödel. However I don't investigate that here.

In this chapter I indicate how the CB-process for frames has more potential. For a given frame A the operation ***der*** on A can be lifted up to each finite level of the assembly tower of A (and probably can be lifted even further). This lifting is described in Sect. 9.5, and then in remaining sections I show how it can be used to measure a certain 'pathology' of the tower. I believe there is much more that can be done here.

Of course, each space gives a frame, its topology, and so this point-free analysis can be applied to the point-sensitive case. In [22, 23] I look at this application for certain rather simple spaces. In [23] I show there are spaces S such that

$$\mathcal{O}S \quad N\mathcal{O}S \quad N^2\mathcal{O}S \quad N^3\mathcal{O}S$$

are different, and $N^3\mathcal{O}S$ turns out to be the power set $\mathcal{P}S$ of the carrying set. I haven't been able to determine the structure of $N^2\mathcal{O}S$. This seems to depend on a certain problem in infinite combinatorics.

What is more important is that I have almost no information about what happens beyond level three. There is much to be investigated here.

There is also another direction that I have taken. By slightly weakening the Frame Distributive Law we obtain a larger class of complete lattices. For each module M over an arbitrary ring the lattice of sub-modules of M is a concrete example in this larger class. We then find that certain ranking techniques for modules can be investigated by methods analogous to the ones described here. These include the socle length, the Gabriel dimension, and a less well-know dimension due to Boyle. This last is nothing more than the CB-dimension of the assembly of the parent lattice. This idea first appeared in [20, 21], but a fuller account will appear in [24].

References

1. Beazer R, Macnab D (1979) Modal extensions of Heyting algebras. Colloq Math 41:1–12
2. Bezhanishvili G, Ghilardi S (2007) An algebraic approach to subframe logics. Intuitionistic case. Ann Pure Appl Logic 147:84–100
3. Dowker C, Papert D (1966) Quotient frame and subspaces. Proc Lond Math Soc 16(3):275–296
4. Esakia L (1981) Diagonal constructions, Löb's formula and Cantor's scattered spaces. In: Mikeladze Z (ed) Logical and semantical studies. Academic Press, Tbilisi, pp 124–143 (Russian)
5. Esakia L (1985) Heyting algebras, Duality theory. Academic Press, Tbilisi (Russian)
6. Esakia L, Jibladze M, Pataria D (2000) Scattered toposes. Ann Pure Appl Logic 103:97–107
7. Esakia L (2004) Intuitionistice logic and modality via topology. Ann Pure Appl Logic 127:155–170
8. Esakia L (2006) The modalized Heyting calculus: a conservative modal extension of the intuitionistic logic. J Appl Non-Class Logics 16:349–366
9. Esakia L (2009) Around provability logic. Ann Pure Appl Logic 161:174–184
10. Esakia L (2012) Full bibliography. Studia Logica 100:10–15
11. Isbell J (1972) Atomless parts of spaces. Math Scand 31:5–32
12. Johnstone P (1982) Stone spaces. Cambridge University Press, Cambridge
13. Plewe T (2000) Higher order dissolutions and Boolean coreflections of locales. J Pure Appl Algebra 154:273–293
14. Pultr A (2003) Frames. Handb Algebra 3:791–857. In: Hazelwinkel M (ed) Elsevier

15. Simmons H, Macintyre A (1973) Gödel's diagonalization technique and related properties of theories. Colloquium Math 28:165–180
16. Simmons H (1975) Topological aspects of suitable theories. Proc Edinb Math Soc 19(2):383–391
17. Simmons H (1980) The Cantor-Bendixson analysis of a frame, Séminaire de mathematique pure, no. 92, Université Catholique de Louvain, 42 p
18. Simmons H (1980) Spaces with Boolean assemblies. Colloq Math 43:23–39
19. Simmons H (1982) An algebraic version of Cantor-Bendixson analysis. In: Categorical aspects of topology and analysis, Springer, Lecture notes in mathematics 915:310–323
20. Simmons H (1989) Near-discreteness of modules and spaces as measured by Gabriel and Cantor. J Pure Appl Algebra 56:119–162
21. Simmons H (1989) Separating the discrete from the continuous by iterating derivatives. Bull Soc Math Belg 41:417–463
22. Simmons H (2005) The extended Cantor-Bendixson analysis of trees. Algebra Universalis 52:439–468
23. Simmons H, Examples of higher level assemblies (to appear in the ArXiv)
24. Simmons H, Ranking technique for idioms (a series of documents to appear in the ArXiv)
25. Simmons H, Web page at www.cs.man.ac.uk/~hsimmons
26. Wilson J (1994) The assembly tower and some categorical and algebraic aspects of frame theory. PhD thesis, Carnegie Mellon University

Chapter 10
Topological Interpretations of Provability Logic

Lev Beklemishev and David Gabelaia

In memory of Leo Esakia

Abstract Provability logic concerns the study of modality $\Box$ as provability in formal systems such as Peano Arithmetic. A natural, albeit quite surprising, topological interpretation of provability logic has been found in the 1970s by Harold Simmons and Leo Esakia. They have observed that the dual $\Diamond$ modality, corresponding to consistency in the context of formal arithmetic, has all the basic properties of the topological derivative operator acting on a scattered space. The topic has become a long-term project for the Georgian school of logic led by Esakia, with occasional contributions from elsewhere. More recently, a new impetus came from the study of polymodal provability logic **GLP** that was known to be Kripke incomplete and, in general, to have a more complicated behavior than its unimodal counterpart. Topological semantics provided a better alternative to Kripke models in the sense that **GLP** was shown to be topologically complete. At the same time, new fascinating connections with set theory and large cardinals have emerged. We give a survey of the results on topological semantics of provability logic starting from first contributions by Esakia. However, a special emphasis is put on the recent work on topological models of polymodal provability logic. We also include a few results that have not been published so far, most notably the results of Sect. 10.4 (due to the second author) and Sects. 10.7, 10.8 (due to the first author).

L. Beklemishev (✉)
V.A. Steklov Mathematical Institute, RAS, Moscow, Russia

Moscow M.V. Lomonosov State University, Moscow, Russia

National Research University Higher School of Economics, Moscow, Russia
e-mail: bekl@mi.ras.ru

D. Gabelaia
TSU Andrea Razmadze Mathematical Institute, Tbilisi, Georgia

Free University of Tbilisi, Tbilisi, Georgia
e-mail: gabelaia@gmail.com

G. Bezhanishvili (ed.), *Leo Esakia on Duality in Modal and Intuitionistic Logics*, Outstanding Contributions to Logic 4, DOI: 10.1007/978-94-017-8860-1_10,

Keywords Provability logic · Scattered spaces · GLP · Ordinal topologies

10.1 Provability Logics and Magari Algebras

Provability logics and algebras emerge from, respectively, a modal logical and an algebraic point of view on the proof-theoretic phenomena around Gödel's incompleteness theorems. These theorems are usually perceived as putting fundamental restrictions on what can be formally proved in a given axiomatic system (satisfying modest natural requirements). For the sake of a discussion, we call a formal theory T *gödelian* if

- T is a first order theory in which the natural numbers along with the operations $+$ and $\cdot$ are interpretable;
- T proves some basic properties of these operations and a modicum of induction (it is sufficient to assume that T contains Elementary Arithmetic EA, see [7]);
- T has a recursively enumerable (r.e.) set of axioms.

The Second Incompleteness Theorem of Kurt Gödel (G2) states that a gödelian theory T cannot prove its own consistency provided it is indeed consistent. More accurately, for any r.e. presentation of such a theory T, Gödel has shown how to write down an arithmetical formula $\mathsf{Prov}_T(x)$ expressing that *x is (a natural number coding) a formula provable in T*. Then the statement $\mathsf{Con}(T) := \neg\mathsf{Prov}_T(\ulcorner\bot\urcorner)$ naturally expresses that the theory T is consistent. G2 states that $T \nvdash \mathsf{Con}(T)$ provided T is consistent.

Provability logic emerged from the question of what properties of formal provability Prov_T can be verified in T, even if the consistency of T cannot. Several such properties have been stated by Gödel himself [33]. Hilbert and Bernays [36] and then Löb [44] stated them in the form of conditions any adequate formalization of a provability predicate in T must satisfy. After Gödel's and Löb's work it was clear that the formal provability predicate calls for a treatment as a modality. It led to the formulation of the Gödel–Löb provability logic **GL** and eventually to the celebrated arithmetical completeness theorem due to Solovay [55].

Independently, Macintyre and Simmons [45] and Magari [46] took a very natural algebraic perspective on the phenomenon of formal provability which led to the concept of *diagonalizable algebra*. Such algebras are now more commonly called *Magari algebras*. This point of view is more convenient for our present purposes.

Recall that the Lindenbaum–Tarski algebra of a theory T is the set of all T-sentences Sent_T modulo provable equivalence in T, that is, the structure $\mathscr{L}_T = \mathrm{Sent}_T/{\sim_T}$ where, for all $\varphi, \psi \in \mathrm{Sent}_T$,

$$\varphi \sim_T \psi \iff T \vdash (\varphi \leftrightarrow \psi).$$

Since we assume T to be based on classical propositional logic, $\mathscr{L}_T$ is a boolean algebra with operations $\wedge, \vee, \neg$. Constants $\bot$ and $\top$ are identified with the sets of refutable and provable sentences of T, respectively. The standard ordering on $\mathscr{L}_T$ is defined by

$$[\varphi] \leq [\psi] \iff T \vdash \varphi \rightarrow \psi \iff [\varphi \wedge \psi] = [\varphi],$$

where $[\varphi]$ denotes the equivalence class of φ.

It is well known that for consistent gödelian theories T all such algebras are isomorphic to the unique countable atomless boolean algebra. (This is a consequence of a strengthening of Gödel's First Incompleteness Theorem due to Rosser.) We obtain more interesting algebras by enriching the structure of the boolean algebra $\mathscr{L}_T$ by additional operation(s).

Gödel's consistency formula induces a unary operator $\Diamond_T$ acting on $\mathscr{L}_T$:

$$\Diamond_T : [\varphi] \longmapsto [\mathsf{Con}(T + \varphi)].$$

The sentence $\mathsf{Con}(T + \varphi)$ expressing the consistency of T extended by φ can be defined as $\neg\mathsf{Prov}_T(\ulcorner\neg\varphi\urcorner)$. The dual operator is $\Box_T : [\varphi] \longmapsto [\mathsf{Prov}_T(\ulcorner\varphi\urcorner)]$, thus $\Box_T x = \neg\Diamond_T\neg x$ for all $x \in \mathscr{L}_T$.

Hilbert–Bernays–Löb derivability conditions ensure that $\Diamond_T$ is correctly defined on the equivalence classes of the Lindenbaum–Tarski algebra of T. Moreover, it satisfies the following identities (where we write $\Diamond_T$ simply as $\Diamond$ and the variables range over arbitrary elements of $\mathscr{L}_T$):

M1. $\Diamond\bot = \bot$; $\Diamond(x \vee y) = \Diamond x \vee \Diamond y$;
M2. $\Diamond x = \Diamond(x \wedge \neg\Diamond x)$.

Notice that Axiom M2 is a formalization of G2 stated for the theory $T' = T + \varphi$, where $[\varphi] = x$. In fact, the left hand side states that T' is consistent, whereas the right hand side states that $T' + \neg\mathsf{Con}(T')$ is consistent, that is, $T' \nvdash \mathsf{Con}(T')$. The dual form of Axiom M2, $\Box(\Box x \rightarrow x) = \Box x$, expresses the formalization of Löb's theorem [44].

A Boolean algebra with an operator $\mathscr{M} = (M, \Diamond)$ satisfying M1, M2 is called *Magari algebra*. Thus, the main example of a Magari algebra is the structure $(\mathscr{L}_T, \Diamond_T)$ for any consistent gödelian theory T.

Notice that M1 induces $\Diamond$ to be monotone: if $x \leq y$ then $\Diamond x \leq \Diamond y$. The *transitivity* inequality $\Diamond\Diamond x \leq \Diamond x$ is often postulated as an additional axiom of Magari algebras, however, as discovered independently by de Jongh, Kripke and Sambin in the 1970s, it follows from M1 and M2.

Proposition 1. *In any Magari algebra $\mathscr{M}$ it holds that $\Diamond\Diamond x \leq \Diamond x$ for all $x \in M$.*

Proof Given any $x \in M$, consider $y := x \vee \Diamond x$. On the one hand, we have

$$\Diamond\Diamond x \leq (\Diamond x \vee \Diamond\Diamond x) = \Diamond y.$$

On the other hand, since $\Diamond x \wedge \neg\Diamond y = \bot$ we obtain

$$\Diamond y \leq \Diamond(y \wedge \neg\Diamond y) \leq \Diamond((x \vee \Diamond x) \wedge \neg\Diamond y) = \Diamond(x \wedge \neg\Diamond y) \vee \Diamond\bot \leq \Diamond x.$$

Hence, $\Diamond\Diamond x \leq \Diamond x$. □

In general, we call an *identity* of an algebraic structure $\mathcal{M}$ a formula of the form $t(\mathbf{x}) = u(\mathbf{x})$, where t, u are terms, such that $\mathcal{M} \models \forall\mathbf{x}\ (t(\mathbf{x}) = u(\mathbf{x}))$. Identities of Maragi algebras can be described in terms of modal logic as follows. Any term (built from the variables using boolean operations and $\Diamond$) is naturally identified with a formula in the language of propositional logic with a new unary connective $\Diamond$. If $\varphi(\mathbf{x})$ is such a formula and $\mathcal{M}$ a Magari algebra, we write $\mathcal{M} \models \varphi$ iff $\forall\mathbf{x}\ (t_\varphi(\mathbf{x}) = \top)$ is valid in $\mathcal{M}$, where t_φ is the term corresponding to φ. Since any identity in Magari algebras can be equivalently written in the form $t = \top$ for some term t, the axiomatization of identities of $\mathcal{M}$ amounts to axiomatizing modal formulas valid in $\mathcal{M}$. The *logic of* $\mathcal{M}$, $\mathrm{Log}(\mathcal{M})$, is the set of all modal formulas valid in $\mathcal{M}$, that is, $\mathrm{Log}(\mathcal{M}) := \{\varphi : \mathcal{M} \models \varphi\}$, and the logic of a class of modal algebras is defined similarly.

One of the main parameters of a Magari algebra $\mathcal{M}$ is its *characteristic* $\mathrm{ch}(\mathcal{M}) := \min\{k \in \omega : \Diamond^k\top = \bot\}$ and $\mathrm{ch}(\mathcal{M}) := \infty$ if no such k exists. If T is arithmetically sound, that is, if the arithmetical consequences of T are valid in the standard model, then $\mathrm{ch}(\mathcal{L}_T) = \infty$. Theories (whose algebras are) of finite characteristics are, in a sense, close to being inconsistent and may be considered a pathology.

Solovay [55] proved that any identity valid in the structure $(\mathcal{L}_T, \Diamond_T)$ follows from the boolean identities together with M1–M2, provided T is arithmetically sound. This has been generalized by Visser [58] to arbitrary theories of infinite characteristic.

Theorem 1. (Solovay, Visser) *Suppose* $\mathrm{ch}(\mathcal{L}_T, \Diamond_T) = \infty$. *An identity holds in* $(\mathcal{L}_T, \Diamond_T)$ *iff it holds in all Magari algebras.*

Apart from the equational characterization by M1, M2 above, the identities of Magari algebras can be axiomatized modal-logically. In fact, the logic of all Magari algebras, and by the Solovay theorem the logic $\mathrm{Log}(\mathcal{L}_T, \Diamond_T)$ of the Magari algebra of T, for any fixed theory T of infinite characteristic, coincides with the familiar Gödel–Löb logic **GL**. Abusing the language we will often identify **GL** with the set of identities of Magari algebras.[1]

A Hilbert-style axiomatization of **GL** is usually given in the modal language where $\Box$ rather than $\Diamond$ is taken as basic and the latter is treated as an abbreviation for $\neg\Box\neg$. The axioms and inference rules of **GL** are as follows.

Axiom schemata:

L1. All instances of propositional tautologies;
L2. $\Box(\varphi \to \psi) \to (\Box\varphi \to \Box\psi)$;

[1] For normal modal logics, going from an equational to a Hilbert-style axiomatization and back is automatic, as they are known to be strongly finitely algebraizable (see [19, 31]). We do not assume the reader's familiarity with algebraic logic and prefer to give explicit axiomatizations for the systems at hand.

L3. $\Box(\Box\varphi \to \varphi) \to \Box\varphi$.

Rules: $\varphi,\ \varphi \to \psi/\psi$ (modus ponens), $\varphi/\Box\varphi$ (necessitation).

By a well-known result of Segerberg [51], **GL** is sound and complete w.r.t. the class of all transitive and upwards well-founded Kripke frames. In fact, it is sufficient to restrict the attention to frames that are finite irreflexive trees. Thus, summarizing various characterizations above, we have

Theorem 2. *Let T be a gödelian theory of infinite characteristic. For any modal formula φ, the following statements are equivalent:*

(i) $\mathbf{GL} \vdash \varphi$;
(ii) φ *is valid in all Magari algebras;*
(iii) $(\mathscr{L}_T, \Diamond_T) \vDash \varphi$;
(iv) φ *is valid in all finite irreflexive tree-like Kripke frames.*

10.2 Topological Interpretation

A natural, albeit quite surprising, topological interpretation of provability logic was found by Simmons [53]. He observed that the topological derivative operator acting on a scattered topological space satisfies all the identities of Magari algebras. Esakia [28], working independently, considered a more general problem of set-theoretic interpretations of Magari algebras.

Let X be a nonempty set and let $\mathscr{P}(X)$ the boolean algebra of subsets of X. Consider any operator $\delta : \mathscr{P}(X) \to \mathscr{P}(X)$ and the structure $(\mathscr{P}(X), \delta)$. Can $(\mathscr{P}(X), \delta)$ be a Magari algebra and, if yes, when? Esakia [28] found what may be called a canonical answer to this question (Theorem 4 below).

Let (X, τ) be a topological space, where τ denotes the set of open subsets of X, and let $A \subseteq X$. Topological *derivative* $d_\tau(A)$ *of* A is the set of limit points of A:

$$x \in d_\tau(A) \iff \forall U \in \tau\ (x \in U \Rightarrow \exists y \neq x\ (y \in U \cap A)).$$

Notice that $c_\tau(A) := A \cup d_\tau(A)$ is the closure of A and $iso_\tau(A) := A \setminus d_\tau(A)$ is the set of isolated points of A.

The classical notion of a scattered topological space is due to Georg Cantor. (X, τ) is called *scattered* if every nonempty subspace $A \subseteq X$ has an isolated point.

Theorem 3. (Simmons, Esakia) *The following statements are equivalent:*

(i) (X, τ) *is scattered;*
(ii) $(\mathscr{P}(X), d_\tau)$ *is a Magari algebra, that is, for all* $A \subseteq X$, $d_\tau(A) = d_\tau(A \setminus d_\tau(A))$.

Notice that $d_\tau(A) = d_\tau(A \setminus d_\tau(A))$ means that each limit point of A is a limit point of its isolated points. The algebra of the form $(\mathscr{P}(X), d_\tau)$ associated with a topological space (X, τ) will be called *the derivative algebra of* X. Thus, this theorem states that the derivative algebra of (X, τ) is Magari iff (X, τ) is scattered.

Proof Suppose (X, τ) is scattered, $A \subseteq X$ and $x \in d_\tau(A)$. Consider any open neighborhood U of x. Since $(U \cap A) \setminus \{x\}$ is nonempty, it has an isolated point $y \neq x$. Since U is open, y is an isolated point of A, that is, $y \in A \setminus d_\tau(A)$. Hence, $x \in d_\tau(A \setminus d_\tau(A))$. The inclusion $d_\tau(A \setminus d_\tau(A)) \subseteq d_\tau(A)$ follows from the monotonicity of d_τ. Therefore Statement (ii) holds.

Suppose that (ii) holds and let $A \subseteq X$ be nonempty. We show that A has an isolated point. If $d_\tau A$ is empty, we are done. Otherwise, take any $x \in d_\tau A$. Since x is a limit of isolated points of A, there must be at least one such point. □

We notice that the transitivity principle $d_\tau d_\tau A \subseteq d_\tau A$ topologically means that the set $d_\tau A$, for any $A \subseteq X$, is closed. We recall the following standard equivalent characterization an easy proof of which we shall omit.

Proposition 2. *For any topological space (X, τ), the following statements are equivalent:*

(i) Every $x \in X$ is an intersection of an open and a closed set;
(ii) For each $A \subseteq X$, the set $d_\tau A$ is closed.

Topological spaces satisfying either of these conditions are called *T_d-spaces*. Condition (i) shows that T_d is a weak separation property located between T_0 and T_1. Thus, Proposition 1 yields, as a corollary, the modal proof of the following well-known fact.

Corollary 1. *All scattered spaces are T_d.*

We have seen in Theorem 3 that each scattered space equipped with a topological derivative operator is a Magari algebra. The following result by Esakia [28] shows that any Magari algebra on $\mathcal{P}(X)$ can be described in this way.

Theorem 4. (Esakia) *If $(\mathcal{P}(X), \delta)$ is a Magari algebra, then X bears a unique topology τ for which $\delta = d_\tau$. Moreover, τ is scattered.*

Proof We first remark that if $(\mathcal{P}(X), \delta)$ is a Magari algebra, then the operator $c(A) := A \cup \delta A$ satisfies the Kuratowski axioms of the topological closure: $c\varnothing = \varnothing$, $c(A \cup B) = cA \cup cB$, $A \subseteq cA$, $ccA = cA$. This defines a topology τ on X in which a set A is τ-closed iff $A = c(A)$ iff $\delta A \subseteq A$. If ν is any topology such that $\delta = d_\nu$, then ν has the same closed sets, that is, $\nu = \tau$. So if the required topology exists, it is unique. To show that $\delta = d_\tau$ we need an auxiliary lemma. □

Lemma 1. *Suppose $(\mathcal{P}(X), \delta)$ is Magari. Then, for all $x \in X$,*

(i) $x \notin \delta(\{x\})$;
(ii) $x \in \delta A \iff x \in \delta(A \setminus \{x\})$.

Proof (i) By Axiom M2 we have $\delta\{x\} \subseteq \delta(\{x\} \setminus \delta\{x\})$. If $x \in \delta\{x\}$ then $\delta(\{x\} \setminus \delta\{x\}) = \delta\varnothing = \varnothing$. Hence, $\delta\{x\} = \varnothing$, a contradiction.

(ii) $x \in \delta A$ implies $x \in \delta((A \setminus \{x\}) \cup \{x\}) = \delta(A \setminus \{x\}) \cup \delta\{x\}$. By (i), $x \notin \delta\{x\}$, hence $x \in \delta(A \setminus \{x\})$. The other implication follows from the monotonicity of δ. □

Lemma 2. *Suppose* $(\mathscr{P}(X), \delta)$ *is Magari and* τ *is the associated topology. Then* $\delta = d_\tau$.

Proof Let $d = d_\tau$; we show that for any set $A \subseteq X$ $dA = \delta A$. Notice that for any B, $cB = dB \cup B = \delta B \cup B$. Assume $x \in \delta A$. Then $x \in \delta(A \setminus \{x\}) \subseteq c(A \setminus \{x\}) \subseteq d(A \setminus \{x\}) \cup (A \setminus \{x\})$. Since $x \notin A \setminus \{x\}$, we obtain $x \in d(A \setminus \{x\})$. By the monotonicity of d, $x \in dA$. Similarly, if $x \in dA$ then $x \in d(A \setminus \{x\})$. Hence, $x \in c(A \setminus \{x\}) = \delta(A \setminus \{x\}) \cup (A \setminus \{x\})$. Since $x \notin A \setminus \{x\}$ we obtain $x \in \delta A$. □

From this lemma and Theorem 3 we also infer that τ is a scattered topology.

Theorem 4 shows that to study a natural set-theoretic interpretation of provability logic means to study the semantics of $\Diamond$ as a derivative operation on a scattered topological space. Derivative semantics of modality was first suggested in the fundamental paper by McKinsey and Tarski [48]. See [43] for a detailed survey of such semantics for arbitrary topological spaces. The emphasis in this chapter is on the logics related to formal provability and scattered topological spaces.

10.3 Topological Completeness Theorems

Natural examples of scattered topological spaces come from orderings. Two examples will play an important role below.

Let $(X, \prec)$ be a strict partial ordering. The *left topology* or the *downset topology* $\tau_\leftarrow$ on $(X, \prec)$ is given by all sets $A \subseteq X$ such that $\forall x, y\ (y \prec x \in A \Rightarrow y \in A)$. We obviously have that $(X, \prec)$ is well-founded iff $(X, \tau_\leftarrow)$ is scattered. The *right topology* or the *upset topology* is defined similarly.

The left topology is, in general, non-Hausdorff. More natural is the *interval topology* on a linear ordering $(X, <)$, which is generated by all open intervals $(\alpha, \beta) = \{x \in X \mid \alpha < x < \beta\}$ such that $\alpha, \beta \in X \cup \{\pm\infty\}$ and $\alpha < \beta$. The interval topology refines both the left topology and the right topology and is scattered on any ordinal [52].

Given a topological space (X, τ), we denote the logic of its derivative algebra $(\mathscr{P}(X), d_\tau)$ by $\mathrm{Log}(X, \tau)$, and we let $\mathrm{Log}(\mathscr{C})$ denote the logic of (the class of derivative algebras associated with) a class $\mathscr{C}$ of topological spaces. Thus, if $\mathscr{C}$ is a class of scattered spaces, $\mathrm{Log}(\mathscr{C})$ is a normal modal logic extending **GL**.

Esakia [28] has noted that the completeness theorem for **GL** w.r.t. its Kripke semantics (see [22, 51]) implies that **GL** is the modal logic of scattered spaces. In fact, if $(X, \prec)$ is a strict partial ordering, then the modal algebra associated with the Kripke frame $(X, \prec)$ is the same as the derivative algebra of (X, τ) where τ is its upset topology. This implies that any modal logic of a class of strict partial orders, including **GL**, is complete w.r.t. topological derivative semantics.

We can also note that **GL** is the logic of a single countable scattered space. Abashidze [1] and Blass [18] independently proved a stronger completeness result.

Theorem 5. (Abashidze, Blass) *Let* $\alpha \geq \omega^\omega$ *be any ordinal equipped with the interval topology. Then* $\mathrm{Log}(\alpha) = \mathbf{GL}$.

Thus, **GL** is complete w.r.t. a natural scattered topological space. The rest of this section is devoted to a new proof of this result. We need some technical prerequisites that will be also useful later in this chapter.

Ranks and d-maps. An equivalent characterization of scattered spaces is often given in terms of the following transfinite *Cantor–Bendixson sequence* of subsets of a topological space (X, τ):

- $d^0_\tau X = X; \quad d^{\alpha+1}_\tau X = d_\tau(d^\alpha_\tau X)$ and
- $d^\alpha_\tau X = \bigcap_{\beta<\alpha} d^\beta_\tau X$ if α is a limit ordinal.

It is easy to show by transfinite induction that for any (X, τ), all sets $d^\alpha_\tau X$ are closed and that $d^\alpha_\tau X \supseteq d^\beta_\tau X$ whenever $\alpha \leq \beta$.

Theorem 6. (Cantor) *(X, τ) is scattered iff $d^\alpha_\tau X = \varnothing$ for some ordinal α.*

Proof Let $d = d_\tau$. If (X, τ) is scattered then we have $d^\alpha X \supset d^{\alpha+1} X$ for each α such that $d^\alpha X \neq \varnothing$. By cardinality arguments this yields an α such that $d^\alpha X = \varnothing$.

Conversely, suppose $A \subseteq X$ is nonempty. Let α be the least ordinal such that $A \not\subseteq d^\alpha X$. Obviously, α cannot be a limit ordinal, hence $\alpha = \beta + 1$ for some β and there is an $x \in A \setminus d^{\beta+1} X$. Since $A \subseteq d^\beta X$, we also have $x \in d^\beta X$. Since $x \notin d^{\beta+1} X = d(d^\beta X)$, x is isolated in the relative topology of $d^\beta X$, and hence in the relative topology of $A \subseteq d^\beta X$. □

Call the least α such that $d^\alpha_\tau X = \varnothing$ the *Cantor–Bendixson rank* of X and denote it by $\rho_\tau(X)$. Let On denote the class of all ordinals. Then the *rank function* $\rho_\tau : X \to$ On is defined by

$$\rho_\tau(x) := \min\{\alpha : x \notin d^{\alpha+1}_\tau(X)\}.$$

Notice that ρ_τ maps X onto $\rho_\tau(X) = \{\alpha : \alpha < \rho_\tau(X)\}$. Also, $\rho_\tau(x) \geq \alpha$ iff $x \in d^\alpha_\tau X$. We omit the subscript τ whenever there is no danger of confusion.

Example 1. For an ordinal equipped with its *left topology*, $\rho(\alpha) = \alpha$ for all α. When the same ordinal is equipped with its *interval topology*, ρ is the function ℓ defined by $\ell(0) = 0$; $\ell(\alpha) = \beta$ if $\alpha = \gamma + \omega^\beta$ for some γ, β. By the Cantor normal form theorem for any $\alpha > 0$, such a β is uniquely determined, thus ℓ is well-defined. Notice that $\ell(\alpha) = 0$ iff α is a non-limit ordinal.

Let (X, τ_X) and (Y, τ_Y) be topological spaces, and let d_X, d_Y denote the corresponding derivative operators. A map $f : X \to Y$ is called a *d-map* if f is continuous, open and *pointwise discrete*, that is, $f^{-1}(y)$ is a discrete subspace of X for each $y \in Y$. d-maps are well known to satisfy the properties expressed in the following lemma (see [16]).

Lemma 3.

(i) $f^{-1}(d_Y(A)) = d_X(f^{-1}(A))$ *for any* $A \subseteq Y$;
(ii) $f^{-1} : (\mathscr{P}(Y), d_Y) \to (\mathscr{P}(X), d_X)$ *is a homomorphism of derivative algebras;*

(iii) If f is onto, then $\mathrm{Log}(X, \tau_X) \subseteq \mathrm{Log}(Y, \tau_Y)$.

Property (i) is easy to check directly; (ii) follows from (i), and (iii) follows from (ii). Each of the conditions (i) and (ii) is equivalent to f being a d-map.

A proof of the following lemma can be found in [5].

Lemma 4. *Let Ω be the ordinal $\rho_\tau(X)$ taken with its left topology. Then*

(i) $\rho_\tau : X \twoheadrightarrow \Omega$ *is an onto d-map;*

(ii) If $f : X \to \lambda$ is a d-map, where λ is an ordinal with its left topology, then $f(X) = \Omega$ and $f = \rho_\tau$.

An immediate corollary is that the rank function is preserved under d-maps.

The d-sum construction. The constructions of summing up structures, in particular, topological spaces or orderings 'along' another structure play an important role in various branches of logic and mathematics (see, e.g., [34]). Here we present another construction of this type, called *d-sum*, which can be used to recursively build both finite trees and ordinals. Given a tree T, one can construct a new tree by 'plugging in' other trees in place of the leaves of T. Similarly, given an ordinal α, one can 'plug in' new ordinals α_i for each isolated point $i \in \alpha$ to obtain another ordinal. The d-sum construction turned out to be rather useful for proving topological completeness theorems. Its particular case called *d-product* serves as a tool in the proof of topological completeness of **GLP** in [5].

Definition 1 Let X be a topological space and let $\{Y_j \mid j \in iso(X)\}$ be a collection of spaces indexed by the set $iso(X)$ of isolated points of X. We uniquely extend it to the collection $\{Y_j \mid j \in X\}$ by letting $Y_j = \{j\}$ for all $j \in dX$.

We define the *d-sum* (Z, τ_Z) of $\{Y_j\}$ over X (denoted $\sum^d_{j\in X} Y_j$) as follows. The base set is the disjoint union $Z := \bigsqcup_{j\in X} Y_j$. Define the map $\pi : Z \to X$ by putting $\pi(y) = j$ whenever $y \in Y_j$. Now let the topology τ_Z consist of the sets $V \cup \pi^{-1}(U)$ where V is open in the topological sum $\bigsqcup_{j\in iso(X)} Y_j$ and U is open in X. It is not difficult to check that τ_Z qualifies for a topology.

Example 2. (trees) Consider finite irreflexive trees equipped with the upset topology. Note that the leaves of a tree are the isolated points in the topology. Therefore, taking the d-sum of trees T_i over a tree T simply means plugging in T_i's in place of the leaves of T.

Let us call an *n-fork* a tree $\mathfrak{F}_n = (W_n, R_n)$, where $W_n = \{r, w_0, w_1, \ldots, w_{n-1}\}$ and $R_n = \{(r, w_i) \mid 0 \leq i < n\}$. Observe that any finite tree is either an irreflexive point, or an n-fork, or can be obtained (possibly in several ways) as a d-sum of trees of smaller depth.

Example 3. (ordinals) Consider ordinals equipped with the interval topology. If $(\alpha_i)_{i\in\beta}$ is a family of ordinals such that $\alpha_i = 1$ for limit i, then the d-sum $\sum^d_{i\in\beta} \alpha_i$ is homeomorphic to the ordinal sum $\sum_{i\in\beta} \alpha_i$. This can be checked directly by examining the descriptions of neighborhoods in respective spaces. Thus, a d-sum of ordinals along another ordinal is homeomorphic to an ordinal.

The following lemma shows that d-sums, in a way, commute with d-maps.

Lemma 5. *Let X and X' be two spaces and let $\{Y_j \mid j \in iso(X)\}$ and $\{Y'_k \mid k \in iso(X')\}$ be collections of spaces indexed by $iso(X)$ and $iso(X')$, respectively. Suppose further that $f : X \to X'$ is an onto d-map, and for each $j \in iso(X)$ there is an onto d-map $f_j : Y_j \to Y'_{f(j)}$. Then there exists an onto d-map $g : \sum^d_{j\in X} Y_j \to \sum^d_{k\in X'} Y'_k$.*

Proof First note that since f is a d-map, $f(j)$ is isolated in X' iff j is isolated in X. Indeed, by openness of f, if $\{j\} \in \tau$, then $\{f(j)\} \in \tau'$. Conversely, if $f(j)$ is isolated, then $f^{-1}f(j)$ is both open and discrete by continuity and pointwise discreteness of f. Hence, any point in $f^{-1}f(j)$, and j in particular, is isolated in X. For convenience, let us denote $f_* \equiv f\restriction_{d_\tau X}$ and $f^* \equiv f\restriction_{iso(X)}$. It follows that $f^* : iso(X) \to iso(X')$ and $f_* : d_\tau X \to d_{\tau'} X'$ are well-defined onto maps and $f = f^* \cup f_*$. Thus, in particular, the space $Y'_{f(j)}$ in the formulation of the theorem is well-defined.

Take g to be the set-theoretic union $g = f_* \cup \bigcup_{j\in iso(X)} f_j$. We show that g is a d-map. Let π and π' be the 'projection' maps associated with $\sum^d_{j\in X} Y_j$ and $\sum^d_{k\in X'} Y'_k$, respectively. To show that g is open, take $W = V \cup \pi^{-1}(U) \in \tau_Z$. Then $g(W) = g(V) \cup g(\pi^{-1}(U))$. That $g(V)$ is open in the topological sum of Y'_k is clear from the openness of the maps f_j. Moreover, from the definition of g and the fact that all f_j are onto it can be easily deduced that $g(\pi^{-1}(U)) = \pi'^{-1}(f(U))$. Since f is an open map, it follows that $g(W)$ is open in τ'_Z. To see that g is continuous, take $W' = V' \cup \pi'^{-1}(U') \in \tau'_Z$. Then $g^{-1}(W') = g^{-1}(U') \cup g^{-1}(\pi'^{-1}(U'))$. Again, the openness of $g^{-1}(U')$ is trivial. It is also easily seen that $g^{-1}(\pi'^{-1}(U')) = \pi^{-1}(f^{-1}(U'))$. It follows that $g^{-1}(W')$ is open in τ_Z. To see that g is pointwise discrete is straightforward, given that f and all the f_j are pointwise discrete. □

The following lemma is crucial for a proof of Theorem 5.

Lemma 6. *For each finite irreflexive tree T there exists a countable ordinal $\alpha < \omega^\omega$ and an onto d-map $f : \alpha \twoheadrightarrow T$.*

Proof The proof proceeds by induction on the depth of T. It is clear that the claim is true for a one-point tree. If T is an n-fork $\mathfrak{F}_n$ we define a d-map $f : \omega + 1 \twoheadrightarrow \mathfrak{F}_n$ by letting $f(x) := w_{x \bmod n}$ for $x < \omega$ and $f(\omega) := r$.

Now consider a tree T of depth $n > 1$ and suppose the claim is true for all trees of depth less than n. Clearly T can be presented as a d-sum of trees of strictly smaller depth in various ways. Using the induction hypothesis, each of the smaller trees is an image of a countable ordinal under a d-map. Applying Lemma 5 and observing that a countable d-sum of countable ordinals is a countable ordinal produces a countable ordinal α and an onto d-map $f : \alpha \twoheadrightarrow T$. Since the rank function is preserved under d-maps, the rank of α is equal to the rank of T, that is, to n. It follows that $\alpha < \omega^\omega$, which completes the proof. □

Now we prove Theorem 5.

Proof Take a non-theorem φ of **GL**. Then φ can be refuted on a finite irreflexive tree T by theorem 2. By Lemma 6, there exists an ordinal $\beta < \omega^\omega$ that maps onto T via a d-map. By Lemma 3 (iii), φ can be refuted on β. But β is an open subspace of α. It follows that φ can be refuted on α. □

Another, perhaps the simplest, proof of Theorem 5 appeared recently in [17, Theorem 3.5]. It relied on a direct proof of Lemma 6 rather than on Lemma 5. However, we believe that our approach illuminates the underlying recursive mechanism and may lead to additional insights in more complicated situations (see [5]).

10.4 Topological Semantics of Linearity Axioms

For a gödelian theory T consider the 0-generated subalgebra $\mathcal{L}_T^0$ of $(\mathcal{L}_T, \Diamond_T)$, that is, the subalgebra generated by $\top$. If $\mathrm{ch}(\mathcal{L}_T, \Diamond_T) = \infty$, then also $\mathrm{ch}(\mathcal{L}_T^0, \Diamond_T) = \infty$. In fact, the modal logic of the Magari algebra $(\mathcal{L}_T^0, \Diamond_T)$ is known (see [37]) to be **GL.3** which is obtained from **GL** by adding the following axiom:

$$(.3) \qquad \Diamond p \wedge \Diamond q \to \Diamond(p \wedge q) \vee \Diamond(p \wedge \Diamond q) \vee \Diamond(\Diamond p \wedge q).$$

This is the so called 'linearity axiom' and, as the name suggests, its finite rooted Kripke frames are precisely the finite strict linear orders. Since **GL.3** is Kripke complete (see, e.g., [24]), its topological completeness is immediate. However, it is not immediately clear what kind of scattered spaces does the linearity axiom isolate. To characterize GL.3-spaces, let us first simplify the axiom (.3). Consider the following formula:

$$(lin) \qquad \Box(\Box^+ p \vee \Box^+ q) \to \Box p \vee \Box q,$$

where $\Box^+\varphi$ is a shorthand for $\varphi \wedge \Box\varphi$.

Lemma 7. *In* ***GL*** *the schema* (.3) *is equivalent to* (lin).

Proof To show that $(lin) \vdash_{\mathbf{GL}}$ (.3), witness the following syntactic argument. Observe that the dual form of (lin) looks as follows:

$$\Diamond p \wedge \Diamond q \to \Diamond(\Diamond^+ p \wedge \Diamond^+ q) \qquad (*)$$

where $\Diamond^+\varphi := \varphi \vee \Diamond\varphi$. Furthermore, an instance of the **GL** axiom looks as follows:

$$\Diamond(\Diamond^+ p \wedge \Diamond^+ q) \to \Diamond(\Diamond^+ p \wedge \Diamond^+ q \wedge \Box(\Box^+\neg p \vee \Box^+\neg q)).$$

By the axiom (lin) we also have: $\Box(\Box^+\neg p \vee \Box^+\neg q) \to (\Box\neg p \vee \Box\neg q)$. So, using the monotonicity of $\Diamond$ we obtain:

$$\Diamond p \wedge \Diamond q \rightarrow \Diamond(\Diamond^+ p \wedge \Diamond^+ q \wedge (\Box\neg p \vee \Box\neg q)).$$

By boolean logic

$$\Diamond^+ p \wedge \Diamond^+ q \leftrightarrow (p \wedge q) \vee (p \wedge \Diamond q) \vee (\Diamond p \wedge q) \vee (\Diamond p \wedge \Diamond q) \qquad (**)$$

and

$$(\Box\neg p \vee \Box\neg q) \leftrightarrow \neg(\Diamond p \wedge \Diamond q).$$

Using these, together with the monotonicity of $\Diamond$ we finally arrive at:

$$\Diamond p \wedge \Diamond q \rightarrow \Diamond((p \wedge q) \vee (p \wedge \Diamond q) \vee (\Diamond p \wedge q)),$$

which is equivalent to (.3) since $\Diamond$ distributes over $\vee$.

To show the converse, we observe that (.3) implies (lin) even in the system **K**. Indeed, the formula $(*)$, which is the dual form of (lin), can be rewritten, using $(**)$ and the distribution of $\Diamond$ over $\vee$ as follows:

$$\Diamond p \wedge \Diamond q \rightarrow \Diamond(p \wedge q) \vee \Diamond(p \wedge \Diamond q) \vee \Diamond(\Diamond p \wedge q) \vee \Diamond(\Diamond p \wedge \Diamond q),$$

which is clearly a weakening of (.3). Therefore (.3) $\vdash_{\mathbf{GL}}$ (lin). □

It follows that a scattered space is a GL.3-space iff it validates (lin). To characterize such spaces, consider the following definition.

Definition 2 Call a scattered space *primal* if for each $x \in X$ and $U, V \in \tau$, $\{x\} \cup U \cup V \in \tau$ implies $\{x\} \cup U \in \tau$ or $\{x\} \cup V \in \tau$.

It can be shown that X is primal iff the collection of punctured open neighborhoods of each non-isolated point is a prime filter in the Heyting algebra τ.

Theorem 7. *Let X be a scattered space. Then $X \vDash (lin)$ iff X is primal.*

Proof Let X be a scattered space together with a valuation ν. Let $P := \nu(p)$ and $Q := \nu(q)$ denote the truth-sets of p and q, respectively. Then the truth sets of $\Box^+ p$ and $\Box^+ q$ are $I_\tau P$ and $I_\tau Q$, where I_τ is the interior operator of X. We write $x \vDash \varphi$ for $X, x \vDash_\nu \varphi$.

Suppose X is primal and for some valuation $x \vDash \Box(\Box^+ p \vee \Box^+ q)$. Then there exists an open neighborhood W of x such that $W \setminus \{x\} \vDash \Box^+ p \vee \Box^+ q$. In other words, $W \setminus \{x\} \subseteq I_\tau P \cup I_\tau Q$. Let $U = W \cap I_\tau P \in \tau$ and $V = W \cap I_\tau Q \in \tau$. Then $\{x\} \cup U \cup V = W \in \tau$. It follows that either $\{x\} \cup U \in \tau$ or $\{x\} \cup V \in \tau$. Hence $x \vDash \Box p$ or $x \vDash \Box q$. This proves that $X \vDash (lin)$.

Suppose now X is not primal. Then there exist $x \in X$ and $U, V \in \tau$ such that $\{x\} \cup U \cup V \in \tau$, but $\{x\} \cup U \notin \tau$ and $\{x\} \cup V \notin \tau$. Take a valuation such that $P = U$ and $Q = V$. Then clearly $x \vDash \Box(\Box^+ p \vee \Box^+ q)$. However, neither $x \vDash \Box p$ nor $x \vDash \Box q$ is true. Indeed, if, for example, $x \vDash \Box p$, then there exists an open neighborhood W of x such that $W \setminus \{x\} \subseteq P = U$. But then $\{x\} \cup U = W \cup U \in \tau$, which is a contradiction. This shows that $X \not\models (lin)$. □

Example 4. (primal spaces) The left topology of any well-founded linear order is clearly primal. To give an example of a primal space not coming from order, consider any countable set A, a point $b \notin A$ and a free ultrafilter $\mathbf{u}$ over A. Then the set $A \cup \{b\}$ with the topology $\wp(A) \cup \{U \cup \{b\} \mid U \in \mathbf{u}\}$ is easily seen to be primal. This space is homeomorphic to a subspace of the Stone-Čech compactification of a countable discrete space A defined by $A \cup \{\mathbf{u}\}$.

The primal scattered spaces are closely related to *maximal scattered* spaces of [5]. A scattered space is called *maximal* if it does not have any proper refinements with the same rank function. It is easy to see that each maximal scattered space is primal, but there are primal spaces which are not maximal. The two notions do coincide for the scattered spaces of finite rank. It follows that the logic of maximal scattered spaces is **GL.3**.

10.5 GLP-Algebras and Polymodal Provability Logic

A natural generalization of provability logic **GL** to a language with infinitely many modal diamonds $\langle 0 \rangle$, $\langle 1 \rangle$, ... has been introduced in 1986 by Japaridze [40]. He interpreted $\langle 1 \rangle \varphi$ as an arithmetical statement expressing the *ω-consistency of* φ over a given gödelian theory T.[2] Similarly, $\langle n \rangle \varphi$ was interpreted as the consistency of the extension of $T + \varphi$ by n nested applications of the ω-rule.

While the logic of each of the individual modalities $\langle n \rangle$ over Peano Arithmetic was known to coincide with **GL** by a relatively straightforward extension of the Solovay theorem [20], Japaridze found a complete axiomatization of the *joint* logic of the modalities $\langle n \rangle$ for all $n \in \omega$. This result involved considerable technical difficulties and lead to one of the first genuine extensions of Solovay's arithmetical fixed-point construction. Later, Japaridze's work has been simplified and extended by Ignatiev [39] and Boolos [21]. In particular, Ignatiev showed that **GLP** is complete for more general sequences of 'strong' provability predicates in arithmetic and analyzed the variable-free fragment of **GLP**. Boolos included a treatment of **GLB** (the fragment of **GLP** with just two modalities) in his popular book on provability logic [22].

More recently, **GLP** has found interesting applications in proof-theoretic analysis of arithmetic [2, 6, 7, 9] which stimulated some further interest in the study of modal-logical properties of **GLP** [11, 15, 23, 38]. For such applications, the algebraic language appears to be more natural and a different choice of the interpretation of the provability predicates is needed. The relevant structures have been introduced in [6] under the name of *graded provability algebras*.

---

[2] A gödelian theory U is ω-consistent if its extension by unnested applications of the ω-rule $U' := U + \{\forall x\, \varphi(x) : \forall n\, U \vdash \varphi(\underline{n})\}$ is consistent.

Recall that an arithmetical formula is called Π_n if it can be obtained from a formula containing only bounded quantifiers $\forall x \le t$ and $\exists x \le t$ by a prefix of n alternating blocks of quantifiers starting from $\forall$. Arithmetical Σ_n-formulas are defined dually.

Let T be a gödelian theory. T is called *n-consistent* if T together with all true arithmetical Π_n-sentences is consistent. (Alternatively, T is n-consistent iff every Σ_n-sentence provable in T is true.) Let $n\text{-}\mathsf{Con}(T)$ denote an arithmetical formula expressing the n-consistency of T (it can be defined using the standard Π_n-definition of truth for Π_n-sentences in arithmetic). Since we assume T to be recursively enumerable, it is easy to check that the formula $n\text{-}\mathsf{Con}(T)$ itself belongs to the class Π_{n+1}.

The n-consistency formula induces an operator $\langle n \rangle_T$ acting on the Lindenbaum–Tarski algebra $\mathcal{L}_T$:

$$\langle n \rangle_T : [\varphi] \longmapsto [n\text{-}\mathsf{Con}(T + \varphi)].$$

The dual *n-provability* operators are defined by $[n]_T x = \neg \langle n \rangle_T \neg x$ for all $x \in \mathcal{L}_T$. Since every true Π_n-sentence is assumed to be an axiom for n-provability, we notice that every true Σ_{n+1}-sentence must be n-provable. Moreover, this latter fact is formalizable in T, so we obtain the following lemma (see [54]). (By the abuse of notation we denote by $[n]_T \varphi$ the arithmetical formula expressing the n-provability of φ in T.)

Lemma 8. *For each true Σ_{n+1}-formula $\sigma(x)$, $T \vdash \forall x\, (\sigma(x) \to [n]_T \sigma(\underline{x}))$.*

As a corollary we obtain a basic observation probably due to Smorynski [54].

Proposition 3. *For each $n \in \omega$, the structure $(\mathcal{L}_T, \langle n \rangle_T)$ is a Magari algebra.*

A proof of this fact consists of verifying the Hilbert–Bernays–Löb derivability conditions for $[n]_T$ in T and of deducing from them, in the usual way, an analog of Löb's theorem for $[n]_T$.

The structure $(\mathcal{L}_T, \{\langle n \rangle_T : n \in \omega\})$ is called the *graded provability algebra of T* or the *GLP-algebra of T*. Apart from the identities inherited from the structure of Magari algebras for each $\langle n \rangle$, it satisfies the following principles for all $m < n$:

P1. $\langle m \rangle x \le [n] \langle m \rangle x$;
P2. $\langle n \rangle x \le \langle m \rangle x$.

The validity of P1 follows from Lemma 8 because the formula $\langle m \rangle_T \varphi$, for any φ, belongs to the class Π_{m+1}. P2 holds since $\langle n \rangle_T \varphi$ asserts the consistency of a stronger theory than $\langle m \rangle_T \varphi$ for $m < n$.

In general, we call a *GLP-algebra* a structure $(M, \{\langle n \rangle : n \in \omega\})$ such that each $(M, \langle n \rangle)$ is a Magari algebra and conditions P1, P2 (that are equivalent to identities) are satisfied for all $x \in M$.

At this point it is worth noticing that condition P1 has an equivalent form that has proved to be quite useful in the study of GLP-algebras.

Lemma 9. *Modulo the other identities of GLP-algebras, P1 is equivalent to*

P1'. $\langle n\rangle y \wedge \langle m\rangle x = \langle n\rangle(y \wedge \langle m\rangle x)$ *for all* $m < n$.

Proof First, we prove P1$'$. We have $y \wedge \langle m\rangle x \leq y$, hence $\langle n\rangle(y \wedge \langle m\rangle x) \leq \langle n\rangle y$. Similarly, by P2 and transitivity, $\langle n\rangle(y \wedge \langle m\rangle x) \leq \langle n\rangle\langle m\rangle x \leq \langle m\rangle\langle m\rangle x \leq \langle m\rangle x$. Hence, $\langle n\rangle(y \wedge \langle m\rangle x) \leq \langle n\rangle y \wedge \langle m\rangle x$. In the other direction, by P1, $\langle n\rangle y \wedge \langle m\rangle x \leq \langle n\rangle y \wedge [n]\langle m\rangle x$. However, as in any modal algebra, we also have $\langle n\rangle y \wedge [n]z \leq \langle n\rangle(y \wedge z)$. It follows that $\langle n\rangle y \wedge [n]\langle m\rangle x \leq \langle n\rangle(y \wedge \langle m\rangle x)$. Thus, P1$'$ is proved.

To infer P1 from P1$'$ it is sufficient to prove that $\langle m\rangle x \wedge \neg[n]\langle m\rangle x = \bot$. We have that $\neg[n]\langle m\rangle x = \langle n\rangle\neg\langle m\rangle x$. Therefore, by P1$'$, $\langle m\rangle x \wedge \langle n\rangle\neg\langle m\rangle x = \langle n\rangle(\neg\langle m\rangle x \wedge \langle m\rangle x) = \langle n\rangle\bot = \bot$, as required. □

An equivalent formulation of Japaridze's arithmetical completeness theorem is that any identity of $(\mathcal{L}_T, \{\langle n\rangle_T : n \in \omega\})$ follows from the identities of GLP-algebras [40]. It is somewhat strengthened to the current formulation in [13, 39].

Theorem 8. (Japaridze) *Suppose* T *is gödelian,* T *contains Peano Arithmetic, and* $\mathrm{ch}(\mathcal{L}_T, \langle n\rangle_T) = \infty$ *for each* $n < \omega$. *Then, an identity holds in* $(\mathcal{L}_T, \{\langle n\rangle_T : n \in \omega\})$ *iff it holds in all GLP-algebras.*

We note that the condition $\mathrm{ch}(\mathcal{L}_T, \langle n\rangle_T) = \infty$, for each $n \in \omega$, is equivalent to $T + n\text{-}\mathsf{Con}(T)$ being consistent for each $n \in \omega$, and is clearly necessary for the validity of Japaridze's theorem.

The logic of all GLP-algebras can also be axiomatized as a Hilbert-style calculus (see the footnote in Sect. 10.1). The corresponding system **GLP** was originally introduced by Japaridze. **GLP** is formulated in the language of propositional logic enriched by modalities $[n]$ for all $n \in \omega$. The axioms of **GLP** are those of **GL**, formulated for each $[n]$, as well as the two analogs of P1 and P2 for all $m < n$:

P1. $\langle m\rangle\varphi \to [n]\langle m\rangle\varphi$;
P2. $[m]\varphi \to [n]\varphi$.

The inference rules of **GLP** are modus ponens and $\varphi/[n]\varphi$ for each $n \in \omega$.

We let $\mathbf{GLP}_n$ denote the fragment of **GLP** in the language with the first n modalities; thus **GLB** is $\mathbf{GLP}_2$.

For any modal formula φ, $\mathbf{GLP} \vdash \varphi$ iff the identity $t_\varphi = \top$ holds in all **GLP**-algebras. Hence, **GLP** coincides with the logic of all GLP-algebras as well as with the logic of the GLP-algebra of T for any theory T such that $T + n\text{-}\mathsf{Con}(T)$ is consistent for each $n < \omega$.

10.6 GLP-Spaces

Topological semantics for **GLP** has been first considered in [14]. The main difficulty in the modal-logical study of **GLP** comes from the fact that it is incomplete with respect to its relational semantics; that is, **GLP** is the logic of no class of *frames* [22]. Even though a suitable class of relational *models* for which **GLP** is sound and complete was developed in [11], these models are not so easy to handle. So, it is natural to consider a generalization of the topological semantics we have for **GL**. As it turns out, topological semantics provides another natural class of GLP-algebras which is interesting in its own right, and also due to its analogy with the proof-theoretic GLP-algebras.

As before, we are interested in GLP-algebras of the form $(\mathcal{P}(X), \{\langle n\rangle : n \in \omega\})$, where $\mathcal{P}(X)$ is the boolean algebra of subsets of a given set X. Since each $(\mathcal{P}(X), \langle n\rangle)$ is a Magari algebra, the operator $\langle n\rangle$ is the derivative operator with respect to some uniquely defined scattered topology on X. Thus, we come to the following definition [14].

A polytopological space $(X, \{\tau_n : n \in \omega\})$ is called a *GLP-space* if the following conditions hold for each $n \in \omega$:

D0. (X, τ_n) is a scattered space;
D1. For each $A \subseteq X$, $d_{\tau_n}(A)$ is τ_{n+1}-open;
D2. $\tau_n \subseteq \tau_{n+1}$.

We notice that the last two conditions directly correspond to conditions P1 and P2 of GLP-algebras. By a *GLP$_m$-space* we mean a space $(X, \{\tau_n : n < m\})$ satisfying conditions D0–D2 for the first m topologies.

Proposition 4. *(i) If $(X, \{\tau_n : n \in \omega\})$ is a GLP-space, then the structure $(\mathcal{P}(X), \{d_{\tau_n} : n \in \omega\})$ is a GLP-algebra.*
(ii) If $(\mathcal{P}(X), \{\langle n\rangle : n \in \omega\})$ is a GLP-algebra, then there are uniquely defined topologies $\{\tau_n : n \in \omega\}$ on X such that $(X, \{\tau_n : n \in \omega\})$ is a GLP-space and $\langle n\rangle = d_{\tau_n}$ for each $n < \omega$.

Proof (i) Suppose $(X, \{\tau_n : n \in \omega\})$ is a GLP-space. Let $d_n := d_{\tau_n}$ denote the corresponding derivative operators and let $\tilde{d}_n$ denote its dual $\tilde{d}_n(A) := X \backslash d_n(X \backslash A)$.[3] By Theorem 3 $(\mathcal{P}(X), d_n)$ is a Magari algebra for each $n \in \omega$. Notice that $A \in \tau_n$ iff $A \subseteq \tilde{d}_n A$. If $m < n$, then $d_m A \in \tau_n$, so $d_m A \subseteq \tilde{d}_n d_m A$, hence P1 holds. Since $\tau_n \subseteq \tau_{n+1}$, we have $d_{n+1} A \subseteq d_n A$, thus P2 holds.

(ii) Let $(\mathcal{P}(X), \{\langle n\rangle : n \in \omega\})$ be a GLP-algebra. Since each of the algebras $(\mathcal{P}(X), \langle n\rangle)$ is Magari, by Theorem 4 a scattered topology τ_n on X is defined for which $\langle n\rangle = d_{\tau_n}$. In fact, we have $U \in \tau_n$ iff $U \subseteq [n]U$. We check that conditions D1 and D2 are met.

Suppose A is τ_n-closed, that is, $\langle n\rangle A \subseteq A$. Then $\langle n+1\rangle A \subseteq \langle n\rangle A \subseteq A$ by P2. Hence, A is τ_{n+1}-closed. Thus, $\tau_n \subseteq \tau_{n+1}$.

By P1 for any set A we have $\langle n\rangle A \subseteq [n+1]\langle n\rangle A$. Hence, $d_{\tau_n}(A) = \langle n\rangle A \in \tau_{n+1}$. Thus, $(X, \{\tau_n : n \in \omega\})$ is a GLP-space. □

To obtain examples of GLP-spaces let us first consider the case of two modalities. The following basic example is due to Esakia (private communication, see [14]).

---

[3] There is no conventional name for the dual of the derivative operator. Sometimes it is denoted by t. Here we choose the notation $\tilde{d}$ to emphasize its connection with d.

Example 5. Consider a bitopological space $(\Omega; \tau_0, \tau_1)$, where Ω is an ordinal, τ_0 is its left topology, and τ_1 is its interval topology. Esakia noticed that this space is a model of **GLB**, that is, in our terminology, a GLP_2-space. In fact, for any $A \subseteq \Omega$ the set $d_0(A) = (\min A, \Omega)$ is an open interval, whenever A is not empty. Hence, D1 holds (the other two conditions are immediate). Esakia also noticed that such spaces can never be complete for **GLP** as the linearity axiom (.3) holds for $\langle 0 \rangle$.

In general, to define GLP_n-spaces for $n > 1$, we introduce an operation $\tau \longmapsto \tau^+$ on topologies on a given set X. This operation plays a central role in the study of GLP-spaces.

Given a topological space (X, τ), let τ^+ be the coarsest topology containing τ such that each set of the form $d_\tau(A)$, with $A \subseteq X$, is open in τ^+. Thus, τ^+ is generated by τ and $\{d_\tau(A) : A \subseteq X\}$. Clearly, τ^+ is the coarsest topology on X such that $(X; \tau, \tau^+)$ is a GLP_2-space. Sometimes we call τ^+ the *derivative topology* of (X, τ).

Getting back to Esakia's example, it is easy to verify that, on any ordinal Ω, the derivative topology of the left topology coincides with the interval topology. (In fact, any open interval is an intersection of a downset and an open upset.)

Example 6. Even though we are mainly interested in scattered spaces, the derivative topology makes sense for arbitrary spaces. The reader can check that if τ is the *coarsest* topology on a set X (whose open sets are just X and $\varnothing$), then τ^+ is the *cofinite* topology on X (whose open sets are exactly the cofinite subsets of X together with $\varnothing$). On the other hand, if τ is the cofinite topology, then $\tau^+ = \tau$. We note that the logic of the cofinite topology on an infinite set is **KD45** (see [57]).

For scattered spaces, τ^+ is always strictly finer than τ, unless τ is discrete. We present a proof using the language of Magari algebras.

Proposition 5. *If (X, τ) is scattered, then $d_\tau(X)$ is not open, unless $d_\tau(X) = \varnothing$.*

Proof The set $d_\tau(X)$ corresponds to the element $\Diamond\top$ in the associated Magari algebra; $d_\tau(X)$ being open means $\Diamond\top \leq \Box\Diamond\top$. By M2 we have $\Box\Diamond\top \leq \Box\bot = \neg\Diamond\top$. Hence, $\Diamond\top \leq \neg\Diamond\top$, that is, $\Diamond\top = \bot$. This means $d_\tau(X) = \varnothing$. □

We will see later that τ^+ can be much finer than τ. Notice that if τ is T_d, then each set of the form $d_\tau(A)$ is τ-closed. Hence, it will be clopen in τ^+. Thus, τ^+ is obtained by adding to τ new clopen sets. In particular, τ^+ will be zero-dimensional if so is τ.[4]

Iterating the plus operation yields a GLP-space. Let (X, τ) be a scattered space. Define: $\tau_0 := \tau$ and $\tau_{n+1} := \tau_n^+$. Then $(X, \{\tau_n : n \in \omega\})$ is a GLP-space that will be called the GLP-space *generated from* (X, τ) or simply the *generated GLP-space*.

Thus, from any scattered space we can always produce a GLP-space in a natural way. The question is whether this space will be nontrivial, that is, whether we can guarantee that the topologies τ_n are non-discrete.

[4] Recall that a topological space is zero-dimensional if it has a base of clopen sets.

In fact, the next observation from [14] shows that for many natural τ already the topology τ^+ will be discrete. Recall that a topological space X is *first-countable* if every point $x \in X$ has a countable basis of open neighborhoods.

Proposition 6. *If (X, τ) is Hausdorff and first-countable, then τ^+ is discrete.*

Proof It is easy to see that if (X, τ) is first-countable and Hausdorff, then every point $a \in d_\tau(X)$ is a (unique) limit point of a countable sequence of points $A = \{a_n\}_{n \in \omega}$. Hence, there is a set $A \subseteq X$ such that $d_\tau(A) = \{a\}$. By D1 this means that $\{a\}$ is τ^+-open. □

Thus, if τ is the interval topology on a countable ordinal, then τ^+ is discrete. The same holds, for example, if τ is the (non-scattered) topology of the real line.

We remark that the left topology τ on any countable ordinal $> \omega$ yields an example of a non-Hausdorff first-countable space such that τ^+ is non-discrete. In the following section we will also see that if τ is the interval topology on any ordinal $> \omega_1$, then τ^+ is non-discrete (ω_1 is its least non-isolated point). However, we do not have any topological characterization of spaces (X, τ) such that τ^+ is discrete. (See, however, Proposition 8, which provides a characterization in terms of d-reflection.)

Given an arbitrary scattered topology τ, it is natural to ask about the separation properties of τ^+. In fact, for τ^+ we can infer a bit more separation than for an arbitrary scattered topology. Recall that a topological space X is T_1 if for any two different points $a, b \in X$ there is an open set U such that $a \in U$ and $b \notin U$.

Proposition 7. *Let (X, τ) be any topological space. Then (X, τ^+) is T_1.*

Proof Let $a, b \in X, a \neq b$. Consider the set $B := d_\tau(\{b\})$, which is open in τ^+. We either have $a \in B$ (and $b \notin B$ by definition) or a belongs to the complement of the closure of $\{b\}$. □

The following example shows that, in general, τ^+ need not always be Hausdorff.

Example 7. Let $(X, \prec)$ be a strict partial ordering on $X := \omega \cup \{a, b\}$, where ω is taken with its natural order, a and b are $\prec$-incomparable, and $n \prec a, b$ for all $n \in \omega$. Let τ be the left topology on $(X, \prec)$. Since $\prec$ is well-founded, τ is scattered.

Notice that for any $A \subseteq X$ we have $d_\tau(A) = \{x \in X : \exists y \in A\ y \prec x\}$. Hence, if A intersects ω, then $d_\tau(A)$ contains an end-segment of ω. Otherwise, $d_\tau(A) = \varnothing$. It follows that a base of open neighborhoods of a in τ^+ consists of sets of the form $I \cup \{a\}$, where I is an end-segment of ω. Similarly, sets of the form $I \cup \{b\}$ are a base of open neighborhoods of b. But any two such sets have a non-empty intersection.

10.7 d-Reflection

In the next section we are going to describe in some detail the GLP-space generated from the left topology on the ordinals. Strikingly, we will see that it naturally leads to some of the central notions of combinatorial set theory, such as Mahlo operation and stationary reflection. In fact, part of our analysis can be easily stated using the language of modal logic for arbitrary generated GLP-spaces. In this section we provide a necessary setup and characterize the topologies of a generated GLP-space in terms of what we call *d-reflection*.[5]

Throughout this section we fix a topological space (X, τ) and let $d = d_\tau$.

Definition 3 A point $a \in X$ is called *d-reflexive* if $a \in dX$ and, for each $A \subseteq X$,

$$a \in dA \Rightarrow a \in d(dA).$$

In modal logic terms this means that the formula $\Diamond\top \wedge (\Diamond p \to \Diamond\Diamond p)$ is valid at $a \in X$ for any evaluation of the variable p in (X, τ).

Similarly, a point $a \in X$ is called *m-fold d-reflexive* if $a \in dX$ and for each $A_1, \dots, A_m \subseteq X$,

$$a \in dA_1 \cap \cdots \cap dA_m \Rightarrow a \in d(dA_1 \cap \cdots \cap dA_m).$$

2-fold d-reflexive points will also be called *doubly d-reflexive* points. Expressed with the help of the modal language, $a \in X$ is doubly d-reflexive iff the formula $\Diamond\top \wedge (\Diamond p \wedge \Diamond q \to \Diamond(\Diamond p \wedge \Diamond q))$ is valid at a for any evaluation of p, q.

Lemma 10. *Let (X, τ) be a T_d-space. Each doubly d-reflexive point $x \in X$ is m-fold d-reflexive for any finite m.*

Proof The argument goes by induction on $m \geq 2$. Suppose $x \in dA_1 \cap \cdots \cap dA_{m+1}$, then $x \in dA_1 \cap \cdots \cap dA_m$ and $x \in dA_{m+1}$. By induction hypothesis, $x \in d(dA_1 \cap \cdots \cap dA_m)$ and by 2-fold reflection $x \in d(d(dA_1 \cap \cdots \cap dA_m) \cap dA_{m+1})$. However, by T_d property $d(dA_1 \cap \cdots \cap dA_m) \subseteq dA_1 \cap \cdots \cap dA_m$, hence $x \in d(dA_1 \cap \cdots \cap dA_m \cap dA_{m+1})$, as required. □

Proposition 8. *Let (X, τ) be a T_d-space. A point $x \in X$ is doubly d-reflexive iff x is a limit point of (X, τ^+).*

Proof For the (if) direction, we give an argument in the algebraic format. In fact, it is sufficient to show the following inequality in the algebra of (X, τ) for any elements $p, q \subseteq X$:

$$\langle 1\rangle\top \wedge \langle 0\rangle p \wedge \langle 0\rangle q \leq \langle 0\rangle(\langle 0\rangle p \wedge \langle 0\rangle q).$$

Notice that by Lemma 9, $\langle 1\rangle\top \wedge \langle 0\rangle p = \langle 1\rangle(\top \wedge \langle 0\rangle p) = \langle 1\rangle\langle 0\rangle p$. Hence, using P1$'$ once again, we obtain: $\langle 1\rangle\top \wedge \langle 0\rangle p \wedge \langle 0\rangle q = \langle 1\rangle\langle 0\rangle p \wedge \langle 0\rangle q = \langle 1\rangle(\langle 0\rangle p \wedge \langle 0\rangle q)$. The latter formula can be weakened to $\langle 0\rangle(\langle 0\rangle p \wedge \langle 0\rangle q)$ by P2, as required.

[5] Curiously, the reader may notice that the notion of *reflection principle* as used in provability logic and formal arithmetic matches very nicely the notions such as *stationary reflection* in set theory. (As far as we know, the two terms have evolved completely independently from one another.)

For the (only if) direction, it is sufficient to show that each doubly d-reflexive point of (X, τ) is a limit point of τ^+. Suppose x is doubly d-reflexive. By Lemma 10, x is m-fold d-reflexive. Any basic open subset of τ^+ has the form $U := A_0 \cap dA_1 \cap \cdots \cap dA_m$, where $A_0 \in \tau$. Assume $x \in U$, we have to find a point $y \neq x$ such that $y \in U$.

Since $x \in dA_1 \cap \cdots \cap dA_m$, by m-fold d-reflexivity we obtain $x \in d(dA_1 \cap \cdots \cap dA_m)$. Since A_0 is an open neighborhood of x, there is a $y \in A_0$ such that $y \neq x$ and $y \in dA_1 \cap \cdots \cap dA_m$. Hence, $y \in U$ and $y \neq x$, as required. □

Let d^+ denote the derivative operator associated with τ^+. We obtain the following characterization of derived topology in terms of neighborhoods.

Proposition 9. *Let (X, τ) be a T_d-space. A subset $U \subseteq X$ contains a τ^+-neighborhood of $x \in X$ iff one of the following two cases holds:*

(i) x is not doubly d-reflexive and $x \in U$;
(ii) x is doubly d-reflexive and there is an $A \in \tau$ and a B such that $x \in A \cap dB \subseteq U$.

Proof Since (i) ensures that x is τ^+-isolated by Proposition 8, each condition is clearly sufficient for U to contain a τ^+-neighborhood of x. To prove the converse, assume that U contains a τ^+-neighborhood of x. This means $x \in A \cap dA_1 \cap \cdots \cap dA_m \subseteq U$ for some $A, A_1, \ldots, A_m$ with $A \in \tau$. If x is τ^+-isolated, condition (i) holds. Otherwise, $x \in d^+X$. Let $B := dA_1 \cap \cdots \cap dA_m$. Since B is closed in τ we have $dB \subseteq B$, hence $A \cap dB \subseteq U$. It remains to show that $x \in A \cap dB$. By Lemma 9, $B \cap d^+X = d^+B \subseteq dB$. Hence, $x \in A \cap B \cap d^+X \subseteq A \cap dB$. □

Remark 1. Since in clause (ii) of Proposition 9 the set A is open, we have $A \cap dB = A \cap d(A \cap B)$ for any B. Hence, we may assume $B \subseteq A$.

Corollary 2. *Let (X, τ) be a T_d-space. Then, for all $x \in X$ and $A \subseteq X$, $x \in d^+A$ iff the following two conditions hold:*

(i) x is doubly d-reflexive;
(ii) For all $B \subseteq X$, $x \in dB \Rightarrow x \in d(A \cap dB)$.

Proof The fact that (i) and (ii) are necessary is proved using Proposition 8 and the inequality $d^+A \cap dB = d^+(A \cap dB) \subseteq d(A \cap dB)$. We prove that (i) and (ii) are sufficient. Assume $x \in U \in \tau^+$. By Proposition 9 we may assume that U has the form $V \cap dB$, where $V \in \tau$. By (ii), from $x \in dB$ we obtain $x \in d(A \cap dB)$. Hence, there is a $y \neq x$ such that $y \in V$ and $y \in A \cap dB$. It follows that $y \in A$ and $y \in V \cap dB = U$. ⊣ □

10.8 The Ordinal GLP-Space

Here we discuss the GLP-space generated from the left topology on the ordinals, that is, the GLP-space $(\Omega; \{\tau_n : n \in \omega\})$, where Ω is a fixed ordinal, τ_0 is the left topology on Ω and $\tau_{n+1} = \tau_n^+$ for each $n \in \omega$. The material in this section comes from a so far unpublished manuscript of the first author [10]. Our basic findings are summarized in the following table, to which we provide extended comments below.

The rows of the table correspond to topologies τ_n. The first column contains the name of the topology (the first two are standard, the third one is introduced in [14], the fourth one is introduced here). The second column indicates the first limit point of τ_n, which is denoted θ_n. The last column describes the derivative operator associated with τ_n. We note that θ_3 is a large cardinal which is sometimes referred to as *the first cardinal reflecting for pairs of stationary sets* (see below), but we know no special notation for this cardinal.

	Name	θ_n	$d_n(A)$
τ_0	Left	1	$\{\alpha : A \cap \alpha \neq \varnothing\}$
τ_1	Interval	ω	$\{\alpha \in Lim : A \cap \alpha$ is unbounded in $\alpha\}$
τ_2	Club	ω_1	$\{\alpha : \mathrm{cf}(\alpha) > \omega$ and $A \cap \alpha$ is stationary in $\alpha\}$
τ_3	Mahlo	θ_3	……

We have already seen that the derivative topology of the left topology is exactly the interval topology. Therefore, basic facts related to the first two rows of the table are rather clear. We turn to the next topology τ_2.

Club topology. Recall that the *cofinality* $\mathrm{cf}(\alpha)$ of a limit ordinal α is the least order type of a cofinal subset of α; $\mathrm{cf}(\alpha) := 0$ if $\alpha \notin \mathrm{Lim}$. (We use the words *cofinal in* α and *unbounded in* α as synonyms.) An ordinal α is *regular* if $\mathrm{cf}(\alpha) = \alpha$.

To characterize τ_2 we apply Proposition 9, hence it is useful to see what corresponds to the notion of doubly d-reflexive point of the interval topology.

Lemma 11. *For any ordinal α, α is d_1-reflexive iff α is doubly d_1-reflexive iff $\mathrm{cf}(\alpha) > \omega$.*

Proof d_1-reflexivity of α means that $\alpha \in \mathrm{Lim}$ and, for all $A \subseteq \alpha$, if A is cofinal in α, then $d_1(A)$ is cofinal in α. If $\mathrm{cf}(\alpha) = \omega$, then there is an increasing sequence $(\alpha_n)_{n\in\omega}$ such that $\sup\{\alpha_n : n \in \omega\} = \alpha$. Then, for $A := \{\alpha_n : n \in \omega\}$ we obviously have $d_1(A) = \{\alpha\}$, hence A violates the reflexivity property. Therefore, d_1-reflexivity of α implies $\mathrm{cf}(\alpha) > \omega$.

Now we show that $\mathrm{cf}(\alpha) > \omega$ implies α is doubly d_1-reflexive. Suppose $\mathrm{cf}(\alpha) > \omega$ and $A, B \subseteq \alpha$ are both cofinal in α. We show that $d_1 A \cap d_1 B$ is cofinal in α. Assume $\beta < \alpha$. Using the cofinality of A, B we can construct an increasing sequence $(\gamma_n)_{n\in\omega}$ above β such that $\gamma_n \in A$ for even n, and $\gamma_n \in B$ for odd n. Let $\gamma := \sup\{\gamma_n : n < \omega\}$. Obviously, both A and B are cofinal in γ whence $\gamma \in d_1 A \cap d_1 B$. Since $\mathrm{cf}(\alpha) > \omega$ and $\mathrm{cf}(\gamma) = \omega$, we have $\gamma < \alpha$. □

Corollary 3. *Limit points of τ_2 are exactly the ordinals of uncountable cofinality.*

It turns out that topology τ_2 is strongly related to the well-known concept of a *club filter*, i.e., the filter generated by all clubs on a limit ordinal. Recall that a subset $C \subseteq \alpha$ is called a *club* in α if C is closed in the interval topology of α and unbounded in α.

Proposition 10. *Assume* $\mathrm{cf}(\alpha) > \omega$. *The following statements are equivalent:*

(i) *U contains a* τ_2*-neighborhood of* α*;*
(ii) *There is a* $B \subseteq \alpha$ *such that* $\alpha \in d_1 B \subseteq U$*;*
(iii) $\alpha \in U$ *and U contains a club in* α*;*
(iv) $\alpha \in U$ *and* $U \cap \alpha$ *belongs to the club filter on* α.

Proof Statement (ii) implies (iii) since $\alpha \cap d_1 B$ is a club in α whenever $\alpha \in d_1 B$. Statement (iii) implies (iv) for obvious reasons.

Statement (iv) implies (i). If C is a club in α, then $C \cup \{\alpha\}$ contains a τ_2-neighborhood $d_1 C$ of α. Indeed, $d_1 C$ is τ_2-open, contains α, and $d_1 C \subseteq C \cup \{\alpha\}$ since C is τ_1-closed in α.

Statement (i) implies (ii). Assume U contains a τ_2-neighborhood of α. Since $\mathrm{cf}(\alpha) > \omega$, by Lemma 11 and Proposition 9 there is an $A \in \tau_1$ and a B_1 such that $\alpha \in A \cap d_1 B_1 \subseteq U$. Since A is a τ_1-neighborhood of α, by Proposition 9 again there are $A_0 \in \tau_0$ and B_0 such that $\alpha \in A_0 \cap d_0 B_0$. Since τ_0 is the left topology, we may assume that A_0 is the minimal τ_0-neighborhood $[0, \alpha]$ of α. Besides, we have $\alpha \in d_0 B_0 \cap d_1 B_1 = d_1(B_1 \cap d_0 B_0) \subseteq U$. Since $[0, \alpha]$ is τ_1-clopen, $d_1(C \cap \alpha) = [0, \alpha] \cap d_1 C$ for any C, so we can take $B_1 \cap d_0 B_0 \cap \alpha$ for B. □

Corollary 4. τ_2 *is the unique topology on* Ω *such that*

- *If* $\mathrm{cf}(\alpha) \leq \omega$, *then* α *is an isolated point;*
- *If* $\mathrm{cf}(\alpha) > \omega$, *then, for any* $U \subseteq \Omega$, *U contains a neighborhood of* α *iff* $\alpha \in U$ *and U contains a club in* α.

Hence, we may call τ_2 the *club topology.*

The derivative operation for the club topology is also well known in set theory. Recall the following definition for $\mathrm{cf}(\alpha) > \omega$.

A subset $A \subseteq \alpha$ is called *stationary in* α if A intersects every club in α. Observe that this happens exactly when α is a limit point of A in τ_2, so

$$d_2(A) = \{\alpha : \mathrm{cf}(\alpha) > \omega \text{ and } A \cap \alpha \text{ is stationary in } \alpha\}.$$

The map d_2 is usually called the *Mahlo operation* (see [41], where d_2 is denoted Tr). Its main significance is associated with the notion of Mahlo cardinal, one of the basic examples of large cardinals in set theory. Let Reg denote the class of regular cardinals; the ordinals in $d_2(\mathrm{Reg})$ are called *weakly Mahlo cardinals.* Their existence implies the consistency of ZFC, as well as the consistency of ZFC together with the assertion 'inaccessible cardinals exist.'

Now we turn to topology τ_3.

Stationary reflection and Mahlo topology. Since the open sets of τ_3 are generated by the Mahlo operation, we call τ_3 *Mahlo topology.* It turns out to be intrinsically connected with *stationary reflection*, an extensively studied phenomenon in set theory (see [32, Chaps. 1, 15]).

We adopt the following terminology. An ordinal λ is called *reflecting* if $\mathrm{cf}(\lambda) > \omega$ and, whenever A is stationary in λ, there is an $\alpha < \lambda$ such that $A \cap \alpha$ is stationary in

α. Similarly, λ is *doubly reflecting* if $\mathrm{cf}(\lambda) > \omega$ and whenever A, B are stationary in λ there is an $\alpha < \lambda$ such that both $A \cap \alpha$ and $B \cap \alpha$ are stationary in α.

Mekler and Shelah's notion of *reflection cardinal* [49] is somewhat more general than the one given here, however it has the same consistency strength. Reflection for pairs of stationary sets has been introduced by Magidor [47]. Since d_2 coincides with the Mahlo operation, we immediately obtain the following statement.

Proposition 11. *(i) λ is reflecting iff λ is d_2-reflexive;*
(ii) λ is doubly reflecting iff λ is doubly d_2-reflexive;
(iii) λ is a non-isolated point in τ_3 iff λ is doubly reflecting.

Together with the next proposition this yields a characterization of Mahlo topology in terms of neighborhoods.

Proposition 12. *Suppose λ is doubly reflecting. For any subset $U \subseteq \Omega$, the following conditions are equivalent:*

(i) U contains a τ_3-neighborhood of λ;
(ii) $\lambda \in U$ and there is a $B \subseteq \lambda$ such that $\lambda \in d_2B \subseteq U$;
(iii) $\lambda \in U$ and there is a τ_2-closed (in the relative topology of λ) stationary $C \subseteq \lambda$ such that $C \subseteq U$.

Notice that the notion of τ_2-closed stationary C in (iii) is the analog of the notion of club for the τ_2-topology.

Proof Condition (ii) implies (iii). Since λ is reflecting, if $\lambda \in d_2B$, then $\lambda \in d_2d_2B$, that is, $\lambda \cap d_2B$ is stationary in λ. So we may take $C := \lambda \cap d_2B$.

Condition (iii) implies (ii). If C is τ_2-closed and stationary in λ, then $d_2C \subseteq C \cup \{\lambda\} \subseteq U$ and $\lambda \in d_2C$. Thus, $\lambda \cap d_2C$ can be taken for B.

Condition (ii) implies (i). If (ii) holds, U contains a subset of the form d_2B. The latter is τ_3-open and contains λ, thus it is a neighborhood of λ.

For the converse direction, we note that by Proposition 9 U contains a subset of the form $A \cap d_2B$, where $A \in \tau_2$, $B \subseteq A$ and $\lambda \in A \cap d_2B$. Since A is a τ_2-neighborhood of λ, by Proposition 10 there is a set B_1 such that $\lambda \in [0, \lambda] \cap d_1B_1 \subseteq A$. Then

$$\lambda \in [0, \lambda] \cap d_1B_1 \cap d_2B = [0, \lambda] \cap d_2(B \cap d_1B_1).$$

Since $[0, \lambda]$ is clopen, we obtain $\lambda \in d_2C$ with $C := B \cap d_1B_1 \cap \lambda$. □

Reflecting and doubly reflecting cardinals are large cardinals in the sense that their existence implies consistency of ZFC. They have been studied by Mekler and Shelah [49] and Magidor [47] who investigated their consistency strength and related them to some other well-known large cardinals. By a result of Magidor, the existence of a doubly reflecting cardinal is equiconsistent with the existence of a *weakly compact cardinal*.[6] More precisely, the following proposition holds.

[6] Weakly compact cardinals are the same as Π^1_1-indescribable cardinals, see below.

Proposition 13. *(i) If λ is weakly compact, then λ is doubly reflecting.*
(ii) (Magidor) *If λ is doubly reflecting, then λ is weakly compact in L.*

Here, the first item is well known and easy. Magidor originally proved the analog of the second item for $\lambda = \aleph_2$ and stationary sets of ordinals of countable cofinality in $\aleph_2$. However, it has been remarked by Mekler and Shelah [49] that essentially the same proof yields the stated claim.[7]

Corollary 5. *Assertion "τ_3 is non-discrete" is equiconsistent with the existence of a weakly compact cardinal.*

Corollary 6. *If* ZFC *is consistent, then it is consistent with* ZFC *that τ_3 is discrete and hence that* **GLP**$_3$ *is incomplete w.r.t. any ordinal space.*

Recall that θ_n denotes the first non-isolated point of τ_n (in the space of all ordinals). We have: $\theta_0 = 1, \theta_1 = \omega, \theta_2 = \omega_1$, θ_3 is the first doubly reflecting cardinal.

ZFC does not know much about the location of θ_3, however the following facts are interesting.

- θ_3 is regular, but not a successor of a regular cardinal;
- While weakly compact cardinals are non-isolated, θ_3 need not be weakly compact: If infinitely many supercompact cardinals exist, then there is a model, where $\aleph_{\omega+1}$ is doubly reflecting [47];
- If θ_3 is a successor of a singular strong limit cardinal, then it is consistent that infinitely many Woodin cardinals exist, see [56].[8]

Further topologies. Further topologies of the ordinal GLP-space do not seem to have prominently occurred in set-theoretic work. They yield some large cardinal notions, for the statement that τ_n is non-discrete (equivalently, θ_n exists) implies the existence of a doubly reflecting cardinal for any $n > 2$. We do not know whether cardinals θ_n coincide with any of the standard large cardinal notions.

Here we give a sufficient condition for the topology τ_{n+2} to be non-discrete. We show that if there exists a *Π^1_n-indescribable cardinal*, then τ_{n+2} is non-discrete.

Let Q be a class of second order formulas over the standard first order set-theoretic language enriched by a unary predicate R. We assume Q to contain at least the class of all first order formulas (denoted Π^1_0). We shall consider standard models of that language of the form $(V_\alpha, \in, R)$, where α is an ordinal, V_α is the α-th class in the cumulative hierarchy, and R is a subset of V_α.

We would like to give a definition of Q-indescribable cardinals in topological terms. They can then be defined as follows.

Definition 4 For any sentence $\varphi \in Q$ and any $R \subseteq V_\kappa$, let $U_\kappa(\varphi, R)$ denote the set $\{\alpha \leq \kappa : (V_\alpha, \in, R \cap V_\alpha) \vDash \varphi\}$. The *$Q$-describable topology* τ_Q on Ω is generated by a subbase consisting of sets $U_\kappa(\varphi, R)$ for all $\kappa \in \Omega$, $\varphi \in Q$, and $R \subseteq V_\kappa$.

---

[7] The first author thanks J. Cummings for clarifying this.

[8] Stronger results have been announced, see [50].

As an exercise, the reader can check that the intervals $(\alpha, \kappa]$ are open in any τ_Q (consider $R = \{\alpha\}$ and $\varphi = \exists x\ (x \in R)$). The main strength of the Q-describable topology, however, comes from the fact that a second order variable R is allowed to occur in φ. So, all subsets of Ω that can be 'described' in this way are open in τ_Q.

Let d_Q denote the derivative operator for τ_Q. An ordinal $\kappa < \Omega$ is called *Q-indescribable* if it is a limit point of τ_Q. In other words, κ is Q-indescribable iff $\kappa \in d_Q(\Omega)$ iff $\kappa \in d_Q(\kappa)$.

It is not difficult to show that, whenever Q is any of the classes Π^1_n, the sets $U_\kappa(\varphi, R)$ actually form a base for τ_Q. Hence, our definition of Π^1_n-indescribable cardinals is equivalent to the standard one given in [42]: κ is Q-indescribable iff, for all $R \subseteq V_\kappa$ and all sentences $\varphi \in Q$,

$$(V_\kappa, \in, R) \vDash \varphi \Rightarrow \exists \alpha < \kappa\ (V_\alpha, \in, R \cap V_\alpha) \vDash \varphi.$$

It is well known that *weakly compact cardinals* coincide with the Π^1_1-indescribable ones (see [41]). From this it is easy to conclude that the Mahlo topology τ_3 is contained in $\tau_{\Pi^1_1}$. The following more general proposition was suggested to the first author by Philipp Schlicht (see [10]).

Proposition 14. *For any $n \geq 0$, τ_{n+2} is contained in $\tau_{\Pi^1_n}$.*

Proof We shall show that for each n, there is a Π^1_n-formula $\varphi_{n+1}(R)$ such that

$$\kappa \in d_{n+1}(A) \iff (V_\kappa, \in, A \cap \kappa) \vDash \varphi_{n+1}(R). \qquad (**)$$

This implies that for each $\kappa \in d_{n+1}(A)$, the set $U_\kappa(\varphi_{n+1}, A \cap \kappa)$ is a $\tau_{\Pi^1_n}$-open subset of $d_{n+1}(A)$ containing κ. Hence, each $d_{n+1}(A)$ is $\tau_{\Pi^1_n}$-open. Since τ_{n+2} is generated over τ_{n+1} by the open sets of the form $d_{n+1}(A)$ for various A, we have $\tau_{n+2} \subseteq \tau_{\Pi^1_n}$.

We prove (**) by induction on n. For $n = 0$, notice that $\kappa \in d_1(A)$ iff ($\kappa \in \mathrm{Lim}$ and $A \cap \kappa$ is unbounded in κ) iff

$$(V_\kappa, \in, A \cap \kappa) \vDash \forall \alpha\ \exists \beta\ (R(\beta) \wedge \alpha < \beta).$$

For the induction step recall that by Corollary 2, $\kappa \in d_{n+1}(A)$ iff

(i) κ is doubly d_n-reflexive;
(ii) $\forall Y \subseteq \kappa\ (\kappa \in d_n(Y) \to \exists \alpha < \kappa\ (\alpha \in A \wedge \alpha \in d_n(Y))$.

By the induction hypothesis, for some $\varphi_n(R) \in \Pi^1_{n-1}$, we have

$$\alpha \in d_n(A) \iff (V_\alpha, \in, A \cap \alpha) \vDash \varphi_n(R).$$

Hence, part (ii) is equivalent to

$$(V_\kappa, \in, A \cap \kappa) \vDash \forall Y \subseteq \mathrm{On}\ (\varphi_n(Y) \to \exists \alpha\ (R(\alpha) \wedge \varphi_n^{V_\alpha}(Y \cap \alpha))).$$

Here, φ^{V_α} means the relativization of all quantifiers in φ to V_α. We notice that V_α is first order definable, hence the complexity of $\varphi_n^{V_\alpha}$ remains in the class Π^1_{n-1}. So, the resulting formula is Π^1_n.

To treat part (i) we recall that $\kappa < \Omega$ is doubly d_n-reflexive iff $\kappa \in d_n(\Omega)$ and

$$\forall Y_1, Y_2 \subseteq \kappa \, (\kappa \in d_n(Y_1) \cap d_n(Y_2) \to \exists \alpha < \kappa \; \alpha \in d_n(Y_1) \cap d_n(Y_2)).$$

Similarly to the above, using the induction hypothesis this can be rewritten as a Π^1_n-formula. □

Corollary 7. *If there is a Π^1_n-indescribable cardinal $\kappa < \Omega$, then τ_{n+2} has a non-isolated point.*

Corollary 8. *If for each n there is a Π^1_n-indescribable cardinal $\kappa < \Omega$, then all τ_n are non-discrete.*

By the result of Magidor [47] we know that θ_3 need not be weakly compact in some models of ZFC (e.g. in a model, where $\theta_3 = \aleph_{\omega+1}$). Hence, in general, the condition of the existence of Π^1_n-indescribable cardinals is not a necessary one for the nontriviality of the topologies τ_{n+2}. However, Bagaria et al. [4] prove that in L the Π^1_n-indescribable cardinals coincide with the limit points of τ_{n+2}.

10.9 Topological Completeness Results for GLP

As in the case of the unimodal language (cf. Sect. 10.3), one can ask two basic questions: Is **GLP** complete w.r.t. the class of all GLP-spaces? Is **GLP** complete w.r.t. some fixed natural GLP-space?

In the unimodal case, both questions received positive answers due to Esakia and Abashidze–Blass, respectively. Now the situation is more complicated.

The first question was initially studied by Beklemishev et al. in [14], where only some partial results were obtained. It was proved that the bimodal system **GLB** is complete w.r.t. GLP_2-spaces of the form (X, τ, τ^+), where X is a well-founded partial ordering and τ is its left topology. A proof of this result was based on the Kripke model techniques coming from [11].

Already at that time it was clear that these techniques cannot be immediately generalized to GLP_3-spaces since the third topology τ^{++} on such orderings is sufficiently similar to the club topology. From the results of Blass [18] (see Theorem 10 below) it was known that some stronger set-theoretic assumptions would be needed to prove completeness w.r.t. such topologies. Moreover, without any large cardinal assumptions it was not even known whether a GLP-space with a non-discrete third topology could exist at all.

First examples of GLP-spaces in which all topologies are non-discrete are constructed in [5], where also the stronger fact of topological completeness of **GLP** w.r.t. the class of all (countable, Hausdorff) GLP-spaces is established.

Theorem 9. *(i)* $\mathrm{Log}(\mathscr{C}) = \mathbf{GLP}$, *where* $\mathscr{C}$ *is the class of all GLP-spaces.*
(ii) There is a countable Hausdorff GLP-space X *such that* $\mathrm{Log}(X) = \mathbf{GLP}$.

In fact, X is the ordinal $\varepsilon_0 = \sup\{\omega, \omega^\omega, \omega^{\omega^\omega}, \ldots\}$ equipped with a sequence of topologies refining the interval topology. However, these topologies cannot be first-countable and are, in fact, defined using non-constructive methods such as Zorn's lemma.[9] In this sense, it is not an example of a *natural* GLP-space. The proof of this theorem introduces the techniques of maximal and limit-maximal extensions of scattered spaces. It falls outside the present survey (see [5]).

The question whether **GLP** is complete w.r.t. some natural GLP-space is still open. Some partial results concerning the GLP-space generated from the interval topology on the ordinals (in the sense of the plus operation) are described below. Here, we call this space the *ordinal GLP-space*. (The space described in Sect. 10.8 is not an exact model of **GLP** as the left topology validates the linearity axiom.)

As we know from Corollary 6, it is consistent with ZFC that the Mahlo topology is discrete. Hence, it is consistent that **GLP** is incomplete w.r.t. the ordinal GLP-space. However, is it consistent with ZFC that **GLP** is complete w.r.t. the ordinal GLP-space? To this question we do not know a full answer. A pioneering work has been done by Blass [18] who studied the question of completeness of the Gödel–Löb logic **GL** w.r.t. a semantics equivalent to the topological interpretation w.r.t. the *club topology* τ_2. He used the language of filters rather than that of topological spaces as is more common in set theory.

Theorem 10. (Blass)

(i) If $V = L$ *and* $\Omega \geq \aleph_\omega$*, then* **GL** *is complete w.r.t.* (Ω, τ_2).
(ii) If there is a weakly Mahlo cardinal, there is a model of ZFC *in which* **GL** *is incomplete w.r.t.* (Ω, τ_2) *for any* Ω.

A corollary of (i) is that the statement "**GL** is complete w.r.t. τ_2" is consistent with ZFC (provided ZFC is consistent). In fact, instead of $V = L$ Blass used the so-called *square principle* for all $\aleph_n$, $n < \omega$, which holds in L by the results of Ronald Jensen. A proof of (i) is based on an interesting combinatorial construction using the techniques of splitting stationary sets.

A proof of (ii) is much easier. It uses a model of Harrington and Shelah in which $\aleph_2$ is reflecting for stationary sets of ordinals of countable cofinality [35]. Assuming Mahlo cardinals exist, they have shown that the following statement holds in some model of ZFC:

> If S is a stationary subset of $\aleph_2$ such that $\forall \alpha \in S\ \mathrm{cf}(\alpha) = \omega$, then there is a $\beta < \alpha$ (of cofinality ω_1) such that $S \cap \beta$ is stationary in β.

In fact, this statement can be expressed in the language of modal logic. First, we remark that this principle implies its generalization to all ordinals λ of cofinality $\aleph_2$ (consider an increasing continuous function mapping $\aleph_2$ to a club in λ). Second, we

[9] It seems to be interesting to study the question of topological completeness of **GLP** in the absence of the full axiom of choice, possibly with the axiom of determinacy.

remark that for the club topology the formula $\Diamond^n\top$ represents the class of ordinals of cofinality at least $\aleph_n$. This is a straightforward generalization of Lemma 11. Thus, the formula $\Box^3\bot\wedge\Diamond^2\top$ represents the subclass of Ω consisting of ordinals of cofinality ω_2.

Hence, the above reflection principle amounts to the validity of the following modal formula:

$$\Box^3\bot\wedge\Diamond^2\top\wedge\Diamond(p\wedge\Box\bot)\rightarrow\Diamond^2(p\wedge\Box\bot). \qquad (*)$$

In fact, if the antecedent is valid in λ, then $\mathrm{cf}(\lambda)=\omega_2$ and the interpretation of $p\wedge\Box\bot$ is a set S consisting of ordinals of countable cofinality such that $S\cap\lambda$ is stationary in λ. The consequent just states that this set reflects. Thus, formula $(*)$ is valid in (Ω,τ_2) for any Ω. Since this formula is clearly not provable in **GL**, the topological completeness fails for (Ω,τ_2).

Thus, Blass managed to give an exact consistency strength of the statement "**GL** is incomplete w.r.t. τ_2".

Corollary 9. *"**GL** is incomplete w.r.t. τ_2" is consistent iff it is consistent that Mahlo cardinals exist.*

It is possible to generalize these results to the case of bimodal logic **GLB** [12]. The situation remains essentially unchanged, although a proof of Statement (i) of Theorem 10 needs considerable adaptation.

Theorem 11. *If $V=L$ and $\Omega\geq\aleph_\omega$, then **GLB** is complete w.r.t. $(\Omega;\tau_1,\tau_2)$.*

10.10 Topologies for the Variable-Free Fragment of GLP

A natural topological model for the variable-free fragment of **GLP** has been introduced by Icard [38]. It is not a GLP-space and thus it is not a model of the full **GLP** (nor even of **GLB**). However, it is sound and complete for the variable-free fragment of **GLP**. It gives a convenient tool for the study of this fragment, which plays an important role in proof-theoretic applications of the polymodal provability logic. Here we give a simplified presentation of Icard's polytopological space.

Let Ω be an ordinal and let $\ell:\Omega\rightarrow\Omega$ denote the rank function for the interval topology on Ω (see Example 1). We define $\ell^0(\alpha)=\alpha$ and $\ell^{k+1}(\alpha)=\ell\ell^k(\alpha)$.

Icard's topologies υ_n, for each $n\in\omega$, are defined as follows. Let υ_0 be the left topology, and let υ_n be generated by υ_0 and all sets of the form

$$U^m_\beta:=\{\alpha\in\Omega:\ell^m(\alpha)>\beta\}$$

for $m<n$ and $\beta<\Omega$.

Clearly, υ_n is an increasing sequence of topologies. In fact, υ_1 is the interval topology. We let d_n and ρ_n denote the derivative operator and the rank function for υ_n, respectively. We have the following characterizations.

Lemma 12. *(i)* $\ell : (\Omega, \upsilon_{n+1}) \to (\Omega, \upsilon_n)$ *is a d-map;*
(ii) υ_{n+1} *is the coarsest topology* ν *on* Ω *such that* ν *contains the interval topology and* $\ell : (\Omega, \nu) \to (\Omega, \upsilon_n)$ *is continuous;*
(iii) ℓ^n *is the rank function of* υ_n*, that is,* $\rho_n = \ell^n$*;*
(iv) υ_{n+1} *is generated by* υ_n *and* $\{d_n^{\alpha+1}(\Omega) : \alpha < \rho_n(\Omega)\}$.

Proof (i) The map $\ell : (\Omega, \upsilon_{n+1}) \to (\Omega, \upsilon_n)$ is continuous. In fact, $\ell^{-1}[0, \beta)$ is open in the interval topology υ_1 since $\ell : (\Omega, \upsilon_1) \to (\Omega, \upsilon_0)$ is its rank function, hence a d-map. Also, if $m < n$, then $\ell^{-1}(U_\beta^m) = U_\beta^{m+1}$, hence it is open in υ_{n+1}.

The map ℓ is open. Notice that υ_{n+1} is generated by υ_1 and some sets of the form $\ell^{-1}(U)$, where $U \in \upsilon_n$. A base of υ_{n+1} consists of sets of the form $V \cap \ell^{-1}(U)$ for some $V \in \upsilon_1$ and $U \in \upsilon_n$. We have $\ell(V \cap \ell^{-1}(U)) = \ell(V) \cap U$. $\ell(V)$ is υ_0-open since $\ell : (\Omega, \upsilon_1) \to (\Omega, \upsilon_0)$ is a d-map and $V \in \upsilon_1$. Hence, the image of any basic open in υ_{n+1} is open in υ_n.

The map ℓ is pointwise discrete since $\ell^{-1}\{\alpha\}$ is discrete in the interval topology υ_1, hence in υ_{n+1}.

(ii) By (i), $\ell : (\Omega, \upsilon_{n+1}) \to (\Omega, \upsilon_n)$ is continuous, hence $\nu \subseteq \upsilon_{n+1}$. On the other hand, if $\ell : (\Omega, \nu) \to (\Omega, \upsilon_n)$ is continuous, then $\ell^{-1}(U_\beta^m) \in \nu$ for each $m < n$. Therefore, $U_\beta^m \in \nu$ for all m such that $1 \leq m \leq n$. Since ν also contains the interval topology, we have $\upsilon_{n+1} \subseteq \nu$.

(iii) By (i), we have that $\rho_n \circ \ell$ is a d-map from (Ω, υ_{n+1}) to (Ω, υ_0). Hence, it coincides with the rank function for υ_{n+1}, $\rho_{n+1} = \rho_n \circ \ell$. The claim follows by an easy induction on n.

(iv) By (iii),

$$d_n^{\beta+1}(\Omega) = \{\alpha \in \Omega : \rho_n(\alpha) > \beta\} = \{\alpha \in \Omega : \ell^n(\alpha) > \beta\} = U_\beta^n.$$

Obviously, υ_{n+1} is generated by υ_n and U_β^n for all β. Hence, the claim. □

We call an *Icard space* a polytopological space of the form $(\Omega; \upsilon_0, \upsilon_1, \ldots)$. Icard originally considered just $\Omega = \varepsilon_0$. We are going to give an alternative proof of the following theorem [38].

Theorem 12. (Icard) *Let* φ *be a variable-free* **GLP***-formula.*

(i) If **GLP** $\vdash \varphi$*, then* $(\Omega; \upsilon_0, \upsilon_1, \ldots) \vDash \varphi$.
(ii) If $\Omega \geq \varepsilon_0$ *and* **GLP** $\nvdash \varphi$*, then* $(\Omega; \upsilon_0, \upsilon_1, \ldots) \nvDash \varphi$.

Proof Within this proof we abbreviate $(\Omega; \upsilon_0, \upsilon_1, \ldots)$ by Ω. To prove part (i) we first remark that all topologies υ_n are scattered, hence all axioms of **GLP** except for P1 are valid in Ω. Moreover, $\mathrm{Log}(\Omega)$ is closed under the inference rules of **GLP**. Thus, we only have to show that the variable-free instances of axiom P1 are valid in Ω. This is sufficient because any derivation of a variable-free formula in **GLP** can be replaced by a derivation in which only the variable-free formulas occur (replace all the variables by the constant $\top$).

Let φ be a variable-free formula. We denote by φ^* its uniquely defined interpretation in Ω. The validity of an instance of P1 for φ amounts to the fact that $d_m(\varphi^*)$ is open in υ_n, whenever $m < n$. Thus, we have to prove the following proposition.□

Proposition 15. *For any variable-free formula φ, $d_n(\varphi^*)$ is open in υ_{n+1}.*

Let φ^+ denote the result of replacing in φ each modality $\langle n\rangle$ by $\langle n+1\rangle$. We need the following auxiliary claim.

Lemma 13. *If φ is variable-free, then $\ell^{-1}(\varphi^*) = (\varphi^+)^*$.*

Proof This goes by induction on the build-up of φ. The cases of constants and boolean connectives are easy. Suppose $\varphi = \langle n\rangle\psi$. We notice that since $\ell : (\Omega, \upsilon_{n+1}) \to (\Omega, \upsilon_n)$ is a d-map, we have $\ell^{-1}(d_n(A)) = d_{n+1}(\ell^{-1}(A))$ for any $A \subseteq \Omega$. Therefore, $\ell^{-1}(\varphi^*) = \ell^{-1}(d_n(\psi^*)) = d_{n+1}(\ell^{-1}(\psi^*)) = d_{n+1}((\psi^+)^*) = (\varphi^+)^*$, as required. □

We prove Proposition 15 in two steps. First, we show that it holds for a subclass of variable-free formulas called *ordered formulas*. Then we show that any variable-free formula is equivalent in Ω to an ordered one.

A formula φ is called *ordered* if no modality $\langle m\rangle$ occurs within the scope of $\langle n\rangle$ in φ for any $m < n$. The *height of* φ is the index of its maximal modality.

Lemma 14. *If $\langle n\rangle\varphi$ is ordered, then $d_n(\varphi^*)$ is open in υ_{n+1}.*

Proof This goes by induction on the height of $\langle n\rangle\varphi$. If it is 0, then $n = 0$. If $n = 0$, the claim is obvious since $d_0(A)$ is open in υ_1 for any $A \subseteq \Omega$. If $n > 0$, since $\langle n\rangle\varphi$ is ordered, we observe that $\langle n\rangle\varphi$ has the form $(\langle n-1\rangle\psi)^+$ for some ψ. The height of $\langle n-1\rangle\psi$ is less than that of $\langle n\rangle\varphi$. Hence, by the induction hypothesis, $(\langle n-1\rangle\psi)^* \in \upsilon_n$. Since $\ell : (\Omega, \upsilon_{n+1}) \to (\Omega, \upsilon_n)$ is continuous, we conclude that $\ell^{-1}(\langle n-1\rangle\psi)^*$ is open in υ_{n+1}. By Lemma 13, this set coincides with $(\langle n\rangle\varphi)^* = d_n(\varphi^*)$. □

Lemma 15. *Any variable-free formula φ is equivalent in Ω to an ordered one.*

Proof We argue by induction on the complexity of φ. The cases of boolean connectives and constants are easy. Suppose φ has the form $\langle n\rangle\psi$, where we may assume ψ to be in disjunctive normal form $\psi = \bigvee_i \bigwedge_j \pm\langle n_{ij}\rangle\psi_{ij}$. By the induction hypothesis, we may assume all the subformulas $\langle n_{ij}\rangle\psi_{ij}$ (and ψ itself) are ordered. Since $\langle n\rangle$ commutes with disjunction, it will be sufficient to show that for each i the formula $\theta_i := \langle n\rangle \bigwedge_j \pm\langle n_{ij}\rangle\psi_{ij}$ can be ordered.

By Lemma 14 each set $(\langle n_{ij}\rangle\psi_{ij})^*$ is open in υ_n whenever $n_{ij} < n$. Being a derived set, it is also closed in $\upsilon_{n_{ij}}$ and hence in υ_n. Thus, all such sets are clopen.

If U is open, then $d(A \cap U) = d(A) \cap U$ for any topological space. In particular, for any $A \subseteq \Omega$ and $n_{ij} < n$, $d_n(A \cap (\pm\langle n_{ij}\rangle\psi_{ij})^*) = d_n(A) \cap (\pm\langle n_{ij}\rangle\psi_{ij})^*$. This allows us to bring all the conjuncts $\pm\langle n_{ij}\rangle\psi_{ij}$ from under the $\langle n\rangle$ modality in θ_i. The resulting conjunction is ordered. □

This concludes the proof of Proposition 15 and thereby of Part (i).

A variable-free formula A is called a *word* if it is built-up from $\top$ only using connectives of the form $\langle n\rangle$ for any $n \in \omega$. We write $A \vdash B$ for **GLP** $\vdash A \to B$.

To prove Part (ii), we shall rely on the following fundamental lemma about the variable-free fragment of **GLP**. For a proof of this lemma we refer to [6, 8].

Lemma 16. *(i) Every variable-free formula is equivalent in* **GLP** *to a boolean combination of words;*
(ii) For any words A and B, either $A \vdash \langle 0\rangle B$, *or* $B \vdash \langle 0\rangle A$, *or A and B are equivalent;*
(iii) Conjunction of words is equivalent to a word.

We prove Part (ii) of Theorem 12 in a series of lemmas. First, we show that any word is true at some point in Ω provided $\Omega \geq \varepsilon_0$.

Lemma 17. *For any word* A, $\varepsilon_0 \in A^*$.

Proof We know that $\rho_n(\varepsilon_0) = \ell^n(\varepsilon_0) = \varepsilon_0$. Hence, $\varepsilon_0 \in d_n(\Omega)$ for each n. Assume n exceeds all the indices of modalities in A and $A = \langle m\rangle B$. By Proposition 15 the set B^* is open in υ_n. By the induction hypothesis $\varepsilon_0 \in B^*$. Hence, $\varepsilon_0 \in d_n(B^*) \subseteq d_m(B^*) = A^*$. This proves the claim. □

Applying this lemma to the word $\langle 0\rangle A$ we obtain the following corollary.

Corollary 10. *For every word A, there is an* $\alpha < \varepsilon_0$ *such that* $\alpha \in A^*$.

Let $\min(A^*)$ denote the least ordinal $\alpha \in \Omega$ such that $\alpha \in A^*$.

Lemma 18. *For any words A, B, if* $A \nvdash B$, *then* $\min(A^*) \notin B^*$.

Proof If $A \nvdash B$, then, by Lemma 16 (ii), $B \vdash \langle 0\rangle A$. Therefore, by the soundness of **GLP** in Ω, $B^* \subseteq d_0(A^*)$. It follows that for each $\beta \in B^*$ there is an $\alpha \in A^*$ such that $\alpha < \beta$. Thus, $\min(A^*) \notin B^*$. □

Now we are ready to prove Part (ii). Assume φ is variable-free and **GLP** $\nvdash \varphi$. By Lemma 16 (i) we may assume that φ is a boolean combination of words. Writing φ in conjunctive normal form we observe that it is sufficient to prove the claim only for formulas φ of the form $\bigwedge_i A_i \to \bigvee_j B_j$, where A_i and B_j are words. Moreover, $\bigwedge_i A_i$ is equivalent to a single word A.

Since **GLP** $\nvdash \varphi$ we have $A \nvdash B_j$ for each j. Let $\alpha = \min(A^*)$. By Lemma 18 we have $\alpha \notin B_j^*$ for each j. Hence, $\alpha \notin (\bigvee_j B_j)^*$ and $\alpha \notin \varphi^*$. This means that $\Omega \nvDash \varphi^*$.

10.11 Further Results

Topological semantics of polymodal provability logic has been extended to the language with transfinitely many modalities. A logic $\mathbf{GLP}_\Lambda$ having modalities $[\alpha]$ for all ordinals $\alpha < \Lambda$ is introduced in [8]. It was intended for the proof-theoretic analysis of predicative theories and is currently being actively investigated for that purpose.

David Fernandez and Joost Joosten undertook a thorough study of the variable-free fragment of that logic mostly in connection with the arising ordinal notation systems (see [25, 27] for a sample). In particular, they found a suitable generalization of Icard's polytopological space and showed that it is complete for that fragment [26]. Fernandez [30] also proved topological completeness of the full $\mathbf{GLP}_\Lambda$ by generalizing the results of [5].

The ordinal GLP-space is easily generalized to transfinitely many topologies $(\tau_\alpha)_{\alpha<\Lambda}$ by letting τ_0 be the left topology, $\tau_{\alpha+1} := \tau_\alpha^+$ and, for limit ordinals λ, τ_λ be the topology generated by all τ_α such that $\alpha < \lambda$. This space is a natural model of $\mathbf{GLP}_\Lambda$ and has been studied quite recently by Bagaria [3] and further by Bagaria et al. [4]. In particular, the three authors proved that in L the limit points of τ_{n+2} are Π^1_n-indescribable cardinals. The question posed in [14] whether the non-discreteness of τ_{n+2} is equiconsistent with the existence of Π^1_n-indescribable cardinals still appears to be open.

Acknowledgments We wish to thank the referee for many useful comments, which helped to significantly improve the readability of the paper. Thanks are also due to Guram Bezhanishvili both for his detailed comments and his patience with the slow pace this chapter was taking.

The first author was supported by the Russian Foundation for Basic Research (RFBR), Russian Presidential Council for Support of Leading Scientific Schools, and the Swiss–Russian cooperation project STCP–CH–RU "Computational proof theory".

The second author was supported by the Shota Rustaveli National Science Foundation grant #FR/489/5-105/11 and the French–Georgian grant CNRS–SRNSF #4135/05-01.

References

1. Abashidze M (1985) Ordinal completeness of the Gödel-Löb modal system. In: Intensional logics and the logical structure of theories. Metsniereba, Tbilisi, pp 49–73 (Russian)
2. Artemov S, Beklemishev L (2004) Provability logic. In: Gabbay D, Guenthner F (eds) Handbook of philosophical logic, vol 13, 2nd edn. Kluwer, Dordrecht, pp 229–403
3. Bagaria J (2012) Topologies on ordinals and the completeness of polymodal provability logic (in preparation)
4. Bagaria J, Magidor M, Sakai H (2012) Reflection and indescribability in the constructible universe. Manuscript, To appear in Israel Journal of Mathematics
5. Beklemishev L, Gabelaia D (2013) Topological completeness of the provability logic glp. Ann Pure Appl Logic 128(12):1201–1223
6. Beklemishev L (2004) Provability algebras and proof-theoretic ordinals, i. Ann Pure Appl Logic 128:103–123
7. Beklemishev L (2005) Reflection principles and provability algebras in formal arithmetic. Uspekhi Matematicheskikh Nauk 60(2):3–78 (English trans: Russ Math Surv 60(2):197–268)
8. Beklemishev L (2005) Veblen hierarchy in the context of provability algebras. In: Hájek P, Valdés-Villanueva L, Westerståhl D (eds) Proceedings of the 12th international congress on logic, methodology and philosophy of science. Kings College Publications, London, pp 65–78
9. Beklemishev L (2006) The Worm principle. In: Chatzidakis Z, Koepke P, Pohlers W (eds) Lecture notes in logic 27. Logic Colloquium '02. AK Peters, pp 75–95
10. Beklemishev L (2009) On GLP-spaces. Manuscript. http://www.mi.ras.ru/~bekl/Papers/glp-sp.pdf
11. Beklemishev L (2010) Kripke semantics for provability logic glp. Ann Pure Appl Logic 161:756–774
12. Beklemishev L (2011) Ordinal completeness of bimodal provability logic GLB. In: Bezhanishvili N et al (ed) Logic, language and computation, TbiLLC 2009. Lecture notes in artificial intelligence, number 6618, pp 1–15

13. Beklemishev L (2011) A simplified proof of the arithmetical completeness theorem for the provability logic GLP. Trudy Matematicheskogo Instituta imeni V.A.Steklova, 274(3):32–40 (English trans: Proc Steklov Inst Math 274(3):25–33)
14. Beklemishev L, Bezhanishvili G, Icard T (2010) On topological models of GLP. In: Schindler R (ed) Ways of proof theory. Ontos Verlag, Heusendamm bei Frankfurt, Germany, pp 133–152
15. Beklemishev L, Joosten J, Vervoort M (2005) A finitary treatment of the closed fragment of japaridze's provability logic. J Logic Comput 15(4):447–463
16. Bezhanishvili G, Esakia L, Gabelaia D (2005) Some results on modal axiomatization and definability for topological spaces. Studia Logica 81:325–355
17. Bezhanishvili G, Morandi P (2010) Scattered and hereditarily irresolvable spaces in modal logic. Archive for Mathematical Logic 49(3):343–365
18. Blass A (1990) Infinitary combinatorics and modal logic. J Symb Logic 55(2):761–778
19. Block W, Pigozzi D (1989) Algebraizable logics. Memoirs of the AMS 77(396)
20. Boolos G (1979) The unprovability of consistency: an essay in modal logic. Cambridge University Press, Cambridge
21. Boolos G (1993) The analytical completeness of dzhaparidze's polymodal logics. Ann Pure Appl Logic 61:95–111
22. Boolos G (1993) The logic of provability. Cambridge University Press, Cambridge
23. Carlucci L (2005) Worms, gaps and hydras. Math Logic Q 51(4):342–350
24. Chagrov A, Zakharyaschev M (1997) Modal logic. Clarendon Press, Oxford
25. Fernández-Duque D, Joosten J (2012) Kripke models of transfinite provability logic. Adv Modal Logic 9:185–199
26. Fernández-Duque D, Joosten J (2013) Models of transfinite provability logic. J Symb Logic 78(2):543–561
27. Fernández-Duque D, Joosten J (2012) Turing progressions and their well-orders. In: How the world computes. Lecture notes in computer science. Springer, Berlin, pp 212–221
28. Esakia L (1981) Diagonal constructions, löb's formula and cantor's scattered spaces. Studies in logic and semantics. Metsniereba, Tbilisi, pp 128–143 (Russian)
29. Feferman S, Dawson J, Kleene S, Moore G, Solovay R, van Heijenoort J (eds) (1996) Kurt Gödel collected works, vol 1, Publications 1929–1936. Oxford University Press, Oxford
30. Fernández-Duque D (2012) The polytopologies of transfinite provability logic. Preprint arXiv:1207.6595v2 [math.LO]
31. Font J, Jansana R, Pigozzi D (2003) A survey of abstract algebraic logic. Studia Logica 74(1–2):13–97
32. Foreman M, Kanamori A (eds) (2010) Handbook of set theory, vol 1–3. Springer, Berlin
33. Gödel K (1933) Eine Interpretation des intuitionistischen Aussagenkalkuls. Ergebnisse Math. Kolloq., 4:39–40 (English trans: [29], pp 301–303)
34. Gurevich Y (1985) Monadic second-order theories. In Barwise J, Feferman S (eds) Model-theoretical logics. Springer, Berlin, pp 479–506
35. Harrington L, Shelah S (1985) Some exact equiconsistency results in set theory. Notre Dame J Formal Logic 26:178–188
36. Hilbert D, Bernays P (1968) Grundlagen der Mathematik, vols I and II, 2d edn. Springer, Berlin
37. Icard T, Joosten J (2012) Provability and interpretability logics with restricted realizations. Notre Dame J Formal Logic 53(2):133–154
38. Icard T III (2011) A topological study of the closed fragment of glp. J Logic Comput 21(4):683–696
39. Ignatiev K (1993) On strong provability predicates and the associated modal logics. J Symb Logic 58:249–290
40. Japaridze G (1987) Modal-logical means of studying provability. PhD thesis, Moscow (Russian)
41. Jech T (2002) Set theory: the third millenium edition. Springer, Berlin
42. Kanamori A (2009) The higher infinite, 2nd edn. Springer, Berlin
43. Kudinov A, Shehtman V (2014) Derivational modal logics with the difference modality (this volume)
44. Löb M (1955) Solution of a problem of leon henkin. J Symb Logic 20:115–118

45. Macintyre A, Simmons H (1973) Gödel's diagonalization technique and related properties of theories. Colloq Math 28:165–180
46. Magari R (1975) The diagonalizable algebras (the algebraization of the theories which express Theor.:II). Bollettino della Unione Matematica Italiana, Serie 4, 12, 1975. Suppl Fasc 3, 117–125
47. Magidor M (1982) Reflecting stationary sets. J Symb Logic 47:755–771
48. McKinsey J, Tarski A (1944) The algebra of topology. Ann Math 45:141–191
49. Mekler A, Shehlah S (1989) The consistency strength of "every stationary set reflects". Israel J Math 67(3):353–366
50. Sargsyan G (2012) On the strength of PFA, I. http://tinyurl.com/bmknyyp
51. Segerberg K (1971) An essay in classical modal logic. Filosofiska Föreningen och Filosofiska Institutionen vid Uppsala Universitet, Uppsala
52. Semadeni Z (1971) Banach spaces of continuous functions. In: Monografie matematyczne, Tom 55. PWN—Polish Scientific Publishers, Warsaw
53. Simmons H (1975) Topological aspects of suitable theories. Proc Edinb Math Soc 2(19):383–391
54. Smoryński C (1985) Self-reference and modal logic. Springer, Berlin
55. Solovay R (1976) Provability interpretations of modal logic. Israel J Math 28:33–71
56. Steel J (2005) Pfa implies $ad^{l(\searrow)}$. J Symb Logic 70(4):1255–1296
57. Steinsvold C (2007) Topological models of belief logics. PhD thesis, New York
58. Visser A (1984) The provability logics of recursively enumerable theories extending Peano arithmetic at arbitrary theories extending Peano arithmetic. J Philos Logic 13:97–113

Chapter 11
Derivational Modal Logics with the Difference Modality

Andrey Kudinov and Valentin Shehtman

In memory of Leo Esakia

Abstract In this chapter we study modal logics of topological spaces in the combined language with the derivational modality and the difference modality. We give axiomatizations and prove completeness for the following classes: all spaces, T_1-spaces, dense-in-themselves spaces, a zero-dimensional dense-in-itself separable metric space, $\mathbf{R}^n$ ($n \geq 2$). We also discuss the correlation between languages with different combinations of the topological, derivational, universal and difference modalities in terms of definability.

Keywords Modal logic · Topological semantics · Derivational modality · Universal modality · Difference modality

A. Kudinov (✉) · V. Shehtman
Institute for Information Transmission Problems, Russian Academy of Sciences, Moscow, Russia

National Research University Higher School of Economics, Moscow, Russia
e-mail: kudinov@iitp.ru

A. Kudinov
Moscow Institute of Physics and Technology, Moscow, Russia

V. Shehtman
Moscow State University, Moscow, Russia
e-mail: shehtman@netscape.net

G. Bezhanishvili (ed.), *Leo Esakia on Duality in Modal and Intuitionistic Logics*, Outstanding Contributions to Logic 4, DOI: 10.1007/978-94-017-8860-1_11,

11.1 Introduction

Topological modal logic was initiated by the works of A. Tarski and J.C.C. McKinsey in the 1940s. They were first to consider both topological interpretations of the diamond modality: one as closure, and another as derivative

Their studies of closure modal logics were rather detailed and profound. In particular, in the fundamental paper [32] they have shown that the logic of any metric separable dense-in-itself space is **S4**. This remarkable result also demonstrates a relative weakness of the closure operator to distinguish between interesting topological properties.

The derivational interpretation gives more expressive power. For example, the real line can be distinguished from the real plane (the observation made by K. Kuratowski as early as in 1920s, cf. [27]); the real line can be distinguished from the rational line [36]; T_0 and T_D separation axioms become expressible [8, 15]. In [32] McKinsey and Tarski only gave basic definitions for derivational modal logics and stated several open problems which were solved much later.

The derivational semantics also has its limitations (for example, it is still impossible to distinguish $\mathbf{R}^2$ from $\mathbf{R}^3$). Further increase of expressive power can be provided by the well-known methods of adding universal or difference modalities [18, 19]. In the context of topological semantics this approach also has proved fruitful—for example, connectedness is expressible in modal logic with the closure and universal modalities [37], and the T_1 separation axiom in modal logic with the closure and difference modalities [17, 22].

Until the early 1990s, when the connections between topological modal logic and Computer Science were established, the interest in that subject was moderate. Leo Esakia was one of the enthusiasts of the modal logic approach to topology, and he was probably the first to appreciate the role of the derivational modality, in particular, in modal logics of provability [14]. Another strong motivation for further studies of derivational modal logics ('d-logics') were the axiomatization problems left open in [32].[1] In recent years d-logics have been studied rather intensively, a brief summary of results can be found in Sect. 11.3.

In this chapter the first thorough investigation is provided for logics in the most expressive language in this context,[2] namely the derivational modal logics with the difference modality ('dd-logics'). It unifies earlier studies by the first author in closure modal logics with the difference modality ('cd-logics') and by the second author in d-logics.

The diagram in Sect. 11.12 compares the expressive power of different kinds of topomodal logics. Our conjecture is that dd-logics are strictly more expressive than the others, but it is still an open question if the dd-language is stronger than the cd-language. Speaking informally, it is more convenient—for example, Kura-

[1] The early works of the second author in this field were greatly influenced by Leo Esakia.

[2] Some other kinds of topomodal logics arise when we deal with topological spaces with additional structures, e.g. spaces with two topologies, spaces with a homeomorphism etc. (cf. [3]).

towski's axiom for $\mathbf{R}^2$ (Definition 21) is expressible in cd-logic as well, but in a more complicated form [23].

We show that still in many cases properties of dd-logics are similar to those of d-logics: finite axiomatizability, decidability and the finite model property (fmp). Besides specific results characterizing logics of some particular spaces, our goal was to propose some general methods. In fact, nowadays in topomodal logic there are many technical proofs, but few general methods. In this chapter we propose only two simplifying novelties—dd-morphisms (Sect. 11.6) and the Glueing lemma 16, but we hope that much more can be done in this direction, cf. the recent paper [20].

In more detail, the plan of the chapter is as follows. Preliminary Sects. 11.2, 11.3 and 11.4 include standard definitions and basic facts about modal logics and their semantics. Some general completeness results for dd-logics can be found in Sects. 11.5 and 11.7. In Sect. 11.5 we show that every extension of the minimal logic $\mathbf{K4}^\circ\mathbf{D}^+$ by variable-free axioms is topologically complete. In Sect. 11.8 we prove the same for extensions of $\mathbf{DT_1}$ (the logic of dense-in-themselves T_1-spaces); the proof is based on a construction of d-morphisms from the recent paper [9].

In Sect. 11.6 we consider validity-preserving maps from topological to Kripke frames (d-morphisms and dd-morphisms) and prove a modified version of McKinsey–Tarski's lemma on dissectable spaces. In Sect. 11.7 we prove that $\mathbf{DT_1}$ is complete w.r.t. an arbitrary zero-dimensional dense-in-itself separable metric space by the method from [36, 38].

Sections 11.8, 11.9 and 11.10 study the axiom of connectedness *AC* and Kuratowski's axiom *Ku* related to local 1-componency. In particular we prove that the logic $\mathbf{DT_1CK}$ with both these axioms has the fmp. This is a refinement of an earlier result [36, 38] on the fmp of the d-logic $\mathbf{D4} + Ku$ (the new proof uses a simpler construction).

Section 11.11 contains our central result: $\mathbf{DT_1CK}$ is the dd-logic of $\mathbf{R}^n$ for $n > 1$. The proof uses an inductive construction of dd-morphisms onto finite frames of the corresponding logic, and it combines methods from [23, 36, 38], with an essential improvement motivated by [31] and based on the Glueing lemma.

The final section discusses some further directions and open questions. The Appendix contains technical details of some proofs.

11.2 Basic Notions

The material of this section is quite standard, and most of it can be found in [12]. We consider *n-modal (propositional) formulas* constructed from a countable set of propositional variables *PV* and the connectives $\bot$, $\to$, $\Box_1, \dots, \Box_n$. The derived connectives are $\wedge$, $\vee$, $\neg$, $\top$, $\leftrightarrow$, $\Diamond_1, \dots, \Diamond_n$. A formula without occurrences of propositional variables is called *closed*.

A *(normal) n-modal logic* is a set of modal formulas containing the classical tautologies, the axioms $\Box_i(p \to q) \to (\Box_i p \to \Box_i q)$ and closed under the standard inference rules: Modus Ponens $(A,\ A \to B/B)$, Necessitation $(A/\Box_i A)$, and Substitution $(A(p_j)/A(B_j))$. To be more specific, we use the terms '$(\Box_1, \dots, \Box_n)$-modal formula' and '$(\Box_1, \dots, \Box_n)$-modal logic'.

$\mathbf{K}_n$ denotes the minimal n-modal logic (and $\mathbf{K} = \mathbf{K}_1$). An n-modal logic containing a certain n-modal logic Λ is called an *extension* of Λ or a *Λ-logic*. The minimal Λ-logic containing a set of n-modal formulas Γ is denoted by $\Lambda + \Gamma$. In particular,

$$\mathbf{K4} := \mathbf{K} + \Box p \to \Box\Box p, \mathbf{S4} := \mathbf{K4} + \Box p \to p, \mathbf{D4} := \mathbf{K4} + \Diamond\top,$$
$$\mathbf{K4}^\circ := \mathbf{wK4} := \mathbf{K} + p \wedge \Box p \to \Box\Box p.$$

The *fusion* $L_1 * L_2$ of modal logics L_1, L_2 with distinct modalities is the smallest modal logic in the joined language containing $L_1 \cup L_2$.

A (*normal*) *n-modal algebra* is a Boolean algebra with extra n unary operations preserving $\mathbf{1}$ (the unit) and distributing over $\cap$; they are often denoted by $\Box_1, \dots, \Box_n$ as the modal connectives. A *valuation* in a modal algebra $\mathfrak{A}$ is a set-theoretic map $\theta : PV \longrightarrow \mathfrak{A}$. It extends to all n-modal formulas by induction:

$$\theta(\bot) = \varnothing,\ \theta(A \to B) = -\theta(A) \cup \theta(B),\ \theta(\Box_i A) = \Box_i \theta(A).$$

A formula A is *true in* $\mathfrak{A}$ (in symbols: $\mathfrak{A} \vDash A$) if $\theta(A) = \mathbf{1}$ for any valuation θ. The set $\mathbf{L}(\mathfrak{A})$ of all n-modal formulas true in an n-modal algebra $\mathfrak{A}$ is an n-modal logic called the *logic of* $\mathfrak{A}$.

An n-modal *Kripke frame* is a tuple $F = (W, R_1, \dots, R_n)$, where W is a nonempty set (of worlds), R_i are binary relations on W. We often write $x \in F$ instead of $x \in W$. In this chapter (except for Sect. 11.2) all 1-modal frames are assumed to be transitive. The associated n-modal algebra $MA(F)$ is 2^W (the Boolean algebra of all subsets of W) with the operations $\Box_1, \dots, \Box_n$ such that $\Box_i V = \{x \mid R_i(x) \subseteq V\}$ for any $V \subseteq W$.

A *valuation* in F is the same as in $MA(F)$, i.e., it is a map from PV to $\mathscr{P}(W)$ (the power set of W). A *(Kripke) model* over F is a pair $M = (F, \theta)$, where θ is a valuation in F. The notation $M, x \vDash A$ means $x \in \theta(A)$, which is also read as 'A is *true in M at x*'. A (modal) formula A is *true in M* (in symbols: $M \vDash A$) if A is true in M at all worlds. A formula A is called *valid in* a Kripke frame F (in symbols: $F \vDash A$) if A is true in all Kripke models over F; this is obviously equivalent to $MA(F) \vDash A$.

The *modal logic* $\mathbf{L}(F)$ of a Kripke frame F is the set of all modal formulas valid in F, i.e., $\mathbf{L}(MA(F))$. For a class of n-modal frames $\mathscr{C}$, the *modal logic* of $\mathscr{C}$ (or *the modal logic determined* by $\mathscr{C}$) is $\mathbf{L}(\mathscr{C}) := \bigcap\{\mathbf{L}(F) \mid F \in \mathscr{C}\}$. Logics determined by classes of Kripke frames are called *Kripke complete*. An n-modal frame validating an n-modal logic Λ is called a *Λ-frame*. A modal logic has the *finite model property (fmp)* if it is determined by some class of finite frames.

It is well known that $(W, R) \vDash \mathbf{K4}$ iff R is transitive; $(W, R) \vDash \mathbf{S4}$ iff R is reflexive transitive (a *quasi-order*).

A *cluster* in a transitive frame (W, R) is an equivalence class under the relation $\sim_R := (R \cap R^{-1}) \cup I_W$, where I_W is the equality relation on W. A *degenerate cluster* is an irreflexive singleton. A cluster that is a reflexive singleton is called *trivial* or *simple*. A *chain* is a frame (W, R) with R transitive, antisymmetric and linear, i.e., it satisfies $\forall x \forall y\ (xRy \vee yRx \vee x = y)$. A point $x \in W$ is *strictly (R-)minimal* if $R^{-1}(x) = \varnothing$.

A *subframe* of a frame $F = (W, R_1, \ldots, R_n)$ obtained by restriction to $V \subseteq W$ is $F|V := (V, R_1|V, \ldots, R_n|V)$. Then for any Kripke model $M = (F, \theta)$ we have a *submodel* $M|V := (F|V, \theta|V)$, where $(\theta|V)(q) := \theta(q) \cap V$ for each $q \in PV$. If $R_i(V) \subseteq V$ for any i, the subframe $F|V$ and the submodel $M|V$ are called *generated*.

The *union* of subframes $F_j = F|W_j,\ j \in J$, is the subframe $\bigcup_{j \in J} F_j := F| \bigcup_{j \in J} W_j$.

A *generated subframe (cone) with the root* x is $F^x := F|R^*(x)$, where R^* is the reflexive transitive closure of $R_1 \cup \ldots \cup R_n$; so for a transitive frame (W, R), $R^* = R \cup I_W$ is the reflexive closure of R (which is also denoted by $\overline{R}$). A frame F is called *rooted* with the root u if $F = F^u$. Similarly we define a cone M^x of a Kripke model M.

Every finite rooted transitive frame $F = (W, R)$ can be presented as the union $(F|C) \cup F^{x_1} \cup \ldots \cup F^{x_m}$ $(m \geq 0)$, where C is the root cluster, x_i are its successors (i.e., $x_i \notin C,\ \overline{R}^{-1}(x_i) = \sim_R(x_i) \cup C$). If C is non-degenerate, the frame $F|C$ is (C, C^2), which we usually denote just by C. If $C = \{a\}$ is degenerate, $F|C$ is $(\{a\}, \varnothing)$, which we denote by $\breve{a}$.

For the rest of the section we fix the propositional language (and the number n).

Lemma 1 (Generation Lemma)

(1) $\mathbf{L}(F) = \bigcap \{\mathbf{L}(F^x) \mid x \in F\}$.
(2) If F is a generated subframe of G, then $\mathbf{L}(G) \subseteq \mathbf{L}(F)$.
(3) If M is a generated submodel of N, then for any formula A and for any x in M

$$N, x \vDash A \text{ iff } M, x \vDash A.$$

Lemma 2 *For any Kripke complete modal logic Λ,*

$$\Lambda = \mathbf{L}(\textit{all } \Lambda\textit{-frames}) = \mathbf{L}(\textit{all rooted } \Lambda\textit{-frames}).$$

A *p-morphism* from a frame $(W, R_1, \ldots, R_n)$ onto a frame $(W', R'_1, \ldots, R'_n)$ is a surjective map $f : W \longrightarrow W'$ satisfying the following conditions (for any i):

(1) $\forall x \forall y\ (xR_i y \Rightarrow f(x) R'_i f(y))$ (monotonicity);
(2) $\forall x \forall z\ (f(x) R'_i z \Rightarrow \exists y (f(y) = z\ \&\ xR_i y))$ (the lift property).

If $xR_i y$ and $f(x) R'_i f(y)$, we say that $xR_i y$ *lifts* $f(x) R'_i f(y)$.

Note that (1) & (2) is equivalent to

$$\forall x\, f(R_i(x)) = R'_i(f(x)).$$

$f : F \twoheadrightarrow F'$ denotes that f is a p-morphism from F onto F'.

Every set-theoretic map $f : W \longrightarrow W'$ gives rise to the dual morphism of Boolean algebras $2^f : 2^{W'} \longrightarrow 2^W$ sending every subset $V \subseteq W'$ to its inverse image $f^{-1}(V) \subseteq W$.

Lemma 3 (P-morphism Lemma)

(1) $f : F \twoheadrightarrow F'$ *iff* 2^f *is an embedding of* $MA(F')$ *into* $MA(F)$.
(2) $f : F \twoheadrightarrow F'$ *implies* $\mathbf{L}(F) \subseteq \mathbf{L}(F')$.
(3) If $f : F \twoheadrightarrow F'$, *then* $F \models A \Leftrightarrow F' \models A$ *for any closed formula* A.

In this chapter in the proofs of the fmp we will use the well-known filtration method [12]. Let us recall the construction we need.

Let Ψ be a set of modal formulas closed under subformulas. For a Kripke model $M = (F, \varphi)$ over a frame $F = (W, R_1, \ldots, R_n)$, define an equivalence relation on W by

$$x \equiv_\Psi y \Longleftrightarrow \forall A \in \Psi(M, x \models A \Leftrightarrow M, y \models A).$$

Put $W' := W/\equiv_\Psi$; $x^\sim := \equiv_\Psi (x)$ (the equivalence class of x), $\varphi'(q) := \{x^\sim \mid x \in \varphi(q)\}$ for $q \in PV \cap \Psi$ (and let $\varphi'(q)$ be arbitrary for $q \in PV - \Psi$).

Lemma 4 (Filtration Lemma) *Under the above assumptions, consider the relations* $\underline{R}_i, R'_i$ *on* W' *such that*

$$a\underline{R}_i b \textit{ iff } \exists x \in a\, \exists y \in b\, xR_iy,$$

$$R'_i = \begin{cases} \textit{the transitive closure of } \underline{R}_i & \textit{if } R_i \textit{ is transitive,} \\ \underline{R}_i & \textit{otherwise.} \end{cases}$$

Put $M' := (W', R'_1, \ldots, R'_n, \varphi')$. *Then for any* $x \in W,\ A \in \Psi$:

$$M, x \models A \textit{ iff } M', x^\sim \models A.$$

Definition 1 An *m-formula* is a modal formula in propositional variables $\{p_1, \ldots, p_m\}$. For a modal logic Λ we define the *m-weak* (or *m-restricted) canonical frame* $F_{\Lambda\lceil m} := (W, R_1, \ldots, R_n)$ *and canonical model* $M_{\Lambda\lceil m} := (F_{\Lambda\lceil m}, \varphi)$, where W is the set of all maximal Λ-consistent sets of m-formulas, xR_iy iff for any m-formula A ($\Box_i A \in x \Rightarrow A \in y$),

$$\varphi(p_i) := \begin{cases} \{x \mid p_i \in x\} & \text{if } i \leq m, \\ \varnothing & \text{if } i > m. \end{cases}$$

Λ is called *weakly canonical* if $F_{\Lambda\lceil m} \models \Lambda$ for any finite m.

Proposition 1 *For any m-formula A and a modal logic* Λ

(1) $M_{\Lambda\lceil m}, x \models A$ *iff* $A \in x$;
(2) $M_{\Lambda\lceil m} \models A$ *iff* $A \in \Lambda$;
(3) if Λ *is weakly canonical, then it is Kripke complete.*

Corollary 1 *If for any m-formula A,* $M_{\Lambda\lceil m}, x \models A \Leftrightarrow M_{\Lambda\lceil m}, y \models A$, *then* $x = y$.

Definition 2 A cluster C in a transitive frame (W, R) is called *maximal* if $\overline{R}(C) = C$.

Lemma 5 *Let $F_{\Lambda\lceil m} = (W, R_1, \dots, R_n)$ and suppose $\Lambda \vdash \Box_1 p \to \Box_1\Box_1 p$ (i.e., R_1 is transitive). Then every generated subframe of (W, R_1) contains a maximal cluster.*

The proof is based on the fact that the general Kripke frame corresponding to a canonical model is descriptive; cf. [12, 16] for further details.[3]

11.3 Derivational Modal Logics

We denote topological spaces by $\mathfrak{X}, \mathfrak{Y}, \dots$ and the corresponding sets by $X, Y, \dots$.[4] The interior operation in a space $\mathfrak{X}$ is denoted by $\mathbf{I}_X$ and the closure operation by $\mathbf{C}_X$, but we often omit the subscript X. A set S is a *neighbourhood* of a point x if $x \in \mathbf{I}S$; then $S - \{x\}$ is called a *punctured neighbourhood* of x.

Definition 3 Let $\mathfrak{X}$ be a topological space, $V \subseteq X$. A point $x \in X$ is said to be *limit* for V if $x \in \mathbf{C}(V - \{x\})$; a non-limit point of V is called *isolated*.

The *derived set of V* (denoted by $\mathbf{d}V$, or by $\mathbf{d}_X V$) is the set of all limit points of V. The unary operation $V \mapsto \mathbf{d}V$ on $\mathscr{P}(X)$ is called *the derivation* (in $\mathfrak{X}$).

A set without isolated points is called *dense-in-itself.*

Lemma 6 [28] *For a subspace $\mathscr{Y} \subseteq \mathfrak{X}$ and $V \subseteq X$, $\mathbf{d}_Y(V \cap Y) = \mathbf{d}_X(V \cap Y) \cap Y$; if Y is open, then $\mathbf{d}_Y(V \cap Y) = \mathbf{d}_X V \cap Y$.*

Definition 4 The *derivational algebra of a topological space* $\mathfrak{X}$ is $DA(X) := (2^X, \tilde{\mathbf{d}})$, where 2^X is the Boolean algebra of all subsets of X, $\tilde{\mathbf{d}}V := -\mathbf{d}(-V)$.[5] The *closure algebra of a space* $\mathfrak{X}$ is $CA(\mathfrak{X}) := (2^X, \mathbf{I})$.

Remark 1 In [32] the derivational algebra of $\mathfrak{X}$ is defined as $(2^X, \mathbf{d})$, and the closure algebra as $(2^X, \mathbf{C})$, but here we adopt equivalent dual definitions.

It is well known that $CA(\mathfrak{X})$, $DA(\mathfrak{X})$ are modal algebras, $CA(\mathfrak{X}) \vDash \mathbf{S4}$ and $DA(\mathfrak{X}) \vDash \mathbf{K4}^\circ$ (the latter is due to Esakia).

Every Kripke **S4**-frame $F = (W, R)$ is associated with a topological space $N(F)$ on W, with the *Alexandrov* (or *right*) *topology* $\{V \subseteq W \mid R(V) \subseteq V\}$. In $N(F)$ we have $\mathbf{C}V = R^{-1}(V)$, $\mathbf{I}V = \{x \mid R(x) \subseteq V\}$; thus $MA(F) = CA(N(F))$.

Definition 5 A modal formula A is called *d-valid in a topological space* $\mathfrak{X}$ (in symbols, $\mathfrak{X} \vDash^d A$) if it is true in the algebra $DA(\mathfrak{X})$. The logic $\mathbf{L}(DA(\mathfrak{X}))$ is called the *derivational modal logic* (or the *d-logic*) of $\mathfrak{X}$ and denoted by $\mathbf{Ld}(\mathfrak{X})$.

A formula A is called *c-valid* in $\mathfrak{X}$ (in symbols, $\mathfrak{X} \vDash^c A$) if it is true in $CA(\mathfrak{X})$. $\mathbf{Lc}(\mathfrak{X}) := \mathbf{L}(CA(\mathfrak{X}))$ is called the *closure modal logic*, or the *c-logic* of $\mathfrak{X}$.

---

[3] For the 1-modal case this lemma has been known as folklore since the 1970s; the second author learned it from Leo Esakia in 1975.

[4] Sometimes we neglect this difference.

[5] There is no common notation for this operation; some authors use τ.

Definition 6 For a class of topological spaces $\mathscr{C}$ we also define the *d-logic* $\mathbf{Ld}(\mathscr{C}) := \bigcap\{\mathbf{Ld}(\mathfrak{X}) \mid \mathfrak{X} \in \mathscr{C}\}$ and the *c-logic* $\mathbf{Lc}(\mathscr{C}) := \bigcap\{\mathbf{Lc}(\mathfrak{X}) \mid \mathfrak{X} \in \mathscr{C}\}$. Logics of this form are called *d-complete* (respectively, *c-complete*).

Definition 7 A *valuation* in a topological space $\mathfrak{X}$ is a map $\varphi : PV \longrightarrow \mathscr{P}(\mathfrak{X})$. Then $(\mathfrak{X}, \varphi)$ is called a *topological model* over $\mathfrak{X}$.

So valuations in $\mathfrak{X}$, $CA(\mathfrak{X})$, and $DA(\mathfrak{X})$ are the same. Every valuation φ can be prolonged to all formulas in two ways, according either to $CA(\mathfrak{X})$ or $DA(\mathfrak{X})$. The corresponding maps are denoted respectively by φ_c or φ_d. Thus

$$\varphi_d(\Box A) = \tilde{\mathbf{d}}\varphi_d(A), \qquad \varphi_d(\Diamond A) = \mathbf{d}\varphi_d(A),$$
$$\varphi_c(\Box A) = \mathbf{I}\varphi_c(A), \qquad \varphi_c(\Diamond A) = \mathbf{C}\varphi_c(A).$$

A formula A is called *d-true* (respectively, *c-true*) in $(\mathfrak{X}, \varphi)$ if $\varphi_d(A) = X$ (respectively, $\varphi_c(A) = X$). So A is d-valid in $\mathfrak{X}$ iff A is d-true in every topological model over $\mathfrak{X}$, and similarly for c-validity.

Definition 8 A modal formula A is called *d-true at a point* x in a topological model $(\mathfrak{X}, \varphi)$ if $x \in \varphi_d(A)$.

Instead of $x \in \varphi_d(A)$, we write $x \vDash^d A$ if the model is clear from the context. Similarly we define the c-truth at a point and use the corresponding notation. From the definitions we obtain.

Lemma 7 [15] *For a topological model over a space* $\mathfrak{X}$

- $x \vDash^d \Box A$ *iff* $\exists U \ni x$ *(*U *is open in* $\mathfrak{X}$ *&* $\forall y \in U - \{x\}\ y \vDash^d A$*);*
- $x \vDash^d \Diamond A$ *iff* $\forall U \ni x$ *(*U *is open in* $\mathfrak{X} \Rightarrow \exists y \in U - \{x\}\ y \vDash^d A$*).*

Definition 9 A *local* T_1*-space* (or a T_D*-space* [5]) is a topological space in which every point is *locally closed*, i.e, closed in some neighbourhood.

Note that a point x in an Alexandrov space $N(W, R)$ is closed iff it is minimal (i.e., $R^{-1}(x) = \{x\}$); x is locally closed iff $R(x) \cap R^{-1}(x) = \{x\}$. Thus $N(F)$ is local T_1 iff F is a poset.

Lemma 8 [15] *For a topological space* $\mathfrak{X}$

(1) $\mathfrak{X} \vDash^d \mathbf{K4}$ *iff* $\mathfrak{X}$ *is local* T_1*;*
(2) $\mathfrak{X} \vDash^d \Diamond\top$ *iff* $\mathfrak{X}$ *is dense-in-itself.*

Definition 10 A Kripke frame (W, R) is called *weakly transitive* if $R \circ R \subseteq \overline{R}$.

It is obvious that the weak transitivity of R is equivalent to the transitivity of $\overline{R}$.

Proposition 2 [15] *(1)* $(W, R) \models$ **K4**$^\circ$ *iff* (W, R) *is weakly transitive;*
(2) **K4**$^\circ$ *is Kripke-complete.*

Lemma 9 [15] *(1) Let* $F = (W, R)$ *be a Kripke* **S4***-frame, and let* $R^\circ := R - I_W$, $F^\circ := (W, R^\circ)$. *Then* $\mathbf{Ld}(N(F)) = \mathbf{L}(F^\circ)$.
(2) Let $F = (W, R)$ *be a weakly transitive irreflexive Kripke frame, and let* $\overline{F} =: (W, \overline{R})$ *be its reflexive closure. Then* $\mathbf{Ld}(N(\overline{F})) = \mathbf{L}(F)$.
(3) If $\Lambda = \mathbf{L}(\mathscr{C})$, *for some class* $\mathscr{C}$ *of weakly transitive irreflexive Kripke frames, then* Λ *is d-complete.*

Definition 11 For a 1-modal formula A we define $A^\sharp$ as the formula obtained by replacing every occurrence of every subformula $\Box B$ with $\overline{\Box} B := \Box B \wedge B$. For a 1-modal logic Λ its *reflexive fragment* is ${}^\sharp\Lambda := \{A \mid \Lambda \vdash A^\sharp\}$.

Proposition 3 [6] *(1) If* Λ *is a* **K4**$^\circ$*-logic, then* ${}^\sharp\Lambda$ *is an* **S4***-logic.*
(2) For any topological space X, $\mathbf{Lc}(\mathfrak{X}) = {}^\sharp\mathbf{Ld}(\mathfrak{X})$,
(3) For any weakly transitive Kripke frame F, $\mathbf{L}(\overline{F}) = {}^\sharp\mathbf{L}(F)$.

Let us give some examples of d-complete logics.

(1) **Ld**(all topological spaces) = **K4**$^\circ$. This was proved by L. Esakia in the 1970s and published in [15].
(2) **Ld**(all local T_1-spaces) = **K4**. This is also a result from [15].
(3) **Ld**(all T_0-spaces) $=$ **K4**$^\circ + p \wedge \Diamond(q \wedge \Diamond p) \to \Diamond p \vee \Diamond(q \wedge \Diamond q)$. This result is from [8].
(4) Esakia [14] also proved that Gödel - Löb logic $\mathbf{GL} := \mathbf{K} + \Box(\Box p \supset p) \supset \Box p$ is the derivational logic of the class of all scattered spaces (a space is *scattered* if each of its nonempty subsets has an isolated point).
(5) The papers [1, 2, 10] give a complete description of d-logics of ordinals with the interval topology: $\mathbf{Ld}(\alpha)$ is either **GL** (if $\alpha \geq \omega^\omega$), or $\mathbf{GL} + \Box^n\bot$ (if $\omega^{n-1} \leq \alpha < \omega^n$). In particular, $\mathbf{Ver} := \mathbf{K} + \Box\bot$ is the d-logic of any finite ordinal (and of any discrete space).
(6) The well-known "difference logic" [13, 35] $\mathbf{DL} := \mathbf{K4}^\circ + \Diamond\Box p \supset p$ is determined by Kripke frames with the difference relation: $\mathbf{DL} = \mathbf{L}(\{(W, \neq_W) \mid W \neq \varnothing\})$, where $\neq_W := W^2 - I_W$; hence by Lemma 9, **DL** is the d-logic of the class of all trivial topological spaces. However, for any particular trivial space $\mathfrak{X}$, $\mathbf{Ld}(\mathfrak{X}) \neq \mathbf{DL}$. Moreover, $\mathbf{Ld}(\mathfrak{X})$ is not finitely axiomatizable for any infinite trivial $\mathfrak{X}$ [26]; this surprising result is easily proved by a standard technique using Jankov formulas (cf. [25]).
(7) In [38] it was proved that **Ld**(all zero-dimensional separable metric spaces) = **K4**. All these spaces are embeddable in **R** [28].
(8) In [38] it was also proved that for any dense-in-itself zero-dimensional separable metric space $\mathfrak{X}$, $\mathbf{Ld}(\mathfrak{X}) = \mathbf{D4}$; this was a generalization of an earlier proof [36] for $\mathfrak{X} = \mathbf{Q}$. A more elegant proof for **Q** is in [30].

(9) Every extension of **K4** by a set of closed axioms is a d-logic of some subspace of **Q** [9]. This gives us a continuum of d-logics of countable metric spaces.
(10) In [36] $\mathbf{Ld}(\mathbf{R}^2)$ was axiomatized and it was also proved that the d-logics of $\mathbf{R}^n$ for $n \geq 2$ coincide. We will simplify and extend that proof in the present chapter.
(11) **Ld**(**R**) was described in [38]; for a simpler completeness proof cf. [31].
(12) **Ld**(all Stone spaces) = **K4** and **Ld**(all weakly scattered Stone spaces) = $\mathbf{K4} + \Diamond\top \to \Diamond\Box\bot$, cf. [7].
(13) d-logics of special types of spaces were studied in [6, 9, 30]. They include submaximal, perfectly disconnected, maximal, weakly scattered and some other spaces.

However, not all extensions of $\mathbf{K4}^\circ$ are d-complete. In fact, the formula $p \supset \Diamond p$ never can be d-valid because $\mathbf{d}Y = \varnothing$ for a singleton Y. So every extension of **S4** is d-incomplete, and thus Kripke completeness does not imply d-completeness.

Proposition 4 *Let $F = (\omega^*, \prec)$ be the "standard irreflexive transitive tree", where ω^* is the set of all finite sequences in ω; $\alpha \prec \beta$ iff α is a proper initial segment of β. Then*

$$\mathbf{D4} = \mathbf{L}(F) = \mathbf{Ld}(N(\overline{F})) = \mathbf{Ld}(\mathscr{D}),$$

where $\mathscr{D}$ denotes the class of all dense-in-themselves local T_1-spaces.

Proof The first equality is well known [39]; the second one holds by Lemma 9. By Lemma 8, **D4** is d-valid exactly in spaces from $\mathscr{D}$. So $N(\overline{F}) \in \mathscr{D}$, $\mathbf{D4} \subseteq \mathbf{Ld}(\mathscr{D})$, and the third equality follows. □

11.4 Adding the Universal Modality and the Difference Modality

Recall that the *universal modality* $[\forall]$ and the *difference modality* $[\neq]$ correspond to Kripke frames with the universal and difference relations. So (under a valuation in a set W) these modalities are interpreted in the standard way:

$$x \vDash [\forall]A \text{ iff } \forall y \in W\ y \vDash A; \qquad x \vDash [\neq]A \text{ iff } \forall y \in W\ (y \neq x \Rightarrow y \vDash A).$$

The corresponding dual modalities are denoted by $\langle\exists\rangle$ and $\langle\neq\rangle$.

Definition 12 For a $[\forall]$-modal formula A we define the $[\neq]$-modal formula A^u by induction:

$$A^u := A \text{ for } A \text{ atomic}, \quad (A \supset B)^u := A^u \supset B^u, \quad ([\forall]B)^u := [\neq]B^u \wedge B^u.$$

We can consider 2-modal topological logics obtained from $\mathbf{Lc}(\mathfrak{X})$ or $\mathbf{Ld}(\mathfrak{X})$ by adding the universal or difference modalities.[6] Thus for a topological space $\mathfrak{X}$ we

[6] So we extend the definitions of the d-truth or the c-truth by adding the item for $[\forall]$ or $[\neq]$.

obtain four 2-modal logics : $\mathbf{Lc}_\forall(\mathfrak{X})$ (the *closure universal (cu-) logic*), $\mathbf{Ld}_\forall(\mathfrak{X})$ (the *derivational universal (du-) logic*), $\mathbf{Lc}_{\neq}(\mathfrak{X})$ (the *closure difference (cd-) logic*), $\mathbf{Ld}_{\neq}(\mathfrak{X})$ (the *derivational difference (dd-) logic*). Similar notations ($\mathbf{Lc}_\forall(\mathscr{C})$ etc.) are used for logics of a class of spaces $\mathscr{C}$, and we can define four kinds of topological completeness (cu-, du-, cd-, dd-) for 2-modal logics.

cd-logics were first studied in [17], cu-logics in [37], du-logics in [31], but dd-logics have never been addressed so far.

For a $\Box$-modal logic $\mathbf{L}$ we define the 2-modal logics

$$\mathbf{LD} := \mathbf{L} * \mathbf{DL} + [\neq]p \wedge p \to \Box p, \qquad \mathbf{LD}^+ := \mathbf{L} * \mathbf{DL} + [\neq]p \to \Box p,$$
$$\mathbf{LU} := \mathbf{L} * \mathbf{S5} + [\forall]p \to \Box p.$$

Here we suppose that **S5** is formulated in the language with $[\forall]$ and **DL** in the language with $[\neq]$. The following is checked easily:

Lemma 10 *For any topological space $\mathfrak{X}$,*

$$\mathbf{Lc}_\forall(\mathfrak{X}) \supseteq \mathbf{S4U}, \quad \mathbf{Ld}_\forall(\mathfrak{X}) \supseteq \mathbf{K4^\circ U}, \quad \mathbf{Lc}_{\neq}(\mathfrak{X}) \supseteq \mathbf{S4D}, \quad \mathbf{Ld}_{\neq}(\mathfrak{X}) \supseteq \mathbf{K4^\circ D^+}.$$

Definition 13 For a 1-modal Kripke frame $F = (W, R)$ we define 2-modal frames $F_\forall := (F, W^2)$, $F_{\neq} := (F, \neq_W)$ and modal logics $\mathbf{L}_\forall(F) := \mathbf{L}(F_\forall)$, $\mathbf{L}_{\neq}(F) := \mathbf{L}(F_{\neq})$.

Sahlqvist theorem [12] implies

Proposition 5 *The logics* **S4U**, **K4°U**, **S4D**, **K4°D**$^+$ *are Kripke complete.*

Using the first-order equivalents of the modal axioms for these logics (in particular, Proposition 2) we obtain

Lemma 11 *For a rooted Kripke frame $G = (W, R, S)$*

(1) $G \vDash \mathbf{S4U}$ *iff R is a quasi-order & $S = W^2$,*
(2) $G \vDash \mathbf{K4^\circ U}$ *iff R is weakly transitive & $S = W^2$,*
(3) $G \vDash \mathbf{S4D}$ *iff R is a quasi-order & $\overline{S} = W^2$,*
(4) $G \vDash \mathbf{K4^\circ D^+}$ *iff R is weakly transitive & $\overline{S} = W^2$& $R \subseteq S$.*

Also note that $\overline{S} = W^2$ iff $\neq_W \subseteq S$.

Definition 14 A rooted Kripke **K4°D**$^+$-frame described by Lemma 11 (4) is called *basic*. The class of these frames is denoted by $\mathfrak{F}_0$.

Next, we easily obtain the 2-modal analogue to Lemma 9.

Lemma 12 *(1) Let F be an* **S4***-frame. Then*

$$\mathbf{Ld}_{\neq}(N(F)) = \mathbf{L}_{\neq}(F^\circ), \; \mathbf{Ld}_\forall(N(F)) = \mathbf{L}_\forall(F^\circ).$$

(2) Let F be a weakly transitive irreflexive Kripke frame. Then

$$\mathbf{Ld}_{\neq}(N(\overline{F})) = \mathbf{L}_{\neq}(F),\ \mathbf{Ld}_{\forall}(N(\overline{F})) = \mathbf{L}_{\forall}(F).$$

(3) Let $\mathscr{C}$ be a class of weakly transitive irreflexive Kripke 1-frames. Then $\mathbf{L}_{\neq}(\mathscr{C})$ is dd-complete, $\mathbf{L}_{\forall}(\mathscr{C})$ is du-complete.

Let us extend the translations $(-)^\sharp$, $(-)^u$ to 2-modal formulas.

Definition 15 $(-)^u$ translates $(\Box, [\forall])$-modal formulas to $(\Box, [\neq])$-modal formulas so that $([\forall]B)^u = [\neq]B^u \wedge B^u$ and $(-)^u$ distributes over the other connectives.

Similarly, $(-)^\sharp$ translates $(\Box, [\neq])$-modal formulas and $(\Box, [\forall])$-modal formulas to formulas of the same kind, so that $(\Box B)^\sharp = \Box B^\sharp \wedge B^\sharp$ and $(-)^\sharp$ distributes over the other connectives.

$^u\Lambda := \{A \mid A^u \in \Lambda\}$ for a $(\Box, [\forall])$-modal logic Λ (the *universal fragment*),

$^\sharp\Lambda := \{A \mid A^\sharp \in \Lambda\}$ for a $(\Box, [\neq])$- or a $(\Box, [\forall])$-modal Λ (the *reflexive fragment*),

$^{\sharp u}\Lambda := {}^\sharp({}^u\Lambda)$ for a $(\Box, [\neq])$-modal Λ(the *reflexive universal fragment*).

Proposition 6 *(1) The map $\Lambda \mapsto {}^\sharp\Lambda$ sends* **K4°D**$^+$*-logics to* **S4U**-*logics.*
(2) The map $\Lambda \mapsto {}^u\Lambda$ sends **K4°D**$^+$*-logics to* **K4°U**-*logics and* **S4D**-*logics to* **S4U**-*logics.*
(3) The map $\Lambda \mapsto {}^{\sharp u}\Lambda$ sends **K4°D**$^+$*-logics to* **S4U**-*logics.*
(4) For a topological space $\mathfrak{X}$

$$\mathbf{Lc}_{\neq}(\mathfrak{X}) = {}^\sharp\mathbf{Ld}_{\neq}(\mathfrak{X}),\ \mathbf{Ld}_{\forall}(\mathfrak{X}) = {}^u\mathbf{Ld}_{\neq}(\mathfrak{X}),\ \mathbf{Lc}_{\forall}(\mathfrak{X}) = {}^u\mathbf{Lc}_{\neq}(\mathfrak{X}) = {}^\sharp\mathbf{Ld}_{\forall}(\mathfrak{X}).$$

(5) For a weakly transitive Kripke frame F

$$\mathbf{L}_{\neq}(\overline{F}) = {}^\sharp\mathbf{L}_{\neq}(F),\ \mathbf{L}_{\forall}(F) = {}^u\mathbf{L}_{\neq}(F),\ \mathbf{L}_{\forall}(\overline{F}) = {}^\sharp\mathbf{L}_{\forall}(F) = {}^{\sharp\, u}\mathbf{L}_{\neq}(F).$$

Proposition 6 (4) implies that dd-logics are the most expressive of all the kinds of logics that we consider.

Corollary 2 *If $\mathbf{Ld}_{\neq}(\mathfrak{X}) = \mathbf{Ld}_{\neq}(\mathfrak{Y})$ for spaces $\mathfrak{X}, \mathfrak{Y}$, then all the other logics (du-, cu-, cd-, d-, c-) of these spaces coincide.*

Let

$$AT_1 := [\neq]p \to [\neq]\Box p, \quad AC := [\forall]\,(\Box p \vee \Box\neg p) \to [\forall]\,p \vee [\forall]\,\neg p.$$

Proposition 7 *For a topological space $\mathfrak{X}$*

(1) $\mathfrak{X} \models^d \Diamond\top$ iff $\mathfrak{X}$ is dense-in-itself;
(2) $\mathfrak{X} \models^d AT_1$ iff $\mathfrak{X} \models^c AT_1$ iff $\mathfrak{X}$ is a T_1-space;
(3) $\mathfrak{X} \models^d AC^\sharp$ iff $\mathfrak{X} \models^c AC$ iff $\mathfrak{X}$ is connected.

Proof (1) and the first equivalence in (2) are trivial. The first equivalence in (3) follows from Proposition 6 (4). The remaining ones are checked easily, cf. [23, 37]. □

For a □-modal logic **L** put

$$\mathbf{LD^+T_1} := \mathbf{LD^+} + AT_1, \quad \mathbf{LD^+T_1C} := \mathbf{LD^+} + AT_1 + AC^{\sharp u}.$$

Also put

$$\mathbf{KT_1} := \mathbf{K4D^+T_1}, \ \mathbf{DT_1} := \mathbf{D4D^+T_1}, \ \mathbf{DT_1C} := \mathbf{D4D^+T_1C}.$$

Proposition 8 [23] *If $F = (W, R, R_D)$ is basic, then $F \models AT_1$ iff all R_D-irreflexive points are strictly R-minimal iff $R_D \circ R \subseteq R_D$.*

Remark 2 Density-in-itself is expressible in cd-logic and dd-logic by the formula $DS := [\neq]p \supset \Diamond p$. So for any space $\mathfrak{X}$, $\mathfrak{X} \models^c DS$ iff $\mathfrak{X} \models^d DS$ iff $\mathfrak{X} \models^d \Diamond\top$. It is known that DS axiomatizes dense-in-themselves spaces in cd-logic [23]. However, in dd-logic this axiom is insufficient: $\mathbf{Ld}_{\neq}$(all dense-in-themselves spaces) $= \mathbf{D4^\circ D^+} = \mathbf{K4^\circ D^+} + \Diamond\top$, and it is *stronger* than $\mathbf{K4^\circ D^+} + DS$. (To see the latter, consider a singleton Kripke frame, which is R_D-reflexive, but R-irreflexive.) Therefore, $\mathbf{K4^\circ D^+} + DS$ is dd-incomplete.

Remark 3 Every T_1-space is a local T_1-space, so the dd-logic of all T_1-spaces contains $\Box p \to \Box\Box p$. However, $\mathbf{K4^\circ D^+T_1} \not\vdash \Box p \to \Box\Box p$. In fact, consider a 2-point frame $F := (W, \neq_W, W^2)$. It is clear that $F \models \mathbf{K4^\circ D^+}$. Also $F \models AT_1$ by Proposition 8, but $F \not\models \Box p \to \Box\Box p$ since $\neq_W$ is not transitive.

It follows that $\mathbf{K4^\circ D^+T_1}$ is dd-incomplete; T_1-spaces are actually axiomatized by $\mathbf{KT_1}$ (Corollary 4).

Let us give some examples of du-, cu- and cd-complete logics.

(1) $\mathbf{Lc}_\forall$(all spaces) $= \mathbf{S4U}$.
(2) $\mathbf{Lc}_\forall$(all connected spaces) $= \mathbf{Lc}_\forall(\mathbf{R}^n) = \mathbf{S4U} + AC$ for any $n \geq 1$ [37].[7]
(3) $\mathbf{Ld}_\forall$(all spaces) $= \mathbf{S4D}$ [13].
(4) $\mathbf{Lc}_{\neq}(\mathfrak{X}) = \mathbf{S4DT_1} + \mathbf{DS}$, where $\mathfrak{X}$ is a zero-dimensional separable metric space [23].
(5) $\mathbf{Lc}_{\neq}(\mathbf{R}^n)$ for any $n \geq 2$ is finitely axiomatized in [22]; all these logics coincide.
(6) $\mathbf{Ld}_\forall(\mathbf{R})$ is finitely axiomatized in [31].

[7] Shehtman [37] contains a stronger claim: $\mathbf{Lc}_\forall(\mathfrak{X}) = \mathbf{S4U} + AC$ for any connected dense-in-itself separable metric $\mathfrak{X}$. However, recently we found a gap in the proof of Lemma 17 from that paper. Now we state the main result only for the case $\mathfrak{X} = \mathbf{R}^n$; a proof can be obtained by applying the methods of the present chapter, but we are planning to publish it separately.

11.5 dd-Completeness of $\mathbf{K4^\circ D^+}$ and Some of its Extensions

This section contains some simple arguments showing that there are many dd-complete bimodal logics.

All formulas and logics in this section are $(\Box, [\neq])$-modal. An arbitrary Kripke frame for $(\Box, [\neq])$-formulas is often denoted by (W, R, R_D).

Lemma 13 *(1) Every weakly transitive Kripke 1-frame is a p-morphic image of some irreflexive weakly transitive Kripke 1-frame.*
(2) Every rooted $\mathbf{K4^\circ D^+}$*-frame is a p-morphic image of some R- and* R_D*-irreflexive rooted* $\mathbf{K4^\circ D^+}$*-frame.*

Proof (1) Cf. [15].

(2) Similar to the proof of (1). For $F = (W, R, R_D) \in \mathfrak{F}_0$ put

$$W_r := \{a \mid aR_D a\},\ W_i = W - W_r,\quad \tilde{W} := W_i \cup (W_r \times \{0, 1\}).$$

Then we define the relation $\tilde{R}$ on $\tilde{W}$ such that

$$(b, j)\tilde{R}a \ \text{ iff } \ bRa, \qquad a\tilde{R}(b, j) \ \text{ iff } \ aRb,$$
$$(b, j)\tilde{R}(b', k) \ \text{ iff } \ bRb' \,\&\, b \neq b' \vee b = b' \,\&\, j \neq k, \qquad a\tilde{R}a' \ \text{ iff } \ aRa'.$$

Here $a, a' \in W_i$; $b, b' \in W_r$; $j, k \in \{0, 1\}$. So we duplicate all R_D-reflexive points making them irreflexive (under both relations). It follows that $\tilde{F} := (\tilde{W}, \tilde{R}, \neq_{\tilde{W}}) \in \mathfrak{F}_0$ and $\tilde{R}$ is irreflexive; the map $f : \tilde{W} \to W$ sending (b, j) to b and a to itself (for $b \in W_r$, $a \in W_i$) is a p-morphism $\tilde{F} \twoheadrightarrow F$. □

Proposition 9 *Let* Γ *be a set of closed 2-modal formulas,* $\Lambda := \mathbf{K4^\circ D^+} + \Gamma$. *Then*

(1) Λ *is Kripke complete.*
(2) Λ *is dd-complete.*

Proof (1) $\mathbf{K4^\circ D^+}$ is axiomatized by Sahlqvist formulas. One can easily check that (in the minimal modal logic) every closed formula is equivalent to a positive formula, so we can apply Sahlqvist theorem.

(2) Suppose $A \notin \Lambda$. By (1) and the Generation lemma there exists a rooted Kripke 2-frame F such that $F \vDash L$ and $F \nvDash A$. Then by Lemma 13, for some irreflexive weakly transitive 1-frame $G = (W, R)$ there is a p-morphism $(G, \neq_W) \twoheadrightarrow F$. By the p-morphism lemma $(G, \neq_W) \nvDash A$ and $(G, \neq_W) \vDash \Lambda$ (since Γ consists of closed formulas). Hence by Lemma 12, $\Lambda \subseteq \mathbf{Ld}_{\neq}(N(\overline{G}))$, $A \notin \mathbf{Ld}_{\neq}(N(\overline{G}))$. □

Remark 4 Using Proposition 9 and the construction from [9], one can prove that there is a continuum of dd-complete logics. Such a claim is rather weak because Proposition 9 deals only with Alexandrov spaces. In Sect. 11.7 we will show how to construct many dd-complete logics of metric spaces.

11.6 d-Morphisms and dd-Morphisms; Extended McKinsey–Tarski's Lemma

In this section we recall the notion of a d-morphism (a validity-preserving map for d-logics) and introduce dd-morphisms, the analogues of d-morphisms for dd-logics. This is the main technical tool in the present chapter. Two basic lemmas are proved here, an analogue of McKinsey–Tarski's lemma on dissectability for d-morphisms and the Glueing lemma.

The original McKinsey–Tarski's lemma [32] states the existence of a c-morphism (cf. Remark 5) from an arbitrary separable dense-in-itself metric space onto a certain quasi-tree of depth 2. The separability condition is actually redundant [33, Chap. 3] (note that the latter proof is quite different from [32][8]). But c-morphisms preserve validity only for c-logics, and unfortunately, the constructions by McKinsey–Tarski and Rasiowa–Sikorski cannot be used for d-morphisms. So we need another construction to prove a stronger form of McKinsey–Tarski's lemma.

Definition 16 Let $\mathfrak{X}$ be a topological space, $F = (W, R)$ a transitive Kripke frame. A map $f : X \longrightarrow W$ is called a *d-morphism* from $\mathfrak{X}$ to F if f is open and continuous as a map $\mathfrak{X} \longrightarrow N(\overline{F})$ and also satisfies

$$\text{r-density}: \ \forall w \in W(wRw \Rightarrow f^{-1}(w) \subseteq \mathbf{d}f^{-1}(w)),$$
$$\text{i-discreteness}: \ \forall w \in W(\neg wRw \Rightarrow f^{-1}(w) \cap \mathbf{d}f^{-1}(w) = \varnothing).$$

If f is surjective, we write $f : \mathfrak{X} \twoheadrightarrow^d F$.

Proposition 10 [6] *(1) f is a d-morphism from $\mathfrak{X}$ to F iff 2^f is a homomorphism from $MA(F)$ to $DA(\mathfrak{X})$.*
(2) If $f : \mathfrak{X} \twoheadrightarrow^d F$, then $\mathbf{Ld}(\mathfrak{X}) \subseteq \mathbf{L}(F)$.

Corollary 3 [36] *A map f from a topological space $\mathfrak{X}$ to a finite transitive Kripke frame F is a d-morphism iff*

$$\forall w \in W \ \mathbf{d}f^{-1}(w) = f^{-1}(R^{-1}(w)).$$

Remark 5 For a space $\mathfrak{X}$ and a Kripke **S4**-frame $F = (W, R)$ one can also define a *c-morphism* $\mathfrak{X} \longrightarrow F$ just as an open and continuous map $f : \mathfrak{X} \longrightarrow N(F)$. So every d-morphism to an **S4**-frame is a c-morphism. It is well known [33] that $f : X \longrightarrow W$ is a c-morphism iff 2^f is a homomorphism $MA(F) \longrightarrow CA(\mathfrak{X})$. Again for a finite F this is equivalent to

$$\forall w \in W \ \mathbf{C}f^{-1}(w) = f^{-1}(R^{-1}(w)).$$

Lemma 14 *If $f : \mathfrak{X} \twoheadrightarrow^d F$ for a finite frame F and $\mathscr{Y} \subseteq \mathfrak{X}$ is an open subspace, then $f|Y$ is a d-morphism.*

---

[8] Recently Kremer [21] has showed that **S4** is *strongly complete* w.r.t. any dense-in-itself metric space. His proof uses much of the construction from [33].

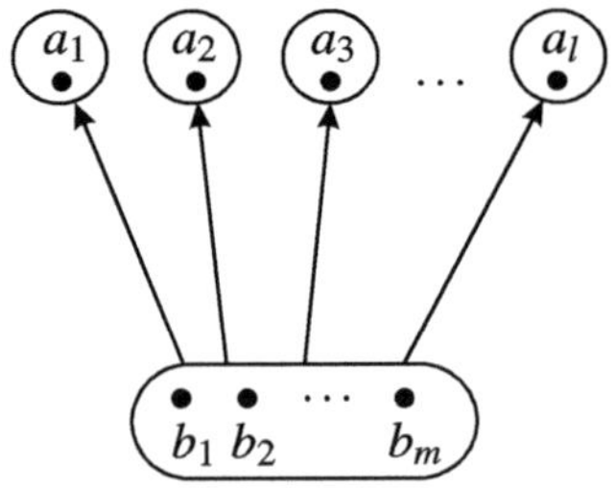

Fig. 11.1 Frame Φ_{ml}

Proof We apply Proposition 10. Note that $f|Y$ is the composition $f \cdot j$, where $j : Y \hookrightarrow X$ is the inclusion map. Then $2^{f|Y} = 2^j \cdot 2^f$. Since 2^f is a homomorphism $MA(F) \longrightarrow DA(\mathfrak{X})$, it remains to show that 2^j is a homomorphism $DA(\mathfrak{X}) \longrightarrow DA(\mathcal{Y})$, i.e., it preserves the derivation: $j^{-1}(\mathbf{d}V) = \mathbf{d}_Y j^{-1}(V)$, or $\mathbf{d}V \cap Y = \mathbf{d}_Y(V \cap Y)$, which follows from Lemma 6. □

Definition 17 A set γ of subsets of a topological space $\mathfrak{X}$ is called *dense* at $x \in X$ if every neighbourhood of x contains a member of γ.

Proposition 11 *For $m, l > 0$ let Φ_{ml} be a "quasi-tree" of height 2, with singleton maximal clusters and an m-element root cluster (Fig. 11.1). For $l = 0,\ m > 0,\ \Phi_{ml}$ denotes an m-element cluster.*

Let $\mathfrak{X}$ be a dense-in-itself separable metric space, $B \subset X$ a closed nowhere dense set. Then there exists a d-morphism $g : \mathfrak{X} \twoheadrightarrow^d \Phi_{ml}$ with the following properties:

(1) $B \subseteq g^{-1}(b_1)$;
(2) every $g^{-1}(a_i)$ (for $i \leq l$) is a union of a set α_i of disjoint open balls, which is dense at any point of $g^{-1}(\{b_1, \dots, b_m\})$.

For the proof see Appendix.

Lemma 15 *Assume that*

(1) $\mathfrak{X}$ is a dense-in-itself separable metric space,
(2) $B \subset X$ is closed nowhere dense,
(3) $F = C \cup F_1 \cup \dots \cup F_l$ is a **D4***-frame, where $C = \{b_1, \dots, b_m\}$ is a non-degenerate root cluster, $F_1, \dots, F_l$ are the subframes generated by the successors of C,*
(4) for any nonempty open ball U in $\mathfrak{X}$, for any $i \in \{1, \dots, l\}$ there exists a d-morphism $f_i^U : U \twoheadrightarrow^d F_i$.

Then there exists $f : \mathfrak{X} \twoheadrightarrow^d F$ such that $f(B) = \{b_1\}$.

Proof First, we construct $g : \mathfrak{X} \twoheadrightarrow^d \Phi_{ml}$ according to Proposition 11. Then $B \subseteq g^{-1}(b_1)$ and $A_i = g^{-1}(a_i)$ is the union of a set α_i of disjoint open balls. Put

$$f(x) := \begin{cases} g(x) & \text{if } g(x) \in C, \\ f_i^U(x) & \text{if } x \in U,\ U \in \alpha_i. \end{cases} \tag{11.1}$$

Since g and f_i^U are surjective, the same holds for f. So let us show

$$\mathbf{d}f^{-1}(a) = f^{-1}(R^{-1}(a))$$

(R is the accessibility relation on F). First suppose $a \in C$. Then (since g is a d-morphism)

$$\mathbf{d}f^{-1}(a) = \mathbf{d}g^{-1}(a) = g^{-1}(C) = f^{-1}(C) = f^{-1}(R^{-1}(a)).$$

Now suppose $a \notin C$, $I = \{i | a \in F_i\}$ and let R_i be the accessibility relation on F_i. We have:

$$f^{-1}(a) = \bigcup_{i \in I} \bigcup_{U \in \alpha_i} (f_i^U)^{-1}(a), \qquad R^{-1}(a) = C \cup \bigcup_{i \in I} R_i^{-1}(a),$$

and so $f^{-1}(R^{-1}(a)) = g^{-1}(C) \cup \bigcup_{i \in I} \bigcup_{U \in \alpha_i} (f_i^U)^{-1}(R_i^{-1}(a))$. Since f_i^U is a d-morphism,

$$f^{-1}(R^{-1}(a)) = g^{-1}(C) \cup \bigcup_{i \in I} \bigcup_{U \in \alpha_i} \mathbf{d}_U((f_i^U)^{-1}(a)) \subseteq g^{-1}(C) \cup \mathbf{d}f^{-1}(a). \quad (11.2)$$

Let us show that

$$g^{-1}(C) \subseteq \mathbf{d}f^{-1}(a). \quad (11.3)$$

Let $x \in g^{-1}(C)$. Since α_i is dense at x, every neighbourhood of x contains some $U \in \alpha_i$. Since f_i^U is surjective, $f(u) = f_i^U(u) = a$ for some $u \in U$. Therefore, $x \in \mathbf{d}f^{-1}(a)$.

Equations (11.2) and (11.3) imply $f^{-1}(R^{-1}(a)) \subseteq \mathbf{d}f^{-1}(a)$. Let us prove the converse:

$$\mathbf{d}f^{-1}(a) \subseteq f^{-1}(R^{-1}(a)). \quad (11.4)$$

We have $A_j \cap f^{-1}(a) = \varnothing$ for $j \notin I$ and A_j is open, hence $A_j \cap \mathbf{d}f^{-1}(a) = \varnothing$. Thus $\mathbf{d}f^{-1}(a) \subseteq g^{-1}(C) \cup A_i$. Now $g^{-1}(C) \subseteq f^{-1}(R^{-1}(a))$ by (11.2), so it remains to show that for any $i \in I$

$$\mathbf{d}f^{-1}(a) \cap A_i \subseteq f^{-1}(R^{-1}(a)). \quad (11.5)$$

To check this, consider any $x \in \mathbf{d}f^{-1}(a) \cap A_i$. Then $x \in U$ for some $U \in \alpha_i$, and thus by Lemma 6 and (11.2), $x \in \mathbf{d}f^{-1}(a) \cap U = \mathbf{d}_U(f^{-1}(a) \cap U) = \mathbf{d}_U(f_i^U)^{-1}(a) \subseteq f^{-1}(R^{-1}(a))$. This implies (11.5) and completes the proof of (11.4). □

Recall that ∂ denotes the boundary of a set in a topological space: $\partial A := \mathbf{C}A - \mathbf{I}A$.

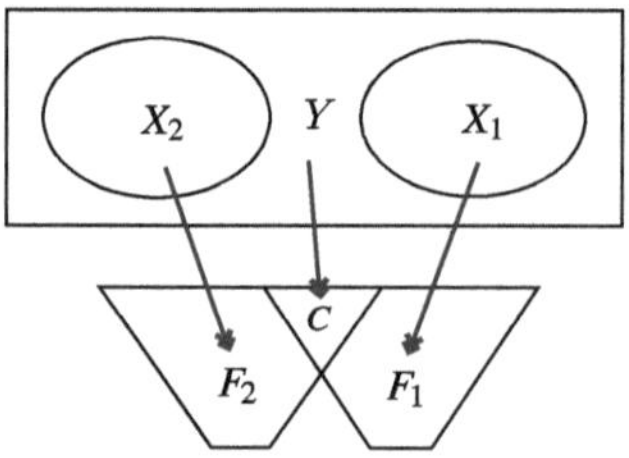

Fig. 11.2 Case (a)

Lemma 16 (Glueing lemma) *Let $\mathfrak{X}$ be a local T_1-space satisfying*
(a) $X = X_1 \cup Y \cup X_2$ for closed nonempty subsets X_1, Y, X_2 such that

- $X_1 \cap X_2 = X_1 \cap \mathbf{I}Y = X_2 \cap \mathbf{I}Y = \varnothing$,
- $\partial X_1 \cup \partial X_2 = \partial Y$,
- $\mathbf{dI}Y = Y$ *(i.e., Y is regular and dense in-itself).*

or
(b) $X = X_1 \cup X_2$ is a nontrivial closed partition.

Let $F = (W, R)$ be a finite **K4***-frame, $F_1 = (W_1, R_1)$, $F_2 = (W_2, R_2)$ its generated subframes such that $W = W_1 \cup W_2$ and suppose there are d-morphisms $f_i : \mathfrak{X}_i \twoheadrightarrow^d F_i$, $i = 1, 2$, where $\mathfrak{X}_i$ is the subspace of $\mathfrak{X}$ corresponding to X_i.*

In case (a) we also assume that F_1, F_2 have a common maximal cluster C, $f_i(\partial X_i) \subseteq R^{-1}(C)$ for $i = 1, 2$ and there is $g : \mathbf{I}Y \twoheadrightarrow^d C$ (where C is regarded as a frame with the universal relation, $\mathbf{I}Y$ as a subspace of $\mathfrak{X}$). Then $f_1 \cup f_2 \cup g : \mathfrak{X} \twoheadrightarrow^d F$ in case (a), $f_1 \cup f_2 : \mathfrak{X} \twoheadrightarrow^d F$ in case (b).[9]

Proof Let $f := f_1 \cup f_2 \cup g$ (or $f = f_1 \cup f_2$), $F_i = (W_i, R_i)$, $\mathbf{d} := \mathbf{d}_X$, $\mathbf{d}_i := \mathbf{d}_{X_i}$. For $w \in W$ there are four options.

(1) $w \in W_1 - W_2$. Then $\mathbf{d}f^{-1}(w) = \mathbf{d}f_1^{-1}(w) = \mathbf{d}_1 f_1^{-1}(w) = f_1^{-1}(R_1^{-1}(w))$ (since X_1 is closed and f_1 is a d-morphism). It remains to note that $R_1^{-1}(w) = R^{-1}(w) \subseteq W_1 - W_2$, and thus $f_1^{-1}(R_1^{-1}(w)) = f^{-1}(R^{-1}(w))$.

(2) $w \in W_2 - W_1$. Similar to case (1).

(3) $w \in (W_1 \cap W_2) - C$ in case (a) or $w \in W_1 \cap W_2$ in case (b). Then $f^{-1}(w) = f_1^{-1}(w) \cup f_2^{-1}(w)$, so as in (1),

$$\mathbf{d}f^{-1}(w) = \mathbf{d}_1 f_1^{-1}(w) \cup \mathbf{d}_2 f_2^{-1}(w) = f_1^{-1}(R_1^{-1}(w)) \cup f_2^{-1}(R_2^{-1}(w)) = f^{-1}(R^{-1}(w)).$$

(4) $w \in C$ in case (a). First note that $\mathbf{d}g^{-1}(w) = Y$. Indeed, g is a d-morphism onto the cluster C, so $\mathbf{d}_{\mathbf{I}Y} g^{-1}(w) = g^{-1}(C) = \mathbf{I}Y$. Hence $\mathbf{I}Y \subseteq \mathbf{d}g^{-1}(w) \subseteq \mathbf{dI}Y = Y$, and thus

$$Y = \mathbf{dI}Y \subseteq \mathbf{dd}g^{-1}(w) \subseteq \mathbf{d}g^{-1}(w)$$

[9] $f_1 \cup f_2$ is the map f such that $f|X_i = f_i$; similarly for $f_1 \cup f_2 \cup g$ (Fig. 11.2).

by Lemma 8(2). Next, since X_1, X_2 are closed and f_1, f_2 are d-morphisms we have

$$\begin{aligned}\mathbf{d}f^{-1}(w) &= \mathbf{d}f_1^{-1}(w) \cup \mathbf{d}f_2^{-1}(w) \cup \mathbf{d}g^{-1}(w) = \mathbf{d}_1 f_1^{-1}(w) \cup \mathbf{d}_2 f_2^{-1}(w) \cup Y \\ &= f_1^{-1}(R_1^{-1}(w)) \cup f_2^{-1}(R_2^{-1}(w)) \cup Y = f^{-1}(R^{-1}(w)).\end{aligned}$$

□

Case (b) of the previous lemma can be generalized as follows.

Lemma 17 *Suppose a topological space $\mathfrak{X}$ is the disjoint union of open subspaces: $\mathfrak{X} = \bigsqcup_{i \in I} \mathfrak{X}_i$. Suppose a Kripke* **K4***-frame F is the union of its generated subframes: $F = \bigcup_{i \in I} F_i$ and suppose $f_i : \mathfrak{X}_i \twoheadrightarrow^d F_i$. Then $\bigcup_{i \in I} f_i : \mathfrak{X} \twoheadrightarrow^d F$.*

Definition 18 Let $\mathfrak{X}$ be a topological space, $F = (W, R, R_D)$ be a frame. Then a surjective map $f : X \longrightarrow W$ is called a *dd-morphism* (in symbols, $f : \mathfrak{X} \twoheadrightarrow^{dd} F$) if

(1) $f : \mathfrak{X} \twoheadrightarrow^d (W, R)$ is a d-morphism;
(2) $f : (X, \neq_X) \twoheadrightarrow (W, R_D)$ *is a p-morphism of Kripke frames.*

Lemma 18 *If $f : \mathfrak{X} \twoheadrightarrow^{dd} F$, then $\mathbf{Ld}_{\neq}(\mathfrak{X}) \subseteq \mathbf{L}(F)$ and for any closed 2-modal A*

$$\mathfrak{X} \vDash A \Leftrightarrow F \vDash A.$$

Proof Similar to Proposition 10 and Lemma 3. □

Definition 19 A set-theoretic map $f : X \longrightarrow Y$ is called *n-fold at $y \in Y$* if $|f^{-1}(y)| = n$[10]; f is called *manifold at y* if it is n-fold for some $n > 1$.

Proposition 12 *(1) Let $G = (X, \neq_X)$, $F = (W, S)$ be Kripke frames such that $\overline{S} = W^2$, and let $f : X \longrightarrow W$ be a surjective function. Then*

$$f : G \twoheadrightarrow F \quad \textit{iff} \quad f \textit{ is manifold exactly at } S\textit{-reflexive points of } F.$$

(2) Let $\mathfrak{X}$ be a T_1-space, $F = (W, R, R_D)$ a rooted **KT₁***-frame, $f : \mathfrak{X} \twoheadrightarrow^d (W, R)$. Then $f : \mathfrak{X} \twoheadrightarrow^{dd} F$ iff for any strictly R-minimal v*

$$vR_Dv \Leftrightarrow f \textit{ is manifold at } v.$$

(3) If $\mathfrak{X}$ is a T_1-space, $f : \mathfrak{X} \twoheadrightarrow^d F = (W, R)$ and $R^{-1}(w) \neq \varnothing$ for any $w \in W$, then $f : \mathfrak{X} \twoheadrightarrow^{dd} F_\forall$, where $F_\forall := (W, R, W^2)$.

Proof (1) Note that f is a p-morphism iff for any $x \in X$

$$f(X - \{x\}) = S(f(x)) = \begin{cases} W & \text{if } f(x)Sf(x), \\ W - \{f(x)\} & \text{otherwise.} \end{cases}$$

10 $|\ldots|$ denotes the cardinality.

(2) By (1), $f : \mathfrak{X} \twoheadrightarrow^{dd} F$ iff

$$\forall v \in W(vR_D v \Leftrightarrow \left|f^{-1}(v)\right| > 1).$$

The latter equivalence holds whenever $R^{-1}(v) \neq \varnothing$. Indeed, by Corollary 3, $\mathbf{d}f^{-1}(v) = f^{-1}(R^{-1}(v)) \neq \varnothing$, and thus $f^{-1}(v)$ is not a singleton (since $\mathfrak{X}$ is a T_1-space). $R^{-1}(v) \neq \varnothing$ also implies $vR_D v$ by Proposition 8.

(3) follows from (2). □

After we have proved the main technical results, in the next sections we will study dd-logics of specific spaces.

11.7 D4 and $\mathbf{DT_1}$ as Logics of Zero-Dimensional Dense-in-Themselves Spaces

In this section we will prove d-completeness of **D4** and dd-completeness of $\mathbf{DT_1}$ w.r.t. zero-dimensional spaces. The proof follows rather easily from the previous section and an additional technical fact (Proposition 13) similar to the McKinsey–Tarski lemma.

Recall that a (nonempty) topological space $\mathfrak{X}$ is *zero-dimensional* if clopen sets constitute its open base [4]. Zero-dimensional T_1-spaces with a countable base are subspaces of the Cantor discontinuum or of the set of irrationals [28].

Lemma 19 *Let $\mathfrak{X}$ be a zero-dimensional dense-in-itself Hausdorff space. Then for any n there exists a nontrivial open partition $\mathfrak{X} = \mathfrak{X}_1 \sqcup \ldots \sqcup \mathfrak{X}_n$, in which every $\mathfrak{X}_i$ is also a zero-dimensional dense-in-itself Hausdorff space.*

Proof It is sufficient to prove the claim for $n = 2$ and then apply induction. A dense-in-itself space cannot be a singleton, so there are two different points $x, y \in X$. Since $\mathfrak{X}$ is T_1 and zero-dimensional, there exists a clopen set U such that $x \in U$, $y \notin U$. So $X = U \cup (X - U)$ is a nontrivial open partition. The Hausdorff property, density-in-itself, zero-dimensionality are inherited for open subspaces. □

Proposition 13 *Let $\mathfrak{X}$ be a zero-dimensional dense-in-itself metric space, $y \in X$. Let Ψ_l be the frame consisting of an irreflexive root b and its reflexive successors $a_0, \ldots, a_{l-1}$ (Fig. 11.3). Then there exists $f : \mathfrak{X} \twoheadrightarrow^{d} \Psi_l$ such that $f(y) = b$ and for every i there is an open partition of $f^{-1}(a_i)$ which is dense at y.*

Proof Let $O(a, r) := \{x \in X \mid \rho(a, x) < r\}$, where ρ is the distance in $\mathfrak{X}$. There exist clopen sets $Y_0, Y_1, \ldots$ such that

$$\{y\} \subset \ldots \subset Y_{n+1} \subset Y_n \subset \ldots Y_1 \subset Y_0 = X$$

and $Y_n \subseteq O(y, 1/n)$ for $n > 0$. These Y_n can be easily constructed by induction. Then

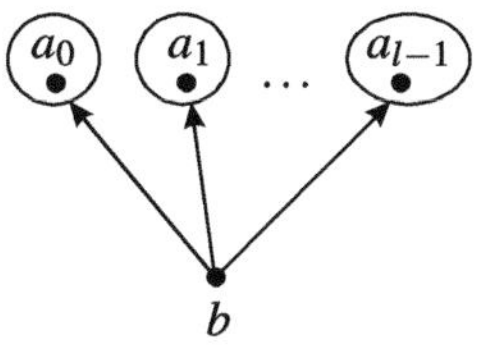

Fig. 11.3 Frame Ψ_l

$$\bigcap_n Y_n = \{y\} \text{ and } X - \{y\} = \bigsqcup_n X_n,$$

where $X_n = Y_n - Y_{n+1}$. Note that the X_n are nonempty and open, $X_n \subseteq O(y, 1/n)$ for $n > 0$.

Now define a map $f : X \longrightarrow \Psi_l$ as follows:

$$f(x) = \begin{cases} a_{r(n)} & \text{if } x \in X_n; \\ b & \text{if } x = y, \end{cases}$$

where $r(n)$ is the remainder of dividing n by l; it is clear that f is surjective.

Let us show that for any x,

$$x \in \mathbf{d}f^{-1}(u) \text{ iff } f(x)Ru. \qquad (*)$$

(i) Assume that $u = a_j$. Then $f^{-1}(u) = \bigcup_n X_{nl+j}$, and

$$f(x)Ru \text{ iff } (f(x) = b \text{ or } f(x) = u).$$

To prove 'if' in (*), consider two cases.

1. Suppose $f(x) = u$, $x \in X_{nl+j}$. Since X_{nl+j} is nonempty and open, it is dense-in-itself, and thus $x \in \mathbf{d}X_{nl+j} \subseteq \mathbf{d}f^{-1}(u)$.

2. Suppose $f(x) = b$, i.e. $x = y$. Then $x \in \mathbf{d}f^{-1}(u)$ since $X_{nl+j} \subseteq O(y, 1/n)$.

The previous argument also shows that $\{X_{nl+j} \mid n \geq 0\}$ is an open partition of $f^{-1}(a_j)$ which is dense at y.

To prove 'only if', suppose $f(x)Ru$ is not true. Then $f(x) = a_k$ for some $k \neq j$, and so for some n, $x \in X_n$, $X_n \cap f^{-1}(u) = \varnothing$. Since X_n is open, $x \notin \mathbf{d}f^{-1}(u)$.

(ii) Assume that $u = b$. Then $f^{-1}(u) = \{y\}$, and so $\mathbf{d}f^{-1}(u) = \varnothing = f^{-1}(R^{-1}(u))$. □

Proposition 14 *Let $\mathfrak{X}$ be a zero-dimensional dense-in-itself separable metric space, F a finite rooted* **D4***-frame. Then there exists a d-morphism $\mathfrak{X} \twoheadrightarrow^d F$ which is 1-fold at the root of F if this root is irreflexive.*

Proof By induction on the size of F.

(i) If F is a finite cluster, the claim follows from Proposition 11.

(ii) If $F = C \cup F_1 \cup \dots \cup F_l$, where $C = \{b_1, \dots, b_m\}$ is a non-degenerate root cluster, $F_1, \dots, F_l$ are the subframes generated by the successors of C, we can apply Lemma 15. In fact, every open ball U in $\mathfrak{X}$ is zero-dimensional and dense-in-itself.

(iii) Suppose $F = \breve{b} \cup F_0 \cup \dots \cup F_{l-1}$, where b is an irreflexive root of F, F_i are the subframes generated by the successors of b. There exists $g : X \twoheadrightarrow^d \Psi_l$ by Proposition 13, with an arbitrary $y \in X$. Then $g^{-1}(a_i)$ is a union of a set α_i of disjoint open sets, and α_i is dense at y. If $U \in \alpha_i$, then by IH, there exists $f_i^U : U \twoheadrightarrow^d F_i$. Put

$$f(x) = \begin{cases} b & \text{if } x = y; \\ f_i^U(x) & \text{if } x \in U,\ U \in \alpha_i. \end{cases}$$

Then as in Lemma 15 it follows that $f : X \twoheadrightarrow^d F$.

Finally note that if the root of F is irreflexive, the first step of the construction is case (iii), so the preimage of the root is a singleton. □

Theorem 1 *If $\mathfrak{X}$ is a zero-dimensional dense-in-itself separable metric space, then* $\mathbf{Ld}(\mathfrak{X}) = \mathbf{D4}$.

Proof By Propositions 14 and 10, $\mathbf{Ld}(\mathfrak{X}) \subseteq \mathbf{L}(F)$ for any finite rooted $\mathbf{D4}$-frame F, thus $\mathbf{Ld}(\mathfrak{X}) \subseteq \mathbf{D4}$ since $\mathbf{D4}$ has the fmp. By Lemma 8, $\mathbf{D4} \subseteq \mathbf{Ld}(\mathfrak{X})$. □

Lemma 20 *Let $\mathfrak{X}$ be a zero-dimensional dense-in-itself separable metric space, F a finite* $\mathbf{D4}$*-frame. Then there exists a d-morphism $\mathfrak{X} \twoheadrightarrow^d F$ which is 1-fold at all strictly minimal points.*

Proof $F = F_1 \cup \dots \cup F_n$ for different finite rooted $\mathbf{D4}$-frames F_i. By Lemma 19, $\mathfrak{X} = \mathfrak{X}_1 \sqcup \dots \sqcup \mathfrak{X}_n$ for zero-dimensional dense-in-themselves subspaces $\mathfrak{X}_i$, which are also metric and separable. By Proposition 14, we construct $f_i : \mathfrak{X}_i \twoheadrightarrow^d F_i$. Then by Lemma 17, $\bigcup_{i=1}^{n} f_i : \mathfrak{X} \twoheadrightarrow^d F$. Every strictly minimal point of F is an irreflexive root of a unique F_i, so its preimage is a singleton. □

Proposition 15 *Let $\mathfrak{X}$ be a zero-dimensional dense-in-itself separable metric space, $F \in \mathfrak{F}_0$ a finite* $\mathbf{DT_1}$*-frame. Then there exists a dd-morphism $\mathfrak{X} \twoheadrightarrow^{dd} F$.*

Proof We slightly modify the proof of the previous lemma. Let $F = (W, R, R_D)$, $G = (W, R)$. Then $G = G_1 \cup \dots \cup G_n$ for different cones G_i. We call G_i *special* if its root is strictly R-minimal and R_D-reflexive. We may assume that exactly $G_1, \dots, G_m$ are special. Then we count them twice and present G as $G_1 \cup G_1' \cup \dots \cup G_m \cup G_m' \cup G_{m+1} \cup \dots \cup G_n$, where $G_i' = G_i$ for $i \leq m$ (or as $G_1 \cup G_1' \cup \dots \cup G_m \cup G_m'$ if $m = n$).

Now we can argue as in the proof of Lemma 20. By Lemma 19, $\mathfrak{X} = \mathfrak{X}_1 \sqcup \mathfrak{X}_1' \sqcup \dots \sqcup \mathfrak{X}_m \sqcup \mathfrak{X}_m' \sqcup \mathfrak{X}_{m+1} \sqcup \dots \sqcup \mathfrak{X}_n$ for zero-dimensional dense-in-itself separable metric $\mathfrak{X}_i$, $\mathfrak{X}_i'$. By Proposition 14, we construct the maps $f_i : \mathfrak{X}_i \twoheadrightarrow^d G_i$, $f_i' : \mathfrak{X}_i' \twoheadrightarrow^d G_i'$, which are 1-fold at irreflexive roots; hence by Lemma 17, $f : \mathfrak{X} \twoheadrightarrow^d G$ for $f := \bigcup_{i=1}^{n} f_i \cup \bigcup_{i=1}^{m} f_i'$.

Every strictly minimal point $a \in G$ is an irreflexive root of a unique G_i. If a is R_D-irreflexive, then G_i is not special, so $f^{-1}(a) = f_i^{-1}(a)$ is a singleton. If a is

R_D-reflexive, then G_i is special, so $f^{-1}(a) = f_i^{-1}(a) \cup (f_i')^{-1}(a)$, and thus f is 2-fold at a. Therefore, $f : \mathfrak{X} \twoheadrightarrow^{dd} F$ by Proposition 12. □

Lemma 21 *Let $M = (W, R, R_D, \varphi)$ be a rooted Kripke model over a basic frame*[11] *validating AT_1, Ψ a set of 2-modal formulas closed under subformulas. Let $M' = (W', R', R'_D, \theta')$ be a filtration of M through Ψ described in Lemma 4.*[12] *Then the frame (W', R', R'_D) is also basic and validates AT_1.*

Proof Clearly R' is transitive by definition. For any two different $a, b \in W'$ we have $aR'_D b$ since $xR_D y$ for any $x \in a,\ y \in b$ (as $F \in \mathfrak{F}_0$).

Next, note that if a is R'_D-irreflexive, then $a = \{x\}$ for some R_D-irreflexive x. In this case, since $(W, R, R_D) \vDash AT_1$, there is no y such that yRx (Proposition 8), hence $(R')^{-1}(a) = \varnothing$, and thus $(W', R', R'_D) \vDash AT_1$.

Finally, $R' \subseteq R'_D$. Indeed, all different points in F' are R'_D-related, so it remains to show that every R'_D-irreflexive point is R'-irreflexive. As noted above, such a point is a singleton class $x^{\sim} = \{x\}$, where x is R_D-irreflexive. Then x is R-minimal, so in W' there is no loop of the form $x^{\sim}\underline{R}x_1\underline{R}\ldots\underline{R}x^{\sim}$, and thus $x^{\sim}$ is R'-irreflexive. □

By a standard argument, Lemma 21 implies

Theorem 2 *Every logic of the form* $\mathbf{KT_1}+A$, *where A is a closed 2-modal formula, has the finite model property.*

Theorem 3 *Let $\mathfrak{X}$ be a zero-dimensional dense-in-itself separable metric space. Then* $\mathbf{Ld}_{\neq}(\mathfrak{X}) = \mathbf{DT_1}$.

Proof For any finite $\mathbf{DT_1}$-frame F we have $\mathbf{Ld}_{\neq}(\mathfrak{X}) \subseteq \mathbf{L}(F)$ by Proposition 15 and Lemma 18. By the previous theorem, $\mathbf{DT_1}$ has the fmp, so $\mathbf{Ld}_{\neq}(\mathfrak{X}) \subseteq \mathbf{DT_1}$. Since $\mathfrak{X} \vDash^d \mathbf{DT_1}$ (Proposition 7), it follows that $\mathbf{Ld}_{\neq}(\mathfrak{X}) = \mathbf{DT_1}$. □

Proposition 16 [9, Lemma 3.1] *Every countable*[13] *rooted* **K4**-*frame is a d-morphic image of a subspace of* **Q**.

To apply this proposition to the language with the difference modality, we need to examine the preimage of the root for the constructed morphism. Fortunately, in the proof of Lemma 3.1 in [9] the preimage of a root r is a singleton iff r is irreflexive.

Lemma 22 *Let F be a countable* **K4**-*frame. Then there exists a d-morphism from a subspace of* **Q** *onto F which is 1-fold at all strictly minimal points.*

Proof Similar to Lemma 20. We can present F as a countable union of different cones $\bigcup_{i \in I} F_i$ and **Q** as a disjoint union $\bigsqcup_{i \in I} \mathfrak{X}_i$ of spaces homeomorphic to **Q**. By Proposition 16 (and the discussion after it), for each i there exists $f_i : \mathcal{Y}_i \twoheadrightarrow^d F_i$ for

[11] Basic frames were defined in Sect. 11.4.

[12] Recall that R' is the transitive closure of $\underline{R}$, $R'_D = \underline{R_D}$.

[13] In this chapter, as well as in [9], 'countable' means 'of cardinality at most $\aleph_0$'.

some subspace $\mathcal{Y}_i \subseteq \mathfrak{X}_i$ such that f_i is 1-fold at the root r_i of F_i if r_i is irreflexive. Now by Lemma 17 $f := \bigcup_{i \in I} f_i : \bigsqcup_{i \in I} \mathcal{Y}_i \twoheadrightarrow^d F$, and f is 1-fold at all strictly minimal points of F (i.e., the irreflexive r_i)—since every r_i belongs only to F_i, so $f^{-1}(r_i) = f_i^{-1}(r_i)$. □

Proposition 17 *Let F be a countable* $\mathbf{KT_1}$*-frame. Then there exists a dd-morphism from a subspace of* $\mathbf{Q}$ *onto F.*

Proof Similar to Proposition 15. If $F = (W, R, R_D)$, the frame $G = (W, R)$ is a countable union of different cones. There are two types of cones: non-special G_i $(i \in I)$ and special (with strictly R-minimal and R_D-reflexive roots) H_j $(j \in J)$:

$$G = \bigcup_{i \in I} G_i \cup \bigcup_{j \in J} H_j.$$

Then we duplicate all special cones

$$G = \bigcup_{i \in I} G_i \cup \bigcup_{j \in J} H_j \cup \bigcup_{j \in J} H'_j$$

and as in the proof of 16, construct $f : \bigsqcup_{i \in I} \mathcal{Y}_i \sqcup \bigsqcup_{j \in J} \mathcal{Z}_j \sqcup \bigsqcup_{j \in J} \mathcal{Z}'_j \twoheadrightarrow^d F$. This map is 1-fold exactly at all R_D-irreflexive points, so it is a dd-morphism onto F. □

Corollary 4 $\mathbf{Ld}_{\neq}$*(all T_1-spaces)* $= \mathbf{KT_1}$.

Proof Note that $\mathbf{KT_1}$ is complete w.r.t. countable frames and every subspace of $\mathbf{Q}$ is T_1. □

Proposition 18 *Let $\Lambda = \mathbf{KT_1} + \Gamma$ be a consistent logic, where Γ is a set of closed formulas. Then Λ is dd-complete w.r.t. subspaces of* $\mathbf{Q}$.

Proof Since every closed formula is canonical, Λ is Kripke complete. So for every formula $A \notin \Lambda$ there is a frame F_A such that $F_A \models \Lambda$ and $F_A \not\models A$. By Proposition 17, there is a subspace $\mathfrak{X}_A \subseteq \mathbf{Q}$ and $f_A : \mathfrak{X}_A \twoheadrightarrow^{dd} F_A$. Then $\mathfrak{X}_A \not\models A$, $\mathfrak{X}_A \models \Lambda$ by Lemma 18. Therefore, $\mathbf{Ld}_{\neq}(\mathcal{K}) = \Lambda$ for $\mathcal{K} := \{\mathfrak{X}_A \mid A \notin \Lambda\}$. □

Remark 6 A logic of the form described in Proposition 18 is dd-complete w.r.t. a set of subspaces of $\mathbf{Q}$. This set may be non-equivalent to a single subspace. For example, there is no subspace $\mathfrak{X} \subseteq \mathbf{Q}$ such that $\mathbf{KT_1} = \mathbf{Ld}_{\neq}(\mathfrak{X})$. Indeed, consider

$$A := [\neq]\Box\bot \wedge \Box\bot.$$

Then A is satisfiable in $\mathfrak{X}$ iff $\mathfrak{X} \models^d A$ iff $\mathfrak{X}$ is discrete. So A is consistent in $\mathbf{KT_1}$. Now if $\mathbf{KT_1} = \mathbf{Ld}_{\neq}(\mathfrak{X})$, then A must be satisfiable in $\mathfrak{X}$, hence $\mathfrak{X} \models^d A$; but $\mathbf{KT_1} \not\vdash A$, and so we have a contradiction.

11.8 Connectedness

Connectedness was the first example of a property expressible in cu-logic, but not in c-logic. The corresponding connectedness axiom from [37] will be essential for our further studies. In this section we show that it is weakly canonical, i.e., valid in weak canonical frames—a fact not mentioned in [37].

Lemma 23 [37] *A topological space $\mathfrak{X}$ is connected iff $\mathfrak{X} \vDash^c AC$, where*

$$AC := [\forall](\Box p \vee \Box\neg p) \to [\forall]p \vee [\forall]\neg p.$$

For the case of Alexandrov topology there is an equivalent definition of connectedness in relational terms.

Definition 20 For a transitive Kripke frame $F = (W, R)$ we define the *comparability* relation $R^{\pm} := R \cup R^{-1} \cup I_W$. F is called *connected* if the transitive closure of $R^{\pm}$ is universal. A subset $V \subseteq W$ is called *connected* in F if the frame $F|V$ is connected.

A 2-modal frame (W, R, S) is called *(R)-connected* if (W, R) is connected.

Thus F is connected iff every two points x, y can be connected by a *non-oriented path* (which we call just a *path*), a sequence of points $x_0x_1 \dots x_n$ such that $x = x_0R^{\pm}x_1 \dots R^{\pm}x_n = y$.

From [37] and Proposition 6 we obtain

Lemma 24 *(1) For an* **S4***-frame F, the associated space $N(F)$ is connected iff F is connected.*

(2) For a **K4***-frame F, $F_{\forall} \vDash AC^{\sharp u}$ iff F is connected.*

Lemma 25 *Let $M = (W, R, R_D, \theta)$ be a rooted generated submodel of the m-weak canonical model for a modal logic $\Lambda \supseteq$* **K4D**$^+$*. Then*

(1) Every R-cluster in M is finite of cardinality at most 2^m.
(2) (W, R) has finitely many R-maximal clusters.
(3) For each R-maximal cluster C in M there exists an m-formula $\beta(C)$ such that:

$$\forall x \in M\ (M, x \vDash \beta(C) \Leftrightarrow x \in \overline{R}^{-1}(C)).$$

The proof is similar to [12, Sect. 8.6].

Lemma 26 *Every rooted generated subframe of a weak canonical frame for a logic $\Lambda \supseteq$* **K4D**$^+$ *$+ AC^{\sharp u}$ is connected.*

Proof Let M be a weak canonical model for Λ, M_0 its rooted generated submodel with the frame $F = (W, R, R_D)$, and suppose F is disconnected. Then there exists a nonempty proper clopen subset V in the space $N(W, \overline{R})$. Let Δ be the set of all R-maximal clusters in V and put

$$B := \bigvee_{C \in \Delta} \beta(C).$$

Then B defines V in M_0, i.e., $V = \overline{R}^{-1}(\bigcup \Delta)$. Indeed, $\bigcup \Delta \subseteq V$ implies $\overline{R}^{-1}(\bigcup \Delta) \subseteq V$ since V is closed. The other way round, $V \subseteq \overline{R}^{-1}(\bigcup \Delta)$ since for any $v \in V$, $\overline{R}(v)$ contains an R-maximal cluster $C \in \Delta$, and $\overline{R}(v) \subseteq V$ as V is open.

So $w \models B$ for any $w \in V$, and since V is open, $w \models \overline{\Box}B$. By the same reason, $w \models \overline{\Box}\neg B$ for any $w \notin V$. Hence

$$M_0 \models [\forall]\,(\overline{\Box}B \vee \overline{\Box}\neg B).$$

By Proposition 1, all substitution instances of AC are true in M_0. So we have

$$M_0 \models [\forall]\,(\overline{\Box}B \vee \overline{\Box}\neg B) \to [\forall]\,B \vee [\forall]\,\neg B,$$

and thus

$$M_0 \models [\forall]\,B \vee [\forall]\,\neg B.$$

This contradicts the fact that V is a nonempty proper subset of W. □

In d-logic instead of connectedness we can express some its local versions; they will be considered in the next section.

11.9 Kuratowski Formula and Local 1-Componency

In this section we briefly study the Kuratowski formula distinguishing $\mathbf{R}$ from $\mathbf{R}^2$ in d-logic. Here the main proofs are similar to the previous section, so most of the details are left to the reader.

Definition 21 We define the *Kuratowski formula* as

$$Ku := \quad \Box(\overline{\Box}p \vee \overline{\Box}\neg p) \to \Box p \vee \Box\neg p.$$

The spaces validating Ku are characterized as follows [31].

Lemma 27 *For a topological space $\mathfrak{X}$, $\mathfrak{X} \models^d Ku$ iff for any $x \in X$ and any open neighbourhood U of x, if $U - \{x\}$ is a disjoint union $V_1 \cup V_2$ of sets open in the subspace $U - \{x\}$, then there exists a neighbourhood[14] $V \subseteq U$ of x such that $V - \{x\} \subseteq V_1$ or $V - \{x\} \subseteq V_2$.*

[14] In [31] neighbourhoods are assumed to be open, but this does not matter here since every neighbourhood contains an open neighbourhood.

Definition 22 A topological space $\mathfrak{X}$ is called *locally connected* if every neighbourhood of any point x contains a connected neighbourhood of x. Similarly, $\mathfrak{X}$ is called *locally 1-component* if every punctured neighbourhood of any point x contains a connected punctured neighbourhood of x.

It is well known [4] that in a locally connected space every neighbourhood U of any point x contains a connected *open* neighbourhood of x (e.g. the connected component of x in $\mathbf{I}U$).

Lemma 28 *If $\mathfrak{X}$ is locally 1-component, then $\mathfrak{X} \vDash^d Ku$.*

The proof is straightforward, and we leave it to the reader.

Lemma 29 *(1) Every space d-validating Ku has the following* non-splitting property*:*

(NSP) If an open set U is connected, $x \in U$ and $U - \{x\}$ is open, then $U - \{x\}$ is connected.

(2) Suppose $\mathfrak{X}$ is locally connected and local T_1. Then (NSP) holds in $\mathfrak{X}$ iff $\mathfrak{X}$ is locally 1-component iff $\mathfrak{X} \vDash^d Ku$.

Proof (1) We assume $\mathfrak{X} \vDash^d Ku$ and check (NSP). Suppose U is open and connected, $U^\circ := U - \{x\}$ is open, and consider a partition $U^\circ = U_1 \cup U_2$ for open U_1, U_2. By Lemma 27 there exists an open $V \subseteq U$ containing x such that $V \subseteq \{x\} \cup U_1$ or $V \subseteq \{x\} \cup U_2$. Consider the first case (the second one is similar). We have a partition

$$U = (\{x\} \cup U_1) \cup U_2,$$

and $\{x\} \cup U_1 = V \cup U_1$, so $\{x\} \cup U_1$ is open. Hence by connectedness, $U = \{x\} \cup U_1$, i.e., $U^\circ = U_1$. Therefore, U° is connected.

(2) It suffices to show that (NSP) implies the local 1-componency. Consider $x \in X$ and its neighbourhood U_1. Since $\mathfrak{X}$ is local T_1, U_1 contains an open neighborhood U_2 in which x is closed, i.e., $\mathbf{C}\{x\} \cap U_2 = \{x\}$. By the local connectedness, U_2 contains a connected open neighbourhood U_3, and again $\mathbf{C}\{x\} \cap U_3 = \{x\}$; thus $U_3 - \{x\}$ is open. By (NSP), $U_3 - \{x\}$ is connected. □

Remark 7 The *(n-th) generalized Kuratowski formula* is the following formula in variables $p_0, \ldots, p_n$

$$Ku_n := \Box \bigvee_{k=0}^{n} \overline{\Box} Q_k \to \bigvee_{k=0}^{n} \Box \neg Q_k,$$

where $Q_k := p_k \wedge \bigwedge_{j \neq k} \neg p_j$.

The formula Ku_1 is related to the equality found by Kuratowski [27]:

$$(*) \quad \mathbf{d}((x \cap \mathbf{d}(-x)) \cup (-x \cap \mathbf{d}x)) = \mathbf{d}x \cap \mathbf{d}(-x),$$

which holds in every algebra $DA(\mathbf{R}^n)$ for $n > 1$, but not in $DA(\mathbf{R})$. This equality corresponds to the modal formula

$$Ku' := \quad \Diamond((p \wedge \Diamond\neg p) \vee (\neg p \wedge \Diamond p)) \leftrightarrow \Diamond p \wedge \Diamond\neg p,$$

and one can show that $\mathbf{D4} + Ku' = \mathbf{D4} + Ku_1 = \mathbf{D4} + Ku$.

Remark 8 The class of spaces validating Ku_n is described in [31]. In particular, it is valid in all locally n-component spaces defined as follows.

A neighbourhood U of a point x in a topological space is called *n-component at x* if the punctured neighbourhood $U - \{x\}$ has at most n connected components. A topological space is called *locally n-component* if the n-component neighbourhoods at each of its points constitute a local base (i.e., every neighbourhood contains an n-component neighbourhood).

Lemma 30 [31] *For a transitive Kripke frame* (W, R)

$(W, R) \vDash Ku$ *iff for any R-irreflexive x, the subset* $R(x)$ *is connected (in the sense of Definition 20).*

Theorem 4 *The logics* $\mathbf{K4} + Ku$, $\mathbf{D4} + Ku$ *are weakly canonical, and thus Kripke complete.*

A proof of Theorem 4 based on Lemma 30 and a 1-modal version of Lemma 30 is straightforward, cf. [36] or [31] (the latter paper proves the same for Ku_n).

Hence we obtain

Theorem 5 *The logic* $\mathbf{DT_1K} := \mathbf{DT_1} + Ku$ *is weakly canonical, and thus Kripke complete.*

Proof (Sketch.) For the axiom Ku the argument from the proof of Theorem 4 is still valid due to the definability of all maximal clusters (Lemma 25). The remaining axioms are Sahlqvist formulas. □

Theorem 6 *The logic* $\mathbf{DT_1CK} := \mathbf{DT_1K} + AC^{\sharp u}$ *is weakly canonical, and thus Kripke complete.*

Proof We can apply the previous theorem and Lemma 26. □

Completeness theorems from this section can be refined: in the next section we will prove the fmp for the logics considered above.

11.10 The Finite Model Property of D4K, DT$_1$K, and DT$_1$CK

For the logic $\mathbf{D4} + Ku$ the first proof of the fmp was given in [36]. Another proof (also for $\mathbf{D4} + Ku_n$) was proposed by M. Zakharyaschev [40]; it is based on a general and powerful method.

In this section we give a simplified version of the proof from [36]. It is based on a standard filtration method, and the same method is also applicable to 2-modal logics $\mathbf{DT_1K}$, $\mathbf{DT_1CK}$.

Theorem 7 *The logics* $\mathbf{DT_1K}$ *and* $\mathbf{DT_1CK}$ *have the finite model property.*

Proof Let Λ be one of these logics. Consider an m-formula $A \notin \Lambda$. Take a generated submodel $M = (W, R, R_D, \varphi)$ of the m-restricted canonical model of Λ such that $M, u \not\models A$ for some u. As we know, its frame is basic and its R-maximal clusters are definable (Lemma 25).

Put

$$\Psi_0 := \{\beta(C) \mid C \text{ is an } R\text{-maximal cluster in } M\},$$
$$\Psi_1 := \{A\} \cup \{\Box\gamma \mid \gamma \text{ is a Boolean combination of formulas from } \Psi_0\},$$
$$\Psi := \text{the closure of } \Psi_1 \text{ under subformulas.}$$

The set Ψ is obviously finite up to equivalence in Λ.

Take the filtration $M' = (W', R', R'_D, \varphi')$ of M through Ψ as in Lemma 21. By that lemma, $F' := (W', R', R'_D) \models \mathbf{KT_1}$. The seriality of R' easily follows from the seriality of R.

Next, if $\Lambda = \mathbf{DT_1CK}$, the frame (W, R, R_D) is connected by Lemma 26. So for any $x, y \in W$ there is an R-path from x to y. aRb implies $a^{\sim}R'b^{\sim}$, so there is an R'-path from $x^{\sim}$ to $y^{\sim}$ in F'. Therefore, $F' \models AC^{\sharp u}$. It remains to show that $F' \models Ku$. Consider an R'-irreflexive point $x^{\sim} \in W'$ and assume that $R'(x^{\sim})$ is disconnected. Let V be a nonempty proper connected component of $R'(x^{\sim})$. Consider

$$\Delta := \{C \mid \exists y(y^{\sim} \in V \ \&\ C \subseteq R(y) \ \&\ C \text{ is an } R\text{-maximal cluster in } M)\};$$
$$B := \bigvee_{C \in \Delta} \beta(C),$$

where $\beta(C)$ is from Lemma 25. Note that

(1) $z \in C \ \&\ C \in \Delta \Rightarrow z^{\sim} \in V.$

Indeed, if $C \in \Delta$, then for some $y^{\sim} \in V$ we have yRz; hence $y^{\sim}R'z^{\sim}$, so $z^{\sim} \in V$ by connectedness of V. Let us show that for any $y^{\sim} \in R'(x^{\sim})$

(2) $M', y^{\sim} \models B$ iff $M, y \models B$ iff $y^{\sim} \in V,$

i.e., B defines V in $R'(x^{\sim})$.

The first equivalence holds by the Filtration Lemma since $B \in \Psi_1$. Let us prove the second equivalence. To show 'if', suppose $y^{\sim} \in V$. By Lemma 5, in the restricted canonical model there is a maximal cluster C R-accessible from y; then $M, y \models \beta(C)$. We have $C \in \Delta$, and thus $M, y \models B$.

To show 'only if', suppose $y^\sim \notin V$, but $M, y \models B$. Then $M, y \models \beta(C)$ for some $C \in \Delta$, hence $C \subseteq R(y)$, i.e., yRz for some (and for all) $z \in C$; so it follows that $y^\sim R' z^\sim$. Thus $y^\sim$ and $z^\sim$ are in the same connected component of $R'(x^\sim)$, which implies $z^\sim \notin V$. However, $z^\sim \in V$ by (1), leading to a contradiction.

By Proposition 1, all substitution instances of Ku are true in M. So

$$M \models Ku(B) := \Box(\overline{\Box}B \vee \overline{\Box}\neg B) \to \Box B \vee \Box\neg B.$$

Consider an arbitrary $y \in R(x)$. Then for any $z \in R(y)$, $y^\sim$ and $z^\sim$ are in the same connected component of $R'(x^\sim)$. Thus $y^\sim$ and $z^\sim$ are both either in V or not in V, and so by (2), both of them satisfy either B or $\neg B$. Hence $M, y \models \overline{\Box}B \vee \overline{\Box}\neg B$. Therefore, x satisfies the premise of $Ku(B)$. Consequently, x must satisfy the conclusion of $Ku(B)$. Thus $M, x \models \Box B$ or $M, x \models \Box\neg B$. Since $\Box B, \Box\neg B \in \Psi_1$, the Filtration Lemma implies $M', x^\sim \models \Box B$ or $M', x^\sim \models \Box\neg B$. By (2), $V = R'(x^\sim)$ or $V = \varnothing$, which contradicts the assumption about V.

To conclude the proof, note that $A \in \Psi$, so by the Filtration Lemma $M', u^\sim \not\models A$. As we have proved, $F' \models \Lambda$. Therefore, Λ has the fmp. □

Theorem 8 *The logic* **D4K** *has the finite model property.*

Proof Use the argument from the proof of Theorem 7 without the second relation. □

Thanks to the fmp, we have a convenient class of Kripke frames for the logic $\mathbf{DT_1CK}$. This will allow us to prove the topological completeness result in the next section.

11.11 The dd-Logic of $\mathbf{R}^n$, $n \geq 2$

This section contains the main result of the chapter. The proof is based on the fmp theorem from the previous section and a technical construction of a dd-morphism presented in the Appendix.

In this section $\|\cdot\|$ denotes the standard norm in $\mathbf{R}^n$, i.e. for $x \in \mathbf{R}^n$

$$\|x\| = \sqrt{x_1^2 + \ldots + x_n^2}.$$

We begin with some simple observations on connectedness. For a path $\alpha = w_0 w_1 \ldots w_n$ in a **K4**-frame (W, R) we use the notation $\overline{R}(\alpha) := \bigcup_{i=0}^{n} \overline{R}(w_i)$. A path α is called *global* (in F) if $\overline{R}(\alpha) = W$.

Lemma 31 *Let $F = (W, R)$ be a finite connected* **K4**-*frame, $w, v \in W$. Then there exists a global path from w to v.*

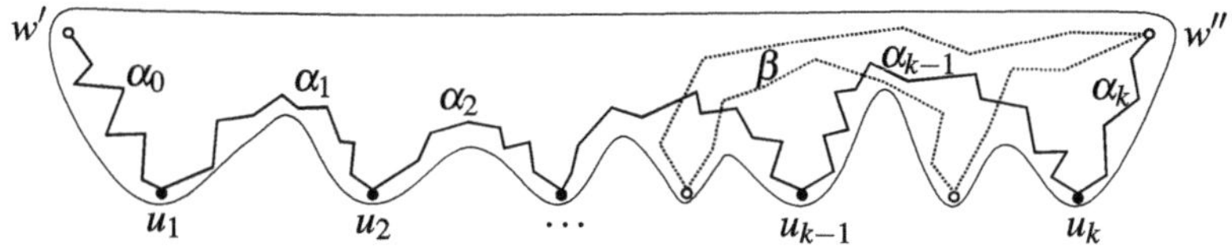

Fig. 11.4 Path α

Proof In the finite connected graph $(W, R^{\pm})$ the vertices w, v can be connected by a path visiting all the vertices (possibly several times). □

Lemma 32 *Let $F = (W, R, R_D)$ be a finite rooted* **DT$_1$CK**-*frame. Then the set of all R_D-reflexive points in F is connected.*

Proof Let x, y be two R_D-reflexive points. Since (W, R) is connected, there exists a path connecting x and y. Consider such a path α with the minimal number n of R_D-irreflexive points, and let us show that $n = 0$.

Suppose not. Take an R_D-irreflexive point z in α; then $\alpha = x \dots uzv \dots y$ for some u, v, and it is clear that zRu, zRv since z is strictly R-minimal. By Lemma 30, $R(z)$ is connected, so u, v can be connected by a path β in $R(z)$. Thus in α we can replace the part uzv with β, and the combined path $x \dots \beta \dots y$ contains $(n - 1)$ R_D-irreflexive points, which contradicts the minimality of n. □

Lemma 33 *Let $F = (W, R, R_D)$ be a finite rooted* **DT$_1$CK**-*frame and let $w', w'' \in W$ be R_D-reflexive. Then there is a global path $\alpha = w_0 \dots w_n$ in (W, R) such that $w' = w_0$, $w_n = w''$ and all R_D-irreflexive points occur only once in α.*

Proof Let $\{u_1, \dots, u_k\}$ be the R_D-irreflexive points. By connectedness, there exist paths $\alpha_0, \dots, \alpha_k$ from w' to u_1, from u_1 to u_2, …, and from u_k to w'', respectively.

By Lemma 32, the set $W' := W - \{u_1, \dots, u_k\}$ is connected. Hence we may assume that each α_i does not contain R_D-irreflexive points except its ends. Also there exists a loop β in $F' := F|W'$ from w'' to w'' such that $W - \bigcup_{i=1}^{k-1} \overline{R}(\alpha_i) \subseteq \overline{R}(\beta)$.

Then we can define α as the joined path $\alpha_0 \dots \alpha_k \beta$ (Fig. 11.4). □

Proposition 19 *For a finite rooted* **DT$_1$CK**-*frame $F = (W, R, R_D)$ and R-reflexive points $w', w'' \in W$, the following holds.*

(a) If $X = \{x \in \mathbf{R}^n \mid ||x|| \le r\}$, $n \ge 2$, then there exists $f : X \twoheadrightarrow^{dd} F$ such that $f(\partial X) = \{w'\}$;

(b) If $0 \le r_1 < r_2$ and

$$X = \{x \in \mathbf{R}^n \mid r_1 \le ||x|| \le r_2\},$$
$$Y' = \{x \in \mathbf{R}^n \mid ||x|| = r_1\},\ Y'' = \{x \in \mathbf{R}^n \mid ||x|| = r_2\},$$

then there exists $f : X \twoheadrightarrow^{dd} F$ such that $f(Y') = \{w'\}, f(Y'') = \{w''\}$.

For the proof see Appendix.

Theorem 9 *For $n \geq 2$, the dd-logic of* $\mathbf{R}^n$ *is* $\mathbf{DT_1CK}$.

Proof Since $\mathbf{R}^n$ is a locally 1-component connected dense-in-itself metric space, $\mathbf{R}^n \models^d \mathbf{DT_1CK}$.

Now consider a formula $A \notin \mathbf{DT_1CK}$. Due to the fmp (Theorem 7) there exists a finite rooted Kripke frame $F = (W, R, R_D) \models \mathbf{DT_1CK}$ such that $F \not\models A$. By Proposition 19 there exists $f : \mathbf{R}^n \twoheadrightarrow^{dd} F$. Hence $\mathbf{R}^n \not\models^d A$ by Lemma 18.

□

11.12 Concluding Remarks

Hybrid logics. Logics with the difference modality are closely related to hybrid logics. The paper [29] describes a validity-preserving translation from the language with the topological and difference modalities into the hybrid language with the topological modality, nominals and the universal modality.

Apparently a similar translation exists for dd-logics considered in our chapter. There may be an additional option—to use 'local nominals', propositional constants that may be true not in a single point, but in a discrete set. Perhaps one can also consider 'one-dimensional nominals' naming 'lines' or 'curves' in the main topological space; there may be many other similar options.

Definability. Among several types of topological modal logics considered in this chapter dd-logics are the most expressive. The correlation between all the types are shown in Fig. 11.5. A language $\mathscr{L}_1$ is *reducible* to $\mathscr{L}_2$ ($\mathscr{L}_1 \leq \mathscr{L}_2$) if every $\mathscr{L}_1$-definable class of spaces is $\mathscr{L}_2$-definable; $\mathscr{L}_1 < \mathscr{L}_2$ if $\mathscr{L}_1 \leq \mathscr{L}_2$ and $\mathscr{L}_2 \not\leq \mathscr{L}_1$. The non-strict reductions 1–7 in Fig. 11.5 are rather obvious. Let us explain why 1–6 are strict.

The relations 1 and 2 are strict since the c-logics of $\mathbf{R}$ and $\mathbf{Q}$ coincide [32], while the cu- and d-logics are different [15, 37].

The relation 3 is strict since in d-logic without the universal modality we cannot express connectedness (this follows from [15]). The relations 4 and 6 are strict, since the cu-logics of $\mathbf{R}$ and $\mathbf{R}^2$ are the same [37], while the cd- and du-logics are different [17, 31].

In cd- and dd-logics we can express *global 1-componency*: the formula

$$[\neq](\overline{\Box} p \vee \overline{\Box} \neg p) \to [\neq] p \vee [\neq] \neg p$$

is c-valid in a space $\mathfrak{X}$ iff the complement of any point in $\mathfrak{X}$ is connected. So we can distinguish the line $\mathbf{R}$ and the circle $\mathbf{S^1}$. In du- (and cu-) logic this is impossible, since there is a local homemorphism $f(t) = e^{it}$ from $\mathbf{R}$ onto $\mathbf{S^1}$. It follows that the relation 5 is strict. Our conjecture is that the relation 7 is strict as well.

Axiomatization. There are several open questions about axiomatization and completeness of certain dd-logics.

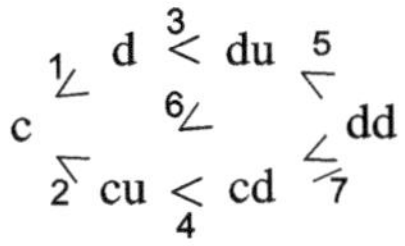

Fig. 11.5 Correlation between topomodal languages

1. The first group of questions is about the logic of **R**. On the one hand, in [25] it was proved that $\mathbf{Lc}_{\neq}(\mathbf{R})$ is not finitely axiomatizable. Probably the same method can be applied to $\mathbf{Ld}_{\neq}(\mathbf{R})$. On the other hand, $\mathbf{Lc}_{\neq}(\mathbf{R})$ has the fmp [24], and we hope that the same holds for the dd-logic. The decidability of $\mathbf{Ld}_{\neq}(\mathbf{R})$ follows from [11], since this logic is a fragment of the universal monadic theory of **R**; and by a result from [34] it is PSPACE-complete. However, constructing an explicit infinite axiomatization of $\mathbf{Lc}_{\neq}(\mathbf{R})$ or $\mathbf{Ld}_{\neq}(\mathbf{R})$ might be a serious technical problem.

2. A 'natural' semantical characterization of the logic $\mathbf{DT_1C} + Ku_2$ (which is a proper sublogic of $\mathbf{Ld}_{\neq}(\mathbf{R})$) is not quite clear. Our conjecture is that it is complete w.r.t. 2-dimensional cell complexes, or more exactly, adjunction spaces obtained from finite sets of 2-dimensional discs and 1-dimensional segments.

3. We do not know any syntactic description of dd-logics of 1-dimensional cell complexes (i.e., unions of finitely many segments in $\mathbf{R}^3$ that may have only endpoints as common). Their properties are probably similar to those of $\mathbf{Ld}_{\neq}(\mathbf{R})$.

4. It may be interesting to study topological modal logics with the graded modalities $[\neq]_n A$ with the following semantics: $x \models [\neq]_n A$ iff there are at least n points $y \neq x$ such that $y \models A$.

5. The papers [32] and [21] prove completeness and strong completeness of **S4** w.r.t. any dense-in-itself metric space. The corresponding result for d-logics is completeness of **D4** w.r.t. an arbitrary dense-in-itself separable metric space. Is separabilty essential here? Does strong completeness hold in this case? Similar questions make sense for dd-logics.

6. Gabelaia [17] presents a 2-modal formula cd-valid exactly in T_0-spaces. However, the cd-logic (and the dd-logic) of the class of T_0-spaces is still unknown. Note that the d-logic of this class has been axiomatized in [8]; probably the same technique is applicable to cd- and dd-logics.

7. In footnote 7 we have mentioned that there is a gap in the paper [37]. Still we can prove that for any connected, locally connected metric space $\mathfrak{X}$ such that the boundary of any ball is nowhere dense, $\mathbf{Lc}_{\forall}(\mathfrak{X}) = \mathbf{S4U} + AC$. But for an arbitrary connected metric space $\mathfrak{X}$ we do not even know if $\mathbf{Lc}_{\forall}(\mathfrak{X})$ is finitely axiomatizable.

8. Is it possible to characterize finitely axiomatizable dd-logics that are complete w.r.t. Hausdorff spaces? metric spaces? Does there exist a dd-logic complete w.r.t. Hausdorff spaces, but incomplete w.r.t. metric spaces?

9. Suppose we have a c-complete modal logic L, and let $\mathscr{K}$ be the class of all topological spaces where L is valid. Is it always true that $\mathbf{Lc}_{\forall}(\mathscr{K}) = \mathbf{LU}$? and $\mathbf{Lc}_{\neq}(\mathscr{K}) = \mathbf{LD}$? Similar questions can be formulated for d-complete modal logics and their du- and dd-extensions.

10. An interesting topic not addressed in this chapter is the complexity of topo-modal logics. In particular, the complexity is unknown for the d-logic (and the dd-logic) of $\mathbf{R}^n$ $(n > 1)$.

Acknowledgments We would like to thank the referee who has helped us to improve the first version of the manuscript. The work on this chapter was supported by RFBR grants 11-01-00281-a, 11-01-00958-a, 11-01-93107-CNRS-a and the Russian President's grant NSh-5593.2012.1.

Appendix

Finally let us give technical details of the proofs of Propositions 11, 19.

Proposition 11 *Let $\mathfrak{X}$ be a dense-in-itself separable metric space, $B \subset X$ a closed nowhere dense set. Then there exists a d-morphism $g : \mathfrak{X} \twoheadrightarrow^d \Phi_{ml}$ with the following properties:*

(1) $B \subseteq g^{-1}(b_1)$;
(2) *every $g^{-1}(a_i)$ (fori $\leq l$) is a union of a set α_i of disjoint open balls, which is dense at any point of $g^{-1}(\{b_1, \dots, b_m\})$.*

The frame Φ_{ml} is shown in Fig. 11.1.

Proof Let $X_1, \dots, X_n, \dots$ be a countable base of $\mathfrak{X}$ consisting of open balls. We construct sets A_{ik}, B_{jk} for $1 \leq i \leq l$, $1 \leq j \leq m$, $k \in \omega$, with the following properties:

(1) A_{ik} is the union of a finite set α_{ik} of nonempty open balls whose closures are disjoint;
(2) $\mathbf{C}A_{ik} \cap \mathbf{C}A_{i'k} = \varnothing$ for $i \neq i'$;
(3) $\alpha_{ik} \subseteq \alpha_{i,k+1}$; $A_{ik} \subseteq A_{i,k+1}$;
(4) B_{jk} is finite;
(5) $B_{jk} \subseteq B_{j,k+1}$;
(6) $A_{ik} \cap B_{jk} = \varnothing$;
(7) $X_{k+1} \subseteq \bigcup_{i=1}^{l} A_{ik} \Rightarrow \alpha_{i,k+1} = \alpha_{ik},\ B_{j,k+1} = B_{jk}$;
(8) if $X_{k+1} \not\subseteq \bigcup_{i=1}^{l} A_{ik}$, there are closed nontrivial balls $P_1, \dots, P_l$ such that for any i, j

$$P_i \subseteq X_{k+1} - A_{ik},\ \alpha_{i,k+1} = \alpha_{ik} \cup \{\mathbf{I}P_i\},\ (B_{j,k+1} - B_{jk}) \cap X_{k+1} \neq \varnothing;$$

(9) $A_{ik} \subseteq X - B$;
(10) $B_{jk} \subseteq X - B$;
(11) $j \neq j' \Rightarrow B_{j'k} \cap B_{jk} = \varnothing$.

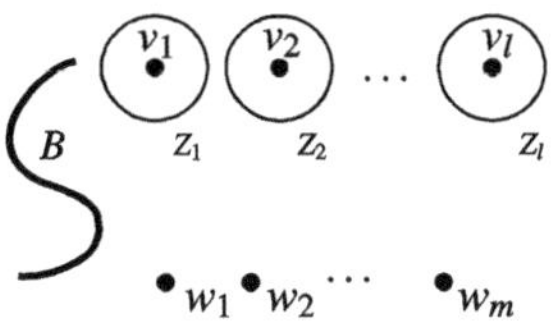

Fig. 11.6 Case $k = 0$

We carry out both the construction and the proof by induction on k.

Let $k = 0$; $(X - B)$ is infinite since it is nonempty and open in a dense-in-itself $\mathfrak{X}$. Take distinct points $v_1, \dots, v_l \notin B$ and disjoint closed nontrivial balls $Z_1, \dots, Z_l \subset X - B$ with centers at $v_1, \dots, v_l$, respectively (see Fig.11.6).

Put

$$\alpha_{i0} := \{\mathbf{I}Z_i\};\ A_{i0} := \mathbf{I}Z_i;$$

then $Z_i = \mathbf{C}A_{i0}$. As above, since $(X - B) - \bigcup_{i=1}^{l} Z_i$ is nonempty and open, it is infinite. Pick distinct $w_1, \dots, w_m \in X - B$ and put $B_{j0} := \{w_j\}$. Then the required properties hold for $k = 0$.

At the induction step we construct $A_{i,k+1}, B_{j,k+1}$. Put $Y_k := \bigcup_{i=1}^{l} A_{ik}$ and consider two cases.

(a) $X_{k+1} \subseteq Y_k$. Then put:

$$\alpha_{i,k+1} := \alpha_{ik};\ A_{i,k+1} := A_{ik};\ B_{j,k+1} := B_{jk}.$$

(b) $X_{k+1} \not\subseteq Y_k$. Then $X_{k+1} \not\subseteq \mathbf{C}Y_k$. Indeed, $X_{k+1} \subseteq \mathbf{C}Y_k$ implies $X_{k+1} \subseteq \mathbf{IC}Y_k = Y_k$ since X_{k+1} is open and by (1) and (2). So we put

$$W_0 := X_{k+1} - \mathbf{C}Y_k - \bigcup_{j=1}^{m} B_{jk},\ W := W_0 - B.$$

Since $(X_{k+1} - \mathbf{C}Y_k)$ is nonempty and open and every B_{jk} is finite by (4), W_0 is also open and nonempty (by the density of $\mathfrak{X}$). By the assumption of Proposition 11, B is closed, and thus W is open.

W is also nonempty. Otherwise $W_0 \subseteq B$, and then $W_0 \subseteq \mathbf{I}B = \varnothing$ (since B is nowhere dense by the assumption of Proposition 11).

Now we argue similarly to the case $k = 0$. Take disjoint closed nontrivial balls $P_1, \dots, P_l \subset W$. Then $W - \bigcup_{i=1}^{l} P_i$ is infinite, so we pick distinct $b_{1,k+1}, \dots, b_{m,k+1}$ in this set and put

$$B_{j,k+1} := B_{jk} \cup \{b_{j,k+1}\},\ \alpha_{i,k+1} := \alpha_{ik} \cup \{\mathbf{I}P_i\},\ A_{i,k+1} := A_{ik} \cup \mathbf{I}P_i.$$

In case (a) all the required properties hold for $(k+1)$ by the construction.

In case (b) we have to check only (1), (2), (6), (8)–(11).

(8) holds since by construction we have

$$P_i \subset W \subset X_{k+1} - \mathbf{C}Y_k \subset X_{k+1} - A_{ik};$$
$$b_{j,k+1} \in W \subseteq X_{k+1},\ b_{j,k+1} \in (B_{j,k+1} - B_{jk}).$$

(1): From IH it is clear that $\alpha_{i,k+1}$ is a finite set of open balls and their closures are disjoint; note that $P_i \cap \mathbf{C}A_{ik} = \varnothing$ since $P_i \subseteq W \subseteq -\mathbf{C}A_{ik}$.

(2): We have

$$\begin{aligned}\mathbf{C}A_{i,k+1} \cap \mathbf{C}A_{i',k+1} &= (\mathbf{C}A_{ik} \cup P_i) \cap (\mathbf{C}A_{i'k} \cup P_{i'})\\ &= (\mathbf{C}A_{ik} \cap \mathbf{C}A_{i'k}) \cup (\mathbf{C}A_{ik} \cap P_{i'}) \cup (\mathbf{C}A_{i'k} \cap P_i) \cup (P_i \cap P_{i'})\\ &= \mathbf{C}A_{ik} \cap \mathbf{C}A_{i'k} = \varnothing\end{aligned}$$

by IH and by the construction; note that $P_i, P_i' \subseteq W \subseteq -\mathbf{C}Y_k$.

(6): We have

$$A_{i,k+1} \cap B_{j,k+1} = (A_{ik} \cap B_{jk}) \cup (\mathbf{I}P_i \cap \{b_{j,k+1}\}) \cup (A_{ik} \cap \{b_{j,k+1}\}) \cup (\mathbf{I}P_i \cap B_{jk}) = \varnothing$$

by IH and since $b_{j,k+1} \notin P_i,\ b_{j,k+1} \in W \subseteq X - Y_k, P_i \subset W \subseteq X - B_{jk}$.

(9): We have $A_{i,k+1} = A_{ik} \cup \mathbf{I}P_i \subseteq -B$ since $A_{ik} \subseteq -B$ by IH, and $P_i \subset W \subseteq -B$ by the construction.

Likewise, (10) follows from $B_{jk} \subseteq -B$ and $b_{j,k+1} \in W \subseteq -B$.

To check (11), assume $j \neq j'$. We have $B_{j',k+1} \cap B_{j,k+1} = B_{j'k} \cap B_{jk}$ since $b_{j',k+1} \neq b_{j,k+1},\ b_{j,k+1} \in W \subseteq -B_{j'k}$ and $b_{j',k+1} \in W \subseteq -B_{jk}$. Then apply IH.

Therefore, the required sets A_{ik}, B_{jk} are constructed. Now put

$$\alpha_i := \bigcup_k \alpha_{ik},\ A_i := \bigcup \alpha_i = \bigcup_k A_{ik},\ B_j := \bigcup_k B_{jk},$$

$$B_1' := X - (\bigcup_i A_i \cup \bigcup_j B_j),$$

and define a map $g : X \longrightarrow \Phi_{ml}$ as follows:

$$g(x) := \begin{cases} a_i & \text{if } x \in A_i,\\ b_j & \text{if } x \in B_j,\ j \neq 1,\\ b_1 & \text{otherwise (i.e., for } x \in B_1'). \end{cases}$$

By (2), (3), (5), (6), (11), g is well defined; by (9), (10), $B \subseteq g^{-1}(b_1)$.

To prove that g is a d-morphism, we check some other properties.

(12) $X - \bigcup_{i=1}^{l} A_i \subseteq \mathbf{d}B_j$.

Indeed, take an arbitrary $x \notin \bigcup_{i=1}^{l} A_i$ and show that $x \in \mathbf{d}B_j$, i.e.,

(13) $(U - \{x\}) \cap B_j \neq \varnothing$

for any neighbourhood U of x. First assume that $x \notin B_j$. Take a basic open X_{k+1} such that $x \in X_{k+1} \subseteq U$. Then $X_{k+1} \nsubseteq \bigcup_{i=1}^{l} A_i$, and (8) implies $B_{j,k+1} \cap X_{k+1} \neq \varnothing$. Thus $B_j \cap U \neq \varnothing$. So we obtain (13).

Suppose $x \in B_j$; then $x \in B_{jk}$ for some k. Since $\mathfrak{X}$ is dense-in-itself and $\{X_1, X_2, \dots\}$ is its open base, $\{X_{s+1} \mid s \geq k\}$ is also an open base (note that every ball in $\mathfrak{X}$ contains a smaller ball). So $x \in X_{s+1} \subseteq U$ for some $s \geq k$. Since $x \notin \bigcup_{i=1}^{l} A_i$, we have $X_{s+1} \nsubseteq \bigcup_{i=1}^{l} A_i$, and so $(B_{j,s+1} - B_{js}) \cap X_{s+1} \neq \varnothing$ by (8); thus $(B_j - B_{js}) \cap U \neq \varnothing$. Now $x \in B_{jk} \subseteq B_{js}$ implies (13).

(14) $\mathbf{d}B_j \subseteq X - \bigcup_{i=1}^{l} A_i$.

Indeed, $B_j \subseteq -A_i$ by (3), (5), (6). So $\mathbf{d}B_j \subseteq \mathbf{d}(-A_i) \subseteq -A_i$ since A_i is open.

Similarly we obtain

(15) $\mathbf{d}B'_1 \subseteq X - \bigcup_{i=1}^{l} A_i, \qquad \mathbf{d}A_i \subseteq X - \bigcup_{r \neq i} A_r.$

Also note that

(16) $A_i \subseteq \mathbf{d}A_i$

since A_i is open, $\mathfrak{X}$ is dense-in-itself. As in (12) we have

(17) α_i is dense at every point of B_j, B'_1 (and thus $B_j,\ B'_1 \subseteq \mathbf{d}A_i$).

To conclude that g is a d-morphism, note that

$$g^{-1}(a_i) = A_i,\ g^{-1}(b_j) = B_j \text{ (for } j \neq 1),\ g^{-1}(b_1) = B'_1,$$

and so by (15), (16), (17)

$$\mathbf{d}g^{-1}(a_i) = \mathbf{d}A_i = X - \bigcup_{r \neq i} A_r = g^{-1}(R^{-1}(a_i)),$$

and by (12), (14), (15)

$$\mathbf{d}g^{-1}(b_j) = \mathbf{d}B_j = X - \bigcup_{i=1}^{l} A_i = g^{-1}(R^{-1}(b_j)) \text{ (for } j \neq 1),$$

$$\mathbf{d}g^{-1}(b_1) = \mathbf{d}B_1' = X - \bigcup_{i=1}^{l} A_i = g^{-1}(R^{-1}(b_1)). \qquad \square$$

Proposition 19 *For a finite rooted* $\mathbf{DT_1CK}$*-frame* $F = (W, R, R_D)$ *and* R*-reflexive points* $w', w'' \in W$*, the following holds.*

(a) *If* $X = \{x \in \mathbf{R}^n \mid ||x|| \leq r\}$, $n \geq 2$, *then there exists* $f : X \twoheadrightarrow^{dd} F$ *such that* $f(\partial X) = \{w'\}$;
(b) *If* $0 \leq r_1 < r_2$ *and*

$$X = \{x \in \mathbf{R}^n \mid r_1 \leq ||x|| \leq r_2\},$$
$$Y' = \{x \in \mathbf{R}^n \mid ||x|| = r_1\},\ Y'' = \{x \in \mathbf{R}^n \mid ||x|| = r_2\},$$

then there exists $f : X \twoheadrightarrow^{dd} F$ *such that* $f(Y') = \{w'\}, f(Y'') = \{w''\}$.

Proof By induction on $|W|$. Let us prove (a) first. There are five cases:

(a1) $W = R(b)$ (and hence bRb) and $b = w'$. Then there exists $f : X \twoheadrightarrow^d (W, R)$. Indeed, let C be the cluster of b (as a subframe of (W, R)). Then $(W, R) = C$ or $(W, R) = C \cup F_1 \cup \ldots \cup F_l$, where the F_i are generated by the successors of C. If $(W, R) = C$, we apply Proposition 11; otherwise we apply Lemma 15 and IH.

By Proposition 8 it follows that R_D is universal. And so by Proposition 12(3) f is a dd-morphism.

(a2) $W = R(b)$ and not $w'Rb$. We may assume that $r = 3$. Put

$$X_1 := \{x \mid ||x|| \leq 1\},\ Y := \{x \mid 1 \leq ||x|| \leq 2\},\ X_2 := \{x \mid 2 \leq ||x|| \leq 3\}.$$

By the case (a1), there is $f_1 : X_1 \twoheadrightarrow^{dd} F$ with $f_1(\partial X_1) = \{b\}$. Let C be a maximal cluster in $R(w')$. By Proposition 11 there is $g : \mathbf{I}Y \twoheadrightarrow^d C$. Since $R(w') \neq W$, we can apply IH to the frame $F' := F_{\forall}^{w'}$ and construct a dd-morphism $f_2 : X_2 \twoheadrightarrow^{dd} F'$ with $f_2(\partial X_2) = \{w'\}$. Now since $f_i(\partial X_i) \subseteq R^{-1}(C)$, the Glueing lemma 16 is applicable. Thus $f : X \twoheadrightarrow^d F$ for $f := f_1 \cup f_2 \cup g$ [See Fig. 11.7, Case (a2)]. Note that $\partial X \subset \partial X_2$, so $f(\partial X) = f_2(\partial X) = \{w'\}$.

As in case (a1), f is a dd-morphism by Proposition 12.

(a3) (W, R) is not rooted. By Lemma 33 there is a global path α in F with a single occurrence of every R_D-irreflexive point. We may assume that $\alpha = b_0c_0b_1c_1 \ldots c_{m-1}b_m$, $b_m = w'$ and for any $i < m$, $c_i \in C_i \subseteq R(b_i) \cap R(b_{i+1})$,

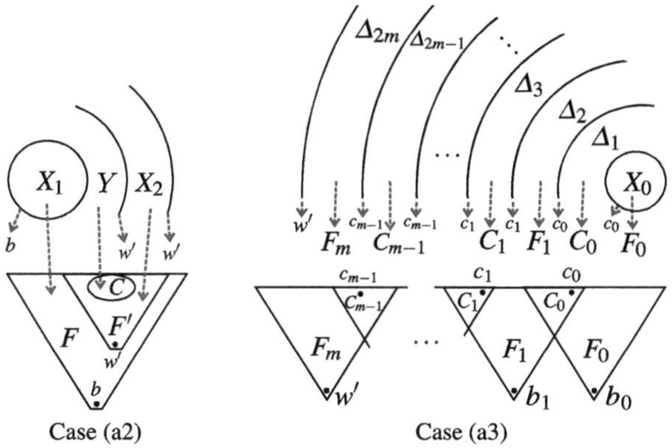

Fig. 11.7 dd-morphism f

where C_i is an R-maximal cluster. Such a path is called *reduced*. For $0 \leq j \leq m$ we put $F_j := F|\overline{R}(b_j)$.

Since (W, R) is not rooted, each F_j is of smaller size than F, so we can apply the induction hypothesis to F_j. We may assume that

$$X = \{x \mid ||x|| \leq 2m + 1\}, \ Y = \{x \mid ||x|| = 2m + 1\}.$$

Then put

$$X_i := \{x \mid ||x|| \leq i + 1\} \text{ for } 0 \leq i \leq 2m,$$

$$Y_i := \partial X_i, \ \Delta_i := \mathbf{C}(X_i - X_{i-1}) \text{ for } 0 \leq i \leq 2m.$$

By IH and Proposition 11 there exist

$$f_0 : X_0 \twoheadrightarrow^{dd} F_0 \text{ such that } f_0(Y_0) = \{c_0\},$$
$$f_{2j} : \Delta_{2j} \twoheadrightarrow^{dd} F_j \text{ such that } f_{2j}(Y_{2j}) = \{c_j\}, \ f_{2j}(Y_{2j-1}) = \{c_{j-1}\} \text{ for } 1 \leq j \leq m,$$
$$f_{2j-1} : \mathbf{I}\Delta_{2j+1} \twoheadrightarrow^{d} C_j \text{ for } 0 \leq j \leq m-1.$$

One can check that $f : X \twoheadrightarrow^{dd} F$ for $f := \bigcup_{j=0}^{2m} f_j$ (Fig. 11.7).

(a4) $W = \overline{R}(b)$, $\neg bR_D b$ (and so $\neg bRb$). We may assume that

$$X = \{x \mid ||x|| \leq 2\}, \ Y = \{x \mid ||x|| = 2\}.$$

Then similar to case (a3) put

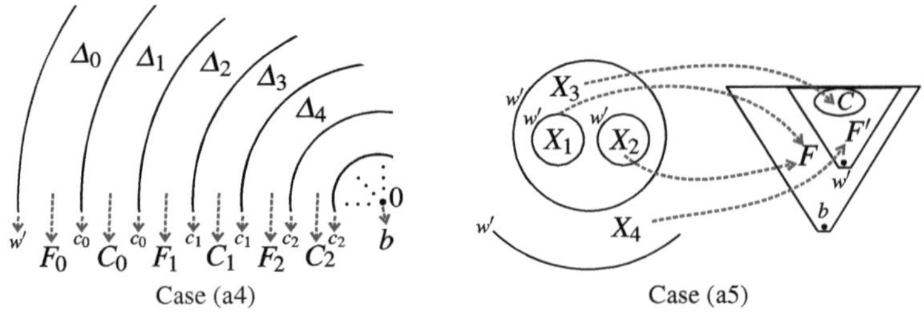

Fig. 11.8 dd-morphism f

$$X_0 := X,\ Y_0 := Y,\ X_i := \left\{x \mid ||x|| \leq \frac{1}{i}\right\},\ Y_i := \partial X_i,\ \Delta_i := \mathbf{C}(X_i - X_{i+1}),\ (i > 0).$$

Consider the frame $F'\!: = F|W'$, where $W' = W - \{b\}$. Note that $w' \in W'$ since $w'Rw'$ by the assumption of Proposition 19. By Lemma 30 F' is connected, and thus $F' \vDash \mathbf{DT_1CK}$. By Lemma 33 there is a reduced global path $\alpha = a_1 \dots a_m$ in F' such that $a_1 = w'$. Let

$$\gamma = a_1 a_2 \dots a_{m-1} a_m a_{m-1} \dots a_2 a_1 a_2 \dots$$

be an infinite path shuttling back and forth through α. Rename the points in γ:

$$\gamma = b_0 c_0 b_1 c_1 \dots b_m c_m b_{m+1} \dots \tag{11.6}$$

Again as in case (a3) we put $F_j\!: = F|\overline{R}(b_j)$, and assume that $c_j \in C_j$ and C_j is an R-maximal cluster. By IH there exist

$$f_0 : \Delta_0 \twoheadrightarrow^{dd} F_0 \text{ such that } f_0(Y_0) = \{b_0\} = \{w'\},\ f_1(Y_1) = \{c_0\},$$
$$f_{2j} : \Delta_{2j} \twoheadrightarrow^{dd} F_j \text{ such that } f_{2j}(Y_{2j}) = \{c_{j-1}\},\ f_{2j}(Y_{2j+1}) = \{c_j\} \text{ for } j > 0,$$

and by Proposition 11 there exist $f_{2j+1} : \mathbf{I}\Delta_{2j+1} \twoheadrightarrow^{d} C_j$. Put

$$f(x) := \begin{cases} b & \text{if } x = \mathbf{0}, \\ f_{2j}(x) & \text{if } x \in \Delta_{2j}, \\ f_{2j+1}(x) & \text{if } x \in \mathbf{I}\Delta_{2j+1}, \end{cases}$$

One can check that f is a d-morphism (Fig. 11.8).

(a5) $W = \overline{R}(b)$, $\neg bRb$ and $bR_D b$. Then R_D is universal, $w' \neq b$. Put

$$X' := \{x \mid ||x|| < 1\},\ X_4 := \{x \mid 1 \leq ||x|| \leq 2\},$$

and let X_1, X_2 be two disjoint closed balls in X', $X_3 := X' - X_1 - X_2$.

Let C be a maximal cluster in $R(w')$, $F' := F|R(w')$. Then there exist:

$$f_i : X_i \twoheadrightarrow^d (W, R) \text{ for } i = 1, 2 \text{ such that } f_i(\partial X_i) = \{w'\}, \text{ by case (a4)},$$
$$f_3 : X_3 \twoheadrightarrow^d C, \text{ by Proposition 11},$$
$$f_4 : X_4 \twoheadrightarrow^{dd} F' \text{ such that } f_4(\partial X_4) = \{w'\}, \text{ by the induction hypothesis.}$$

Put $f := f_1 \cup f_2 \cup f_3 \cup f_4$ (Fig. 11.8). Then $f(\partial X) = \{w'\}$.

By Lemma 16 (b), $f_1 \cup f_2 : X_1 \cup X_2 \twoheadrightarrow^d F$, and hence $f : X \twoheadrightarrow^d F$ by Lemma 16 (a). f is manifold at b, thus it is a dd-morphism by Lemma 18.

Now we prove (b). There are three cases.

(b1) $w' = w'' = b$ and $W = R(b)$. The argument is the same as in case (a1), using Proposition 11, Lemma 15, the induction hypothesis, and Proposition 12.

(b2) $w' = w'' = b$, but $W \neq R(b)$. Consider a maximal cluster $C \subseteq R(b)$. Since all spherical shells for different r_1 and r_2 are homeomorphic, we assume that $r_1 = 1$, $r_2 = 4$. Consider the sets

$$X_1 := \{x \mid 1 \leq ||x|| \leq 2\}, \quad X' := \{x \mid 2 < ||x|| < 3\}, \quad X_3 := \{x \mid 3 \leq ||x|| \leq 4\},$$

and let $X_0 \subset X'$ be a closed ball, $X_2 := X' - X_0$. Let $F' := F|R(b)$. There exist

$$f_1 : X_1 \twoheadrightarrow^{dd} F' \text{ such that } f_1(\partial X_1) = \{b\}, \text{ by case (b1)},$$
$$f_2 : X_2 \twoheadrightarrow^d C, \text{ by Proposition 11},$$
$$f_3 : X_3 \twoheadrightarrow^{dd} F' \text{ such that } f_3(\partial X_3) = \{b\}, \text{ by case (b1)},$$
$$f_0 : X_0 \twoheadrightarrow^{dd} F \text{ such that } f_4(\partial X_0) = \{b\}, \text{ by statement (a) for } F.$$

One can check that $f : X \twoheadrightarrow^{dd} F$ for $f := f_0 \cup f_1 \cup f_2 \cup f_3$.

(b3) $w' \neq w''$ and for some $b \in W$, $W = R(b)$, so F has an R-reflexive root. Let

$$F_1 := F|R(w'), \; F_2 := F|R(w''),$$

and let C_i be an R-maximal cluster in F_i for $i \in \{1, 2\}$.

We assume that $r_1 = 1$, $r_2 = 6$ and consider the sets

$$X_i := \{x \mid i \leq ||x|| \leq i + 1\}, \; i \in \{1, \ldots, 5.\}$$

By case (b1) and Proposition 11 we have

$$f_1 : X_1 \twoheadrightarrow^{dd} F_1 \text{ such that } f_1(\partial X_1) = \{w'\}, \qquad f_2 : \mathbf{I}X_2 \twoheadrightarrow^d C_1,$$
$$f_3 : X_3 \twoheadrightarrow^{dd} F \text{ such that } f_3(\partial X_3) = \{b\}, \qquad f_4 : \mathbf{I}X_4 \twoheadrightarrow^d C_2,$$
$$f_5 : X_5 \twoheadrightarrow^{dd} F_2 \text{ such that } f_1(\partial X_5) = \{w''\}.$$

One can check that $f : X \twoheadrightarrow^{dd} F$ for $f := \bigcup_{i=1}^{5} f_i$ [Fig. 11.9, Case (b3)].

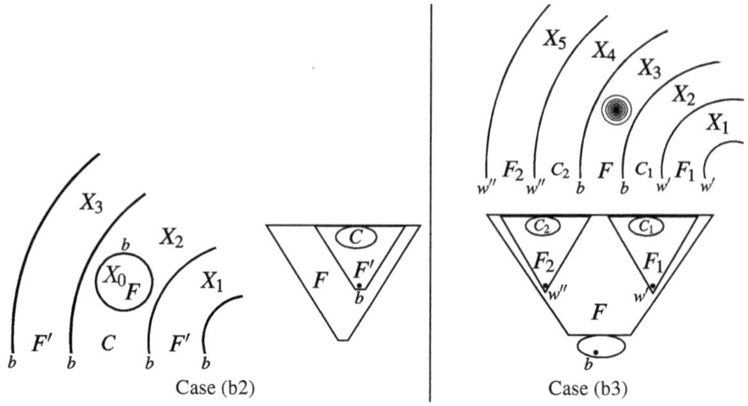

Fig. 11.9 dd-morphism f

(b4) $w' \neq w''$ and $W \neq R(b)$ for any $b \in W$. By Lemma 32 there is a reduced path $\alpha = b_0c_0b_1 \dots c_{m-1}b_m$ from $b_0 = w'$ to $b_m = w''$ that does not contain R_D-irreflexive points, $c_i \in C_i$, where C_i is an R-maximal cluster. We may also assume that

$$\overline{R}(b_i) \neq W \text{ for any } i \in \{1, \dots, m-1\}. \tag{11.7}$$

Indeed, if the frame (W, R) is not rooted, then (11.7) obviously holds. If (W, R) is rooted, then its root r is irreflexive and by Lemma 30, $R(r)$ is connected, so there exists a path α in $R(r)$ satisfying (11.7). Put

$$F_0 := F,\ F_j := F|R(b_j),\ 1 \leq j \leq m.$$

Assuming that $r_1 = 1$, $r_2 = 2m + 1$ we define

$$X_i := \{x \mid ||x|| \leq i + 1\},\ Y_i := \partial X_i \text{ (for } 0 \leq i \leq 2m + 1),$$
$$\Delta_i := \mathbf{C}(X_{i+1} - X_i) \text{ (for } 0 \leq i \leq 2m).$$

By cases (b2), (b1), Proposition 11, and the induction hypothesis there exist

$f_0 : \Delta_0 \twoheadrightarrow^{dd} F = F_0$ such that $f_0(Y_0) = f_0(Y_1) = \{w'\}$;
$f_{2j} : \Delta_{2j} \twoheadrightarrow^{dd} F_j$ such that $f_{2j}(Y_{2j+1}) = \{c_j\}$, $f_{2j}(Y_{2j}) = \{c_{j-1}\}$ $(1 \leq j \leq m)$;
$f_{2j-1} : \mathbf{I}\Delta_{2j-1} \twoheadrightarrow^{d} C_{j-1}$ $(1 \leq j \leq m)$,
$f_{2m} : \Delta_{2m} \twoheadrightarrow^{dd} F_m$ such that $f_{2m}(Y_{2m}) = \{c_m\}$, $f_{2m}(Y_{2m+1}) = \{w''\}$.

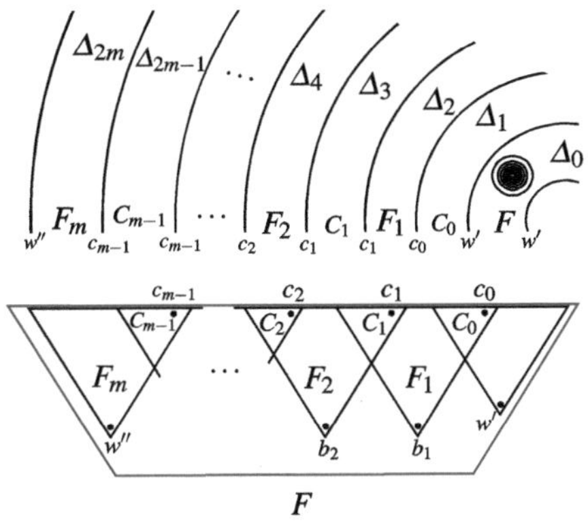

Fig. 11.10 dd-morphism f, case (b4)

We claim that $f : X \twoheadrightarrow^{dd} F$ for $f := \bigcup_{i=0}^{2m} f_i$ (Fig. 11.10). First, we prove by induction using Lemma 16 (see the previous cases) that f is a d-morphism. Note that $f(Y') = f(Y_0) = \{w'\}$ and $f(Y'') = f(Y_{2m+1}) = \{w''\}$.

Second, there are no R_D-irreflexive points in α, so all preimages of R_D-irreflexive points are in Δ_0; since f_0 is a dd-morphism, f is 1-fold at any R_D-irreflexive point and manifold at all the others. Thus f is a dd-morphism by Proposition 12. □

References

1. Abashidze M (1987) Ordinally complete normal extensions of the logic of provability. In: Logic, methodology and philosophy of science, abstracts, vol 5, part 1. Moscow, pp 9–10
2. Abashidze M (1988) Ordinal completeness of modal Gödel-Löb system (Russian). In: Intensional logics and logical structure of theories (Telavi, 1985), Tbilisi, pp 49–73
3. Aiello M, Pratt-Hartmann I, van Benthem J (eds) (2007) Handbook of spatial logics. Springer, Berlin
4. Aleksandrov P (1977) Introduction to set theory and general topology. Nauka, Moscow
5. Aull C, Thron W (1962) Separation axioms between T_0 and T_1. Indag Math 24:26–37
6. Bezhanishvili G, Esakia L, Gabelaia D (2005) Some results on modal axiomatization and definability for topological spaces. Studia Logica 81:325–355
7. Bezhanishvili G, Esakia L, Gabelaia D (2010) The modal logic of Stone spaces: diamond as derivative. Rev Symbolic Logic 3:26–40
8. Bezhanishvili G, Esakia L, Gabelaia D (2011) Spectral and T_0-spaces in d-semantics. In: Bezhanishvili N, Löbner S, Schwabe K, Spada L (eds) Lecture notes in artificial intelligence. Springer, Berlin, pp 16–29
9. Bezhanishvili G, Lucero-Bryan J (2012) More on d-logics of subspaces of the rational numbers. Notre Dame J Formal Logic 53:319–345
10. Blass A (1990) Infinitary combinatorics and modal logic. J Symbolic Logic 55:761–778
11. Burgess J (1985) The decision problem for linear temporal logic. Notre Dame J Formal Logic 26:115–128
12. Chagrov A, Zakharyaschev M (1997) Modal logic. Oxford University Press, Oxford

13. de Rijke M (1993) Extending modal logic. PhD Thesis. ILLC Dissertation Series, Amsterdam
14. Esakia L (1981) Diagonal constructions, Löb formula, and scattered Cantor spaces. In: Logical and semantical investigations, Tbilisi, pp 128–143 (Russian)
15. Esakia L (2001) Weak transitivity—a restitution. In: Logical investigations, vol 8. Nauka, Moscow, pp 244–245 (Russian)
16. Fine K (1985) Logics containing K4, Part II. J Symbolic Logic 50:619–651
17. Gabelaia D (2001) Modal definability in topology. Master's thesis, ILLC
18. Gargov G, Goranko V (1991) Modal logic with names. In: Colloquium on modal logic. Dutch network for language, logic and information, Amsterdam, pp 81–103
19. Goranko V, Passy S (1992) Using the universal modality: gains and questions. J Logic Comput 2:5–30
20. Hodkinson I (2012) On the Priorean temporal logic with 'around now' over the real line. Submitted
21. Kremer P (2013) Strong completeness of S4 for any dense-in-itself metric space. Rev Symbolic Logic 6:545–570
22. Kudinov A (2005) Difference modality in topological spaces. Algebraic and topological methods in non-classical logics II, Barcelona, abstracts, pp 50–51
23. Kudinov A (2006) Topological modal logics with difference modality. In: Advances in modal logic, vol 6. College Publications, London, pp 319–332
24. Kudinov A (2011) Modal logic of R with the difference modality. In: Proceedings of the conference on information technologies and systems. Moscow, pp 335–339 (Russian)
25. Kudinov A (2008) Topological modal logic of R with inequality. Russ Math Surv 63:163–165
26. Kudinov A, Shapirovsky I (2010) Some examples of modal logics without a finite axiomatisation. In: Proceedings of the conference on information technologies and systems. Moscow, pp 258–262 (Russian)
27. Kuratowski K (1922) Sur l'opération $\bar{A}$ de l'Analysis Situs. Fundam Math 3:181–199
28. Kuratowski K (1966) Topology, vol 1. Academic Press, London
29. Litak T (2006) Isomorphism via translation. Advances in modal logic, vol 6. College Publications, London, pp 333–351
30. Lucero-Bryan J (2011) The d-logic of the rational numbers: a fruitful construction. Studia Logica 97:265–295
31. Lucero-Bryan J (2013) The d-logic of the real line. J Logic Comput 23:121–156
32. McKinsey J, Tarski A (1944) The algebra of topology. Ann Math 45:141–191
33. Rasiowa H, Sikorski R (1963) The mathematics of metamathematics. Monografie matematyczne, Tom 41. PWN—Polish Scientific Publishers, Warsaw
34. Reynolds M (2010) The complexity of temporal logic over the reals. Ann Pure Appl Logic 161:1063–1096
35. Segerberg K (1980) A note on the logic of elsewhere. Theoria 46:183–187
36. Shehtman V (1990) Derived sets in Euclidean spaces and modal logic. ITLI Prepublication Series, X-90-05, University of Amsterdam
37. Shehtman V (1999) "Everywhere" and "here". J Appl Non-classical Logics 9:369–380
38. Shehtman V (2000) Modal logics of topological spaces. Habilitation Thesis. Moscow (Russian).
39. van Benthem J (1983) Modal logic and classical logic. Bibliopolis, Napoli
40. Zakharyaschev M (1993) A sufficient condition for the finite model property of modal logics above K4. Bull IGPL 1:13–21

MIX
Papier aus verantwortungsvollen Quellen
Paper from responsible sources
FSC® C105338

If you have any concerns about our products,
you can contact us on
ProductSafety@springernature.com

In case Publisher is established outside the EU,
the EU authorized representative is:
Springer Nature Customer Service Center GmbH
Europaplatz 3, 69115 Heidelberg, Germany

Printed by Libri Plureos GmbH
in Hamburg, Germany